Excel® for Chemists

Excel® for Chemists

A Comprehensive Guide

Third Edition

E. Joseph Billo

A JOHN WILEY & SONS, INC., PUBLICATION

Library of Congress Cataloging-in-Publication Data:

Billo, E. Joseph.
  Excel for chemists : a comprehensive guide / E. Joseph Billo. — 3rd ed.
    p. cm.
  Includes index.
  ISBN 978-0-470-38123-6 (pbk.)
  1. Electronic spreadsheets. 2. Chemistry—Data processing. 3. Microsoft Excel (Computer file) I. Title.
  QD39.3.S67B55 2011
  542'.85—dc22   2011010945

Printed in the United States of America.

oBook: 978-1-118-09395-5
ePDF: 978-1-118-09393-1
ePub: 978-1-118-09394-8

V10016238_121619

Summary of Contents

Contents

PART I   THE BASICS

Chapter 2   Working with Excel 2007 or Excel 2010

Chapter 4   Excel 2007/2010 Charts

Chapter 5   Excel 2003 Charts

PART II   ADVANCED SPREADSHEET TOPICS

Chapter 6   Advanced Worksheet Formulas

Chapter 7   Array Formulas

Chapter 8   Advanced Charting Techniques

PART III   SPREADSHEET MATHEMATICS

Chapter 13   Mathematical Methods for Spreadsheet Calculations

Chapter 14   Linear Regression and Curve Fitting

Chapter 23 Custom Toolbars and Toolbuttons

Preface to the Third Edition

Since the publication of the second edition of *Excel for Chemists* in 2001, three new versions of Excel for the PC have appeared: Excel 2003, Excel 2007 and Excel 2010. With Excel 2007, Microsoft introduced a radically new user interface: the Ribbon, which replaced the familiar menus of earlier versions. The change was so extreme that many users have opted to stay with the "tried-and-true" version, Excel 2003, as long as possible. For this reason, this edition covers Excel 2003 as well as the newer Excel 2007/2010.

There are several new chapters in this edition. Two chapters discuss the features of the new Excel 2007/2010 (plus a new appendix, "What's Where in Excel 2007/2010"), a chapter covers automatic procedures—macros that run automatically—in more detail than was covered in the previous edition, and there is a completely new chapter that discusses how to handle documents created in other language versions of Excel.

Much of the material in this book has been incorporated in courses entitled "Excel for Scientists and Engineers" and "Excel Visual Basic Macros for Scientists and Engineers" which have been presented to over 2500 scientists in the United States, Canada and Europe—not only chemists, but also scientists in many other disciplines. Many changes in this edition were made in light of the experience gained in teaching these courses.

Acknowledgments

Dr. Faith A Morrison, Department of Chemical Engineering, Michigan Technological University, for her "Guidelines on Graphing" used in Chapter 4.

Dr. Lev Zompa, University of Massachusetts–Boston, for spectrophotometric data used in Chapter 19.

Dr. Steve Bell, for NMR data used in Chapter 20.

E. Joseph Billo
May 2011

Before You Begin

Which Version of Excel Are You Using?

This book is for users of Excel 2003 for Windows or Excel 2007/2010 for Windows.

The second edition of this book covered both Excel for the PC and Excel for the Macintosh, but since Excel 2008, the Macintosh version corresponding to Excel 2007, doesn't support Visual Basic for Applications (VBA), I decided not to include Mac-specific instructions in this third edition.

Typographic Conventions

As you read through this book, you'll see several different fonts and capitalization styles within the text. Here are the conventions that I've used.

- Excel 2003 menu headings and menu commands are in boldface type, e.g.:
  File, Format, Delete....

- Excel 2007/2010 Ribbon tab names, icon names and menu command names are in non-bold, e.g.,
  "the Insert icon in the Cells group in the Home tab", "click on Paste Special..."

- For clarity, the titles of some dialog boxes and options are enclosed in quotes, e.g.:
  "When Creating New Workbooks"

- Cell references are in Arial font, e.g.:
  "In cell A9 ..."

- Worksheet functions are in Arial font, e.g.:
  SUM, LINEST

- Excel's built-in argument names (i.e., placeholder arguments) in functions are in Arial italic; required arguments are in bold italic, e.g.:
  LINEST(***known_y's***, ***known_x's***, *const*, *stats*)

- User-defined range names are in Arial, not italic, e.g.:
    {=LINEST(YValues, XValues,TRUE,TRUE)}

- Visual Basic statements are in Arial; VBA keywords are bold, e.g.:
    For Counter = Start **To** End **Step** Increment.

Special Features in this Book

This book has a number of features that you should find useful and helpful. There are over 50 **Excel Tips** to simplify and improve the way you use Excel. For example:

*Excel Tip. To **Fill Down** a value or formula to the same row as an adjacent column of values, select the source cell and double-click on the Fill Handle.*

Throughout the book you'll see **"How-To" Boxes** that outline, in a clear and systematic manner, how to accomplish certain complex tasks. For example:

To Create a Chart with a Secondary Y Axis
(two different Y Axis scales and the same X Axis)

1. Select all three data series to be plotted (the X Axis data series, two Y Axis data series).
2. Create an XY chart.
3. Click on the data series whose axis you want to change.
4. **Excel 2007/2010**. Click on the Format tab. Click on Format Selection in the Current Selection group to display the Format Data Series dialog box. Click on Series Options and press the Secondary Axis option button.

    Excel 2003. Choose **Selected Data Series...** from the **Format** menu and choose the Axis tab (see Figure 8-18).
5. Press the Secondary Axis button. A preview of the combination chart will be displayed. If the chart is suitable, press the OK button.

The CD-ROM

The CD-ROM that accompanies this book contains most of the worksheets that are discussed in the book. The files are in Excel 2003 format, so that they can be opened using either Excel 2003 or Excel 2007/2010.

A complete list of all files on the CD-ROM, with short descriptions, is in Appendix K.

PART I

THE BASICS

1

Working with Excel 2007 or Excel 2010

Extensive changes were made in the Excel 2007 user interface. Excel's familiar drop-down menus with their familiar commands were replaced by the Ribbon. (The Excel 2010 Ribbon is virtually unchanged from the Excel 2007 version.) If you're a long-time Excel user, you'll probably find the ribbon interface confusing and infuriating, although some people profess to like it.

At the end of this chapter I suggest several ways to make the change to Excel 2007 or 2010 easier for Excel 2003 users.

What's New in Excel 2007 and 2010

Let's look at the changes that were introduced in Excel 2007. They're listed in decreasing order of what I consider to be their importance for the average user.

The Ribbon

The Ribbon is the major component of what's officially called the Office Fluent user interface; it replaces the familiar drop-down menus and toolbars of earlier versions of Excel. The Ribbon is essentially a multi-row toolbar. Clicking on one of a series of tabs located on something similar to the Excel 2003 menu bar displays the Ribbon for that particular tab. The primary tab captions are Home, Insert, Page Layout, Formulas, Data, Review, and View. Figure 1-1 shows the Home tab of the Ribbon. (The appearance of the Ribbon may be different, depending on your screen width and/or resolution.)

Other tabs may be displayed depending on context. If a chart is the active window, the Design, Layout and Format tabs are displayed. The Developer tab, for working with VBA procedures, is only displayed if the user previously opted

3

to display it. The Add-Ins tab appears only if custom menu commands or toolbuttons have been added to the Ribbon; these appear in the Menu Commands and Toolbar Commands groups, respectively.

Figure 1-1. The Excel 2007 Home tab.

New File Formats

Microsoft has moved to new XML-based file formats. Instead of the familiar .xls file extension, indicating an Excel workbook, there are now a number of workbook file types. Table 1-1 shows the most important of these new file types. The filename extension now consists of four letters. The first two characters, xl, identify the document as an Excel file. The third character, either s, t or a, identifies the file as spreadsheet, template or add-in. The final character, either x or m, identifies the file as either macro-free or macro-enabled. In addition to these file types, the .xlsb file format identifies a workbook in binary format, while the .crtx file format is used to save chart templates.

Table 1-1. New File Formats in Excel 2007/2010

.xlsx	A workbook containing no macros (the default Excel 2007/2010 file format)
.xlsm	A workbook containing one or more macros (either VBA or Excel 4 macro language)
.xlsb	The Excel 2007/2010 Binary file format.
.xlam	An Add-in workbook containing either VBA procedures or Excel 4.0 macros).
.xltx	An Excel template, without VBA or Excel 4 macro language code.
.xltm	An Excel template with VBA macro code or Excel 4 macro sheets.
.crtx	An Excel chart template.

A Much Larger Worksheet

Worksheets now have more rows and columns: 1,048,576 rows by 16,384 columns for a total of 17,179,869,184 cells in each worksheet. (The column designation now ends at XFD instead of IV.) Granted, Excel 2003's 256 columns was sometimes not enough – many people wanted to have at least 365 columns, one for each day of the year – but seventeen billion cells in each worksheet seems a bit much.

Larger Limits for Some Features

There are larger limits for a number of Excel features. Excel 2003 allowed three levels of conditional formatting; in Excel 2007/2010, the number of conditional formats is limited by available memory. The number of levels of nested functions in a formula in Excel 2003 was seven; the new limit is 64 levels. The number of sort levels in Excel 2003 was three; the new limit is 64 levels. A more complete list of Excel 2007/2010 specifications can be found at the end of this chapter.

New Worksheet Functions

There are more built-in functions, but this is largely because the 89 Engineering functions, loaded in Excel 2003 with the Analysis ToolPak Add-In, are now a permanent part of the list of functions. Apart from the Engineering functions, only five new functions have been added: SUMIFS, AVERAGEIF, AVERAGEIFS, COUNTIFS and IFERROR. These are discussed in Chapter 3, "Excel Formulas and Functions".

The Downside

There are some changes that, in my opinion, are not for the better. Since there are no menus, VBA code that installs a new menu command in, for example, the **Tools** menu of Excel 2003 doesn't work as it did before. If you're lucky, the custom menu command will appear in the Add-Ins tab. The same is true of custom toolbuttons; they also appear in the Add-Ins Tab. If you're not lucky, the VBA code that installs a custom menu command or custom toolbutton may have to be modified.

What's New in Excel 2010

There have been a few changes in Excel 2010 from the previous version, Excel 2007. Among the more substantive changes are: changes in some functions (see "Changes to Functions in Excel 2010" in Chapter 3), the Equation Tool to display an equation in a text box (see "Entering an Equation in a Text Box" later in this chapter), and the ability to customize the Ribbon, not just the Quick Access Toolbar (see "Customizing the Ribbon" in Chapter 23).

The Excel 2007/2010 Document Window

An Excel workbook is a *document* that appears in its own *document window*. Although you can have several workbooks open at the same time and can see several displayed on the screen simultaneously, only one workbook can be the *active workbook*. The default Excel 2007/2010 workbook contains three worksheets; only one worksheet in the active workbook can be the *active worksheet*.

Figure 1-2 shows the Excel 2007 document window. The Excel 2010 window is essentially identical. In Figure 1-2, reading from the top down you'll see the *application title bar*, the *Ribbon tabs* (Home, Insert, Page Layout, etc.), the *Ribbon* (the Home tab of the Ribbon is displayed), the *Quick Access Toolbar* (with New, Open, Save, etc. toolbuttons), the *formula bar*, the rows and columns of cells, and, at the bottom, the *sheet tabs*, the *horizontal scroll bar* and the *status bar*. To the left of the formula bar is the Name Box or *cell reference area*, displaying the cell reference of the currently selected cell. Depending on your monitor, your screen may show a different number of rows or columns and a different view of the Ribbon.

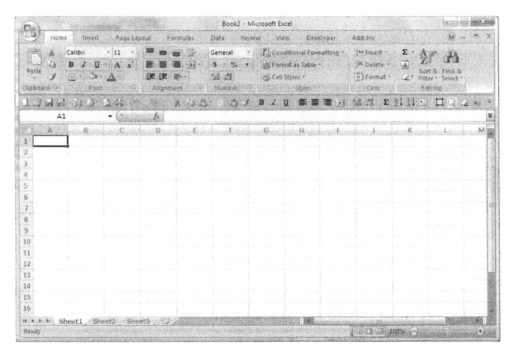

Figure 1-2. The Excel 2007 document window, showing the Home tab.

button ⬇ *that allows you to increase the size of the formula bar to accommodate long formulas. You can also right-click in the formula bar to display a shortcut menu that allows you to expand the formula bar.*

Hiding, Moving or Resizing a Document Window

If you click on the Minimize button (the ⬚ button in the upper right corner of the document) the window will be closed and only the title bar will appear in the tray. To restore the window, click on the document in the tray.

If you click on the Restore Window button (the ⬚ button in the upper right corner of the document), the window size will be reduced so that it no longer completely fills the document window. To restore it to its full size, click the Restore Window button again.

To manually change the size of a window, click and drag any of its borders or corners; the mouse pointer changes shape when you click on a border or corner. You can adjust the document to any size you desire.

To change the position of a document within the Excel window, click on the title bar and drag the document. It can even extend off-screen.

Working with Excel 2007/2010

In my view, some of the most confusing changes that were made in Excel 2007 were those that corresponded to commands in the **File** menu of Excel 2003, and in the **Options** command in the **Tools** menu. These two changes will be discussed first.

The Office Button (Excel 2007)

 The Office Button, shown at the left, is located at the left of the Ribbon tabs. Pressing this button displays the Office Button window, shown in Figure 1-3.

The Office Button window is the approximate equivalent of the **File** menu in Excel 2003. In addition to the commands found in the Excel 2003 **File** menu (**New, Open, Close, Save, Save As, Print, Exit,** etc.), clicking on the Excel Options button at the bottom of the window displays the Options window. The Options window contains the options that were located in the Excel 2003 **Options** command of the **Tools** menu and is described in a following section.

Figure 1-3. The Excel 2007 Office Button window

The File Tab (Excel 2010)

The Excel 2007 Office Button was confusing to new users. (W*hy* did Microsoft disguise the former **File** menu as a simple logo?) Excel 2010 got rid of the Office Button and replaced it with the File tab, shown in Figure 1-4.

The File tab of the Ribbon contains the following buttons: Save, Save As, Open, Close, Info, Recent, New, Print, Save & Send, Help, Options and Exit.

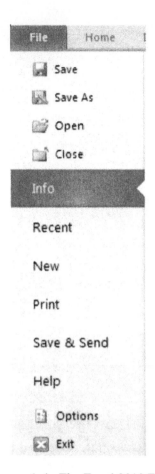

Figure 1-4. The Excel 2010 File tab.

The Excel Options Window

Pressing the Excel Options button in the Office Button window (Excel 2007) or the Options button in the File tab of the Ribbon (Excel 2010) displays the Excel Options window. The menu of options, shown in Figure 1-5, is in the left pane of the window. There are nine options available: Five of these (Popular, Formulas, Save, Advanced, Customize) contain most of the options that are found in the Excel 2003 **Options** command in the **Tools** menu. Figure 1-5 shows the Excel 2010 General options window; the Excel 2007 window is similar but called the Popular Options window.

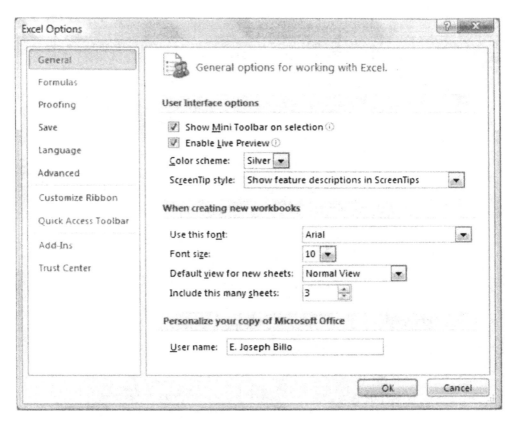

Figure 1-5. The Excel 2010 Options window, showing General options.

One of the five Options windows, the Advanced Options window, is so large that it requires that you scroll down to view all the options. One of the most frustrating things about Excel 2007/2010, even for experienced Excel users, is the difficulty of finding the location of a desired option. Table 1-2 is a guide to help Excel 2003 users navigate their way to Excel 2007/2010 options.

The Ribbon: An Overview

Commands on the Ribbon are represented by icons. Related icons are organized in groups. For example, the Cut, Copy and Paste icons are located in the Clipboard group in the Home tab of the Ribbon. Many of the icons on the Ribbon are identical to Excel 2003 toolbuttons. Some icons have a drop-down button that, when clicked, displays a menu of options.

The appearance and arrangement of icons on the Ribbon depends on the size of the document window. If the window fills the complete screen, the icons in the Ribbon are usually arranged in two rows. If the size of the window is made smaller, the icons may be arranged in three rows, and some groups may not

Table 1-2. A Guide to Excel 2007/2010 Options

Excel 2003 Option Tab	Option	Where to Find It in Excel 2007/ 2010 Options
General	R1C1 reference style	Formulas
	Number of recently used files	Advanced: Display
	Number of sheets in new workbook	General (2010), Popular (2007)
	Standard font and size	General (2010), Popular (2007)
View	Show Formula Bar, Status bar, etc	Advanced: Display
	Show Comment indicator, etc	Advanced: Display
	Show scroll bars, sheet tabs	Advanced: Display options for this workbook
	Show gridlines, row &column headers, page breaks, etc.	Advanced: Display options for this worksheet
Edit	Allow Edit Directly in Cell	Advanced: Editing options
	Allow Drag-and-Drop Editing	Advanced: Editing options
	Automatic % entry	Advanced: Editing options
	Move selection after Enter	Advanced: Editing options
	Extend formats and formulas	Advanced: Editing options
	Fixed decimal	Advanced: Editing options
	Enable AutoComplete	Formula
Calculation	Calculation: Automatic, manual	Formula
	Iteration	Formula
	Update remote references	Advanced: When calculating this workbook
	Precision as displayed	Advanced: When calculating this workbook
	1904 date system	Advanced: When calculating this workbook
	Save external link values	Advanced: When calculating this workbook
Custom Lists		General (2010), Popular (2007)
Chart	Plot empty cells as	(see Chapter 4)
International	Number handling	Advanced: Editing options
Save	AutoRecover every ? Minutes	Save
Error Checking	Enable background error checking	Formula
Color	Standard and custom colors	(Use color palette for object)
	Error-checking rules	Formula
Spelling	Dictionary	Proofing
	AutoCorrect Options	Proofing
Security	File encryption settings	Prepare → Encrypt Document
	File sharing settings	Prepare → Restrict Permission
	Digital signatures	Prepare → Digital Signature
Transition	Menu key, formula evaluation, entry	Advanced: Lotus compatibility

display all available icons. For the screen shots of the Ribbon that are shown in this chapter, windows were resized so that icons were displayed in three rows. For that reason, the images may not look exactly like what you see on your computer screen.

The appearance of a command provides information about its form or availability:

- A command with an ellipsis (...), such as Paste Special..., indicates that the command opens a dialog box to obtain user input.

- Many buttons or commands display submenus, indicated by the ▾ symbol at the bottom of the button or the ▶ symbol at the right edge of the command.

- Some commands are dimmed (i.e., appear as gray characters) when the command is unavailable. Some tabs appear on the Ribbon only when they are available.

- Some commands change the text of their command depending on circumstances. For example, you click on the Add Comment button in the Review tab of the Ribbon to add a comment to a cell; if you select a cell that already has a comment, the button text changes to Edit Comment so that you can edit the text of the comment.

- Some commands are preceded by a check mark if the option has been selected previously. To remove the selection, depending on the command you either click on the check mark or select the command again.

Shortcut Menus

Although Excel's menus have been replaced by the Ribbon, Excel still provides "context-sensitive" shortcut menus. For example, if you press the right mouse button while you select a worksheet element with the mouse pointer, a menu is displayed containing commands that apply to the selection. If you select a column with the right mouse button, a shortcut menu is displayed that contains editing and formatting commands appropriate for a column. In Excel 2007/2010, many of these shortcut menus are accompanied by the Mini Toolbar, described in a following section.

Keyboard Access to the Ribbon

You can access icons on the Ribbon by using the keyboard rather than the mouse. First, press the Alt key to display the KeyTips for the Ribbon icons. Figure 1-6 shows some KeyTips in the Formulas tab of the Ribbon.

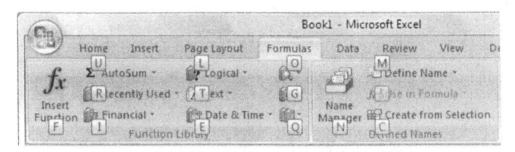

Figure 1-6. KeyTips displayed in the Home tab.

To activate a particular icon, press the keyboard letter that appears in the KeyTip over the icon. Depending on which letter you press, additional KeyTips may appear, and you may have to press additional keys.

To remove the KeyTips, press the Alt key a second time.

The Home Tab

Perhaps this tab, shown in Figure 1-1, should have been named the Format tab. This tab of the Ribbon contains the following groups: Clipboard, Font, Alignment, Number, Styles, Cells and Editing. Many of the icons in the Home tab of the Ribbon correspond to toolbuttons on the Excel 2003 Standard toolbar (Cut, Copy, Paste and the Format Painter) or to toolbuttons on the Formatting toolbar (Font and Font Size, Bold, Italic and Underline, Align Left, Center and Align Right, etc.), as well as a number of icons that perform actions that are in the **Edit, Insert** or **Format** menus of Excel 2003 (Insert or Delete cells, rows or columns, Find or Replace, Format cells, rows, columns or sheet).

Navigating Around the Workbook

Use the Switch Windows icon in the Window group in the View tab of the Ribbon to switch between one workbook and another. All open workbooks are listed in the drop-down menu; the active workbook is indicated with a check mark.

Use the Hide icon to hide a workbook. Most commonly I use Hide with workbooks that contain macros. A macro is still available for use even when it's hidden.

Inserting or Deleting Worksheets

The default Excel 2007/2010 workbook contains three worksheets. You can change the default, so that all new workbooks will have, for example, only one worksheet. In Excel 2010, click on the File tab, press the Options button, click on the General button, and in the "When Creating New Workbooks" group, change the "Include this many sheets" option. In Excel 2007, click on the Office Button, press the Excel Options button, click on the Popular options button, and in the "When Creating New Workbooks" group, change the "Include this many sheets" option.

To insert an additional worksheet in an existing workbook, press the Insert Worksheet button (located to the right of the sheet tabs, see Figure 1-7). A nice feature of this button is that the sheet is added to the right of the existing sheets, instead of to the left of the selected sheet, so that Sheet 4 follows Sheet 3, for example.

Excel Tip. *You can right-click the sheet tab of a worksheet to display the shortcut menu, and then choose Insert or Delete.*

In Excel 2003, if you increase the number of sheets in a workbook, not all of them may be visible. In Excel 2007/2010, as you increase the number of sheets, the area that displays the sheet tabs widens, permitting more sheet tabs to be visible, but the horizontal scroll bar narrows correspondingly. You can change the size of the sheet tab area by dragging the small button between the sheet tabs and the scroll bar.

Figure 1-7. Excel 2007/2010 sheet tabs, with the Insert Worksheet button.

To select a worksheet, simply click on the sheet tab. If the workbook contains a large number of worksheets, the tab for the sheet that you want to select may not be visible. Use the tab scroll buttons, located to the left of the sheet tabs, to scroll through the sheet tabs. From left to right, these four buttons allow you to jump to the first sheet tab, scroll toward the first sheet tab, scroll toward the last sheet tab, or jump to the last sheet tab. When the desired sheet tab is visible, click on it.

You can select a specific sheet by right-clicking on any of the scroll buttons, to display the list of all visible sheets. Choose the desired sheet from the shortcut menu, as illustrated in Figure 1-8.

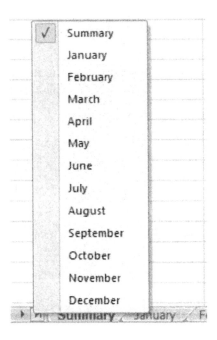

Figure 1-8. Right-click to display the list of all sheets in a workbook.

Changing the Name of a Worksheet

When you create a new workbook, the sheet tabs have the default names Sheet1, Sheet2, etc. To rename a sheet, double-click on the sheet tab, or right-click to display the shortcut menu and choose Rename. The sheet name will be highlighted and you can enter a more descriptive name, as, for example, in Figure 1-9. Click outside the sheet tab to exit from edit mode.

A worksheet name can have a maximum of 31 characters.

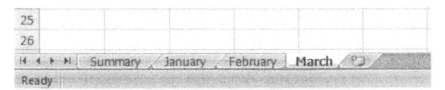

Figure 1-9. Descriptive sheet names are helpful.

Rearranging the Order of Sheets in a Workbook

To move a sheet tab, just click and drag it. The mouse pointer shape becomes an icon showing a sheet at the end of the arrow pointer (Figure 1-10). An arrow above the sheet tab indicates where the tab will be inserted.

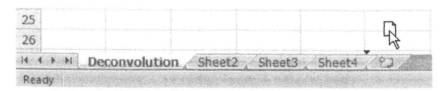

Figure 1-10. Moving a sheet tab with the mouse.

To make a copy of a worksheet, hold down the Ctrl key while dragging the sheet tab. A small + sign appears in the icon (Figure 1-11).

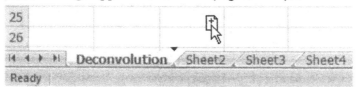

Figure 1-11. Copying a sheet with the mouse.

Selecting Multiple Worksheets: [Group] Mode

To make a non-adjacent selection of worksheets (e.g., Sheet1 and Sheet3), hold down the Ctrl key while clicking on each sheet tab. To select a contiguous range of sheets, click on the sheet tab at one end of the range, then hold down the Shift key and click on the sheet tab at the other end of the range. All the sheet tabs in the range will be selected. To select all the sheets in a workbook, you can right-click on any sheet tab; this displays a shortcut menu that will allow you to Select All Sheets.

You are now in Group Edit mode; [Group] appears in the title bar at the top of the worksheet, and the sheet tabs of all grouped sheets are shown as selected (Figure 1-12).

Once you have grouped sheets, any changes you make in one of them, such as entering or editing data or applying formatting, will be applied to all sheets in the group.

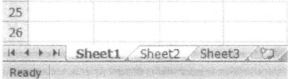

Figure 1-12. Appearance of sheet tabs in [Group] mode.

To Ungroup sheets, click on any ungrouped sheet tab. If all sheets have been selected, right-click on any sheet tab and choose **Ungroup Sheets** from the submenu.

You can also copy, move, delete or print a selection of sheets when in [Group] mode.

Excel Tip. *If you have grouped a number of sheets, choosing Insert Sheet from the Cells group on the Home tab will insert an equal number of worksheets.*

Changing the Color of Sheet Tabs

You can color-code sheet tabs as an aid in organizing your work. Right-click on the sheet tab and then choose Tab Color from the shortcut menu. Alternatively, you can click on the Format button in the Cells group in the Home tab of the Ribbon, and then click on Tab Color.

Using Move or Copy Sheet or Delete Sheet

To move or copy sheets within a workbook, or from one workbook to another, use the Move or Copy Sheet... command (Figure 1-13) in the Format submenu in the Cells group in the Home tab. Since you can easily move sheets within a workbook by using the mouse, the more important use of Move or Copy Sheet is to send a sheet, or a copy of a sheet, from one workbook to another.

Figure 1-13. The Move or Copy Sheet dialog box.

Navigating Around the Worksheet

You can move around a worksheet either by means of the mouse or by using keystrokes.

Use the arrows in the vertical and horizontal scroll bars (the gray bars on the right edge and at the bottom of the window) to scroll through the worksheet. A single click of the mouse on an arrow moves the worksheet one row or column. The position of the scroll box (the white square in the gray bar) indicates the position of the window relative to the worksheet. You can also scroll through the worksheet by clicking on an arrow and holding down the mouse button, by dragging the scroll box with the mouse, or by clicking in the gray space on either side of the scroll box. Table 1-3 lists keystroke commands for cursor movement.

Table 1-3. Keys for Cursor Movement

Arrow keys	Move left, right, up, down one cell
Enter	Move down one cell
Tab	Move right one cell
Home	Move to the beginning of a row
End	Move to the end of a row
Page Up	Move to the top of the window
Page Down	Move to the bottom of the window
Alt + Page Up	Move to the left of the window
Alt + Page Down	Move to the right of the window

Selecting a Range of Cells

You can select a range of cells on the worksheet in several ways:

- Click on the cell in one corner of the range, hold down the mouse button and drag to the cell in the opposite corner of the range. The range of cells will be highlighted. The size of the selection (e.g., $10R \times 3C$) is displayed in the reference area (the Name Box) of the formula bar, until you release the mouse button.

- Click on the cell in one corner of the range, then hold down the Shift key and click on the cell in the diagonally opposite corner of the range. The range of cells will be highlighted.

- Select a complete row or column of cells by clicking on the row or column heading. The row or column will be highlighted.

Selecting Non-Adjacent Ranges

To select non-adjacent ranges, select the first range, then hold down the Ctrl key while selecting the second range. Both cell ranges will be highlighted (Figure 1-14).

	A	B	C	D
	time			
1	(sec)	$[A]_t$	$[B]_t$	$[C]_t$
2	0	5.00E-03	0.00E+00	0.00E+00
3	0.2	4.72E-03	2.60E-04	1.71E-05
4	0.6	4.21E-03	6.50E-04	1.37E-04
5	1.0	3.76E-03	9.04E-04	3.38E-04
6	1.4	3.35E-03	1.06E-03	5.90E-04
7	1.8	2.99E-03	1.14E-03	8.72E-04
8	2.2	2.67E-03	1.16E-03	1.17E-03
9	2.6	2.38E-03	1.16E-03	1.46E-03
10	3.0	2.12E-03	1.12E-03	1.76E-03
11	4.0	1.60E-03	9.76E-04	2.43E-03
12	5.0	1.20E-03	8.04E-04	3.00E-03
13	6.0	9.02E-04	6.41E-04	3.46E-03
14	7.0	6.78E-04	5.02E-04	3.82E-03
15	8.0	5.10E-04	3.87E-04	4.10E-03
16	9.0	3.83E-04	2.97E-04	4.32E-03
17	10.0	2.88E-04	2.26E-04	4.49E-03
18				

Figure 1-14. Selecting non-adjacent ranges.

Extending a Selection

To extend the range of a cell selection you just made, hold down the Shift key, select the last cell in the selection, and drag to include the additional cells. Alternatively, hold down the Shift key and use an arrow key to extend the selection. You can also decrease the number of cells in the selection in the same way.

Selecting a Block of Cells

A *block of cells* is a range of cells containing values and bounded by empty cells. There are several shortcuts for selecting cells within a block:

- Use Ctrl + Shift + (arrow key) to select in the appropriate direction.
- Select a cell at a boundary of the block (at the top, bottom or side of the block). Move the mouse pointer over the edge of the selected cell

until the pointer changes to the arrow pointer (Figure 1-15 Left). Hold down the Shift key and double-click on the bottom edge of the selected cell to select all cells in the column from the top to the bottom of the block, as shown in Figure 1-15 Right. You can select cells from top to bottom, from bottom to top, from left to right or from right to left within a block. You can also select multiple columns or rows in the same way.

	A	B			A	B
1	X	Y		1	X	Y
2	0.100	-2.303		2	0.100	-2.303
3	0.200	-1.609		3	0.200	-1.609
4	0.300	-1.204		4	0.300	-1.204
5	0.400	-0.916		5	0.400	-0.916
6	0.500	-0.693		6	0.500	-0.693
7	0.600	-0.511		7	0.600	-0.511
8	0.700	-0.357		8	0.700	-0.357
9	0.800	-0.223		9	0.800	-0.223
10	0.900	-0.105		10	0.900	-0.105

Figure 1-15. Using the mouse pointer to select a block of data.
Left: Selecting a cell edge. Right: Selecting the block of data
by double-clicking while holding down the Shift key.

Entering Data in a Worksheet

As you type a value in a cell, the characters appear in the formula bar and the active cell. You can complete the entry in several ways.

- Press the Enter key. This moves the selection to the cell below (although you can change the default option so that the selection is not moved). This is the usual way for entering data in cells.

- Press the Enter button ✖ in the formula bar. The cell remains selected. This method is useful if you want to examine a value or formula after entering.

To cancel the entry and revert to the original contents of the cell, press the Cancel button ✔ or the Esc key.

Excel Tip. To enter the same formula or value in a range of cells, select the range of cells, type the value, then press Ctrl + Enter.

Entering Numbers

Excel has a remarkable ability to recognize the type of value that you have entered: a number, a percent, a debit value, as currency, in scientific notation, as a date or time, or even as a fraction. The number will be displayed in the cell in the proper format. Table 1-4 illustrates number formats recognized by Excel.

Table 1-4. Number Formats Recognized by Excel

Type	As Entered at Keyboard	As Displayed in Cell	As Displayed in Formula Bar	As Used in Calculation
Percent	15%	15%	15%	0.15
Scientific	2e-3	2.00E-03	0.002	0.002
Currency	$50	$50	50	50
Currency	$20000	$20,000	20000	20000
Debit	(5000)	-5000	-5000	-5000
Fraction	2 5/8	2 5/8	2.625	2.625
Date	7/4	4-Jul	7/4/2010*	**
Date***	8/3/28	8/3/2028	8/3/2028	**
Date***	8/3/38	8/3/1938	8/3/1938	**
Time	4:30	4:30	4:30:00 AM	**
Time	16:00	16:00	4:00:00 PM	**
Time	4 p	4:00 PM	4:00:00 PM	**

* Enters current year.

** See Chapter 3 for a discussion of date and time calculations.

*** A 2-digit year in the range 00-29 is assumed to be in the 21st century.

Since the slash character indicates either a date or a fraction, if you enter what you intended to be a fraction, such as 1/3, it will be interpreted as a date, specifically 3-Jan. To prevent Excel from converting the entry to a date, enter a zero and a space before the fraction (0 1/3). The zero indicates that the entry is a number, and the value will appear in the formula bar as 0.333333333333333.

How Excel Stores and Displays Numbers

Excel can accept numbers in the range from $\pm$1E-307 to $\pm$9.99999999999999E+307.

Excel stores numbers with 15-significant-figure accuracy. These are displayed in the formula bar and used in all calculations, no matter what number formatting has been applied. Thus the fraction 1/3 appears in the formula bar as 0.333333333333333, and π as 3.14159265358979.

Excel switches between floating-point and scientific notation for best display of values. The formula bar can display numbers up to 21 characters, including the decimal point. Thus 1E-19 entered on the keyboard will be displayed as 0.0000000000000000001 (21 characters) in the formula bar, while 1E-20 will appear as 1E-20. Similarly, 1E20 is displayed as 100000000000000000000, while 1E21 appears as 1E21. Since a total of 21 characters can be displayed, the number of significant figures determines the magnitude of a number less than 1 that can be displayed in non-E format in the formula bar. Thus 1.2345E-15 appears as 0.0000000000000012345, while 1.23456E-15 is displayed as 1.23456E-15.

Entering Text

If you enter text characters (any character other than the digits 0 – 9, the decimal point, or the characters +, -, *, /, ^, $, %) in a cell, Excel will recognize the entry as text. For example, Chestnut Hill MA 02167-3860 is a text entry. Each cell can hold up to 32,767 characters (but only 1,024 will display in a cell). You can distinguish text entries from number entries in the following way: text entries are left-aligned, and numbers are right-aligned. Of course, if you format the alignment of a cell to be right-aligned, its value will be right-aligned whether the value is a number or text.

You can format individual characters in a cell using Bold, Italic, Underlined, etc., or with different fonts, by highlighting the character(s) in the formula bar and then applying the formatting.

> **Excel Tip.** *Sometimes it is necessary to enter a number or a date as a text value. To do this, begin the entry with a single quote.*

Entering Formulas

Instead of entering a number in a cell, you can enter an equation (called a *formula* in Microsoft Excel) that will calculate and display a result. Usually formulas refer to the contents of other cells by using *cell references*, such as A2, a reference to a cell, or B5:B12, a reference to a range of cells. The value displayed in a cell containing a formula will be automatically updated if values elsewhere in the worksheet are changed. Formulas can contain values, arithmetic operators and other operators, cell references, the wide range of Excel's worksheet functions, and parentheses.

The rules for writing formulas (the *syntax*) are as follows:

- A formula must begin with the equal sign (=).
- The *arithmetic operators* are addition (+), subtraction (-), multiplication (*), division (/) and exponentiation (^). Other types of operator are described in Chapter 3.

- Parentheses are used in the usual algebraic fashion to prevent errors caused by the *hierarchy of arithmetic operations* (multiplication or division is performed before addition or subtraction, for example).

Some examples of simple formulas:

=A1+273.15	Adds 273.15 to the value in cell A1.
=A2^2+13*A2-5	Evaluates the function $x^2 + 13x - 5$, where the value of x is stored in cell A2.
=SUM(B3:B47)	Sums the values contained in cells B3 through B47.
=(-C3+SQRT(C3^2-4*C2*C4))/(2*C2)	Finds one of the roots of the quadratic equation whose coefficients a, b and c are stored in cells C2, C3 and C4, respectively.

Excel formulas are discussed in much greater detail in Chapters 3 and 6.

There are some techniques that you can use for entering worksheet formulas.

- Type formulas in lowercase to facilitate detection of typographical errors. When you enter a formula, Excel converts functions and cell references to uppercase. If you type the formula =offset(d1,5,1), Excel will convert it to =OFFSET($D$1,5,1) when you enter the formula; but if you type "ofsett" instead of "offset," Excel won't recognize it and will display the error message #NAME?. When you examine the formula, you'll easily see that the incorrect function name remained in lowercase letters.

- Enter cell or range references in formulas by selecting, not by typing. This makes it less likely that you will enter an incorrect reference and also makes entering complicated references (such as external references) much easier.

- If formulas contain terms identical to those used in other cells, you can Copy that part of the formula from the formula bar and Paste it into the new formula.

Excel Tip. Formulas that return the wrong result because of errors in the hierarchy of calculation are common. When in doubt, use parentheses.

Editing Cell Entries

When you select a cell that contains an entry, the contents of the cell appear in the formula bar. As soon as you begin to enter a new value, the old value disappears. To make minor editing changes in the old entry, place the mouse pointer in the text in the formula bar at the point where you want to edit the entry. The mouse pointer becomes the vertical insertion-point cursor. You can now edit the text in the formula bar using the Copy, Cut, Paste or Delete commands or keys. Complete the entry using the Enter button in the formula bar, or by pressing the Enter key on the keyboard.

Alternatively, you can use Excel's Edit Directly In Cell feature: use function key F2 or double-click on the cell to enter edit mode. You can use the right and left arrow keys to move through the formula, or Ctrl+(arrow key) to jump to the next element of the formula, or Ctrl+Shift+(arrow key) to select the next element of the formula.

Excel uses colors to show range references in formulas and the corresponding ranges on the worksheet. When you enter Edit mode, by clicking in the formula bar, double-clicking on the cell, or pressing F2, the precedents (inputs) to the formula will appear in color and the corresponding inputs indicated by similarly colored cell or range borders (blue for the first input, green for the second, and so on). Each of these colored outlines has a handle in the bottom right corner; you can change the range of a formula input by dragging its handle.

Excel Tip. *To select (highlight) a word or reference for editing, double-click on it.*

The Order in Which Excel Performs Operations in Formulas

If several operators are combined in a single formula, Excel evaluates the formula in the following order: reference operators (colon, space, comma), negation (minus), percent (%), exponentiation (^), multiplication and division (* and /), addition and subtraction (+ and −), concatenation (&), comparison (=, <, >, <=, >=, <>).

If an expression contains operators with the same precedence (multiplication and division or addition and subtraction), Excel evaluates the operators from left to right.

Expressions enclosed in parentheses are evaluated first, no matter where they appear in the formula.

Adding a Text Box

You can add visible comments or other information to a worksheet by typing them into one or more worksheet cells. Another way to add comments, in a much more flexible form, is by using a text box.

To create a text box, click on the Text Box icon [A] in the Text group in the Insert tab, or click on the Text Box toolbutton if you have added it to the Quick Access Toolbar (see Chapter 23 for instructions on how to customize the Quick Access Toolbar). The mouse pointer will change to a crosshair. Position the crosshair pointer where you want to place the text box, and click and drag to outline it (the text box can be moved and sized later). An empty text box will be displayed with a blinking text cursor. Type the desired text within the box.

Text box input has many features of a simple word processor: you can Cut, Copy or Paste text, make individual portions of text bold, italic or underlined, use different font styles, etc., as shown in Figure 1-16. The text within the box can be formatted with the Alignment toolbuttons or with the Alignment command.

	A	B	C	D	E
1					
2	Text in a text box can be **Bold**, *Italic*,				
3	<u>Underlined</u>, or can contain characters in				
4	different fonts, such as α or β by using				
5	the Symbol font.				
6					

Figure 1-16. A text box.

To move a text box, click on it to select it, place the mouse pointer on the border of the text box and drag it to its new position. To resize a text box, select it (white handles will appear), then place the mouse pointer on one of the handles and click and drag to move the border of the box. If you hold down the Ctrl key while dragging, you make a copy of the text box (a small plus sign appears beside the mouse pointer); if you hold down the Alt key, the text box will align with the cell gridlines; if you hold down the Shift key, the text box can only be dragged in either horizontal or vertical alignment with its original position.

Entering an Equation in a Text Box

In addition to the Equation Editor (see "Using the Equation Editor" in Chapter 2), Excel 2007/2010 provides the Equation Tool that allows you to write and edit equations inside a text box on a worksheet.

The Equation button, shown here on the right, is not active until you use the Text Box button to draw a text box. When you click on the Equation button (located in the Insert tab of the Ribbon in the Symbols group), the text "Type equation here" is displayed in the text box, and the Design tab with Equation Tools is displayed.

$$\pi$$

Equation

The Equation Tools Design tab contains icons for inserting a wide range of equation elements. Or you can press the drop-down button below the Equation button to choose from a menu of built-in equations: area of a circle, binomial theorem, expansion of a sum, Fourier series, Pythagorean theorem, quadratic formula, Taylor expansion and two trigonometric equations.

Once the Equation button is made active, you don't need to draw text boxes for additional equations – just press the Equation button and a new text box will be created, containing the text "Type equation here".

Entering a Cell Comment

You can attach a *cell comment* to a cell, for documentation purposes. A comment appears on the worksheet in a small box similar to a Screen Tip. A small red triangle in the upper right corner of the cell indicates that the cell contains a comment. When the mouse pointer is moved over a cell that contains a cell comment, the cell comment appears.

To add a comment to a cell, click on New Comment in the Comments group of the Review tab. Enter the text of the comment in the box (Figure 1-17). To exit, simply click on any cell outside the comment box.

To edit a comment, select the cell containing the comment, then click on Edit Comment in the Comments group of the Review tab. To delete a comment, select the cell containing the comment, then click on Delete Comment, or click on Clear in the Editing group of the Home tab, and choose Comments from the submenu.

To turn display of comments and/or comment indicators on or off, click on the Office Button, press the Excel Options button, click on Advanced in the list of options, scroll down to the Display group, and press the appropriate button in the Comments category.

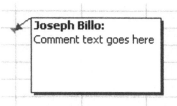

Figure 1-17. A cell comment.

Opening, Closing and Saving Documents

Commands for managing documents are in the File tab of the Ribbon (Excel 2010) or in the Office Button window (Excel 2007). Clicking on the Excel 2007 Office Button displays the Office Button window, shown in Figure 1-3, that contains the commands that were in the Excel 2003 **File** menu: **New, Open, Save, Save As, Print, Close,** and **Exit,** plus a list of recently used documents.

Opening or Creating Workbooks

Use the Open command to locate and open an existing document; or New to create a new document. The New dialog box gives you a choice of opening either a new worksheet or any of the built-in or user-created template sheets.

To open an existing workbook or worksheet from the desktop, simply double-click on it. This will open the document (and will start Excel as well if it isn't already running). If you start Excel first, it will open a new blank workbook.

The List of Recently Used Files

In Excel 2003, a list of recently used files was displayed at the bottom of the **File** menu. In Excel 2007/2010, the list of recent files has an added feature: you can "pin" a file to the list so that it remains even after many other files have been opened.

To keep a file pinned in the recently used files list, click on the Office Button to display the list of Recent Documents (Excel 2007) or click on the File tab and then click the Recent button to display the list of Recent Workbooks (Excel 2010). Click the pin icon next to the filename; click again to unpin the file.

To clear all unpinned files from the list of recently used files (Excel 2010 only), click on the File tab and then click the Recent button to display the list of Recent Workbooks. Right-click any file in the list and select Clear Unpinned Items, then press "Yes".

Using Close or Exit

You can close a document either with the Close button in the File tab (Excel 2010) or the Office Button window (Excel 2007) or by using the Close button on the document title bar. You will be asked if you want to save changes.

When you use the Exit button in the File tab (Excel 2010) or the Office Button window (Excel 2007), you close all open documents (you will be asked if you want to save changes) and then exit from Excel.

Using Save or Save As...

When you save a newly created workbook, the Save dialog box will prompt you to assign a name to the document. Excel 2007/2010 automatically appends a four-letter filename extension (e.g., .xlsx) to identify the file format type.

A document name can contain up to 218 characters (the name includes the complete path to the file, including drive letter, server name, folder path, file name and the filename extension). File names can include spaces but not any of the following characters: slash (/), backslash (\), greater-than sign (>), less-than sign (<), asterisk (*), question mark (?), quotation mark ("), pipe symbol (|), colon (:), semicolon (;). Sheet names can include most of the preceding characters, but can't include question mark (?), colon (:), backslash (\) or asterisk (*) (wildcard characters or file delimiters).

You can use Save As... to create a backup copy of a workbook by giving the copy a different name.

To Save a File for Use in Excel 2003

If you transmit an Excel 2007/2010 workbook to an Excel 2003 user, you may want to send it in Excel 2003 format, since some users of Excel 2003 may be unable to open documents in Excel 2007 format.

Excel 2010. Click on Save As in the File tab of the Ribbon. In the Save As dialog box, choose "Excel 97-2003 Workbook (*.xls)" in the Save As Type list box.

Excel 2007. Click on the Microsoft Office Button, and then click Save As. In the list of file types, choose Excel 97-2003 Workbook (Excel 2007).

Excel features that are specific to Excel 2007/2010 (for example, 16,384 columns) will not be displayed when the workbook is opened in Excel 2003, but they will still be available even if the workbook is edited, saved in Excel 2003 format, and then reopened in Excel 2007/2010.

If you want to save all Excel workbooks in Excel 97-2003 format without going through the above, click on Options in the File tab of the Ribbon (Excel 2010) or click on Excel Options in the Office Button window (Excel 2007) to display the Excel Options window. Click on Save in the list on the left side of the window. In the Save Workbooks category, click on the Save Files In This Format drop-down button to display the list of file formats, and choose Excel 97-2003 Workbook. Now you can just choose Save and your document will be saved in the old format.

Editing a Worksheet

Most commands for editing and formatting a worksheet are located in the Home tab.

Inserting or Deleting Rows or Columns

In Excel 2003, the commands for deleting cells, rows or columns were located in the **Edit** menu, while the corresponding commands for inserting were located in the **Insert** menu. Excel 2007 located all of these commands in a more logical manner, in the same Home tab of the Ribbon.

To insert an entire column of blank cells, click on a column header (the gray rectangle at the top of the column) to select (highlight) an entire column. Then

click on the Insert icon in the Cells group in the Home tab. A new column will
be inserted to the left of the column you selected. Insert a new row in a similar
way; the new row will be inserted to the left of the selected row. Multiple rows
or columns can be inserted in a similar fashion, by selecting as many rows or
columns as you want to insert.

If you click on the drop-down button at the bottom of the Insert icon to
display the Insert submenu, illustrated in Figure 1-18 Left, you can choose to
insert cells, rows, columns, or a new worksheet. Deleting cells, rows, columns or
a worksheet is done in a similar way, using the Delete icon (Figure 1-18 Right).

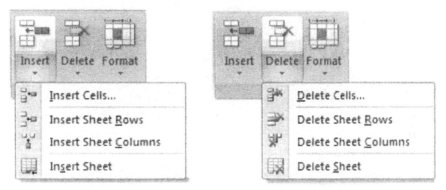

Figure 1-18. The Insert (left) and Delete (right) icons

To insert additional cells within a row or column, select the cell range above
or to the left of which you want to insert cells and click on Insert Cells... in the
Insert menu (Figure 1-18 Left). Excel usually makes a pretty good guess whether
the cells should be shifted to the right or down to make the proper insertion, but
always check to make sure. Then click OK in the dialog box (Figure 1-19).
Deleting cells is done in a similar manner.

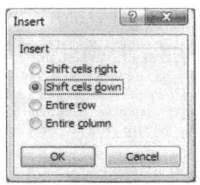

Figure 1-19. The Insert dialog box.

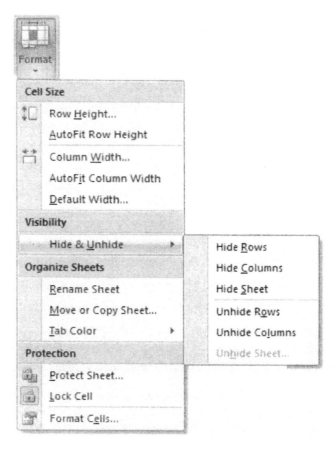

Figure 1-20. The Format Cells menu, showing the Hide/Unhide submenu.

Hiding Rows or Columns

If you have a spreadsheet with many columns or rows of intermediate formulas that lead to a final result, you can hide these intermediate calculations, in order to make the spreadsheet less cluttered. First, select the rows or columns to be hidden. Click on Format in the Cells group in the Home tab of the Ribbon, click on Hide & Unhide in the drop-down menu, and choose Hide Rows or Hide Columns from the submenu (Figure 1-20).

Using Cut, Copy and Paste

Single cells, ranges of cells, or whole rows or columns can be copied or cut from the worksheet and inserted into other locations. In general the destination will be the same size as the range of copied or cut cells, although the contents of a single cell can be copied into a range.

First, select the cell or range that you wish to copy or cut. Then click on Copy or Cut in the Clipboard group of the Home tab, or press Ctrl+C (Copy) or Ctrl+X (Cut). A *marquee* (a dashed line) will appear around the selected cells and a copy of the cells will be placed on the Clipboard. Next, select the destination cell or range. You can now transfer the copy to the destination by clicking on the Paste button in the Clipboard group, or by pressing Ctrl +V (Paste).

Excel Tip.   *When performing a Paste operation, instead of selecting a destination range that is the same size as the copied or cut range, you should always select the single cell that is the upper left corner of the destination range.*

You can also copy or cut text in the formula bar and paste it in a worksheet cell. Select the text to be copied or cut, then copy or cut as described previously. Complete the operation by clicking the Enter button in the formula bar. Then paste in the desired cell.

Using Paste Special...

When you copy a cell and then paste it, Excel transfers the cell's contents, format and comment, if present. You can choose to transfer only some of these cell attributes by using Paste Special.... in the drop-down menu of the Paste button. Paste Special permits you to paste only formulas, formats or comments. In addition, you can convert formulas to constants by choosing Values.

Figure 1-21. The Paste menu, showing the Paste Special... command.

To display the Paste Special dialog box, you must first copy cell contents. After copying, select a cell or range in which you want the copied values to be placed. In the Clipboard group in the Home tab, click on the drop-down arrow of the Paste icon to display the Paste options (Figure 1-21) and click on Paste Special... to display the Paste Special dialog box (Figure 1-22).

If you press one of the Operation buttons in the Paste Special dialog box, the value in the destination cell will be added to, subtracted from, multiplied by, or divided by the value in the copied cell.

Figure 1-22. The Paste Special dialog box.

If the cell in either the source or the destination contains a formula, then the formula will be enclosed in parentheses and joined to the contents of the destination cell by the arithmetic operator. You may wish to experiment to see exactly how this works. Relative references in the source will be changed in the same way as in a normal Paste operation. You can also Copy cells that contain formulas and press both the Values button and one of the Operation buttons to either Add, Subtract, Multiply or Divide.

If you check the Skip Blanks check box, only non-blank cells in the source will be pasted.

Using Paste Special to Transpose Rows and Columns

If values in the source range are arranged in rows, you can convert the data to column format, or vice versa, as shown in Figure 1-23.

First, copy the cells, then select a cell or range in which you want the transposed values to be placed. Display the Paste Special dialog box, check the Transpose box, then press OK.

	A	B	C	D
1	pH	6.00	6.20	etc.
2	Absorbance	0.903	0.861	etc.

F	G
pH	Absorbance
6.00	0.903
6.20	0.861
etc.	etc.

Figure 1-23. Rows and columns transposed.
(Left) Before using and (Right) after using Paste Special (Transpose).

One limitation of Transpose: The copied cells cannot be pasted over any part of the source range.

Copying and Pasting a Picture of Cells

In Excel 2003, if you hold down the Shift key while selecting the **Edit** menu, the **Copy** command becomes **Copy Picture...**. This feature is now found in the Paste menu in the Clipboard group in the Home tab of the Ribbon. Choose the "As Picture" command from the submenu to display the Copy as Picture dialog box (Figure 1-24). You can choose the appearance and format of the copied cells.

Figure 1-24. The Copy Picture dialog box.

Using Clear

Pressing the Clear button in the Editing group in the Home tab of the Ribbon displays a submenu with Clear All, Clear Formats, Clear Contents, and Clear Comments. If you choose Clear Formats from the submenu, for example, you can remove only formats from selected cells. Choosing Clear Contents will delete the cell value but not the format.

Excel Tip.   You can add the Clear All toolbutton ⌀ *to the Quick Access Toolbar. See Chapter 23 for instructions on how to customize the Quick Access Toolbar.*

Copy, Cut or Paste Using Drag-and-Drop Editing

You can also copy, cut or paste using Excel's "Drag-and-Drop" method. With this method you cut and paste or copy and paste a selection by using only the mouse pointer.

To use this method, Drag and Drop must be enabled. Click on the File tab and press the Options button (Excel 2010) or click on the Office Button and press the Excel Options button (Excel 2007). Press the Advanced button and check the "Enable fill handle and cell drag-and-drop" box in the Editing Options group.

To cut and paste by Using Drag-and-Drop, select the range of cells to be moved. Position the mouse pointer over a border of the selection (top, bottom or side). The mouse pointer will change to an arrow. Drag the selection toward the desired position. The border of the selection will be indicated as you drag it (Figure 1-25). Finally, position the selection as desired and release the mouse button.

	A	B	C	D	E	F
1	X	Y				
2	0.100	-2.303				
3	0.200	-1.609			D4:E12	
4	0.300	-1.204				
5	0.400	-0.916				
6	0.500	-0.693				
7	0.600	-0.511				
8	0.700	-0.357				
9	0.800	-0.223				
10	0.900	-0.105				
11						
12						
13						

Figure 1-25. Cutting and Pasting cells using Drag-and-Drop editing.

To copy the selection instead of cutting, hold down Ctrl while dragging. A small plus sign will appear near the arrow pointer.

To insert the selection, hold down the Shift key while dragging. The insertion point of the selection will be indicated by a horizontal or vertical bar as you drag (Figure 1-26).

	A	B
1	X	Y
2	0.400	-0.916
3	0.500	-0.693
4	0.600	-0.511
5	0.100	-2.303
6	0.200	-1.609
7	0.300	-1.204
8	0.700	-0.357
9	0.800	-0.223
10	0.900	-0.105
11		

Figure 1-26. Inserting cells using Shift+ Drag-and-Drop.

Duplicating Values or Formulas in a Range of Cells

To copy a value or formula in one cell into a range of cells, highlight the source cell whose value you want to duplicate, plus the destination cells below or to the right of where you want the value duplicated. Then click on the Fill button in the Editing group of the Home ribbon and choose Down, Right, Up or Left from the submenu.

If the cell contains a number or a text label, the value will be duplicated in the rest of the cells. If the cell contains a formula, the formula will be copied into the selected cells, except that Microsoft Excel uses *relative referencing* when formulas are copied. For example, if cell A2 contains the formula =A1+1, and Fill Down is used to copy the formula into a range of cells below cell A2, the formula copied into cell A3 will be =A2+1, and so on.

Cell references are adjusted when you insert or delete rows or columns, too. If you insert a new column to the left of column A in the preceding example, the formula in cell B2, which used to be cell A2, will read =B1+1.

To use the Across Worksheets option in the Fill submenu, you must select multiple sheets beforehand (see "Selecting Multiple Worksheets: [Group] Mode" earlier in this chapter). When you choose Across Worksheets from the submenu, the Fill Across Worksheets dialog box (Figure 1-27) will appear; you can choose to fill Contents, Formats or both.

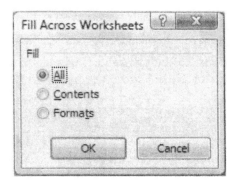

Figure 1-27. The Fill Across Worksheets dialog box.

Absolute, Relative and Mixed References

A relative reference such as A1, becomes A2, A3, etc., as you Fill Down a formula into cells below the original formula. To keep the address of a cell fixed when you use the Fill commands, precede both its letter and number designation by a dollar sign (e.g., \$A\$1). An *absolute reference* such as \$A\$1 remains \$A\$1 as you Fill Down. You will find this absolute cell addressing useful if you wish to use a constant in a formula.

Occasionally it is useful to use *mixed references*. A mixed reference is a reference such as A\$1 or \$A1; the row or the column designation, respectively, will remain constant when you Fill Down or Fill Right.

	A	B
1	X	Y
2	0.100	-2.303
3	0.200	-1.609
4	0.300	-1.204
5	0.400	-0.916
6	0.500	-0.693
7	0.600	-0.511
8	0.700	-0.357
9	0.800	-0.223
10	0.900	-0.105

	A	B
1	X	Y
2	0.100	-2.303
3	0.200	-1.609
4	0.300	-1.204
5	0.400	-0.916
6	0.500	-0.693
7	0.600	-0.511
8	0.700	-0.357
9	0.800	-0.223
10	0.900	-0.105

Figure 1-28. Two views of the same worksheet, showing formulas (left) and values (right). The formula in cell A4 has been filled down into A5:A12.

Relative References When Using Copy or Cut

If you copy and paste a formula, its references will be transferred using relative referencing. Thus, if you copy the formula =A1+1 from cell A2 and paste it in cell A10, the formula in cell A10 will be =A9+1. If you copy the formula from cell A2 and paste it in cell C2, the formula in cell C2 will be =C1+1. (This is probably not the formula you want.)

On the other hand, if you cut the formula in cell A2 and paste it anywhere in the worksheet, it will still be the formula =A1+1.

Thus the difference (with respect to cell references) between copy and paste and cut and paste is that cut adjusts relative references so that they still refer to the original cells, while copy does not adjust relative references, with the result that they refer to different cells.

The best way to copy a formula to a different row and column without altering relative references is to copy it from the formula bar, click the Enter box to complete the copy operation, and then paste in the destination cell.

Using Autofill to Fill Down or Fill Right

Excel's AutoFill feature lets you Fill Down or Fill Right simply by using the mouse pointer. To use AutoFill to Fill Down a formula in a cell, select a cell by clicking on it. You will see a small black square on the lower right corner of the selected cell. Position the mouse pointer exactly over the small black square (the AutoFill handle or simply the *fill handle)*. The mouse pointer becomes a small black cross. Click and drag in the usual way to select a range of cells. If the cell contains a formula, it will be copied into the rest of the range just as if you had used Fill Down or Fill Right; relative references in the formula will be adjusted appropriately. If the cell contains a number or a text label, the value will be duplicated in the rest of the cells. With AutoFill you can also Fill Up and Fill Left.

Excel Tip.   To Fill Down a value or formula to the same row as an adjacent block of values, select the source cell and double-click on the fill handle.

Using Autofill to Create a Series

AutoFill provides an additional feature: you can use it to create a series. There are several ways to create a series. For example, to create the series of integers 1, 2, 3, . in column A, you can either:

- Enter the value 1 in cell A1, enter the formula =A1+1 in cell A2, and then use Fill Down to create the series. You can then use Copy and Paste Special (Values) to convert the formulas to values.

• Use Series... from the Fill submenu of the Editing group in the Home tab. With Series (Figure 1-29) you enter the start value, the end value and the increment.

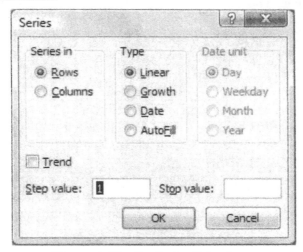

Figure 1-29. The Series dialog box.

• Use AutoFill. This is by far the simplest and most convenient method. If you select a cell containing a number formatted as a date or time, or a text label containing a number, and use AutoFill to fill a range of cells, AutoFill creates a series using the selected cell as the starting value. The value of the series being entered in the active cell is displayed in a Screen Tip box as you drag the AutoFill handle.

If you select a cell containing a date, Excel will create a date series. For example, if you select a cell containing 29-Jan and use AutoFill to Fill Right, the series 29-Jan, 30-Jan, 31-Jan, 1-Feb, 2-Feb... will be created. A cell containing the text "Sunday" will produce the series Sunday, Monday, Tuesday,...; "Sample 1" produces Sample 1, Sample 2, Sample 3,....

If you select two cells, AutoFill will create a series based on the cells you select, as shown in the second and third examples of Figure 1-30.

Excel Tip. *To prevent AutoFill from creating a series, hold down the Ctrl key as you position the black cross pointer over the fill handle. A small plus sign will appear to the right of the black cross pointer. Click and drag in the usual way to fill rather than create a series.*

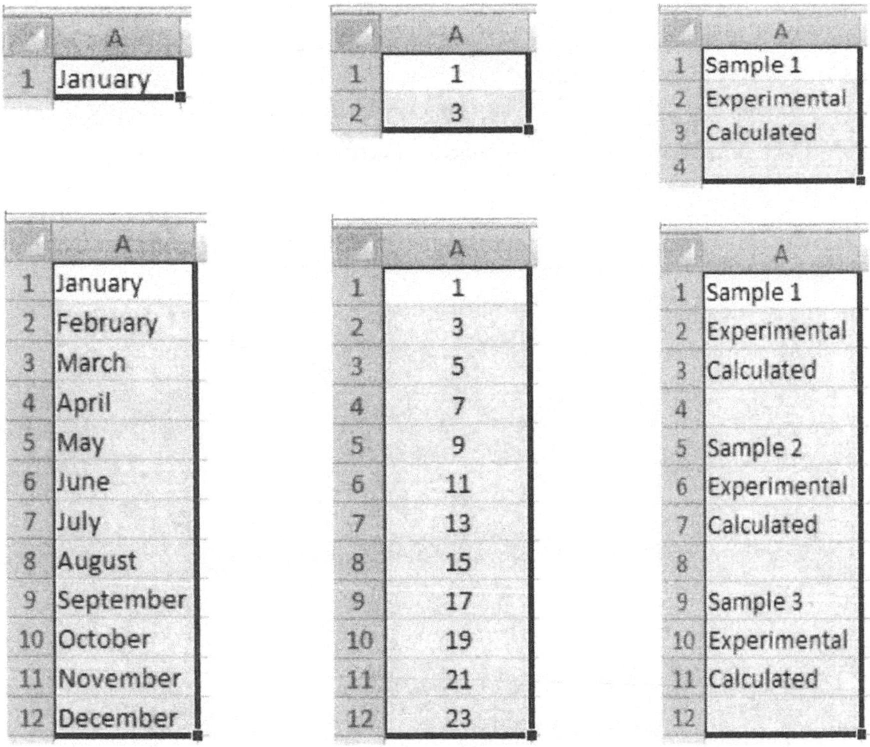

Figure 1-30. Some examples of the use of AutoFill to produce a series. (Above) Cells before using AutoFill. (Below) Series produced by AutoFill.

The AutoFill Shortcut Menu

If you use the right mouse button to drag down the Fill Handle to create a date series, a shortcut menu will be displayed when you release the mouse button (Figure 1-31).

You can then choose to create a series of consecutive dates (Fill Days), a series consisting only of weekdays (Fill Weekdays), a series consisting of a single date in each month (Fill Months) or a series consisting of a single date in each year (Fill Years).

Figure 1-31. The AutoFill shortcut menu.

Formatting Worksheets

You can use commands in the Home tab of the Ribbon to change and improve the appearance of the worksheet and to modify the way number values are displayed.

Using Column Width... and Row Height...

When Excel creates a blank worksheet, all rows are the same default height, all columns the same width. You can change the width of columns, or the height of rows, to improve the appearance of a worksheet or to eliminate wasted space so that you can get more information on a single page. You can also hide rows or columns by reducing their height or width to zero. The data they contain will still be still there, but it will be hidden.

To change the width of a column or the height of a row, click on the Format button in the Cells group in the Home tab of the Ribbon, then choose from the Cell Size group in the drop-down menu. You can enter a value for the height of a row or width of a column (one width unit corresponds to the width of one character of the current font). Column widths and row heights can also be adjusted by choosing the appropriate Autofit option from the submenu. You can adjust the column width or row height to fit a single selected cell or a range of cells, or to be the best fit to the widest entry in a whole column or row.

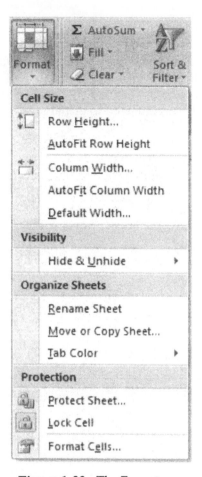

Figure 1-32. The Format menu.

You can also change the column width by using the mouse pointer. Place the cursor on the separator bar between column headings, on the right of the column whose width you want to change. The cursor changes to a double-headed arrow: ✛. Click the mouse button and drag to the right or left to change column width. The column width is displayed in an "InfoBox" as you drag the separator bar.

Excel Tip. To adjust several columns at a time to the same width, select the columns and then perform the column width adjustment with the mouse pointer on any of the selected columns. When you release the mouse button, all the columns will have the adjusted width. You can also get a "best fit" simply by double-clicking on the row or column separator bar. To adjust several rows or columns at once, select the columns and double-click on any row or column separator bar.

Formatting Cells

Formatting allows you to change the appearance of values in cells. You can format cells either by using icons in the Home tab of the Ribbon, by using toolbuttons on the Quick Access Toolbar, by using the Mini Toolbar, or by clicking on the Format Cells... command to display the Format Cells dialog box.

The Mini Toolbar

The Mini Toolbar is a formatting toolbar that can be used to format text. It appears automatically when you select (highlight) the contents of a cell, either in the cell or in the Formula Bar. A faint image of the Mini Toolbar will appear above the text. If you move the mouse toward the Mini Toolbar, the image intensifies; if you move the mouse away, it fades. If you do neither, it disappears after a few seconds.

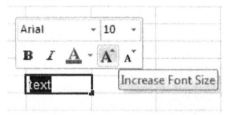

Figure 1-33. The Mini Toolbar appears when formatting text in a cell.

A different Mini Toolbar appears when you select text in a chart.

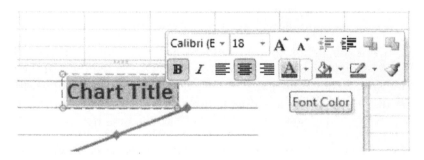

Figure 1-34. The Mini Toolbar appears when formatting text in a chart.

In Excel 2010 only, if you click on the drop-down list of fonts or font sizes and move the mouse downwards to hover over a font or size from the list, the text in the cell will temporarily display the formatting.

Figure 1-35. Using the Mini Toolbar to change font size.

If you right-click on an object – a cell, textbox, row, column, etc. – you will display a shortcut menu with commands appropriate for that object, plus a Mini Toolbar.

To disable the Mini Toolbar in Excel 2010, click on the File tab and click on the Options button. Click on the General options button and uncheck the "Show Mini Toolbar On Selection' box. For Excel 2007, click on the Office Button and click on the Excel Options button. Click on the Popular options button and uncheck the "Show Mini Toolbar On Selection' box.

The Format Cells Dialog Box

The Format Cells dialog box is identical to the dialog box in Excel 2003. To display the dialog box, click on the Format button in the Cells group in the Home tab of the Ribbon. Click on Format Cells in the drop-down menu (Figure 1-32).

The Format Cells dialog box has Font, Alignment and Number tabs to format values within cells, Border and Patterns tabs to format cells, and the Protection tab to set the security of values within cells. The possibilities for Font, Alignment, Border and Patterns are many and varied, and only some of these possibilities will be discussed here. Number formatting is important for scientific spreadsheets; it is discussed in detail in a following section.

Excel Tip.   Click on the small button ▣ at the bottom right of the Font, Alignment or Number groups in the Home tab of the Ribbon to display the complete Format Cells dialog box.

Using Alignment

The Alignment tab in the Format Cells dialog box provides a number of formatting options for the alignment of values in cells. There are option buttons for both horizontal and vertical alignment (Figure 1-36). The Vertical orientation options are useful if you want to add a text label to a narrow column. The Orientation "inclinometer" allows you to display text on any angle.

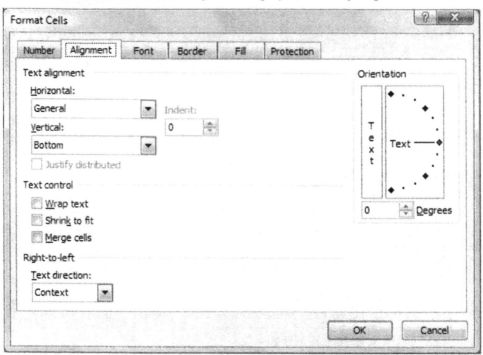

Figure 1-36. The Alignment dialog box.

To use the Merge Cells option, first select the range of cells to be merged, check the Merge Cells box and press OK. You can also use the Merge and Center button in the Alignment group in the Home tab of the Ribbon, or customize the Quick Access Toolbar with the Merge and Center toolbutton (see Chapter 23 for instructions on how to customize the Quick Access Toolbar).

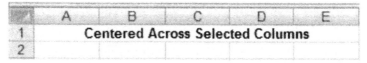

Figure 1-37. Using Merge and Center.

For the most common horizontal alignment options you can use the alignment toolbuttons ▉ ▉ ▉ in the Alignment group in the Home tab of the Ribbon to align text in cells left, centered or right, respectively.

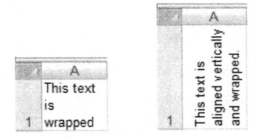

Figure 1-38. Examples of using Wrap Text.

You can also format a text entry in a cell so that the text wraps and is displayed in more than one line (Figure 1-38), by checking the Wrap Text box in the Format Cells dialog box. Excel breaks the text at a space character. Text can be aligned vertically and wrapped.

If you select text in a text box, the Alignment button is not active. To change the alignment of text in a text box, you must press the Orientation button   in the Alignment group. Only vertical orientations are available.

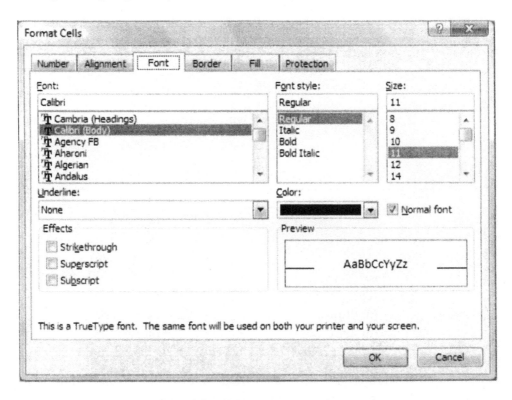

Figure 1-39. The Font dialog box.

Using Font

The Font tab (Figure 1-39) in the Format Cells dialog box allows you to format cells in any of the installed fonts. In addition, you can format individual characters in various font styles or sizes, or as strikethrough, superscript or subscript characters.

You can also use Greek letters, as shown in the text box example in Figure 1-16, by formatting the Roman letter using the Symbol font. Of course, you must know the correspondence between Roman and Greek letters: The Roman S becomes the Greek Σ, for example.

The Alternate Character Set

You can enter symbols that are in the so-called *alternate character set*; an example is shown in Figure 1-40. The characters produced may be different for each different font.

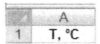

Figure 1-40. A special character (°) typed by using the alternate character set.

The characters are obtained by holding down the Alt key and typing the four-digit ASCII code for the character, *using the numeric keypad*. (If you're using a laptop without a numeric keypad, you'll need to hold down both the Fn and the Alt keys and then press the keys labeled with small digits 1 – 9, color-coded to match the Fn key.)

Table 1-5. Some Useful Alternate Characters

symbol	4-digit code		symbol	4-digit code	
€	0128	(Euro)	ß	0223	(Greek beta)
£	0163	(Pound)	µ	0181	(Greek mu)
¥	0165	(Yen)	Å	0197	(Ångstrom)
•	0149	(bullet)		0160	(unbreakable space)
·	0183	(centered dot)	°	0186	(superscript)
×	0215	(multiply)	¹	0183	(superscript)
÷	0247	(divide)	²	0178	(superscript)
±	0177	(plus or minus)	³	0179	(superscript)
°	0176	(degree)	—	0151	(em dash)

If you use a different font, you'll have to experiment to see what alternate characters are produced.

The range of useful characters obtainable in this way is rather limited. Table 1-5 shows some useful characters obtained using the Arial font. The complete set of alternate characters is listed in Appendix J.

Entering Subscripts and Superscripts

You can enter subscripts and superscripts in text, as in Figure 1-41. First, highlight the characters to be subscripted or superscripted, either in the formula bar or directly in the cell. Choose the Font tab in the Format Cells dialog box, check the Subscript or Superscript check box (see Figure 1-39), press the OK button, and then enter the text by pressing the Enter key or the Enter button.

A
1	$C_{12}H_{22}O_{11}$
2	$BaCl_2 \cdot 2H_2O$
3	$H_2SO_4 + 2NaOH == Na_2SO_4 + 2H_2O$

Figure 1-41. Subscripts in Excel.

Using Border and Patterns

The Border tab in the Format Cells dialog box allows you to place a border around one or more sides of a selected cell or range. This is useful if you want to emphasize comments, instructions or values. The Patterns tab is used to change the background color or pattern of cells.

BOSTON COLLEGE			DATE	6/5/1996
OFFICE OF THE UNIVERSITY REGISTRAR				
CHESTNUT HILL, MASSACHUSETTS 02167			ACADEMIC YEAR	1996
			SEMESTER	96S
COURSE		CH 222 01 INTRO/INORGANIC CHEM		
INSTRUCTOR		BILLO, E JOSEPH		
MEETING PLACE				

STUDENT'S NAME	ID NUMBER	SCHOOL	CLASS	MAJOR	CREDIT	GRADE

Figure 1-42. Using Border to create a custom report form.

The Border option is often used to underline headings, or in a sheet in which the gridlines have been removed, to create a custom form. Figure 1-42 shows a portion of a sheet produced in this way. To remove the existing gridlines, click on Options in the Office Button window or in the File tab, click on Advanced, and in the "Display options for this worksheet" group, and uncheck the Show Gridlines box.

The built-in template sheets provided with Excel (click on New in the File tab in Excel 2010 or click on New in the Office Button window in Excel 2007) are good examples of the use of Border and Patterns to create custom forms.

Using the Format Painter Toolbutton

The Format Painter button in the Clipboard group in the Home tab of the Ribbon, or the Format painter toolbutton 🖌 , copies and pastes formats from one cell or range to another cell or range. To use it, do the following: Select the cell or range with the desired format(s), click the Format Painter toolbutton (this copies the formats), and click on a cell or drag across a range of cells to paste the format(s).

> *Excel Tip.* *To use the Format Painter button to "paint" a format on a series of non-adjacent cells or ranges, select the cell with the desired format, and then double-click on the Format Painter button. This will keep the button in the "pressed" position, allowing you to click on several cells or ranges to paste the format. When you're done, click once on the button to return it to the "unpressed" state.*

Number Formatting

The formatting described in the preceding sections — bold text, italic text, alignment of text, adjusting column widths, etc. — is sometimes referred to as *stylistic formatting*. In addition, it is possible to change the way number values are displayed in cells. This type of formatting, called *number formatting*, is described in the following sections.

Using the Number Formatting Toolbuttons

You can also format number values in cells by using the number formatting toolbuttons shown following.

.00	Increases the number of decimal places.
.00	Decreases the number of decimal places.
%	Formats the number in percent style, with no decimal places.

$ Formats the number in currency style, with two decimal places.

,   Formats a number with commas and two decimal places.

Excel Tip. *There isn't a toolbutton to format number values in scientific format. You can apply scientific format conveniently by using the shortcut key sequence Ctrl+Shift+^. See Appendix E for a complete list of shortcut keys.*

Using Excel's Built-In Number Formats

To change the way a number is displayed in a cell, click on the drop-down list box button (labeled "General") at the top of the Number group in the Home tab of the Ribbon. The list, shown in Figure 1-43, displays most of the available number format categories — Number, Currency, Date, Time, Percentage, Scientific, etc.

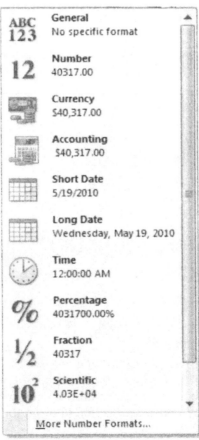

Figure 1-43. The Number Format drop-down menu.

For example, selecting a cell and then choosing Number from the list of categories will display the value in the cell to two decimal places (the default); you can change the number of displayed decimal places using the Decimal Places box. You can also display values as percentages or in exponential notation, for example.

Another route to number formatting is to click on the Format button in the Cells group in the Home tab of the Ribbon and then click on Format Cells in the drop-down menu (Figure 1-32). Click on the Number tab to display the Number Format dialog box (Figure 1-44). The number formats in the Category list correspond to the formats shown in Figure 1-43.

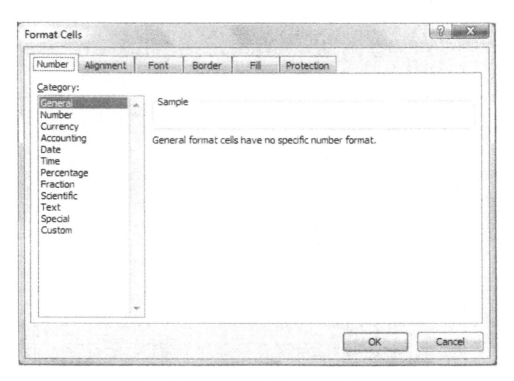

Figure 1-44. The Number Format dialog box.

The appearance of the Number dialog box will be different, depending on the number format category you select.

If you choose the Custom category, you will see the number formatting code that was applied to the cell. If you scroll through the list of codes, you'll see that many of them are quite complex (see Figure 1-45). For the meaning of the built-in number format code symbols, see Table 1-6 or go to "Number Format Codes" in Excel's On-Line Help.

Custom Number Formats

You can create your own custom number formats. First, choose the Custom category from the list of number formats. This will display the list of (so far) built-in number formats. To add a new, user-defined number format (it will be added at the bottom of the list), type the format in the Type box. For example, if you want to display numbers to four decimal places, type 0.0000 in the Type box. The new format will be stored in the list of formats so that you can apply it to other cells; the format is available in all sheets in the workbook.

Table 1-6 lists the formatting symbols you can use to create your own custom formats.

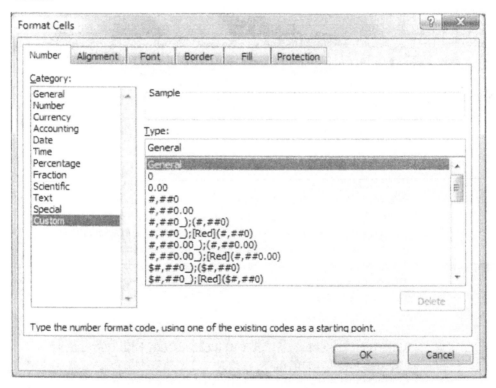

Figure 1-45. The Number Format dialog box showing number formatting codes.

For example, to format a column of telephone numbers, use the custom format (###) ###-####. This will format a cell entry such as 6175523619 in the format (617) 552-3619. The format #.??????? was used to format the table of atomic weight values shown in Figure 1-47, so that they are aligned on the decimal. (Note that, since the format contains seven ? symbols and the atomic weight of Na has only six digits to the right of the decimal point, there is an additional space to the right of the number.)

	A	B
1	H	1.00797
2	O	15.9994
3	Na	22.989768
4	S	32.066

Figure 1-46. Values aligned on the decimal point by using the ? formatting symbol.

You can create some fairly sophisticated number formats. For example, the format $#.0,, (dollar sign, number sign, period, zero, comma, comma) formats financial entries rounded to millions, with one decimal; the value 21180000 is displayed as $21.2.

You can use number formatting to add units to a number value. For example, the format #" g" appends the grams unit g to a number value; the value 50 is displayed as 50 g, as shown in Figure 1-47.

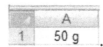

Figure 1-47. Units added to a value by means of number formatting.

Table 1-6. Number Formatting Symbols

#	Placeholder for digit.
0	Placeholder for digit. Displays an extra zero if the number has fewer digits than the number of zeros specified in the format.
?	Placeholder for digit. Same as 0, except that a space character is displayed. Also used when formatting a number as a fraction.
,	Thousands separator (if used with # or 0). Used alone, it rounds and truncates to the thousands place (millions place if two commas are used, etc.)
%	Converts to percent.
E	Converts to scientific format. Use E- to include sign with negative exponents only, E+ to include sign with both positive and negative exponents.
/	Converts to a fraction. Usually used in the form # ??/?? or ##/##. The number of ? or # symbols determines the accuracy of the display.
"*text*"	Text characters can be included in a format by enclosing them in quotes. You can sometimes get away without the quotes, but it's safer to use them.
@	Text placeholder. If the cell contains a text entry, the text is displayed in the format where the @ symbol appears.
[RED]	Displays the characters in the cell in red. You can also use [BLUE], [GREEN], [YELLOW], etc.

Custom Date Formats

You can create custom date formats by using the year, month and day formats listed in Table 1-7. Day or month formats can have one-, two-, three- or four-letter formats; year formats can have either two- or four-letter formats. For example, the number format dddd, mmmm d, yyyy applied to a date entered as 8/3/38 will display Wednesday, August 3, 1938.

Table 1-7. Date Formatting Symbols*

d	Displays the day as a number without leading zeros (1-31)
m	Displays the month as a number without leading zeros (1-12)
dd	Displays the day as a number with leading zeros (01-31)
mm	Displays the month as a number with leading zeros (01-12)
ddd	Displays the day as an abbreviation (Sun-Sat)
mmm	Displays the month as an abbreviation (Jan-Dec)
dddd	Displays the day as a full name (Sunday-Saturday)
mmmm	Displays the month as a full name (January-December)
yy	Displays the year as a two-digit number, e.g., 97
yyyy	Displays the year as a four-digit number, e.g., 1997

* See Chapter 12, "Other Language Versions of Excel", for date formatting symbols for some other languages.

Time Formats

Excel's built-in or custom time formats use the symbols in Table 1-8.

Table 1-8. Time Formatting Symbols

h	Displays the hour without leading zeros (0-23)
hh	Displays the hour with leading zeros (00-23)
m	Displays the minutes with leading zeros (0-59)
mm	Displays the minutes with leading zeros (00-59)
s	Displays the seconds without leading zeros (0-59)
ss	Displays the seconds with leading zeros (00-59)
s.000	Displays seconds to the millisecond
AM/PM	Displays the hour as AM or PM instead of 24-hour time
[h]	Displays elapsed time in hours
[h]:mm	Displays elapsed time in hours and minutes

Variable Number Formats

Different number formats can be applied to positive, negative, zero and text values entered into a cell. A complete format consists of four sections separated by semicolons, for positive, negative, zero and text values, respectively. If only one number format is specified, it applies to all values. If two number formats are specified, then the first one applies to positive numbers and zero, the second to negative numbers. For example, the format $#,###;[Red]$#,### formats positive amounts in black, Excel's default color, and negative amounts in red.

Conditional Number Formats

Conditional number formats can be created by using the syntax [*condition, value*] *format statement.* *Condition* is one of the symbols <, >, =, >=, <=, <>; *value* may be any number. Format statement may be any built-in or custom format. For example, the number format

[>1] "Number too large"

displays any input less than 1 but otherwise issues an error message.

Several conditions may be combined using semicolons. The number format

[>999]#.##,,%;#" ppm"

displays the values 110 and 21560 as 110 ppm and 2.16%, respectively.

Formatting Numbers Using "Precision as Displayed"

To permanently change *all values* stored on a worksheet to their displayed values, use the Precision As Displayed option. Once this command has been invoked, you can't restore the original values.

To apply Precision as Displayed, click on the File tab and press the Options button (Excel 2010) or click on the Office Button and press the Excel Options button (Excel 2007); click on Advanced and check the Set Precision As Displayed box. Because this is an irreversible change, Excel asks you to confirm the change.

To change only a selected range of values to "precision as displayed", use the FIXED worksheet function (see "Text Functions" in Chapter 3).

Excel Tip. You can apply the same formatting to multiple worksheets simply by grouping the sheets (click on the first sheet tab in the range of sheets to be formatted, then hold down the Shift key and click on the last sheet in the range). When you apply the desired formatting to the active sheet, it will be applied to all sheets in the group.

Conditional Formatting (Part I)

The number formatting tools that we have seen described in preceding sections can only change the way the number appears in the cell: as a floating point number, in scientific notation, etc. But none of the number formatting tools allows you to apply stylistic formatting to a cell based on the value in that cell — to display a number in italics if it is negative, for example. Conditional formatting provides this capability.

Conditional formatting allows you to do one of two things: to apply stylistic formatting to a cell based on the value in that cell, or to apply formatting to a cell based on values in several cells (in the same or different worksheets or workbooks). The former — formatting a cell based on the value in that cell — will be described here; the latter will be reserved for Chapter 3, Excel Formulas and Functions.

Conditional formatting has been improved significantly in Excel 2007/2010. In Excel 2003, you can format cells that contain values that satisfy one of eight criteria: between, not between, equal to, not equal to, greater than, less than, greater than or equal to, less than or equal to. In Excel 2007/2010 many more criteria can be applied in addition to the ones listed. Other possibilities include highlighting cells that contain error values, specified text or date values, duplicate or unique values, top N values (e.g., top 10), top Nth percent, above or below average. Figure 1-48 illustrates some of the most popular options provided by Excel 2007/2010 conditional formatting.

As an example of the use of conditional formatting, consider the case of a worksheet that contains a column of values. If any value in the column is above 90 (for example), the measurement is out of range and needs to be examined more closely. You can flag these out-of-range values with conditional formatting. After highlighting the range of cells to be formatted, click on Conditional Formatting in the Styles group of the Home tab, click on Highlight Cells Rules in the drop-down list, and then click on Greater Than... in the submenu, to display the Greater Than dialog box. Enter the value 90 as the criterion, choose Light Red Fill as the format, and press OK. The cells satisfying the criterion will be highlighted, as illustrated in Figure 1-50.

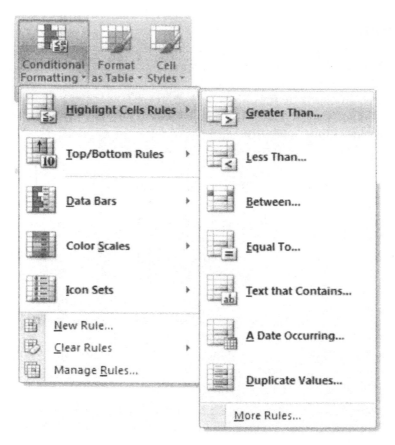

Figure 1-48. Some of the choices for Conditional Formatting.

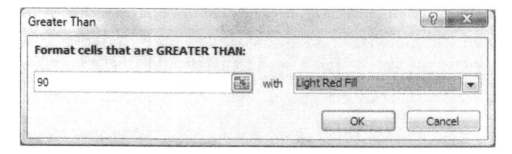

Figure 1-49. The Greater Than dialog box.

	L
1	7.88248
2	86.42525
3	27.06193
4	70.87242
5	99.36292
6	73.5115
7	22.08262
8	74.31941
9	13.85085
10	88.59572
11	11.99503
12	5.40923
13	96.69842
14	75.1717
15	29.25795

Figure 1-50. A portion of a range of cells with conditional formatting.

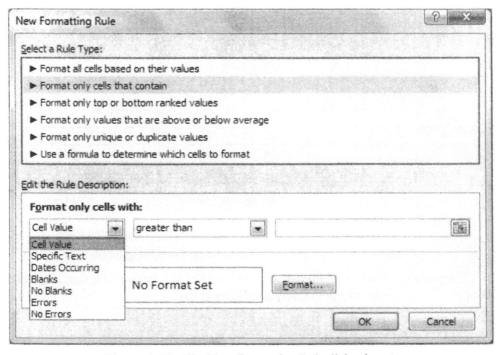

Figure 1-51. The New Formatting Rule dialog box.

Users more comfortable with the Excel 2003 Conditional Formatting dialog box can choose More Rules… in the Highlight Cells Rules submenu (see Figure 1-48) to go directly to the New Formatting Rule dialog box, shown in Figure 1-51. This dialog box provides a greater range of formatting options than the built-in options of, for example, the Greater Than dialog box shown in Figure 1-49.

In addition to formatting cells with font color, cell border or cell background, the Excel 2007/2010 conditional formatting allows you to flag cells with data bars, color scales, or icons. The gallery of icons is shown in Figure 1-52.

Figure 1-52. The gallery of Conditional Formatting icons.

Like other formatting, you can apply a conditional format to a range of cells or apply it to one cell and then Copy it to other cells.

Printing Documents

Commands for printing documents are located in the File tab of the Ribbon (Excel 2010) or in the Office Button window (Excel 2007).

Excel 2010. If you click on the Print button in the File tab, a dialog box is displayed (Figure 1-53) that contains most of the options in the Excel 2003 **Print...** and **Page Setup...** menu commands. You can choose pages to be printed, number of copies, two-sided printing, collated sheets, portrait or landscape orientation, paper size, margins, etc. For users more comfortable with the Excel 2003 Page Setup dialog box, there is a Page Setup hyperlink at the bottom right corner of the dialog box; see the following section for a description of the Page Setup dialog box. The Print Preview view of the document (not shown in Figure 1-53) is also displayed on the right side of the window.

Figure 1-53. The Excel 2010 Print window.

Excel 2007. If you click on the Print button in the Office Button window, the "Print and preview the document" window is displayed (Figure 1-54).

Figure 1-54. The Excel 2007 "Print and preview the document" window.

Choosing Print in the "Print and preview the document" window displays the Print dialog box, identical to the Excel 2003 dialog box (see Figure 2-44 in Chapter 2), which allows you to specify the pages to be printed, and the number of copies. You can also select a printer, if you have more than one printer connected.

Choosing Print Preview in the "Print and preview the document" window activates the Print Preview tab of the Ribbon (Figure 1-55). The Page Setup button displays the Page Setup dialog box, identical to the Excel 2003 dialog box.

Figure 1-55. The Excel 2007 Print Preview tab of the Ribbon.

Print Preview is useful in other ways besides showing what your finished worksheet will look like when printed. If you preview your worksheet and then return to the document window, page breaks will be displayed on the worksheet as dashed lines, to assist you in adjusting column widths, for example, before printing.

Using Page Setup

The Page Setup dialog box, identical to the Excel 2003 version, contains four tabs: Page, Margins, Header/Footer, and Sheet. These dialog boxes are shown in Chapter 2, Figures 2-40, 2-41, 2-42 and 2-43. Use the Page tab to choose Portrait or Landscape orientation. Use the Margins tab to change margins. Use the Header/Footer tab to select header or footer text from a list of built-in options or create a custom header or footer. Use the Sheet tab of the Page Setup dialog box to enter a Print Area (range of cells that will be printed) or Rows To Repeat At Top (row or rows that will be printed at the top of each printed page), or to turn on or turn off the printing of gridlines, row and column headings, etc.

> ***Excel Tip.*** *Click on the small button* ▣ *at the bottom right of the Page Setup or Sheet Options groups in the Page Layout tab to display the Page or Sheet tab of the Page Setup dialog box.*

A new document has no header and no footer; these are the default values. To enter a header or footer, choose the Header/Footer tab in the Page Setup dialog box, which will display list boxes with a wide range of built-in formats for header and footer. You can also create custom headers or footers by pressing the Custom Header... or Custom Footer... button. The Header or Footer dialog boxes are identical; each enables you to enter filename, sheet name, page number, date, time or other information.

To adjust margins, use the Margins tab of the Page Setup dialog box, or click on the Page Layout tab of the Ribbon and click on the Margins icon to display the drop-down menu of margin settings (Figure 1-56).

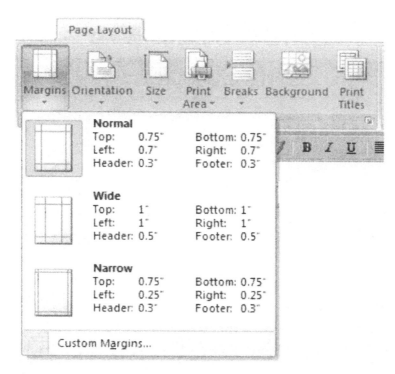

Figure 1-56. The Page Layout tab of the Ribbon, showing the Margins drop-down menu.

To squeeze the maximum amount of worksheet information on a single page, you can decrease the margin widths. The default margin values are 0.75 inch left and right and 0.7 inch top and bottom. If you set the margins to zero, any header and footer information will still be printed, usually right on top of data in your worksheet, so delete the header and/or footer information by choosing "(none)" from the list box.

You can choose Print Row And Column Headings and/or Print Cell Gridlines by choosing the Sheet tab in the Format Cells dialog box. If you de-select Cell Gridlines, they will still be displayed on the screen but they will not be printed.

You may need to use Print Black and White if your worksheet uses color. Colors may be printed as various patterns by your printer; to remove the patterns and produce text in cells in black and white, check the Print Black and White Cells box.

Using Print

If you choose the Print command and simply press the OK button, Excel will print the rectangular array of sheets that includes all filled cells. It's a good idea to use Print Preview before printing; the total number of pages to be printed will be displayed in the status bar. This will tell you whether you can print the whole worksheet, or whether you need to specify a range of pages to be printed.

If you choose Print Preview or Page Setup, Excel displays the automatic page breaks as dashed lines in the worksheet. You can insert a forced page break if you want to print a portion of a worksheet page. To insert a horizontal page break, select an entire row as if you were going to insert a row. Then click on the Breaks button in the Page Setup group in the Page Layout tab of the Ribbon, and choose Insert Page Break from the submenu. The page break will be inserted immediately above the selected row. A forced vertical page break is inserted in a similar fashion; the page break is inserted immediately to the left of the selected column. If you want to insert both a vertical and a horizontal page break, select a single cell within the worksheet; the page breaks will be immediately above and to the left of the cell.

Printing a Selected Range of Cells in a Worksheet

To print a selected range of cells within a worksheet, you must first select (highlight) the range to be printed.

Excel 2010.   Choose Print in the File tab of the Ribbon and press the drop-down button at the right of Print Active Sheets to display the drop-down menu shown in Figure 1-57. Click on Print Selection. Then press the Print button.

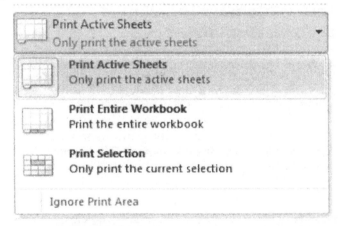

Figure 1-57. The Print Active Sheets drop-down menu.

Excel 2007.   Choose Print from the Office Button window. Press the Selection button in the Print What category box in the lower left corner of the dialog box. Then press the Print button.

Using Set Print Area

To specify a range of cells to be printed each time you choose Print, you must use Set Print Area. You can do this in at least two different ways:

- Display the Page Setup dialog box (see earlier), and choose the Sheet tab. Click in the Print Area text box to select it. Now select the range of cells that you want to print (you can move the dialog box out of the way if necessary), and press the OK button. To cancel a Print Area selection, delete the reference within the Print Area text box.

- First, select the range of cells to be printed, then press the Print Area button in the Page Setup group in the Page Layout tab of the Ribbon, and choose Set Print Area from the submenu. The range to be printed will be indicated by Page Break lines. Choose Remove Print Area from the submenu to cancel the Print Area.

If the Print Area you selected requires more than one page, you can go to Page Setup and change the value in the Reduce/Enlarge box to less than 100%. Sheets printed with values less than about 60% are difficult to read, though. To obtain the appropriate reduction value automatically, after you've selected the area to be printed, choose the Page tab and press the Fit To 1 Pages Wide By 1 Tall button.

Printing Row or Column Headings
for a Multi-Page Worksheet

If you are printing a multi-page worksheet, you can duplicate row or column headings automatically on each printed page. Click on the Print Titles icon in the Page layout tab of the Ribbon, or display the Page Setup dialog box and choose the Sheet tab. Select the Rows To Repeat At Top or the Columns To Repeat At Left text box by clicking the cursor in it. Now select the range of cells that you want to have printed on every page as a title (you can move the dialog box out of the way if necessary). Then click the OK button. The headings will appear at the top or left of each printed page.

Protecting Data in Workbooks

First of all, we should distinguish between security and protection. Security means protecting your computer from viruses. Protection means preventing users from modifying documents or viewing particular workbooks, worksheets or formulas; a number of ways of protecting data will be described in the following sections.

Protecting a Workbook

You can protect the structure of a workbook so that worksheets in the workbook can't be moved, deleted, hidden, unhidden, or renamed, new worksheets can't be inserted, and windows are the same size and in the same position each time the workbook is opened.   Values and formulas in the workbook can still be modified, though; to prevent this, see "Protecting a Worksheet" later in this chapter.

To protect a workbook, click on Protect Workbook in the Protection group in the Review tab of the Ribbon, and choose Protect Structure and Windows from the drop-down menu, to display the dialog box (Figure 1-58).

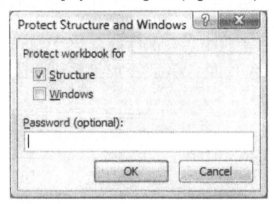

Figure 1-58. The Protect Workbook dialog box.

Check the boxes for Structure and/or Windows, enter a password if necessary (you will be asked to confirm it), and then press OK.

Protecting a Workbook
by Making It a Read-Only Workbook

If you make a workbook read-only, users can view formulas in cells, and change values and formulas, but the changes cannot be saved.

To make a workbook read-only, the document should be closed.   In the Windows Start menu, choose Programs, and then Windows Explorer.   In the Exploring window, open the drive or folder that contains the file and select the document name. Choose Properties from the File menu, choose the General tab, and check the Read-only check box.

Hiding a Worksheet

To hide a worksheet, click on the Format button in the Cells group in the Home tab of the Ribbon, choose Hide & Unhide in the drop-down menu, and choose Hide Sheet in the submenu. However, anyone can view this sheet simply by choosing Unhide from the submenu. You can hide a sheet so that most users can't view it, by using VBA: you set the Visible property of the sheet to VeryHidden, as described in the following paragraph. You may need to read Chapter 16, "Visual Basic for Applications: An Introduction", first.

To make a sheet VeryHidden, switch to the Visual Basic Editor by pressing Alt + F11. If the Project Window is not visible, display it by pressing Ctrl+R. In the hierarchy tree for the desired workbook, click on the name of the sheet you wish to hide.

If the Properties Window is not visible, display it by pressing F4. In the Properties Window, locate the Visible property (at the bottom of the list when the Alphabetic tab is selected). Click on the Visible box; this will cause a drop-down list button to appear in the properties list. Choose the xlSheetVeryHidden property, as shown in Figure 1-60.

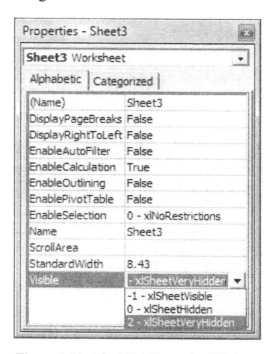

Figure 1-59. The VBA Properties Window.

When you switch back to the Excel workbook, the sheet tab will not be visible and the sheet name will not appear in the Unhide Sheet submenu.

Protecting a Worksheet
by Locking or Hiding Cell Contents

You can lock cells (prevent them from being selected by the user) or hide the contents of the cells. For example, you may want the user to be able to enter values in certain cells while protecting the rest of the worksheet. If you protect cells in this way, before you access the Protect Sheet dialog box you must specify which cells will be protected or unprotected. You do this by using the Format Cells menu.

The process for doing this is somewhat complicated. First you select cells to be locked or unlocked, or cells whose contents will be hidden or visible, and set their status using the Protection tab of the Format Cells dialog box. Then you put the status into effect by choosing Protect Sheet from the Format drop-down menu.

Before you begin, it's important to know that when a new worksheet is opened, the status of *all* cells in the document is Locked. To lock only a limited range of cells in a document (as you will most often want to do), first set the status of all the cells in the document to Unlocked and then select the range of cells that you want to be locked.

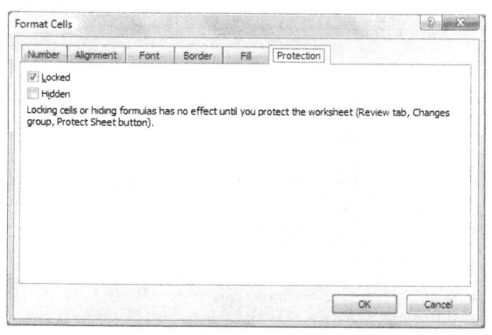

Figure 1-60. The Protection tab of the Format Cells dialog box.

As an example, let's protect a worksheet so that it has only a single unprotected cell. The user will only be able to enter a value in this cell; all other cells in the sheet will be locked. Here's how to do it: First, select the cell you want to be the unlocked cell. Now click on Format in the Cells group in the Home tab of the Ribbon and choose Format Cells from the drop-down menu. Choose the Protection tab to display the dialog box (Figure 1-60). In a new worksheet, by default all cells are locked, so simply uncheck the Locked box so that the cell you selected will be unlocked. Then press OK.

Locking or unlocking cells has no effect unless the worksheet is protected. Activate the worksheet to be protected. Click on Format in the Cells group in the Home tab of the Ribbon and choose Protect Sheet from the submenu. This will display the Protect Sheet dialog box (Figure 1-61). Here you can choose to prevent users from carrying out one or more actions on the whole worksheet (inserting or deleting rows or columns, etc.) or on specified cells. After choosing options from the list, you can enter a password if you desire (you will be asked, in a second dialog box, to confirm the password), in which case users will not be able to unprotect the sheet

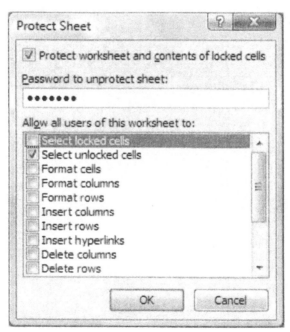

Figure 1-61. The Protect Sheet with Password dialog box.

To Lock a Range of Cells in a New Document

1. Select all cells in the document by clicking on the row/column header button in the upper left corner of the worksheet.

2. Click on Format in the Cells group in the Home tab of the Ribbon and choose Format Cells from the drop-down menu. Choose the Protection tab, uncheck the Locked option, and press the OK button. This un-protects all cells in the worksheet.

3. Now select the range of cells that you want to protect. Click on Format in the Cells group in the Home tab of the Ribbon and choose Format Cells from the drop-down menu, choose the Protection tab, check the Locked option, and press the OK button.

4. Click on Format in the Cells group in the Home tab of the Ribbon and choose Protect Sheet from the submenu, to display the Protect Sheet dialog box. You can enter a password if you wish (Figure 1-61). If you merely want to prevent yourself from making accidental changes, no password is necessary. If you want to protect the document from changes by others, you need a password; make sure that you will be able to retrieve it when you need it.

Controlling the Way Documents Are Displayed

Although only one worksheet at a time can be the active window, Excel provides a number of ways to view data in several different worksheets, or different areas of the same worksheet, at the same time.

Using New Window and Arrange All

If you have more than one document open, you can view several of them simultaneously in a number of ways. One way is to resize and move the documents so that the desired part of each can be seen in the window. Another way is to use the New Window and Arrange All icons in the Window group in the View tab of the Ribbon. The latter method can be used to view multiple documents, or multiple sheets in the same workbook, as described in the following paragraph.

To view multiple worksheets in the active workbook, click on New Window. A second window will be opened for the active workbook. If, for example, the workbook is named Viscosity Data, the windows will be named Viscosity Data:1 and Viscosity Data:2. Activate each window in turn and click on the sheet that you want to display. Now choose Arrange All, Excel displays the Arrange

Windows dialog box (Figure 1-62). You can arrange the windows horizontally (one above the other) or vertically (side by side). If you have created a separate chart sheet from data in a worksheet, Arrange All provides a convenient way to work with a sheet and observe changes in the associated chart. With Arrange All, chart documents are reduced in size so that the whole chart appears in the window; worksheet documents are not reduced in size. Figure 1-63 illustrates a worksheet/chart combination displayed using the Arrange (Vertical) option.

Figure 1-62. The Arrange Windows dialog box.

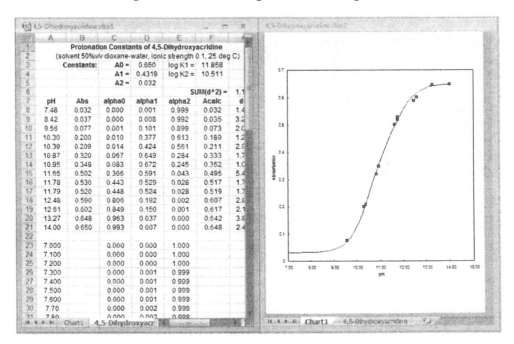

Figure 1-63. Two windows arranged vertically

With three open documents, the Tiled option arranges the documents with the active sheet occupying the left half of the screen; the other two sheets each occupy one-quarter of the screen, one above the other. With four documents Tiled, each occupies one-quarter of the screen. Click on any document to make it the active sheet. Double-click anywhere on the solid border between the windows to undo the arrangement.

Different Views of the Same Worksheet

As your worksheets get larger and more complicated, it becomes impossible to view all of a single worksheet at once, or even all cells in one row or column at one time. Excel provides several convenient ways to display separate portions of a single worksheet on the screen at the same time, so that you can view one part while entering or changing data in another part.

Click on New Window in the Window group of the View tab to display a second window of the active document. Use the Arrange All button to display two windows, either Horizontal or Vertical, of the active sheet.

You can resize and move the windows so that the desired parts of the worksheet can be seen at the same time. Click on the Title Bar at the top of a window and drag it to a suitable position for viewing; click on the side or bottom of a window and drag to resize it. This is useful if you want to Cut or Copy several cell ranges and then Paste them into another area of a worksheet, but the two areas of the worksheet are far apart.

You can set different display options for the two windows. Display values in one window and formulas in another to see the effect of changes to the formulas.

Excel Tip. To remove a workbook view, click on the window you want to remove (the active window has a blue title bar, the inactive one is "grayed out"). Now click on the Close button in the upper right corner of the active window.

Using Split Screens

Use the Split icon in the Window group in the View tab of the Ribbon to split a document window vertically or horizontally into two windows. To split the screen vertically, select an entire column as if you were going to insert a column. Then click on the Split icon. This creates a split in the window, to the left of the selected column, as illustrated in Figure 1-64, with each part of the window displaying the active document. Each part of the document now has its own scroll bar, and you can scroll one part of the document while the other part remains fixed. A horizontal split is accomplished in a similar way.

You can also split the document window by placing the mouse pointer on either split button (the small rectangles at the right end of the horizontal scroll bar and at the top of the vertical scroll bar), and then click and drag the button.

The document window can be split both horizontally and vertically, by first selecting a single worksheet cell, and then clicking on the Split icon.

To remove a split, press the Split icon, or slide the split button back to its original position.

Excel Tip. *To remove a split from a window, it's not necessary to slide the split button back to its original position at the top or left-hand side of the scroll bar. Just place the mouse pointer on the split button and double-click.*

	A	E	F	G	H	I	J	K	M
1	**Grade Sheet**								
2		Hour Exams							
3	Name	#1	#2	#3	Oral report	Paper	Final Exam	Total	Grade
4	CAPICELLA, Jason	15	4	27	17.5	45	23	43.8	D
5	CHUNG, Min-Yin	18	13	44	17.5	45	47	61.5	C+
6	FERREIRO, Kathy	24	16	32	20	45	52	63.0	B-
7	GANGE, Eric	28	13	43	20	40	51	65.0	B-
8	GREALEY, John	22	14	40	17.5	40	56	63.2	C+
9	HAPPERSBACH, Bill	28	12	30	17.5	45	59	63.8	C+
10	HOGAN, Derek	37	17	37	20	50	60	73.7	B
11	LAROZI, Patrick	27	12	38	20	45	58	66.7	B-

Figure 1-64. A document with a split screen.

Using Freeze Panes

Freeze Panes (in the Window group of the View ribbon) can be used to create a similar split document window, but the upper or left part of the window is fixed and cannot be scrolled. Split panes are useful to display fixed row or column headings (or both) while scrolling through the rest of the worksheet.

To use the Freeze Panes feature to split a document window horizontally into two windows, select an entire row as if you were going to insert a row. Then click on the Freeze Panes button in the Window group of the View ribbon, and click on Freeze Panes in the drop-down menu. The portion of the window above the selected row will be frozen. (There are also built-in commands to freeze either the top row or the leftmost column of the sheet.

To split the window both horizontally and vertically, select the cell whose upper left corner defines the location of the split, and click on Freeze Panes.

To unfreeze, simply click on the Freeze Panes button.

	A	B	C	D	E	F	G	H	I	J	K
1	**Grade Sheet**										
2						Hour Exams					
3	Name	ID#	YOG	Major	#1	#2	#3	Oral report	Paper	Final Exam	Total
4	CAPICELLA, Jason	863162407	95	CHEMISTRY	*15*	4	27	17.5	45	23	43.8
5	CHUNG, Min-Yin	325417108	96	CHEMISTRY	18	13	44	17.5	45	47	61.5
6	FERREIRO, Kathy	011083979		EVENING C	24	16	32	20	45	52	63.0
7	GANGE, Eric	436860962	95	CHEMISTRY	28	13	43	20	40	51	65.0
8	GREALEY, John	123757619	96	CHEMISTRY	22	14	40	17.5	40	56	63.2
9	HAPPERSBACH, Bill	761132171	95	CHEMISTRY	28	12	30	17.5	45	59	63.8
10	HOGAN, Derek	547710365	94	BIOCHEMIST	37	17	37	20	50	60	73.7
11	LAROZI, Patrick	533421804	96	CHEMISTRY	27	12	38	20	45	58	66.7

Figure 1-65. A document displayed with Freeze panes.

Using Zoom

You can change the Zoom setting of the worksheet: decrease the zoom setting to see more rows and columns at a reduced size, or increase the setting for easier viewing of small details. To change the zoom setting, click on the Zoom button in the View tab to display the Zoom dialog box. You can choose from several built-in zoom settings or enter a custom setting. If you choose Fit Selection, the zoom setting will be adjusted to display the selected range of rows or columns, or the selected chart, text box, etc., in the window.

You can also use the Zoom Slider, located at the bottom right corner of the window, to change the zoom level.

Figure 1-66. The Excel 2007/2010 Zoom slider.

Excel Tip. *Click on the Zoom Level button to the left of the Zoom Slider to display the Zoom dialog box.*

Easing the Transition from Excel 2003 to Excel 2007/2010

There are some things that Excel 2003 users can do to make the transition to Excel 2007/2010 easier. You can customize the Quick Access Toolbar to look almost exactly like the Excel 2003 toolbar, you can use shortcut keys, or you can install a Classic Menus utility to provide the familiar Excel 2003 menus.

Customize the Quick Access Toolbar

I customized the Quick Access Toolbar (see Chapter 23 for instructions on how to customize it) with familiar toolbuttons to make it look like the Standard and Formatting toolbars of Excel 2003, in the "Toolbars share one row" form. Figure 1-67 shows part of the customized Quick Access Toolbar.

Figure 1-67. A customized Quick Access Toolbar.

Use Shortcut Keys

You can use shortcut keys to carry out many menu or toolbutton commands. Table 1-9 shows a list of the more common ones, and Appendices G and H contain a much more extensive list.

Table 1-9. Shortcut Keys for Excel 2003 Menu Commands

File menu		**Insert** menu	
New	Ctrl + N	**Cells...**	Ctrl + plus
Open	Ctrl + O	**Rows**	Ctrl + plus
Save	Ctrl + S	**Columns**	Ctrl + plus
Edit menu		**Worksheet**	Shift + F11
Undo	Ctrl + Z	**Function...**	Shift + F3
Redo	Ctrl + Y	**Name ▶**	
Cut	Ctrl + X	**Define...**	Ctrl + F3
Copy	Ctrl + C	**Create...**	Ctrl + Shift + F3
Paste	Ctrl + V	**Format** menu	
Fill ▶		**Cells...** (Number tab)	Ctrl + 1
Fill Down	Ctrl + D	**Cells...** (Font tab)	Ctrl + F
Fill Right	Ctrl + R	**Row ▶**	
Clear ▶		**Hide**	Ctrl + zero
Contents	Delete	**Column ▶**	
Delete...	Ctrl + minus	**Hide**	Ctrl + 9
Find	Ctrl + F	**Tools** menu	
Replace...	Ctr l+ H	**Macro ▶**	
Go To...	Ctrl + G	**Macros...**	Alt + F8
		Visual Basic Editor	Alt + F11

Display Classic Menus

The introduction of Excel 2007 was quickly followed by the appearance of several commercial products that, when installed in Excel 2007/2010, provide "classic menus" – that is, menus that look and operate very much like the Excel 2003 menus. You can find several of these products by searching on the web for "excel classic menus" or you can install the Classic Menus utility that is provided on the CD-ROM that accompanies this book.

The Classic Menus workbook contains VBA code that displays a version of the Excel 2003 Worksheet Menu Bar and Chart Menu Bar in Excel 2007/2010. The menus are located in the Add-Ins tab of the Ribbon. Most of the commands found in the Excel 2003 Worksheet Menu Bar are available. Figure 1-68 shows the **Tools** menu with the **Macro** submenu.

A few menu commands, such as the **Customize...** command in the **Tools** menu, are absent from the menus because they are not applicable to Excel 2007/2010.

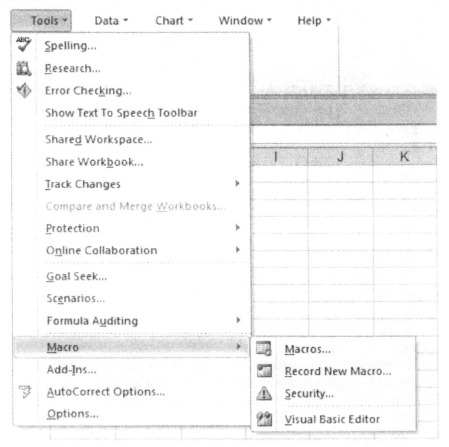

Figure 1-68. Excel 2003 Classic Menus installed in Excel 2010.

The VBA procedure in the Classic Menus workbook runs automatically when you open the workbook. You can install the classic menus by opening the document each time you want to have the menus available, but there are easier approaches. You can save the workbook as an Add-In; this will make the menus available each time you start Excel, provided the Classic Menus Add-In is checked in the Manage Excel Add-Ins list. This allows you to turn off the Classic Menus by unchecking the Add-In.

Excel Tip: *Double-click on the Add-Ins tab to keep the classic menus displayed.*

The following box lists some of the differences between Excel 2003 menus and Classic Menus.

Differences between Excel 2003 Menus and Classic Menus

There is no separate **Chart** menu bar that appears when a chart is selected. The **Chart** menu is located in the "Classic Menus" menu bar; **Chart** menu commands become active when you select a chart.

In all menus:
1.   Letters are not underlined in the menu bar to indicate accelerator keys; for example, **File** instead of **File**.
2.   Shortcut keys are not shown in menu commands, but are still available: for example, Ctrl-C for **Copy**.

In the **File** menu:
1.   Recent files do not have a number preceding the filename for use as a shortcut key.

In the **Edit** menu:
1.   Shift+**Edit** to change **Copy** to **Copy Picture...** has not yet been implemented.

In the **View** menu:
1.   The **Toolbars** command has been omitted, because there are no toolbars to be displayed.

In the **Tools** menu:
1.   The **Customize...** command has been omitted because there are no menus or toolbars to be customized.
2.   In the **Add-Ins...** command, only the **Solver** and **Analysis ToolPak** add-ins will have their menu commands added to the **Tools** menu. To make the command appear in the **Tools** menu, after loading the add-in, click on the **File** menu.

In the **Chart** menu:
1.   The **Chart Options...** command is not available. Use the Layout tab of the Excel 2007/2010 ribbon.

Excel 2007/2010 Workbook and Worksheet Specifications

Worksheet size*	1,048,576 rows by 16,384 columns
Maximum column width	255 characters
Maximum row height	409 points
Number precision	15 digits
Largest allowed positive number	9.99999999999999E+307
Largest allowed negative number	–9.99999999999999E+307
Smallest allowed negative number	–2.225E-308
Smallest allowed positive number	2.225E-308
Levels of Undo*	100
Number of worksheet functions*	341
Maximum number of...	
open workbooks	Limited by available memory
characters in a cell (text)	32,767
characters in a formula	8192
sheets in a workbook	Limited by available memory
colors in a workbook*	16,000,000
custom number formats	Between 200 and 250
windows in a workbook	Limited by system resources
panes in a window	4
arguments in a function*	255
nested levels of functions*	64
sort levels in a single sort	64

* indicates changes from Excel 2003

2

Working with Excel 2003

This chapter covers the basics of working with Excel 2003: navigating around the worksheet, entering values and formulas, and formatting and editing a worksheet. If you are an experienced Excel user, you can probably skip this chapter; however, even experienced users may find a few useful tips in this chapter.

The Excel 2003 Document Window

An Excel workbook is a *document* that appears in its own *document window*. Although you can have several workbooks open at the same time, and can see several displayed on the screen simultaneously, only one workbook can be the *active workbook*. The default Excel 2003 workbook contains three worksheets; only one worksheet in the active workbook can be the *active worksheet*.

An Excel 2003 worksheet consists of 256 columns (labeled A, B, C ... IV) and 65,536 rows (labeled 1, 2, 3, ...). The rows and columns define *cells* (A1, H27, etc.), which constitute the worksheet. Information can be entered into a cell from the keyboard after the cell has been selected, usually with the mouse pointer. The Excel 2003 for Windows document window is shown in Figure 2-1. Depending on your monitor, your screen may show a different number of rows or columns.

In Figure 2-1, reading from the top down you'll see the *application title bar*, the *menu bar* (with **File, Edit, View,** etc. menus), the *Standard toolbar* (with New, Open and Save toolbuttons), the *Formatting toolbar* (with Bold, Italic and Alignment toolbuttons, for example), the *formula bar*, the rows and columns of cells, the *sheet tabs* and the *horizontal scroll bar* and, at the bottom, the *status bar*. The formula bar contains the Name Box or *cell reference area* (displaying the cell reference of the currently selected cell) and the editing area. As you enter values at the keyboard, they appear in the editing area of the formula bar.

The toolbars and other components of your document window may not look the same as in Figure 2-1. See the next section, "Changing What Excel Displays".

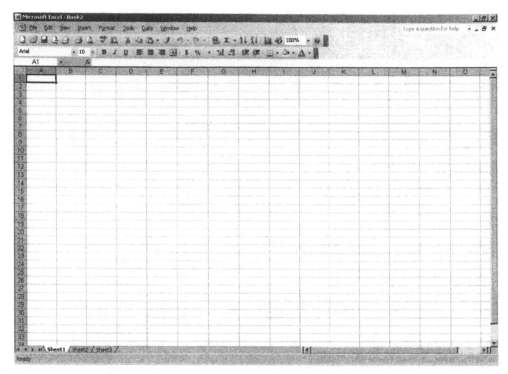

Figure 2-1. The Excel 2003 document window.

In this chapter, the Excel 2003 dialog boxes and other screen shots are shown as they appear when using Windows XP. In subsequent chapters you will see examples from either Windows XP and Windows Vista.

Changing What Excel Displays

You can choose to display or not display most components of the Excel window, such as menu bars, scrollbars, formula bar, gridlines, row and column headers. To turn off the display of these items, or to restore them if they are missing, choose **Options...** from the **Tools** menu to display the Options dialog box (Figure 2-2). Click on the View tab; then check or uncheck the appropriate box. To prevent display of the Startup Task Pane, that annoying "Getting Started" window on the right that appears every time you start Excel, uncheck the Startup Task Pane box.

If you click on the General tab, you can switch from using A1-style references in formulas to R1C1-style references; the labels in the column header row of each worksheet change from A, B, C … to 1, 2, 3 … . You probably won't ever want to use R1C1-style formulas, but there are a few instances where knowing how to create formulas using R1C1-style references can be useful; see "The INDIRECT Function" in Chapter 6.

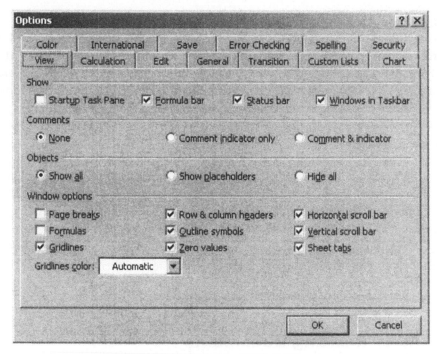

Figure 2-2. The Excel 2003 Options dialog box.

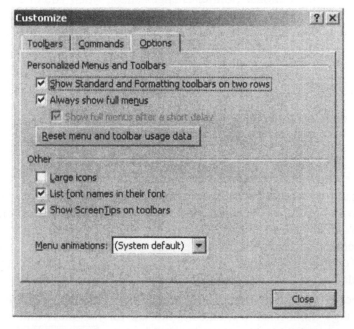

Figure 2-3. The Options tab of the Customize dialog box allows you to specify how toolbars and menus are displayed.

In Excel 2003, the default for menus is that they display recently used commands first, with a drop-down button at the bottom of the menu to display the remaining commands. The default for toolbars is that the Standard and Formatting toolbars share one row. If you prefer to work with "old style" menus and toolbars, choose **Customize...** from the **Tools** menu, choose the Options tab and uncheck the Menus Show Recently Used Commands First and the Standard And Formatting Toolbars Share One Row boxes (Figure 2-3).

Choose **Toolbars** from the **View** menu to display the submenu of available toolbars. The Standard and Formatting toolbars are the default toolbars, but you can display other toolbars by choosing them from the submenu. You can also customize Excel's built-in toolbars or create a custom toobar; to learn about customizing toolbars, see Chapter 23.

If you place the tip of the mouse pointer on one of the toolbuttons, a yellow ScreenTip box appears, describing the button's function. You can deactivate ScreenTips by choosing **Toolbars** from the **View** menu, then **Customize...** from the submenu to display the Customize dialog box. Choose the Options tab and de-select the Show ScreenTips On Toolbars check box.

Moving or Resizing Documents

To change the size of a workbook or worksheet, click and drag any of its borders or corners; the mouse pointer changes shape when you click on a border or corner. You can adjust the document to any size you desire. If you click on the Minimize button (the "underline" symbol in the upper right corner of the document) the document will be minimized so that only the title bar is visible. To restore it to its full size, click the Maximize button (the open square in the upper right corner of the title bar).

To change the position of a document within the Excel window, click on the title bar and drag the document. It can even extend off-screen.

Navigating Around the Workbook

The default Excel 2003 workbook contains three worksheets. If you want a workbook with more than three sheets, you can insert additional worksheets by choosing **Worksheet** from the **Insert** menu. To change the default, so that all new workbooks will have, for example, only one worksheet, choose **Options...** from the **Tools** menu, choose the General tab, change the Sheets In New Workbook default, then create a new workbook.

To select a worksheet, simply click on the sheet tab. If the workbook contains a large number of worksheets, the tab for the sheet that you want to select may not be visible. Use the tab scroll buttons [◄◄ ◄ ► ►►] to the left of the sheet tabs to scroll through the sheet tabs. From left to right, these four buttons allow you to jump to the first sheet tab, scroll toward the first sheet tab, scroll toward the last sheet tab, or jump to the last sheet tab. When the desired sheet tab is visible, click on it.

Excel Tip. *To display a shortcut menu that lists all sheets in the workbook, right-click on any of the tab scroll buttons. You can then select the desired sheet.*

Changing the Name of a Worksheet

When you create a new workbook, the sheet tabs have the default names Sheet1, Sheet2, etc. To rename a sheet, double-click on the sheet tab. The sheet name will be highlighted and you can enter a more descriptive name (the limit is 31 characters), as, for example, in Figure 2-4. Click outside the sheet tab to exit from edit mode.

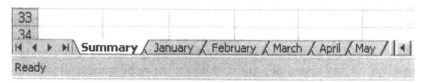

Figure 2-4. Descriptive sheet names are helpful.

Changing the Color of Sheet Tabs

You can color-code sheet tabs: choose **Sheet** from the **Format** menu and choose **Tab Color...** from the submenu. Alternatively, you can right-click on the sheet tab and then choose **Tab Color...** from the shortcut menu.

Rearranging the Order of Sheets in a Workbook

To move a sheet, just click and drag the sheet tab. The mouse pointer shape becomes an icon showing a sheet at the end of the arrow pointer (Figure 2-5). An arrow above the sheet tab indicates where the copy will be inserted.

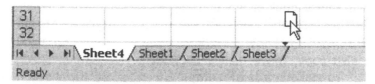

Figure 2-5. Moving a sheet tab with the mouse pointer.

To make a copy of a worksheet, hold down the Ctrl key while dragging the sheet tab. A small + sign appears in the icon (Figure 2-6).

Selecting Non-Adjacent Ranges

To select non-adjacent ranges, select the first range, then hold down the Ctrl key while selecting the second range. Both cell ranges will be highlighted (Figure 2-8).

Extending a Selection

To extend the range of a cell selection you just made, hold down the Shift key, select the last cell in the selection and drag to include the additional cells. Alternatively, hold down the Shift key and use any of the arrow keys to "extend" the selection; i.e., you can also decrease the number of cells in the selection).

	A	B	C	D
1				
2	time, (sec)	$[A]_t$	$[B]_t$	$[C]_t$
3	0.0	5.00E-03	0.00E+00	0.00E+00
4	0.2	4.18E-03	7.91E-04	3.30E-05
5	0.6	2.91E-03	1.83E-03	2.51E-04
6	1.0	2.03E-03	2.37E-03	5.93E-04
7	1.4	1.42E-03	2.59E-03	9.94E-04
8	1.8	9.89E-04	2.60E-03	1.41E-03
9	2.2	6.90E-04	2.49E-03	1.82E-03
10	2.6	4.82E-04	2.31E-03	2.20E-03
11	3.0	3.36E-04	2.11E-03	2.56E-03
12	4.0	1.37E-04	1.57E-03	3.29E-03
13	5.0	5.55E-05	1.12E-03	3.83E-03
14	6.0	2.26E-05	7.76E-04	4.20E-03
15	7.0	9.18E-06	5.31E-04	4.46E-03
16	8.0	3.73E-06	3.60E-04	4.64E-03
17	9.0	1.52E-06	2.43E-04	4.76E-03
18	10.0	6.17E-07	1.64E-04	4.84E-03
19				

Figure 2-8. Selecting non-adjacent ranges.

Selecting a Block of Cells

A *block of cells* is a range of cells containing values and bounded by empty cells. There are several ways to select cells within a block:

- Use Ctrl + Shift + (arrow key) to select in the appropriate direction.
- Select a cell at a boundary of the block (at the top, bottom or side of the block). Move the mouse pointer over the edge of the selected cell until the pointer changes to the arrow pointer (Figure 2-9 Left). Hold

down the Shift key and double-click on the bottom edge of the selected cell to select all cells in the column from the top to the bottom of the block, as shown in Figure 2-9 Right. You can select cells from top to bottom, from bottom to top, from left to right or from right to left within a block. You can also select multiple columns or rows in the same way.

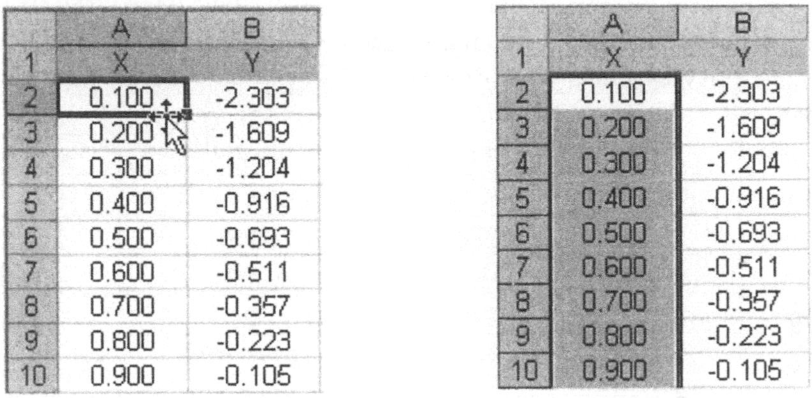

	A	B
1	X	Y
2	0.100	-2.303
3	0.200	-1.609
4	0.300	-1.204
5	0.400	-0.916
6	0.500	-0.693
7	0.600	-0.511
8	0.700	-0.357
9	0.800	-0.223
10	0.900	-0.105

	A	B
1	X	Y
2	0.100	-2.303
3	0.200	-1.609
4	0.300	-1.204
5	0.400	-0.916
6	0.500	-0.693
7	0.600	-0.511
8	0.700	-0.357
9	0.800	-0.223
10	0.900	-0.105

Figure 2-9. Using the mouse pointer to select a block of data. Left: Selecting a cell edge. Right: Selecting the block of data by double-clicking while holding down the Shift key.

Entering Data in a Worksheet

To enter a value in a worksheet cell, select the cell with the mouse pointer, which appears as a large open cross when it passes over cells. Clicking on the desired cell highlights it, indicating that this is the *active cell*, the cell in which you can now enter a value. As you type a value, the characters appear in the formula bar and the active cell. You can complete the entry in several ways.

- Press the Enter button ✓ in the formula bar. The cell remains selected. This method is useful if you want to examine a value or formula after entering.
- Press the Enter key. This moves the selection to the cell below (although you can change the default option so that the selection is not moved). This is the usual way for entering data in successive cells.

To cancel the entry and revert to the original contents of the cell, press the Cancel button ✕ or the Esc key.

Excel Tip. *To enter the same value in a range of cells, select the range of cells, type the value, then press Ctrl + Enter.*

Entering Numbers

Excel has a remarkable ability to recognize the format of the value that you have entered: as a number, a percent, or a debit value; as currency; in scientific notation; as a date or time; or even as a fraction. The number will be displayed in the cell in the proper format, but the number equivalent of the value will appear in the formula bar. Figure 2-10 illustrates number formats recognized by Excel.

Type	As Entered at Keyboard	As Displayed in Cell	As Displayed in Formula Bar	As Used in Calculation
percent	15%	15%	15%	0.15
scientific	2e-3	2.00E-03	0.002	0.002
currency	$50	$50	50	50
currency	$20000	$20,000	20000	20000
debit	(5000)	-5000	-5000	-5000
fraction	2 5/8	2 5/8	2.625	2.625
date	7/4	4-Jul	7/4/2010*	**
date	8/3/28	8/3/2028	8/3/2028	**
date	8/3/38	8/3/1938	8/3/1938	**
time	4:30	4:30	4:30:00 AM	**
time	16:00	16:00	4:00:00 PM	**
time	4 p	4:00 PM	4:00:00 PM	**
* Enters current year				
** See Chapter 3 for a discussion of date and time calculations				

Figure 2-10. Number formats recognized by Excel.

Since the slash character can indicate either a date or a fraction, if you enter a fraction, such as 1/3, it will be interpreted as a date, specifically 3-Jan. To prevent Excel from converting the fraction to a date, enter a zero and a space before the fraction (e. g., 0 1/3). The zero indicates that the entry is a number, and the value will appear in the formula bar as 0.333333333333333.

How Excel Stores and Displays Numbers

Excel can accept numbers in the range from approximately ±1E-307 to ±1E+308 (see "Specifications" at the end of this chapter).

Excel stores numbers with 15-significant-figure accuracy. These are displayed in the formula bar and used in all calculations, no matter what number formatting has been applied. Thus the fraction 1/3 appears in the formula bar as 0.333333333333333, and π as 3.14159265358979.

Excel switches between floating-point and scientific notation for best display of values. The formula bar can display numbers up to 21 characters, including the decimal point. Thus 1E-19 entered on the keyboard will be displayed as 0.0000000000000000001 (21 characters) in the formula bar, while 1E-20 will appear as 1E-20. Similarly, 1E20 is displayed as 100000000000000000000, while 1E21 appears as 1E21. Since a total of 21 characters can be displayed, the number of significant figures determines the magnitude of a number less than 1 that can be displayed in non-E format in the formula bar. Thus 1.2345E-15 appears as 0.0000000000000012345, while 1.23456E-15 is displayed as 1.23456E-15.

Entering Text

If you enter text characters (any character other than numbers, the decimal point, or the characters +, -, *, /, ^, $, %) in a cell, Excel will recognize the entry as text. For example, Chestnut Hill MA 02167-3860 is a text entry. Each cell can hold up to 32,767 characters (but only 1,024 will display in the cell). You can distinguish text entries from number entries in the following way: In a cell that has not been alignment-formatted (e.g., left, centered, right, etc.), text entries are left-aligned, and numbers are right-aligned. Of course, if you format the alignment of a cell to be right-aligned, its value will be right-aligned whether the value is a number or text.

You can format individual characters in a cell using Bold, Italic, Underlined, etc., or with different fonts, by highlighting the character(s) in the formula bar and then applying the formatting.

Excel Tip: Sometimes it is necessary to enter a number or a date as a text value. To do this, begin the entry with a single quote.

Entering Formulas

Instead of entering a number in a cell, you can enter an equation (called a *formula* in Microsoft Excel) that will calculate and display a result. Usually formulas refer to the contents of other cells by using *cell references*, such as A2, a reference to a cell, or B5:B12, a reference to a range of cells. The value displayed in a cell containing a formula will be automatically updated if values elsewhere in the worksheet are changed. Formulas can contain values, arithmetic operators and other operators, cell references, the wide range of Excel's worksheet functions, and parentheses.

The rules for writing formulas (the *syntax*) are as follows:

- A formula must begin with the equal sign (=).
- The *arithmetic operators* are addition (+), subtraction (-), multiplication (*), division (/) and exponentiation (^). Other types of operator are described in Chapter 3.

- Parentheses are used in the usual algebraic fashion to prevent errors caused by the *hierarchy of arithmetic operations* (multiplication or division is performed before addition or subtraction, for example).

Some examples of simple formulas:

=A1+273.15	Adds 273.15 to the value in cell A1.
=A2^2+13*A2-5	Evaluates the function $x^2 + 13x - 5$, where the value of x is stored in cell A2.
=SUM(B3:B47)	Sums the values contained in cells B3 through B47.
=(-C3+SQRT(C3^2-4*C2*C4))/(2*C2)	Finds one of the roots of the quadratic equation whose coefficients a, b and c are stored in cells C2, C3 and C4, respectively.

Excel formulas are discussed in much greater detail in Chapters 3 and 6.

There are some techniques that you can use for entering worksheet formulas.

- Type formulas in lowercase to facilitate detection of typographical errors. When you enter a formula, Excel converts functions and cell references to uppercase. If you type the formula =offset(d1,5,1), Excel will convert it to =OFFSET($D$1,5,1) when you enter the formula; but if you type "ofset" instead of "offset", Excel won't recognize it and will display the error message #NAME?. When you examine the formula, you'll easily see that the incorrect function name remained in lowercase letters.

- Enter cell or range references in formulas by clicking on the cell, not by typing the reference. This makes it less likely that you will enter an incorrect reference (e.g., C25 instead of B25) and also makes entering complicated references much easier.

- If formulas contain terms identical to those used in other cells, you can **Copy** that part of the formula and **Paste** it into the new formula. Here's one method: Before beginning to type the new formula, select the cell containing the formula you want to copy. In the formula bar, highlight the part of the formula you want to copy, **Copy** it, and click the Check Box in the formula bar. Now select the cell into which you want to type the new formula, type the new formula until you reach the part that you've copied, and **Paste** in the formula fragment.

Excel Tip. *Formulas that return the wrong result because of errors in the hierarchy of calculation are common. When in doubt, use parentheses.*

Editing Cell Entries

You can edit cell entries in one of two ways — either in the formula bar or by using the Edit Directly In Cell feature. When you select a cell that contains an entry, the contents of the cell appear in the formula bar. As soon as you begin to enter a new value, the old value disappears. To make minor editing changes in the old entry, place the mouse pointer in the text at the point where you want to edit the entry. The mouse pointer becomes the vertical insertion-point cursor. You can now edit the text in the formula bar using the **Copy**, **Cut**, **Paste** or **Delete** commands or keys. Complete the entry using the Enter button in the formula bar, or by pressing the Enter key on the keyboard.

Alternatively, you can use Excel's Edit Directly In Cell feature: Press function key F2 or double-click on the cell to enter edit mode. You can use the right and left arrow keys to move through the formula, or Ctrl+(arrow key) to jump to the next element of the formula, or Ctrl+Shift+(arrow key) to select the next element of the formula.

Excel uses colors to show range references in formulas and the corresponding ranges on the worksheet. When you enter Edit mode, by clicking in the Formula Bar, double-clicking on the cell, or pressing F2, the references in the formula appear in color, and the corresponding cells or ranges are indicated by similarly colored borders (blue for the first input, green for the second, and so on). Each of these colored outlines has a handle in the bottom right corner; you can change the range of a formula input by dragging its handle.

Excel Tip. To select a word or reference for editing, double-click on it.

Adding a Text Box

You can add notes or comments or other information to a worksheet by typing them into one or more worksheet cells. Another way to add comments, in a much more flexible form, is by using a text box.

To create a text box, press the Text Box toolbutton — it's in the Drawing toolbar, but you can put a copy of it on the Standard or Formatting toolbar (see Chapter 23). The mouse pointer will change to a crosshair. Position the crosshair pointer where you want to place the text box, and click and drag to outline it (the text box can be moved and sized later). An empty text box will be displayed with a blinking text cursor. Type the desired text within the box, and click anywhere outside the text box to complete the entry.

Text box input has many features of a simple word processor: You can Cut, Copy or Paste text, make individual portions of text bold, italic or underlined, use different font styles, etc., as shown in Figure 2-11. The text within the box can be formatted with the Alignment toolbuttons or with the Alignment command.

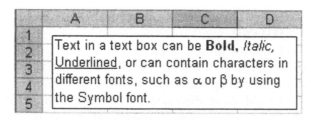

Figure 2-11. A text box.

To move a text box, click on it to activate it, place the mouse pointer on the border of the text box and drag it to its new position. To resize a text box, select it (white handles will appear) and then place the mouse pointer on one of the handles and click and drag to move the border of the box. If you hold down the Ctrl key while dragging, you make a copy of the text box (a small plus sign appears beside the mouse pointer); if you hold down the Alt key, the text box will align with the cell gridlines; if you hold down the Shift key, the text box can only be dragged in either horizontal or vertical alignment with its original position.

Entering a Cell Comment

You can attach a *cell comment* to a specific cell, for documentation purposes. A comment appears on the worksheet in a small box similar to a Screen Tip. A small red triangle in the upper right corner of the cell indicates that the cell contains a comment. When the mouse pointer is moved over a cell that contains a cell comment, the cell comment appears.

To add a comment to a cell, choose **Comment...** from the **Insert** menu. Enter the text of the comment in the box (Figure 2-12). To exit, simply click on any cell outside the comment box. To edit a comment, select the cell containing the comment and then choose **Edit Comment...** from the **Edit** menu. To delete a comment, select the cell containing the comment, choose **Clear** from the **Edit** menu, and then choose **Comments** from the submenu.

You can turn screen display of comments and/or comment indicators on or off by choosing **Options** from the **Tools** menu, choosing the View tab and pressing the appropriate button in the Comments category.

Comment indicators are not printed when you print a worksheet.

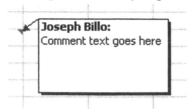

Figure 2-12. A cell comment.

Using the Equation Editor

Sometimes it's useful to be able to display an equation on a spreadsheet. For complicated equations, you can use the Equation Editor; you may already be familiar with its use in Microsoft Word. To use the Equation Editor, choose **Object...** from the **Insert** menu and click on Microsoft Equation 3.0.

The Equation Editor inserts a drawing object that looks like a text box. When you are in Equation Editor mode, the textbox will have a fuzzy border around it, and the Equation Editor toolbar will appear (Figure 2-13).

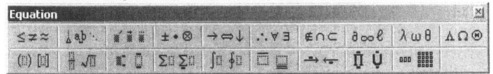

Figure 2-13. The Equation Editor toolbar.

The default sizes of the equation elements, particularly ones such as sub- or superscripts, are quite small. I find it helpful to increase the size of the textbox, which increases the size of the equation elements. To do this, type at least one character of your equation, and then exit from the Equation Editor and return to Microsoft Excel by clicking on an empty cell. Now use the handles around the text box to resize it, as in Figure 2-14.

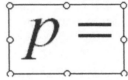

Figure 2-14. The Equation text box resized for easier typing.

Now double-click on the text box to re-enter Equation Editor, and complete the equation. You can select symbols and options from the Equation Editor toolbar.

$$p = Ae^{-B/T}$$

Figure 2-15. A completed equation using the Equation Editor.

To restore the equation textbox to its default size, right-click on the equation object and choose Format Object. Click on the Size tab (Figure 2-16). In either the Height or Width box, enter 100%. Make sure that the Lock Aspect Ratio and Relative To Original Picture Size boxes are checked, and then press OK.

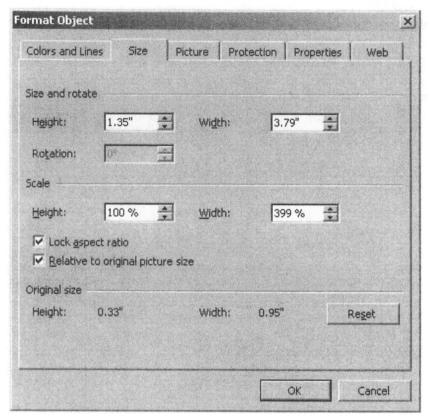

Figure 2-16. Returning a re-sized equation text box to its default size.

Excel's Menus: An Overview

In Excel 2003 for Windows, the Worksheet Menu Bar has the following pull-down menus: **File, Edit, View, Insert, Format, Tools, Data, Window** and **Help.** The **File, Edit, Format** and **Window** menus are discussed in this chapter. Commands in other menus will be discussed in later chapters.

The way in which a command appears in a menu provides information about its form or availability:

- A menu command with an ellipsis (...), such as **Save As...**, indicates that the command opens a dialog box to obtain user input.
- Many Excel menus contain submenus, indicated by the ▶ symbol at the right edge of the menu.
- Some menu commands are dimmed (i.e., appear as gray characters) when the menu command is unavailable. Others appear on the menu only when they are available.

- Some menu commands change the text of their command depending on circumstances. For example, if you use **Comment** to add a comment to a cell, the command changes to **Edit Comment** so that you can edit the text of the comment.
- Some menu commands are preceded by a check mark if the choice has been selected previously. To remove the selection, depending on the command, you either click on the check mark or select the command again.

Shortcut Menus

Excel also provides "context-sensitive" *shortcut menus*. If you press the right mouse button while you select a worksheet element with the mouse pointer, a menu is displayed containing commands that apply to the selection. For example, if you select a column with the right mouse button, a shortcut menu containing editing and formatting commands appropriate for a column appears.

Menu Commands or Toolbuttons?

Many menu commands can be carried out by clicking on a toolbutton. Toolbuttons are more convenient: They often combine a whole series of actions — menu selection plus dialog box options — into a single click of the mouse button.

Some buttons mentioned in this chapter don't appear on either the Standard or Formatting toolbar. To make them available for use, you can display other toolbars, or you can customize a toolbar by adding built-in toolbuttons (see Chapter 23).

Opening, Closing and Saving Documents

Most menu commands for managing documents are in the **File** menu. For the most part, the menu is similar to the **File** menu in other Windows applications, with (among other menu commands) **New...**, **Open...**, **Close**, **Save**, **Save As...**, **Save Workspace...**, **Page Setup...**, **Print Preview...**, **Print...** and **Exit** commands. The **Save Workspace...** command is specific to Excel; see "Using Save Workspace..." later in this chapter.

Opening or Creating Workbooks

Use the **Open...** command to locate and open an existing document; use **New...** to create a new document. **New...** displays a dialog box, docked on the right, in which you have a choice of opening either a new worksheet or any of the built-in or user-created template sheets.

To open an existing workbook or worksheet from the desktop, simply double-click on it. This will open the document (and will start Excel as well if it isn't already running). If you start Excel first, it will open a new blank workbook.

Using Move or Copy Sheet... or Delete Sheet

The default Excel 2003 workbook contains three worksheets, but you can add or remove sheets. Three Excel commands permit you to add or remove sheets from a workbook. The **Delete Sheet** command in the **Edit** menu permanently removes the active sheet from the workbook. To add a worksheet to a workbook, use the **Worksheet** command in the **Insert** menu (you can also insert a chart sheet). To move or copy sheets within a workbook or from one workbook to another, use the **Move or Copy Sheet...** command (Figure 2-17) in the **Edit** menu.

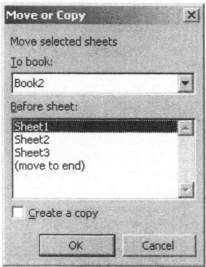

Figure 2-17. The Move or Copy Sheet dialog box.

Using Close or Exit

You can close a document either with the **Close** command from the **File** menu or, more conveniently, by using the Close button on the document title bar. You will be asked if you want to save changes.

If you hold down the Shift key while you pull down the **File** menu, the **Close** command becomes **Close All**. That way you can close all open Excel documents at once.

When you use the **Exit** command, you close all open documents (you will be asked if you want to save changes) and then exit from Excel.

Types of Excel Document

Excel 2003 uses a variety of file formats for its documents. Table 2-2 lists the filename extensions used by Excel for Windows.

Table 2-2. Excel for Windows Filename Extensions

Filename extension	File type
.xla	Add-in macro
.xlb	Toolbar configuration
.xlc	Excel 4.0 chart
.xlm	Excel 4.0 macro
.xls	Excel workbook
.xlt	Template
.xlw	Workspace configuration

Using Save or Save As...

When you **Save** a newly created workbook, the **Save** dialog box will prompt you to assign a name to the document. Excel for Windows automatically appends a three-letter filename extension (e.g., .xls) to identify the file format type.

Earlier versions of Excel for Windows limited document names to a maximum of eight characters, and no spaces were allowed. In Excel 2003, document names can be much longer: The complete path to the file, including drive letter, server name, folder path, file name, and the three-character file name extension, can contain up to 218 characters. File names can include spaces but not any of the following characters: slash (/), backslash (\), greater-than sign (>), less-than sign (<), asterisk (*), question mark (?), quotation mark ("), pipe symbol (|), colon (:), semicolon (;). Sheet names can include most of the preceding characters, but can't include question mark (?), colon (:), backslash (\) or asterisk (*), which are used as wildcard characters or file delimiters.

You can use **Save As...** to create a backup copy of a workbook by giving the copy a different name.

Using Save Workspace...

You can use **Save Workspace** if you have to interrupt your work, when you have several workbooks open at once. The **Save Workspace** command saves the current configuration of the workspace. When you open the .xlw file, the workbooks will be restored to their former size and position on screen.

Editing a Worksheet

You can use commands in the **Edit** menu to modify the arrangement of values in a worksheet.

Inserting or Deleting Rows or Columns

To insert an entire column of blank cells, click on a column header (the gray rectangle at the top of the column), which selects (highlights) an entire column. Then choose **Columns** from the **Insert** menu. A new column will be inserted to the left of the column you selected. Insert a new row in a similar way; the new row will be inserted to the left of the selected row. Multiple rows or columns can be inserted in a similar fashion, by selecting as many rows or columns as you want to insert.

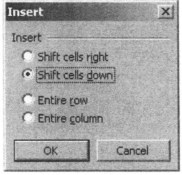

Figure 2-18. The Insert Cells... dialog box.

To insert an extra cell or cells within a row or column, select the cell range above or to the left of which you want to insert cells, and then choose **Insert...**. (Note that **Insert** has become **Insert...** because the Insert dialog box will be displayed.) Excel usually makes a pretty good guess whether the cells should be shifted to the right or down to make the proper insertion, but always check to make sure. Then click OK in the dialog box (Figure 2-18).

Complete rows or columns are deleted in a similar manner. Click on the row or column header to select the rows or columns to be deleted, and then choose **Delete** from the **Edit** menu. To delete cells *within* a row or column, select them with the mouse and then choose **Delete...**. You will be asked whether the cells should be moved up or to the left.

Excel Tip. When inserting or deleting partial rows or columns, take care that other parts of the worksheet do not become misaligned.

Using Cut, Copy and Paste

Single cells, ranges of cells, or whole rows or columns can be copied or cut from the worksheet and inserted or pasted into other locations. In general the destination must be the same size as the copied or cut cells. First, select the cell or range of cells that you wish to copy or cut. Then choose **Copy** or **Cut** from the **Edit** menu, or press the Copy 📋 or Cut ✂ toolbutton, or press Ctrl+C (Copy) or Ctrl+X (Cut). A *marquee* (a dashed line) will appear around the selected cells and a copy of the cells is placed on the Clipboard. Next, select the destination range. You can now transfer the copy to the destination by choosing **Paste** from the **Edit** menu, by pressing the Paste toolbutton 📋, or by pressing Ctrl +V (Paste).

> *Excel Tip. It's much better to select a single cell, and then **Paste**, instead of selecting a destination range that is the same size as the copied or cut range. The selected cell will be the upper left corner of the pasted range of cells.*

You can also **Copy** or **Cut** text in the formula bar and **Paste** it in a worksheet cell. Select the text to be copied or cut, then press the Copy or Cut toolbutton or choose the appropriate command from the **Edit** menu. Complete the operation by clicking the Enter button in the formula bar. Then **Paste** in the desired cell.

> *Excel Tip. Use the Esc key to cancel a **Copy** or **Cut** operation.*

Copying and Pasting Multiple Items

If you need to **Copy** and **Paste** multiple items, you don't need to do them one at a time. You can **Copy** them one-by-one onto the Clipboard, and then **Paste** the collection of items. Use the procedure in the following box.

To Copy and Paste Multiple Items

1. Choose **Office Clipboard...** from the **Edit** menu
2. In the Clipboard window, press the Clear All button to remove anything previously copied.
3. One by one, select and **Copy** the cells, ranges, text, etc., that you want to collect. (You can't include a chart in the collected items to **Paste**.) Up to 24 items can be collected.
4. Select the destination cell and press the Paste All button in the Office Clipboard

---

Using Paste Special...

When you **Copy** a cell and **Paste** it, Excel transfers the cell's contents, format and any comment. You can choose to transfer only some of these cell attributes by using **Paste Special...**. The Paste Special dialog box (Figure 2-19) permits you to paste only Formulas, Formats or Comments. In addition, you can convert formulas to constants by choosing Values.

Excel Tip. *You can use the Paste Values* ⊞ *toolbutton instead of the **Paste Special...** menu command. See Chapter 23 for instructions on how to make this button, and many others, available.*

If you press one of the Operation buttons in the Paste Special dialog box, the value in the destination cell will be added to, subtracted from, multiplied by, or divided by the value in the copied cell.

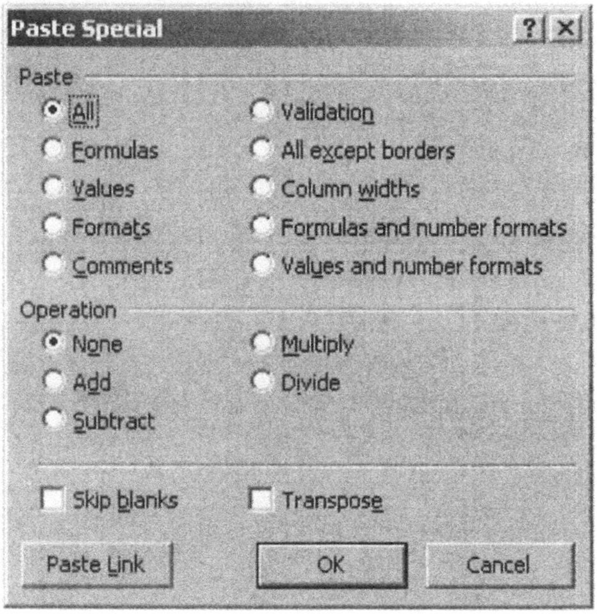

Figure 2-19. The Paste Special dialog box.

If the cell in either the source or the destination contains a formula, then the formula will be enclosed in parentheses and joined to the contents of the destination cell by the arithmetic operator. You may wish to experiment a little to see exactly how this works. Relative references in the source will be changed in the same way as in a normal **Paste** operation. You can also **Copy** cells that contain formulas and press both the Values button and one of the Operation buttons to either Add, Subtract, Multiply or Divide.

If you check the Skip Blanks check box, only non-blank cells in the source will be pasted.

Excel Tip. *Here's a faster way to perform* **Paste Special***. (Values) for a range of cells: Select the range of cells, position the mouse pointer over a border of the selection, hold down the right mouse button, and drag the selection away from its original position, and then back to its original position. When you release the mouse button, a shortcut menu will appear; one of the choices is "Copy here as Values only".*

Using Paste Special to Transpose Rows and Columns

If values in the source range are arranged in rows, you can convert the data to column format, or vice versa, as shown in Figure 2-20.

First **Copy** the cells, and then select a cell or range in which you want the transposed values to be placed. Choose **Paste Special** and check the Transpose box, and then press OK.

	A	B	C	D
1	pH	6.00	6.20	etc.
2	Absorbance	0.903	0.861	etc.

	F	G
	pH	Absorbance
	6.00	0.903
	6.20	0.861
	etc.	etc.

Figure 2-20. Rows and columns transposed.
(Left) Before using and (Right) after using Paste Special (Transpose).

One limitation of Transpose: The copied cells cannot be pasted over any part of the source range.

Using Clear

When you choose **Clear** from the **Edit** menu, a submenu is displayed with All, Formats, Contents, Comments. If you choose Formats from the submenu, for example, you can remove only formats from selected cells. Choosing Contents will delete the cell value but not the format.

Excel Tip. *The easiest way to clear a range of cells is to use the Erase tool*
. To remove only formatting from a cell, use the Clear Formats button
. See Chapter 23 for instructions on how to make these buttons available.

Using the Insert Menu

If you **Copy** a cell or range, then select a cell, range, row or column to specify the destination, then click on the **Insert** menu, the **Cells...** menu command becomes **Copied Cells....** A dialog box will ask whether you want to shift cells down or to the right. If you decide that it's easier to insert new rows or columns first, then **Copy** and **Paste**, you'll have to clear the clipboard before **Copied Cells...** becomes **Cells....** You can do this either (i) by selecting an empty cell and pressing Delete, (ii) by choosing **Clear**... from the **Edit** menu, then exiting from the dialog box by pressing the Cancel button or, most conveniently, (iii) by pressing Esc.

Copy, Cut or Paste Using Drag-and-Drop Editing

You can also **Copy**, **Cut** or **Paste** using Excel's "Drag-and-Drop" method. With this method you **Cut** and **Paste** or **Copy** and **Paste** a selection by using only the mouse pointer.

To Cut and Paste by Using Drag-and-Drop

1. Select the range of cells to be moved.

2. Position the mouse pointer over a border of the selection (top, bottom or side). The mouse pointer will change to an arrow.

3. Drag the selection toward the desired position. The border of the selection will be indicated as you drag it (Figure 2-21).

4. Position the selection as desired and release the mouse button.

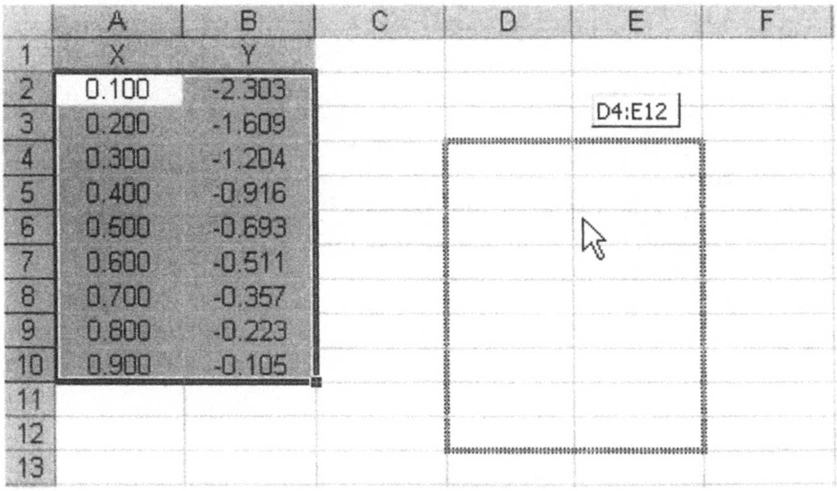

Figure 2-21. Cutting and Pasting cells using Drag-and-Drop editing.

To **Copy** the selection instead of cutting, hold down the Ctrl key while dragging. A small plus sign will appear near the arrow pointer.

To **Insert** the selection, hold down the Shift key while dragging. The insertion point of the selection will be indicated by a horizontal or vertical bar as you drag (Figure 2-22).

To use this method, Drag and Drop must be turned on. Choose **Options** from the **Tools** menu, choose the Edit tab, and check the Allow Cell Drag And Drop box.

	A	B
1	X	Y
2	0.400	-0.916
3	0.500	-0.693
4	0.600	-0.511
5	0.100	-2.303
6	0.200	-1.609
7	0.300	-1.204
8	0.700	-0.357
9	0.800	-0.223
10	0.900	-0.105

Figure 2-22. Inserting cells using Shift+ Drag-and-Drop.

Duplicating Values or Formulas in a Range of Cells

To copy a value or formula in one cell into a range of cells, highlight the cell whose value you want to duplicate, plus cells below or to the right of where you want the value duplicated. Then choose **Fill** from the **Edit** menu and choose **Down**, **Right**, **Up** or **Left** from the submenu.

If the cell contains a number or a text label, the value will be duplicated in the rest of the cells. If the cell contains a formula, the formula will be copied into the selected cells, except that Microsoft Excel uses *relative referencing* when formulas are copied. For example, if cell A2 contains the formula =A1+1, and **Fill Down** is used to copy the formula into a range of cells below cell A2, the formula copied into cell A3 will be =A2+1, and so on.

Cell references are adjusted when you **Insert** or **Delete** rows or columns, too. If you **Insert** a new column to the left of column A in the preceding example, the formula in cell B2, which used to be cell A2, will read =B1+1.

To use the **Across Worksheets** option in the **Fill** submenu, you must have selected multiple sheets beforehand (see "Selecting Multiple Worksheets" earlier in this chapter). When you choose **Across Worksheets** from the submenu, the Fill Across Worksheets dialog box (Figure 2-23) will appear; you can choose to fill Contents, Formats or both.

Figure 2-23. The Fill Across Worksheets dialog box.

Absolute, Relative and Mixed References

A *relative reference* in a formula, such as A1, becomes A2, A3, etc., as you **Fill Down** a formula into cells below the original formula. To keep the address of a cell fixed when you use the **Fill** commands, precede both its letter and number designation by a dollar sign (e.g., $B$1). An *absolute reference* such as $A$1 remains $A$1 as you **Fill Down**. You will find this *absolute cell addressing* useful if you wish to use numerical constants in formulas.

Occasionally it is useful to use *mixed references*. A *mixed reference* is a reference such as A$1 or $A1; the row or the column designation, respectively, will remain constant when you **Fill Down** or **Fill Right**.

	A	B
1		Increment
2		0.001
3	1.000	
4	1.001	
5	1.002	
6	1.003	
7	1.004	
8	1.005	
9	1.006	
10	1.007	
11	1.008	
12	1.009	

	A	B
1		Increment
2		0.001
3	1	
4	=A3+$B$2	
5	=A4+$B$2	
6	=A5+$B$2	
7	=A6+$B$2	
8	=A7+$B$2	
9	=A8+$B$2	
10	=A9+$B$2	
11	=A10+$B$2	
12	=A11+$B$2	

Figure 2-24. Two views of the same worksheet, showing formulas (left) and values (right). The formula in cell A2 has been filled down into A3:A10.

Relative References When Using Copy and Cut

If you **Copy** and **Paste** a formula, its references will be transferred using relative referencing. Thus, if you **Copy** the formula =A1+1 from cell A2 and **Paste** it in cell A10, the formula in cell A10 will be =A9+1. If you **Copy** the formula from cell A2 and **Paste** it in cell C2, the formula in cell C2 will be =C1+1. (This is probably not the formula you want.)

On the other hand, if you **Cut** the formula in cell A2 and **Paste** it anywhere in the worksheet, it will still be the formula =A1+1.

Thus the difference (with respect to cell references) between **Copy** and **Paste** and **Cut** and **Paste** is that **Cut** adjusts relative references so that they still refer to the original cells, while **Copy** does not adjust relative references, with the result that they refer to different cells.

The best way to copy a formula to a different row and column without altering relative references is to **Copy** it from the formula bar, click the Enter box to complete the **Copy** operation, then **Paste** in the destination cell.

Using Autofill to Fill Down or Fill Right

Excel's AutoFill feature lets you **Fill Down** or **Fill Right** simply by using the mouse pointer. To use AutoFill, select a cell by clicking on it. You will see a small black square on the lower right corner of the selected cell. Position the mouse pointer exactly over the small black square (the *fill handle*, or AutoFill handle). The mouse pointer becomes a small black cross. Click and drag in the usual way to select a range of cells. If the cell contains a formula, it will be duplicated to the rest of the range just as if you had used **Fill Down** or **Fill Right**. If the cell contains a number or a text label, the value will be duplicated in the rest of the cells. With AutoFill you can also **Fill Up** and **Fill Left**.

Excel Tip. *To **Fill Down** a value or formula to the same row as an adjacent column of values, select the source cell and double-click on the fill handle.*

Creating a Series

There are three ways to create a series. For example, to create the series of integers 1, 2, 3 ... in column A, you can either:

- Enter the value 1 in cell A1, enter the formula =A1+1 in cell A2, and then use **Fill Down** to create the series. You can then use **Copy** and **Paste Special** (Values) to convert the formulas to values.
- Use **Series**... from the **Fill** submenu of the **Edit** menu. With **Series** (Figure 2-25) you enter the start value, the end value, and the increment.
- The third and most convenient way to create a series is to use AutoFill, described in the following section.

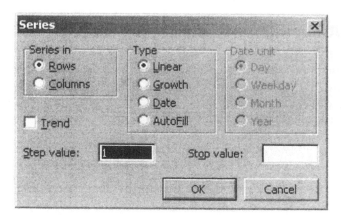

Figure 2-25. The Series dialog box.

Using AutoFill to Create a Series

This is by far the simplest and most convenient method. If you select a cell containing a number formatted as a date or time, or text that contains a number, and use AutoFill to fill a range of cells, AutoFill creates a series using the selected cell as the starting value. The value of the series being entered in the active cell is displayed in a Screen Tip box as you drag the AutoFill handle.

	A
1	January

	B
1	1
2	3

	C
1	Compound 1
2	Experimental
3	Calculated
4	

	A
1	January
2	February
3	March
4	April
5	May
6	June
7	July
8	August
9	September
10	October
11	November
12	December

	B
1	1
2	3
3	5
4	7
5	9
6	11
7	13
8	15
9	17
10	19
11	21
12	23

	C
1	Compound 1
2	Experimental
3	Calculated
4	
5	Compound 2
6	Experimental
7	Calculated
8	
9	Compound 3
10	Experimental
11	Calculated
12	

Figure 2-26. Some examples of the use of AutoFill to produce a series. (Above) Cells before using AutoFill. (Below) Series produced by AutoFill.

If you select two cells, AutoFill will create a series based on the cells you select, as shown in the second and third examples of Figure 2-26.

To create a decreasing-order series, use AutoFill to **Fill Up** or **Fill Left**.

If you select a cell containing a date, Excel will create a date series. For example, if you select a cell containing 29-Jan and use AutoFill to Fill Right, the series 29-Jan, 30-Jan, 31-Jan, 1-Feb, 2-Feb... will be created.

*Excel Tip. To prevent **AutoFill** from creating a series, hold down the Ctrl key as you position the black cross pointer over the fill handle. A small plus sign will appear to the right of the black cross pointer. Click and drag in the usual way to fill rather than create a series.*

The AutoFill Shortcut Menu

If you use the right mouse button to drag down the Fill Handle to create a date series, a shortcut menu will be displayed when you release the mouse button (Figure 2-27).

| Copy Cells |
| Fill Series |
| Fill Formatting Only |
| Fill Without Formatting |
| Fill Days |
| Fill Weekdays |
| Fill Months |
| Fill Years |
| Linear Trend |
| Growth Trend |
| Series... |

Figure 2-27. The AutoFill shortcut menu.

You can then choose to create a series consisting of consecutive dates (Fill Days), weekdays (Fill Weekdays), a single date in each month (Fill Months) or a single date in each year (Fill Years).

Formatting Worksheets

You can use commands from the **Format** menu to change and improve the appearance of the worksheet and to modify the way number values are displayed.

Using Column Width... and Row Height...

When Excel creates a blank worksheet, all rows are the same default height and all columns the same width. You can change the width of columns, or the height of rows, to improve the appearance of a worksheet or to eliminate wasted space so that you can get more information on a single page. You can also hide rows or columns by reducing their height or width to zero. The data they contain will still be still there, but it will be hidden.

To change the width of a cell or column, choose **Column** from the **Format** menu, and then choose **Width** from the submenu. You can enter the new width of the column; one unit corresponds to the width of one character of the current font. Column widths can also be changed by choosing **Autofit Selection** from the submenu. You can adjust the column width to fit a single selected cell or to be the best fit to the widest entry in a whole column.

Row height is adjusted in the same way.

You can also change the column width or row height by using the mouse pointer. To adjust column width, place the cursor on the separator bar between column headings, on the right of the column whose width you want to change. The cursor changes to a double-headed arrow: ↔. Click the mouse button and drag to the right or left to change column width. The column width is displayed in an "InfoBox" as you drag the separator bar.

Double-click on a column separator bar to get "best fit".

Excel Tip. To adjust several columns at a time to the same width, select the columns and then perform the column width adjustment with the mouse pointer on any of the selected columns. When you release the mouse button, all the columns will have the adjusted width. You can also get a "best fit" simply by double-clicking on the row or column separator. To adjust several rows or columns at once, select the columns and then double-click on any row or column separator.

Formatting Cells

The **Cells...** command in the **Format** menu allows you to change the appearance of values in cells. In Excel 2003, the Format Cells dialog box has Font, Alignment and Number tabs for specifying the appearance of values within cells, Border and Patterns tabs for specifying the appearance of cells, and the Protection tab for security of values within cells. The possibilities for Font, Alignment, Border and Patterns are many and varied, and only some of these possibilities will be discussed here. Number formatting is important for scientific spreadsheets; it will be discussed in detail in following sections.

Using Alignment

The **Alignment** command provides a number of formatting options for the alignment of values in cells. Choose **Cells...** from the **Format** menu and choose the Alignment tab. There are option buttons for both horizontal and vertical alignment (Figure 2-28). The Vertical orientation options are useful if you want to add a text label to a narrow data column. The Orientation "inclinometer" allows you to display text on an angle.

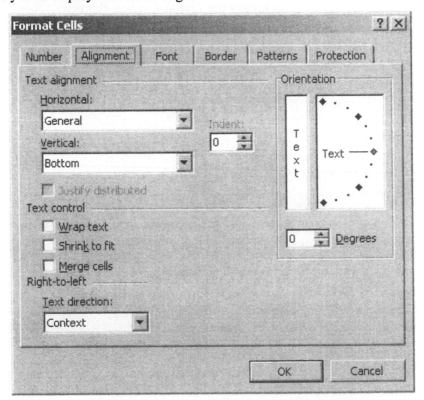

Figure 2-28. The Alignment dialog box.

To use the Merge Cells option, select the columns across which you want the text to be centered, as indicated in Figure 2-29. (The text can be in any of the selected cells.) Check the Merge Cells box and press OK.

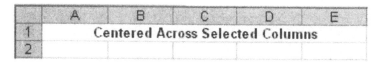

Figure 2-29. Using Merge and Center.

For the most common horizontal alignment options you can use the alignment toolbuttons ▤ ▤ ▤ ▥ to format cells left, centered, right or centered across selected cells, respectively.

You can also format text entries in cells so that the text wraps and is displayed in more than one line (Figure 2-30), by checking the Wrap Text box and then pressing OK. Excel breaks the text at a space character. Text can be both aligned vertically and wrapped.

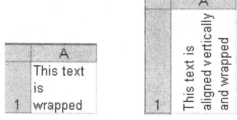

Figure 2-30. Examples of using Wrap Text.

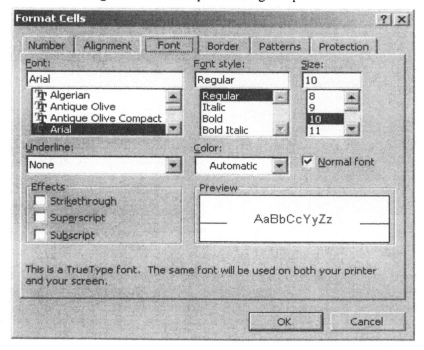

Figure 2-31. The Font dialog box.

When you select a text box, the Alignment dialog box has a slightly different form. The Automatic Size check box, when selected, fits the borders to the text.

Using Font...

The Font tab in the Format Cells dialog box allows you to format cells in any of the installed fonts (Figure 2-31). In addition, you can format individual characters in various font styles or sizes, or as strikethrough, superscript or subscript characters.

You can also use Greek letters, as shown in the text box example in Figure 2-10, by formatting the Roman letter using the Symbol font. Of course, you must know the correspondence between Roman and Greek letters: The Roman S becomes the Greek Σ, for example.

The Alternate Character Set

There is another way to enter some useful characters. From the keyboard, you can enter symbols that are in the so-called *alternate character set*; an example is shown in Figure 2-32. The characters produced may be different for each different font.

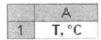

Figure 2-32. A special character (°) typed by using the alternate character set.

The characters are obtained by holding down the Alt key and typing the four-digit ASCII code for the character, *using the numeric keypad.* (If you're using a laptop without a numeric keypad, you'll probably need to hold down both the Fn and the Alt keys, and press the keys labeled with small digits 1–9, color-coded to match the Fn key.)

Table 2-3. Some Useful Alternate Characters

Symbol	4-Digit Code		Symbol	4-Digit Code	
€	0128	(Euro)	ß	0223	(Greek beta)
£	0163	(Pound)	µ	0181	(Greek mu)
¥	0165	(Yen)	Å	0197	(Ångstrom)
•	0149	(bullet)		0160	(unbreakable space)
·	0183	(centered dot)	^{o}	0186	(superscript)
×	0215	(multiply)	^{1}	0183	(superscript)
÷	0247	(divide)	^{2}	0178	(superscript)
±	0177	(plus or minus)	^{3}	0179	(superscript)
°	0176	(degree)	—	0151	(em dash)

If you use a different font, you'll have to experiment to see what alternate characters are produced.

The range of useful characters obtainable in this way is rather limited. Table 2-3 shows some useful characters obtained using the Arial font. The complete set of alternate characters is listed in Appendix J.

Entering Subscripts and Superscripts

You can enter subscripts and superscripts in text, as in Figure 2-33. First, highlight the characters to be subscripted or superscripted. Choose the Font tab in the Format Cells dialog box, check the Subscript or Superscript check box (see Figure 2-31), press the OK button and then enter the text by pressing the Enter key or the Enter button.

A
1 Chemical formulas with subscripts
2 $C_{12}H_{22}O_{11}$
3 $BaCl_2$ $2H_2O$
4 $H_2SO_4 + 2NaOH = Na_2SO_4 + 2H_2O$

Figure 2-33. Subscripts in Excel.

Using Border and Patterns

The Border tab in the Format Cells dialog box allows you to place a border around one or more sides of a selected cell or range. This is useful if you want to emphasize comments, instructions or values. The Patterns tab is used to change the background color or pattern of cells.

BOSTON COLLEGE					
OFFICE OF THE UNIVERSITY REGISTRAR		DATE	6/5/1996		
CHESTNUT HILL, MASSACHUSETTS 02167		ACADEMIC YEAR	1996		
		SEMESTER	96S		
COURSE	CH 222 01 INTRO/INORGANIC CHEM				
INSTRUCTOR	BILLO, E JOSEPH				
MEETING PLACE					

STUDENT'S NAME	ID NUMBER	SCHOOL	CLASS	MAJOR	CREDIT	GRADE

Figure 2-34. Using Border to create a custom report form.

The Border option is often used to underline headings, or in a sheet in which the gridlines have been removed, to create a custom form. Figure 2-34 shows a portion of a template sheet produced in this way. To remove gridlines, choose **Options** from the **Tools** menu and choose the View tab and then uncheck the Gridlines check box.

The built-in template sheets provided with Excel 2003 ("Spreadsheet Solutions") are good examples of the use of Border and Patterns to create custom forms.

Using the Format Painter Toolbutton

The Format Painter toolbutton ✐ copies and pastes formats from one cell or range to another cell or range. To use it, do the following:

- Select the range with the desired format(s).
- Click the Format Painter toolbutton (this copies the format).
- Click on a cell or drag across a range of cells to paste the format(s).

Excel Tip. To use the Format Painter button to "paint" a format on a series of non-adjacent cells or ranges, select the cell with the desired format and then double-click on the Format Painter button. This will keep the button in the "pressed" position, allowing you to click on several cells or ranges to paste the format. When you're done, click once on the button to return it to the "unpressed" state

Number Formatting

The formatting described in the preceding sections — bold text, italic text, alignment of text, adjusting column widths, etc. — is sometimes referred to as *stylistic formatting*. In addition, it is possible to change the way that number values are displayed in cells. This type of formatting, called *number formatting*, is described in the following sections.

Using Excel's Built-In Number Formats

To change the way a number is displayed in a cell, choose **Cells...** from the **Format** menu and choose the Number tab to display the number format categories — Number, Currency, Date, Time, Percentage, Scientific, etc. For example, selecting a cell and then choosing Number from the list of categories will display the value in the cell to two decimal places (the default); you can change the number of displayed decimal places using the Decimal Places box. You can also display values as percentages or in exponential notation, as indicated in Figure 2-35.

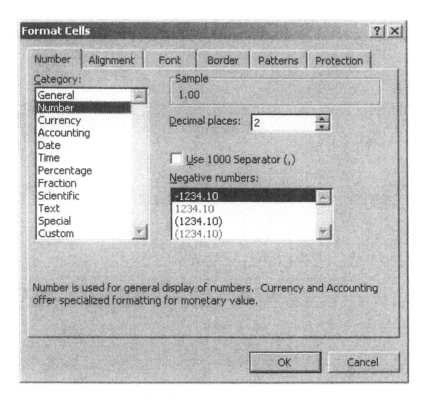

Figure 2-35. The Number Format dialog box.

The appearance of the Number dialog box will be different, depending on the number format category you select.

If you choose the Custom category, you will see the number formatting code that was applied to the cell. If you scroll through the list of codes, you'll see that many of them are quite complex (Figure 2-36). For the meaning of the built-in number format code symbols, see Table 2-4 or go to "Number Format Codes" in Excel's On-Line Help.

Custom Number Formats

You can create custom number formats. First, choose the Custom category from the list of numbers formats. This will display the list of (so far) built-in number formats. To add a new, user-defined number format (it will be added at the bottom of the list), type the format in the Type box. For example, if you want to display numbers to four decimal places, type 0.0000 in the Type box. The new format will be stored in the list of formats so that you can apply it to other cells; the format is available in all sheets in the workbook.

Table 2-4 lists the formatting symbols you can use to create your own custom formats.

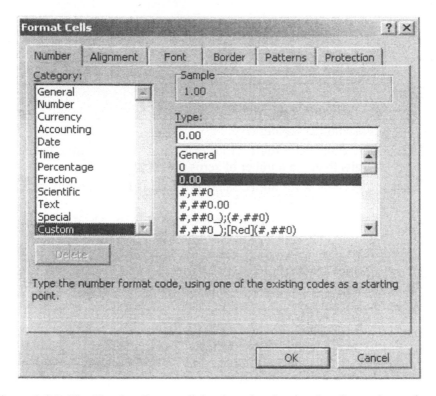

Figure 2-36. The Number Format dialog box showing number formatting codes.

You can create some fairly sophisticated number formats. For example, the format $#.0,, (dollar sign, number sign, period, zero, comma, comma) formats financial entries rounded to millions, with one decimal; the value 21180000 is displayed as $21.2.

The format #.??????? was used to format the table of atomic weight values shown in Figure 2-37, so that they are aligned on the decimal. (Note that, since the format contains seven ? symbols and the atomic weight of Na has only six digits to the right of the decimal point, there is an additional space to the right of the number.)

	A	B
1	H	1.00797
2	O	15.9994
3	Na	22.989768
4	S	32.066

Figure 2-37. Values aligned on the decimal point by using the ? formatting symbol.

You can use number formatting to add units to a number value. For example, the format #" g" appends the grams unit g to a number value; the value 50 is displayed as 50 g, as shown in Figure 2-38.

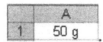

Figure 2-38. Units added to a value by means of number formatting.

Table 2-4. Number Formatting Symbols

#	Placeholder for digit.
0	Placeholder for digit. Adds extra zeros if the number has fewer digits than the number of zeros specified in the format.
?	Placeholder for digit. Same as 0, except that space characters are added.
,	Thousands separator (if used with #, 0 or ?). Used alone, it rounds and truncates to the thousands place (millions place if two commas are used, etc.)
%	Converts a number to percent.
E	Converts a number to scientific format. Use E- to include sign with negative exponents only, E+ to include sign with both positive and negative exponents.
/	Converts to a fraction. Usually used in the form ?/?? or #/## or e.g., #/64. The number of ? or # symbols determines the accuracy of the display.
"text"	Text characters can be included in a format by enclosing them in quotes. You can sometimes get away without the quotes.
@	Text placeholder. If the cell contains a text entry, the text is displayed in the format where the @ symbol appears.
[RED]	Displays the characters in the cell in red. You can also use [BLUE], [GREEN], [YELLOW], etc.

Date Formats

As you saw earlier, Excel recognizes the format of data typed into cells. If you enter a value such as 6/23/00 and then choose **Cells...** from the **Format** menu, choose the Number tab, and choose the Custom category, you'll see that Excel has recognized the value as a date and applied the m/d/yy number format. .

You can create custom date formats by using the year, month and day formats listed in Table 2-5. Day or month formats can have one-, two-, three- or four-letter formats; year formats can have either two- or four-letter formats. For example, the number format dddd, mmmm d, yyyy applied to a date entered as 8/3/38 will display Wednesday, August 3, 1938.

Table 2-5. Date Formatting Symbols*

d	Displays the day as a number without leading zeros (1-31)
m	Displays the month as a number without leading zeros (1-12)
dd	Displays the day as a number with leading zeros (01-31)
mm	Displays the month as a number with leading zeros (01-12)
ddd	Displays the day as an abbreviation (Sun-Sat)
mmm	Displays the month as an abbreviation (Jan-Dec)
dddd	Displays the day as a full name (Sunday-Saturday)
mmmm	Displays the month as a full name (January-December)
yy	Displays the year as a two-digit number, e.g., 97
yyyy	Displays the year as a four-digit number, e.g., 1997

* See Chapter 12 "Other Language Versions of Excel" for date formatting
symbols in some other languages.

Time Formats

Excel's built-in or custom time formats use the symbols in Table 2-6.

Table 2-6. Time Formatting Symbols

h	Displays the hour without leading zeros (0-23)
hh	Displays the hour with leading zeros (00-23)
m	Displays the minutes with leading zeros (0-59)
mm	Displays the minutes with leading zeros (00-59)
s	Displays the seconds without leading zeros (0-59)
ss	Displays the seconds with leading zeros (00-59)
s.000	Displays seconds to the millisecond
AM/PM	Displays the hour as AM or PM instead of 24-hour time
[h]	Displays elapsed time in hours
[h]:mm	Displays elapsed time in hours and minutes

Variable Number Formats

Different number formats can be applied to positive, negative, zero and text
values entered into a cell. A complete format consists of four sections, separated
by semicolons, for positive, negative, zero and text values, respectively. If only
one number format is specified, it applies to all values. If two number formats
are specified, then the first one applies to positive numbers and zero, while the
second one applies to negative numbers.   For example, the format
$#,###;[Red]$#,### formats positive amounts in black, Excel's default color, and
negative amounts in red.

Conditional Number Formats

Conditional number formats can be created by using the syntax [*condition, value*] *format statement*. *Condition* is one of the symbols <, >, =, >=, <=, <>; *value* may be any number. Format statement may be any built-in or custom format. For example, the number format [>1] "Number too large" accepts any input less than 1 but otherwise issues an error message.

Several conditions may be combined using semicolons. The number format

[>999]#.##,,%; #" ppm"

displays the values 110 and 21560 as 110 ppm and 2.16%, respectively.

The maximum number of conditions is three.

Using the Number Formatting Toolbuttons

You can also format number values in cells by using the number formatting toolbuttons shown following.

+.0 .00	Increases the number of decimal places.
.00 →.0	Decreases the number of decimal places.
%	Formats the number in percent style, with no decimal places.
$	Formats the number in currency style, with two decimal places.
,	Formats a number with commas and two decimal places.

Excel Tip. There isn't a toolbutton to format number values in scientific format. You can apply scientific format conveniently by using the shortcut key sequence Ctrl+Shift+^. See Appendix H for a complete list of shortcut keys.

Formatting Numbers Using "Precision as Displayed"

To permanently change *all values* stored on a worksheet to their displayed values, use the Precision as Displayed option. Once this command has been invoked, you can't restore the original values.

To apply Precision as Displayed, choose **Options** from the **Tools** menu and choose the Calculation tab. Check the "Precision as Displayed" box, and then press OK. Because this is an irreversible change, Excel asks you to confirm the change.

To change only a *selected range of values* to "Precision as Displayed", use the FIXED worksheet function (see "Text Functions" in Chapter 3).

Excel Tip. *You can apply the same formatting to multiple worksheets simply by grouping the sheets (click on the first sheet tab in the range of sheets to be formatted, then hold down the Shift key and click on the last sheet in the range). When you apply the desired formatting to the active sheet, it will be applied to all sheets in the group.*

Using Conditional Formatting (Part I)

The number formatting tools that we have seen described in preceding sections can only change the way a number appears in a cell: as a floating point number, in scientific notation, etc. But none of the number formatting tools allows you to apply stylistic formatting to a cell based on the value in that cell, for example, to display a number in italics if it is negative. Conditional formatting provides this capability.

Conditional formatting allows you to do one of two things: to apply stylistic formatting to a cell based on the value in that cell, or to apply formatting to a cell based on the values in several cells (in the same or different worksheets or workbooks). The former — formatting a cell based on the value in that cell — will be described here; the latter will be reserved for Chapter 3, Excel Formulas and Functions.

As an example of the use of conditional formatting, consider the case of a worksheet that contains a column of values. If any value in the column is above 100 (for example), the measurement is out of range and needs to be examined more closely. You can flag these out-of-range values with conditional formatting. The dialog box in Figure 2-39 shows how this is done.

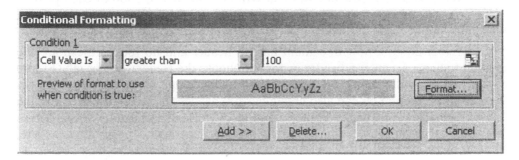

Figure 2-39. The Conditional Formatting dialog box.

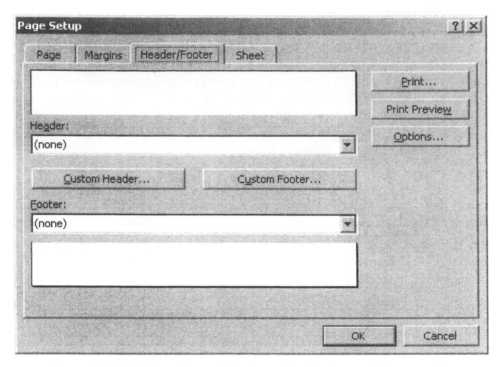

Figure 2-42. The Page Setup dialog box, showing the Header/Footer tab.

To squeeze the maximum amount of worksheet information on a single page, you may want to decrease the margin widths. The default margin values are 0.75 inch left and right and 1 inch top and bottom. If you set the margins to zero, the header and footer information will still be printed, usually right on top of data in your worksheet, so delete the header and/or footer information by choosing "(none)" from the list box.

You can choose Print Row And Column Headings and/or Print Cell Gridlines by choosing the Sheet tab (Figure 2-43). If you de-select Cell Gridlines, they will still be displayed on the screen but they will not be printed. If you turn off screen display of gridlines by choosing the Display command from the Options menu, the Cell Gridlines check box will also be cleared in the Page Setup dialog box and gridlines will not be printed.

You may need to use Print Black and White if your worksheet uses color. Colors may be printed as various patterns by your printer; to remove the patterns and produce text in cells in black and white, check the Print Black and White Cells box.

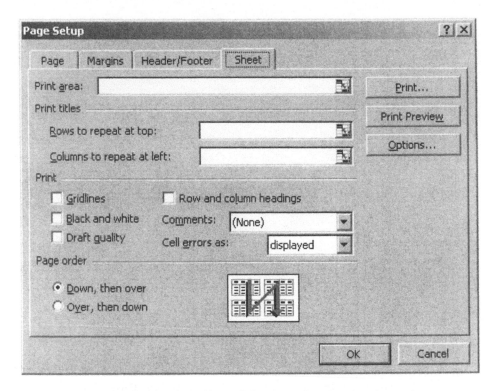

Figure 2-43. The Page Setup dialog box, showing the Sheet tab.

Using Print Preview

Print Preview is useful in other ways besides showing what your finished worksheet will look like when printed. If you preview your worksheet and then return to the document window, page breaks will be displayed on the worksheet as dashed lines, to assist you in adjusting column widths, for example, before printing.

Using Print...

The Print dialog box (Figure 2-44) allows you to specify the pages to be printed and the number of copies. You can also specify a printer, if you have more than one printer connected. You can press the Properties button to display a printer properties dialog box specific to your printer, where you can choose, for example, two-sided printing.

If you choose the **Print** command and simply press the OK button, Excel will print the rectangular array of sheets that includes all filled cells. It's a good idea to use **Print Preview** before printing; the total number of pages to be printed will be displayed in the status bar. This will tell you whether you can print the whole worksheet, or whether you need to specify a range of pages to be printed.

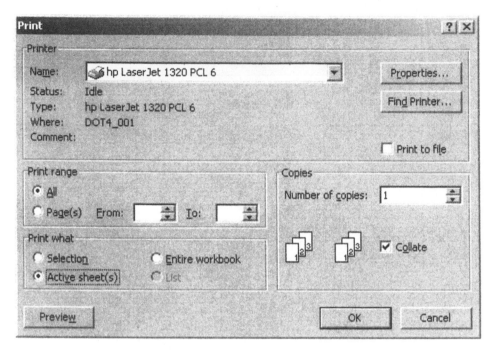

Figure 2-44. The Print dialog box.

If you choose **Print Preview** or **Page Setup**, Excel displays the automatic page breaks as dashed lines in the worksheet. You can also insert a forced page break if you want to print a portion of a worksheet page. To insert a horizontal page break, select an entire row as if you were going to insert a row. Then choose **Page Break** from the **Insert** menu. The page break will be inserted immediately above the row selected. Excel displays forced page breaks as dashed lines that are heavier than the dashed lines used to indicate automatic page breaks. A forced vertical page break is inserted in a similar fashion; the page break is inserted immediately to the left of the selected column. If you want to insert both a vertical and a horizontal page break, select a single cell within the worksheet; the page breaks will be immediately above and to the left of the cell.

Printing a Selected Range of Cells in a Worksheet

To print a selected range of cells within a worksheet, first select (highlight) the range to be printed, then choose **Print...** from the **File** menu. Then press the Selection button in the Print What category box in the lower left corner of the dialog box. The selected range will be printed, but the selection will not be "remembered" the next time you print.

To set a range of cells to be printed each time you choose Print, you must use **Set Print Area**. You can do this in at least two different ways:

- Choose **Page Setup** from the **File** menu, and choose the Sheet tab. Click in the Print Area text box to select it. Now select the range of cells that you want to print (you can move the dialog box out of the way if necessary), and press the OK button. To cancel a Print Area selection, delete the text within the Print Area text box.

- First, select the range of cells to be printed, then choose **Print Area** from the **File** menu, and choose **Set Print Area** from the submenu. The range to be printed will be indicated by Page Break lines. Choose **Remove Print Area** from the submenu to cancel the Print Area.

If the Print Area you selected requires more than one page, you can choose **Page Setup** and change the value in the Reduce/Enlarge box to less than 100%. Sheets printed with values less than about 60% are difficult to read, though. To obtain the appropriate reduction value automatically, after you've selected the area to be printed, choose the Page tab and press the "Fit To 1 Pages Wide By 1 Tall" button.

Printing Row or Column Headings
for a Multi-Page Worksheet

If you are printing a multi-page worksheet, you can duplicate row or column headings automatically on each printed page. Choose **Page Setup...** from the **File** menu, and choose the Sheet tab, shown in Figure 2-43. Select the Rows To Repeat At Top or the Columns To Repeat At Left text box by clicking the cursor in it. Now select the range of cells that you want to print on every page as a title (you can move the dialog box out of the way if necessary). Then click the OK button. The headings will appear at the top or left of each printed page.

Protecting Data in Worksheets

First of all, we should distinguish between security and protection. Security means protecting your computer from viruses. Protection means preventing users from modifying documents or viewing particular workbooks, worksheets or formulas; a number of ways of protecting data will be described in the following sections.

Protecting a Workbook

You can protect the structure of a workbook so that worksheets in the workbook can't be moved, deleted, hidden, unhidden, or renamed, or new worksheets can't be inserted, or windows are the same size and in the same position each time the workbook is opened. Values and formulas in the workbook can still be modified, however; to prevent this, see "Protecting a Worksheet" later in this chapter.

To protect a workbook, choose **Protection** from the **Tools** menu and choose **Protect Workbook...** from the submenu, to display the Protect Workbook dialog box (Figure 2-45).

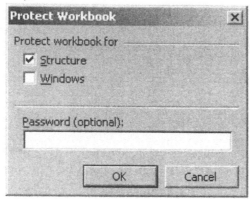

Figure 2-45. The Protect Workbook dialog box.

Check the boxes for Structure and/or Windows, enter a password if necessary (you will be asked to confirm it), and then press OK.

Protecting a Workbook
by Making It a Read-Only Workbook

If you make a workbook read-only, users can view formulas in cells and can change values and formulas, but the changes cannot be saved.

To make a workbook read-only, the document should be closed. In the Windows **Start** menu, choose **Programs**, and then **Windows Explorer**. In the **Exploring** window, open the drive or folder that contains the file and select the document name. Choose **Properties** from the **File** menu, choose the General tab, and check the Read-only check box.

Hiding a Worksheet

You can hide a worksheet by choosing **Sheet** from the **Format** menu, and then **Hide** from the submenu. However, anyone can view this sheet simply by choosing **Unhide** from the submenu.

You can hide a sheet so that most users can't view it, by using VBA – you set the Visible property of the sheet to VeryHidden, as described in the following paragraph. You may need to read Chapter 16 "Visual Basic for Applications: An Introduction" first.

To make a sheet VeryHidden, switch to the Visual Basic Editor by choosing **Macros** from the **Tools** menu and then **Visual Basic Editor** from the submenu, or simply press Alt + F11. If the Project Window is not visible, display it by pressing the Project Explorer toolbutton 🔍 to display it, or simply press Ctrl+R.

In the hierarchy tree for the desired workbook, click on the name of the sheet you wish to hide. If the Properties Window is not visible, display it by pressing the Properties Window toolbutton, or simply press F4. In the Properties Window, locate the Visible property (at the bottom of the list when the Alphabetic tab is selected). Click on the Visible box; this will cause a drop-down list button to appear in the properties list. Choose the xlSheetVeryHidden property, as shown in Figure 2-46. When you switch back to the Excel workbook, the sheet tab will not be visible and the sheet name will not appear in the **Unhide** submenu.

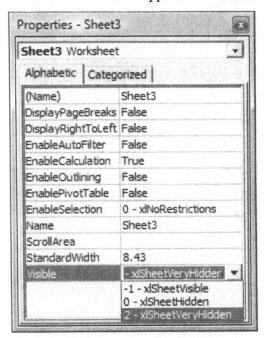

Figure 2-46. The VBA Properties Window.

Protecting a Worksheet
by Locking or Hiding Cell Contents

You can either lock cells (prevent them from being selected by the user) or hide the contents (usually formulas) that the cells contain. For example, you may want the user to be able to enter values in certain cells (perhaps these cells will be the input area) while protecting the rest of the worksheet.

The process for doing this is somewhat complicated. First you select cells to be locked or unlocked, or cells whose contents will be hidden or visible, and set their status using the Protection tab of the **Cells...** command in the **Format** menu. Then you put the status into effect by choosing the **Protection** command in the **Tools** menu.

Before you begin, it's important to know that when a new worksheet is opened, the status of *all* cells in the document is Locked. To lock only a limited range of cells in a document, first set the status of all the cells in the document to Unlocked and then select the range of cells that you want to be locked.

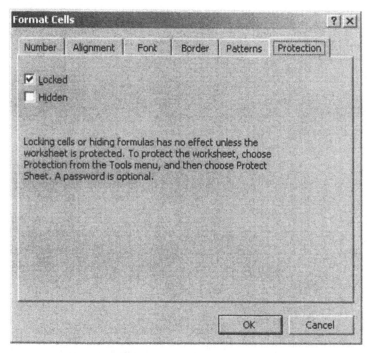

Figure 2-47. The Protection tab of the Format Cells dialog box.

As an example, let's protect a worksheet so that it has only a single unprotected cell. The user will only be able to enter a value in this cell; all other cells in the sheet will be locked. Here's how to do it: first, select the cell you want to be the input area. Now choose **Cells...** from the **Format** menu and choose the Protection tab to display the dialog box (Figure 2-47). In a new worksheet, by default all cells are locked, so simply uncheck the Locked box so that the cell you selected will be unlocked. Then press OK.

Locking or unlocking cells has no effect unless the worksheet is protected. Choose **Protection** from the **Tools** menu, and **Protect Sheet...** from the submenu. This will display the Protect Sheet dialog box (Figure 2-48). Here you can choose to prevent users from carrying out one or more actions on the whole worksheet (inserting or deleting rows or columns, etc.) or on specified cells. After choosing options from the list, you can enter a password if you desire (you will be asked, in a second dialog box, to confirm the password), in which case users will not be able to unprotect the sheet

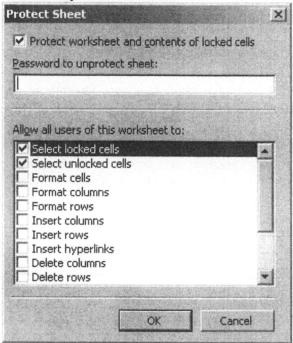

Figure 2-48. The Protect Sheet with Password dialog box.

To Lock a Range of Cells in a New Document

1. Select all cells in the document by clicking on the row/column header button in the upper left corner of the worksheet.

2. Choose **Cells...** from the **Format** menu and choose the Protection tab, uncheck the Locked option, and press the OK button. This un-protects all cells in the worksheet.

3. Now select the range of cells that you want to protect. Choose **Cells...** from the **Format** menu, choose the Protection tab, check the Locked option, and press the OK button.

4. Choose **Protection** from the **Tools** menu and choose **Protect Sheet...** from the submenu. You can enter a password if you wish (Figure 2-48). If you merely want to prevent yourself from making accidental changes, no password is necessary. If you want to protect the document from changes by others, you need a password; make sure that you will be able to retrieve it when you need it.

Controlling the Way Documents Are Displayed

Use the **Window** menu to switch between one Excel document and another. All open documents are listed at the bottom of the **Window** menu; the active document is indicated with a check mark.

Use **Hide** to hide a workbook. Most commonly you'll use **Hide** with workbooks that contain macros. A macro is still "active" even when it's hidden.

To hide a worksheet in a workbook, choose **Sheet** from the **Format** menu and choose **Hide** from the submenu.

Viewing Several Worksheets at the Same Time

Although only one worksheet at a time can be the active window, Excel provides a number of ways to view data in several different worksheets, or different areas of the same worksheet, at the same time.

Using New Window and Arrange...

If you have more than one document open, you can view them simultaneously in a number of ways. One way is to re-size and move the documents so that the desired part of each can be seen at the same time. Another way is to use the **New Window** and **Arrange...** commands in the **Window** menu. The latter method can be used to view multiple documents, or multiple sheets in the same workbook, as described in the following paragraph.

To view multiple worksheets in the active workbook, choose **New Window**. A second window will be opened for the active workbook. If, for example, the workbook is named Viscosity Data, the windows will be named Viscosity Data:1 and Viscosity Data:2. Activate each window in turn and click on the sheet that you want to display. Now choose **Arrange...** from the **Window** menu, Excel displays the Arrange Windows dialog box (Figure 2-49).

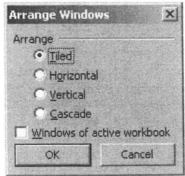

Figure 2-49. The Arrange Windows dialog box.

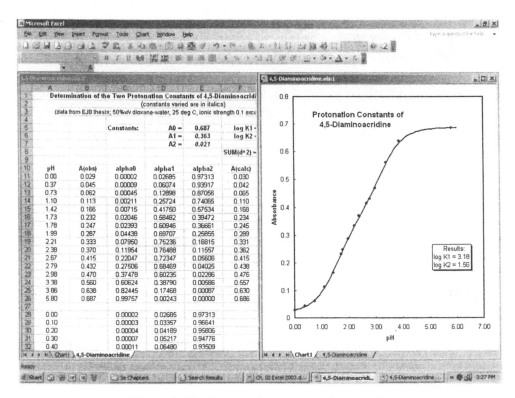

Figure 2-50. Two windows arranged vertically

You can arrange the windows horizontally (one above the other) or vertically (side by side). (The active document will be on top or on the left, respectively.) If you have created a separate chart sheet from data in a worksheet, **Arrange**... provides a convenient way to work with a sheet and observe changes in the associated chart. With **Arrange...**, chart documents are reduced in size so that the whole chart appears in the window; worksheet documents are not reduced in size. Figure 2-50 illustrates a worksheet/chart combination displayed using the **Arrange** (Vertical) option.

With three open documents, the **Tiled** option arranges the documents with the active sheet occupying the left half of the screen; the other two sheets each occupy one-quarter of the screen, one above the other. With four documents **Tiled**, each occupies one-quarter of the screen. Click on any document to make it the active sheet. Double-click anywhere on the solid border between the windows to undo the arrangement.

Different Views of the Same Worksheet

As your worksheets get larger and more complicated, it becomes impossible to view all of a single worksheet at one time, or even all cells in one row or column at one time. Excel provides several convenient ways to display separate portions of a single worksheet on the screen at the same time, so that you can view one part while entering or changing data in another part.

Using New Window

When you choose **New Window**, a second window of the active document is opened. The view name of the original workbook will now be, for example, MyWorkbook:1 and the new one will be MyWorkbook:2.

You can resize and move the windows so that the desired parts of the worksheet can be seen at the same time. This is useful if you want to **Cut** or **Copy** several cell ranges and then **Paste** them into another area of a worksheet, but the two areas of the worksheet are far apart.

You can set different display options for the two windows. Display values in one window and formulas in another to see the effect of changes.

*Excel Tip. To remove the additional workbook views, choose **Arrange...** from the **Window** menu, and choose any one of the arrangements – Horizontal, for example. Click on the window you want to remove (the active window has a blue title bar). Now click on the Close button in the upper right corner of the window to remove the window.*

Using Split Screens

To use the **Split** feature to split a document window horizontally into two windows, select an entire row as if you were going to insert a row. Then choose **Split** from the **Window** menu. This creates a split in the window, above the selected row, with each part of the window displaying the active document. Each part of the document now has its own scroll bar, and you can scroll one part of the document while the other part remains fixed. A vertical split, as illustrated in Figure 2-51, is accomplished in the same way.

You can also split the document window by placing the mouse pointer on either split button (the small rectangles at the right end of the horizontal scroll bar and at the top of the vertical scroll bar), and then click and drag the button.

The document window can be split both horizontally and vertically by first selecting a single worksheet cell, then choosing **Split**.

To remove a split, choose **Remove Split** from the **Window** menu, or slide the split button back to its original position.

Excel Tip. To remove a split from a window, it's not necessary to slide the split button back to its original position at the top or left-hand side of the scroll bar. Just place the pointer on the split button and double-click.

	A	E	F	G	H	I	J	K
1	**Grade Sheet**							
2			**Hour Exams**					
3	**Name**	**#1**	**#2**	**#3**	**Oral report**	**Paper**	**Final Exam**	**Total**
4								
5	CAPICELLA, Jason	*15*	*4*	27	17.5	45	23	43.8
6	CHUNG, Min-Yin	18	13	44	17.5	45	47	61.5
7	FERREIRO, Kathy	24	16	32	20	45	52	63.0
8	GANGE, Eric	28	13	43	20	40	51	65.0
9	GREALEY, John	22	14	40	17.5	40	56	63.2
10	HAPPERSBACH, Bill	28	12	30	17.5	45	59	63.8
11	HOGAN, Derek	37	17	37	20	50	60	73.7

Figure 2-51. A document with a split screen.

Using Freeze Panes

Freeze Panes can be used to create a similar split document window, but the upper or left part of the window is fixed and cannot be scrolled. Split panes are useful to display fixed row or column headings (or both) while scrolling through the rest of the worksheet.

To use the **Freeze Panes** feature to split a document window horizontally into two windows, select an entire row as if you were going to insert a row. Then choose **Freeze Panes** from the **Window** menu. To split the window both horizontally and vertically, select the cell whose upper left corner defines the location of the split, and then choose **Freeze Panes**.

	A	B	C	E	F	G	H	I	J	K
1	Grade Sheet									
2					**Hour Exams**					
3	**Name**	**ID#**	**YOG**	**#1**	**#2**	**#3**	**Oral report**	**Paper**	**Final Exam**	**Total**
4	CAPICELLA, Jason	973124	95	15	4	27	17.5	45	23	43.8
5	CHUNG, Min-Yin	240090	96	18	13	44	17.5	45	47	61.5
6	FERREIRO, Kathy	575994		24	16	32	20	45	52	63.0
7	GANGE, Eric	863790	95	28	13	43	20	40	51	65.0
8	GREALEY, John	662045	96	22	14	40	17.5	40	56	63.2
9	HAPPERSBACH, Bill	965739	95	28	12	30	17.5	45	59	63.8
10	HOGAN, Derek	726781	94	37	17	37	20	50	60	73.7
11	LAROZI, Patrick	952860	96	27	12	38	20	45	58	66.7

Figure 2-52. A document displayed with Freeze panes.

Opening Excel 2007/2010 Documents in Excel 2003

Because Excel 2007/2010 uses a different document format (see Chapter 3 for details), spreadsheets created by using Excel 2007/2010 cannot be opened from Excel 2003. If you are using Excel 2003 and you try to open an Excel 2007/2010 file, you will get a "Windows cannot open this file" error message (Figure 2-53).

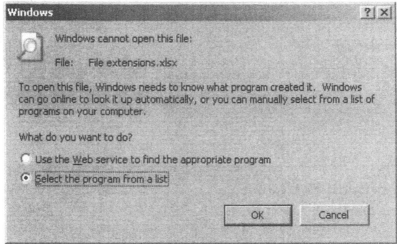

Figure 2-53. Dialog box displayed when trying to open
an Excel 2007/2010 document with Excel 2003.

You can ask the person who sent you the file to save it in Excel 2003 format, or you can follow the instructions in the dialog box to find a file-conversion program.

The best way to handle Excel 2007 files is to download a file conversion utility that will automatically convert Excel 2007/2010 files to Excel 2003-readable files. Search for and download the 2007 Microsoft Office System Compatibility Pack. Once this utility has been downloaded and run, whenever you open an Excel 2007/2010 document, a copy in Excel 97-2003 format will be created, with a filename like Xl0000020. During file conversion process, a message similar to the one shown in Figure 2-54 will be displayed.

Figure 2-54. Dialog box displayed during conversion of
an Excel 2007/2010 document to Excel 2003 format.

Excel 2003 Workbook and Worksheet Specifications

Worksheet size	65,536 rows by 256 columns
Maximum column width	255 characters
Maximum row height	409 points
Number precision	15 digits
Largest allowed positive number	1.79769313486231E+308
Largest allowed negative number	−1E+307
Smallest allowed negative number	−2.2251E-308
Smallest allowed positive number	2.229E-308
Levels of Undo	16
Maximum number of...	
open workbooks	Limited by available memory
characters in a cell (text)	32,767
characters in a formula	1,024
sheets in a workbook	Limited by available memory
colors in a workbook	56
custom number formats	Between 200 and 250
windows in a workbook	Limited by system resources
panes in a window	4
sort levels in a single sort	3

3

Excel Formulas and Functions

This chapter shows you how to use Excel's wide range of worksheet functions to construct sophisticated worksheet formulas. In addition, you'll also learn techniques of formula editing and worksheet troubleshooting.

The Elements of a Worksheet Formula

A worksheet formula consists of some or all of the following: constants, operators, cell references and worksheet functions. This section discusses some of the details of constants, operators and cell references. Subsequent sections describe the more than 300 worksheet functions available in Excel.

Constants

A constant is a fixed value in a formula: a number, a date, or text. Occasionally you will use a constant in a formula, rather than a reference to a cell containing the constant value, but almost always it's better to use a cell reference. If you use a constant instead of a reference (I call this a "hard-wired" formula), you have to modify all formulas that contain the constant, and it's easy to neglect to change some of them.

Operators

There are four classes of worksheet operators: arithmetic, text, logical and reference.

The *arithmetic operator*s +, −, *, / and ^ (for addition, subtraction, multiplication, division and exponentiation, respectively) should already be familiar ones.

There is only one *text operator*: the & (ampersand) symbol. It is used to concatenate text, or text and variables. For example, if cell G256 contains the value 2001, the formula ="Chemical Inventory for "&G256 displays Chemical Inventory for 2001. You'll encounter many other uses of the & operator in the chapters that follow.

The *logical operators* compare two values and produce a logical result, either TRUE or FALSE. The logical operators are the following: = (equal to), > (greater than), < less than), >= (greater than or equal to), <= (less than or equal to) and <> (not equal to). Note that >=, <= and <> must be typed as shown, or Excel will not recognize them as operators and will produce an "Error in formula" message. In worksheet formulas, FALSE can be represented by zero, and TRUE can be represented by any non-zero value.

There are three *reference operators*: the range operator (colon), the union operator (comma) and the intersection operator (space). The *range operator* produces a reference that includes all the cells between and including the two references, e.g., G3:L3. The *union operator* combines multiple references into one reference, e.g., A3,B4 C5,D6. More than one union operator can be used in a single reference. The *intersection operator* produces a reference to the cells common to two references. For example, the reference D2:D4 B3:G3 refers to cell D3, as illustrated in Figure 3-1. The intersection operator is particularly useful when used with *named references* (see "Using Create Names" in Chapter 6).

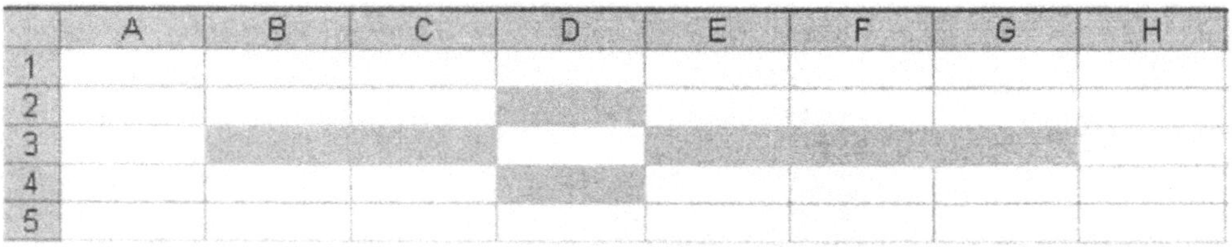

Figure 3-1. A single cell reference (D3) produced by using the intersection operator, in the expression D2:D4 B3:G3.

References: Absolute, Relative and Mixed

Cell references can be absolute, relative or mixed. A *relative reference*, such as A1, becomes B1, C1, and so on, as you **Fill Down** a formula into cells below the original formula. An *absolute reference* such as $A$1 remains $A$1 as you **Fill Down**. A mixed reference is a reference such as A$1 or $A1; use *mixed reference* style if you want to control how the reference is changed as a formula is duplicated using **Fill Down** or **Fill Right**. In the first case the row designation will remain constant when you **Fill Down** or **Fill Right**; in the second case the column designation will remain constant. For example, the reference $D2 becomes $D3 when copied to the cell in the next row down, but remains $D2 when copies to the cell in the next column to the right.

Excel Tip. You can use F4 to toggle between cell reference types. Select the cell reference by double-clicking on it in the formula bar (or just put the insertion point cursor anywhere in the reference), and then press F4 to cycle through the formats, in the sequence: relative (e.g., A1), absolute (e.g., $A$1), mixed (e.g., A$1), mixed (e.g., $A1). If you are typing a formula, you can use F4 immediately after typing the cell reference; if you want to convert a reference after you have completed typing the formula, place the cursor next to or within the typed reference and press F4.

R1C1 Reference Style

Excel can use either of two reference styles: the familiar A1 reference style, such as $A$1, or the R1C1 reference style. In an R1C1-style reference, the location of a cell is specified by an R followed by a row number and a C followed by a column number — for example, R1C5 to indicate cell $E$1.

A reference such as R1C5 is an absolute reference. In R1C1-style, relative references are specified by means of a "pointer" enclosed in square brackets, pointing from the cell containing the formula to the referenced cell.   For example, if the following formula in A1-style is entered in cell B1

    =2*A1

the same formula in R1C1-style would be

    =2*RC[-1]

Some further examples of R1C1-style references:

R[2]C[2]	relative reference to the cell two rows down and two columns to the right. If the reference is in cell A1, it refers to cell C3.
R2C5:R25C5	absolute reference to the range $E$2:$E$25.
C5	absolute reference to column E. The corresponding A1-style reference would be $E:$E.

If you want to see what a formula looks like when displayed in R1C1-style, choose **Options** from the **Tools** menu, and click the General tab.   Under Settings, select the R1C1 reference style check box.

I am sure that almost everyone uses A1-style references in their Excel formulas, but R1C1-style references are useful if you use the INDIRECT worksheet function (See "The INDIRECT Function" in Chapter 6).

Creating and Using External References

External references are used to establish links between documents. You can create an external reference between worksheets within the same workbook, or between worksheets in different workbooks. By linking workbooks, you can utilize the data from one workbook in a formula in another workbook, merge data from several workbooks in a summary sheet, or simplify a complicated model by breaking it up into manageable portions located in different sheets or workbooks.

Linked worksheets are linked dynamically; that is, when you make changes in a *source worksheet*, the changes automatically occur in the *dependent sheet*.

An external reference includes the filename of the source worksheet, separated from the reference or name by an exclamation point, e.g., =SheetName!CellReference if the link is between sheets in the same workbook or =[WorkbookName]SheetName!CellReference if the link is between sheets in different workbooks.

If either the workbook name or the sheet name contain spaces, the [WorkbookName]SheetName must be enclosed in single quotes (e.g., ='[logTplot.xls]Phase diagram data'!$A$1).

There are two ways to enter an external reference in a formula. With the first method, you begin typing the formula in the destination cell and enter the external reference by selecting the appropriate range in the source document. With the second method, you copy the range in the source document and then use **Paste Special...** to paste it into the formula in the destination cell. These two methods are described in the following sections.

Creating an External Reference by Selecting

To enter a reference from a source document into a formula in a dependent document, first select the cell where the value is to be entered and begin typing the formula. At the point in the formula where you want to enter an external reference, click on the sheet tab of the source (switch to the source document if the source sheet is in a different workbook) and select the cell containing the value to be entered (let's say it's in cell H12 of the worksheet Expt #XVIII-32). When you press the Enter button, Excel returns you to the dependent document, and the formula ='Expt #XVIII-32'!$H$12 is entered in the cell.

External references between workbooks, like the preceding example, are absolute by default but you can change the reference to a relative one.

Creating an External Reference by Using Paste Link

To use this method, **Copy** the cell or range in the source document, then switch to the destination. Select the cell or range where the contents are to be pasted, choose Home→Clipboard→Paste→Paste Special (Excel 2007/2010) or **Paste Special...** from the **Edit** menu (Excel 2003), and press the Paste Link button.

This method can be used only to create a link from a cell or range in the source to a cell or range in the destination; it cannot be used to insert an external reference in a formula.

The External Reference Contains the Complete Directory Path

You should never enter an external reference in a formula by typing it. External references are generally long and complicated, and it is easy to make a typing error. You should always enter the reference by selecting the cell or range. Why is this so important? Because an external reference, when entered by selecting, contains the complete directory path to the cell or range. This information is contained in the formula, but is not normally displayed.

To illustrate this, open a workbook (the dependent document) containing a formula with an external reference to another workbook (the source document). Examine the formula containing the external reference (for example, a reference such as [Expt #XVIII-32.xls]Sheet1'!$A$12). Now **Close** the source document, and again examine the formula in the dependent document. The external reference will now display the complete path of the directory or folder containing the referenced cell. e.g., 'C:\Documents and Settings\Owner\Desktop\[Expt #XVIII-32.xls]Sheet1'!$A$12).

Even if the source document is not open, the dependent document can obtain information from cells in the source.

Updating References and Re-establishing Links

If you change a source workbook so that references to cells that provide information to the dependent workbook are changed (by inserting or deleting rows or columns, for example), the cell references in the dependent workbook will be updated, provided that the dependent workbook is open. If the source is not open, the references will not be updated and the calculations will be incorrect. So it's a good idea to always have all source documents open when using a dependent document.

If you move, rename or delete a source document, you'll get a "This workbook contains one or more links that cannot be updated" message the next time you open the dependent document. Unlike cell references, a link (the external reference directory path) is not updated even if the dependent document is open. To re-establish links between documents, see "Re-establishing Links" in Chapter 6, "Advanced Worksheet Formulas".

Creating and Using 3-D References

You can create references that extend across several worksheets in a workbook. If you visualize a workbook as consisting of many worksheets stacked one above the other, a 3-D reference refers to a range of cells extending vertically. For example, you could calculate a total in, e.g., cell D1, on each of 12 sheets, and then calculate a grand total, the sum of the values in cell D1 of all 12 sheets, on a summary sheet, e.g., Sheet13, by using the 3-D formula =SUM(Sheet1:Sheet12!D1). You can enter a 3-D reference by typing it or by selecting.

There are two ways you can enter a 3-D reference by selecting. Follow either of the procedures in the following box.

To Enter a 3-D Reference by Selecting

1. Select the sheet tab of the first sheet in the 3-D reference, e.g., Sheet1.
2. Hold down the Shift key and select the last sheet in the range, e.g., Sheet12.
3. Select the cell or range, e.g., D1.

   or...

1. Select the sheet tab of the first sheet in the 3-D reference, e.g., Sheet1.
2. Select the desired cell or range therein.
3. Hold down the Shift key and select the last sheet in the range, e.g., Sheet12.

Many worksheet functions can be used in a 3-D formula: COUNT, COUNTA, SUM, SUMSQ, PRODUCT, MEDIAN, MAX, MAXA, MIN, MINA, LARGE, SMALL, AVERAGE, AVERAGEA, DEVSQ, STDEV, STDEVA, STDEVP, STDEVPA, VAR, VARA, VARP, VARPA.

Worksheet functions that can't be used include MATCH, INDEX, AND, OR, SUMPRODUCT.

Excel Tip   *If you want to insert rows or columns in a 3-D range such as Sheet1:Sheet12!D1:E10, be sure to group all of the sheets in that range before inserting.*

3-D references cannot be used in array formulas, with the intersection operator, or in formulas that use implicit intersection.

Worksheet Functions: An Overview*

Even though Excel is primarily a business tool, it provides a wide range of functions that are useful for scientific calculations. There are over 300 worksheet functions, organized in ten categories: Database, Date & Time, Engineering, Financial, Information, Logical, Lookup & Reference, Math & Trig Statistical and Text. This chapter provides examples using selected worksheet functions from the Math & Trig, Statistical, Date & Time, Logical, Text, Lookup & Reference and Information function categories. Database functions are described in Chapter 10. Financial functions are not discussed in this text.

The Engineering functions include functions to perform conversions from one number system to another (e.g., decimal to hexadecimal) and functions to operate on complex numbers. In Excel 2003, the Engineering functions are available only when you load the Analysis ToolPak Add-In.

Excel 2007/2010 provides an additional category of worksheet functions, Cube, that operate on OLAP (Online Analytical Processes) databases. Cube functions are not discussed in this text.

Appendix B lists selected worksheet functions in the Database, Date & Time, Information, Logical, Lookup & Reference, Math & Trig, Statistical and Text categories. Appendix C provides an alphabetical list of these worksheet functions along with the required syntax, some comments on the required and optional arguments, one or more examples and a list of related functions.

Function Arguments

Most worksheet functions require one or more *arguments*: the values that the function uses to calculate a return value. The arguments are enclosed in parentheses following the function name, e.g., SQRT(125) or SUM($F$3:$F$28) or SUBSTITUTE(PartNumber, "-1995", "-1996"). A few functions, such as PI() or NOW(), do not require arguments, but the opening and closing parentheses must still be provided.

Function arguments are either required or optional. In accordance with Microsoft's convention, in the following sections of this chapter and in the appendices, required arguments are shown in bold while optional arguments are shown in non-bold text.

When an optional argument is omitted, a default option for that argument is used, so you need to know what is assumed when the argument is omitted. For

---

*The names of the worksheet functions described in this and subsequent chapters are those used in the U.S. version. Different language versions of Excel use function names that are appropriate for the language and country. See Chapter 12, "Other Language Versions of Excel".

example, the worksheet function LOG(**number**, *base*) returns the logarithm of a number to a particular base. If the optional argument *base* is omitted, the function returns the base-10 logarithm.

Most arguments must be of a particular data type (number, text, reference, array, logical or error). Most argument names indicate the data type that is required, by using the words *number, text, reference*, etc., or by appending *_num* or *_number*, etc., to the argument name. For example, the syntax of the SUBSTITUTE function is SUBSTITUTE(**text,** **old_text,** **new_text,** *instance_num*). The first three arguments must be text, the fourth must be a number, or references to cells that contain values of that data type. Some functions can operate on arguments of any data type, indicated by the use of *value* as an argument name.

In every case, a function can use a cell reference as an argument, but the cell must contain a value of the correct data type.

Nested Functions

A function can be used as an argument of another function. This is referred to as *nesting*. A formula can contain up to seven levels of nested functions in Excel 2003, up to 64 in Excel 2007. If function B is used as an argument in function A, function B is a second-level function. If function C uses functions D and E as arguments, both D and E are second-level functions. If function F were nested inside function E as one of its arguments, function F would be a third-level function.

New Functions Introduced in Excel 2007

Five new functions were introduced in Excel 2007: IFERROR in the Logical category, SUMIFS in the Math & Trig category, and AVERAGEIF, AVERAGEIFS and COUNTIFS in the Statistical category. All five of these functions are discussed in the section on Logical functions.

If, while using Excel 2003, you open an Excel 2007/2010 workbook that contains one of these functions, the function will appear like

=_xlfn.COUNTIFS(YOG,97,Major,"biology")

and will return a #NAME? error when recalculated.

Changes to Functions in Excel 2010

For a number of functions, including the MOD function and the RAND function, the algorithm used was modified or changed in order to improve accuracy or performance.

A number of Excel statistical functions have been renamed, "*so that they are more consistent with the function definitions of the scientific community and with other function names in Excel.*" The new function names are recognizable by the inclusion of a period in the name: For example, the BETADIST function is now the BETA.DIST function. For backwards compatibility, the renamed functions remain available with their old names and are now listed in the Compatibility category in the Insert Function dialog box. The list of renamed functions is found in Appendix D

Using Insert Function

Because Excel provides a wide range of functions, it is sometimes difficult to remember them or to enter their arguments correctly. You can use Insert Function to paste a function in a cell, or within a formula that you're typing in the formula bar. To access the Insert Function dialog box, press the f_x button on the formula bar or click on the Formulas tab of the Ribbon (Excel 2007/2010), or choose **Function...** from the **Insert** menu (Excel 2003).

Figure 3-2. The Excel 2007/2010 Formulas tab.

Using Insert Function in Excel 2007/2010. Click on the Formulas tab (Figure 3-2). In the Function Library group, click on the appropriate function category to display a drop-down list of functions in that category (Figure 3-3). Click on the desired function to display the Function Arguments dialog box, or click on **Insert Function...** at the bottom of the list to display the Insert Function dialog box, and choose a function as described for Excel 2003.

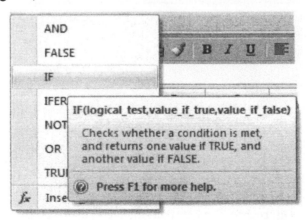

Figure 3-3. Inserting a function with Excel 2007/2010.

Using Insert Function in Excel 2003. Choose **Function...** from the **Insert** menu to display the Excel 2003 Insert Function dialog box (Figure 3-4). You must first select a function category in the "Or select a category" box. Excel will display all the functions in that category in the "Select a function" list box. When you click on a function, its name and syntax appear at the bottom of the dialog box. When you press the OK button, the Function Arguments dialog box will be displayed.   Figure 3-5 shows the Excel 2007/2010 Function Arguments dialog box; the Excel 2003 dialog box is essentially identical.

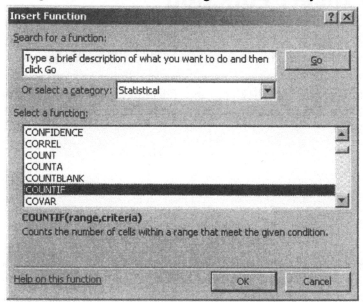

Figure 3-4. The Excel 2003 Insert Function dialog box.

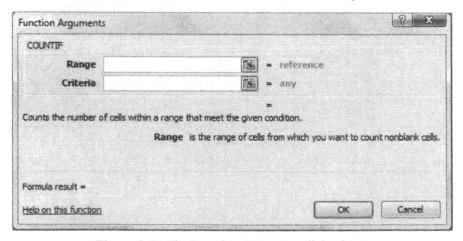

Figure 3-5. The Function Arguments dialog box.

The Function Arguments dialog box provides information about each argument as you enter it: a description of the argument, and whether it is required or optional (the names of required arguments are in bold; optional arguments are non-bold).

As you enter each argument, its value is displayed in the text box on the right. Use the Tab key to move to the next argument. If you need information about a particular argument (the effect of entering either TRUE or FALSE for a logical argument, for example), press the Help button. When you press OK or Enter, the function is entered into the worksheet cell.

Excel Tip. When selecting functions in the Paste Function dialog box, as long as the All category is selected, if you type a letter, the first function beginning with that letter is selected from the list of functions. For example, if you type the letter D, the DATE function is selected. You can type several letters in succession to zero in on the function you want. If you type R, the RAND function is selected, but if you type R-O-W (rapidly), you will select the ROW function. If you type a string of letters that doesn't correspond to any function, you'll get a beep.

As you become familiar with the range of functions provided by Excel, you will probably type most of them directly, rather than using the Function Wizard.

A Shortcut to a Function

Most often you'll know what function you want to enter, in which case it's much faster to type in the function and its arguments instead of using Paste Function.

Occasionally you will not be sure of the arguments and their proper order. If you type the function name in a worksheet cell and then press Ctrl+A, you will go directly to the Step 2 dialog box. You can then enter values for the function's arguments.

If you type the function name in a worksheet cell and then press Ctrl+Shift+A, Excel will paste dummy arguments, called *placeholder arguments,* and add the closing parenthesis. For example, after entering =LINEST, press Ctrl+Shift+A. The function will be completed and will appear as follows:

=LINEST(known_y's,known_x's,const,stats)

The first placeholder argument is selected (highlighted) so that you can enter a value. After entering it, double-click on the next argument to select it.

Using Formula AutoComplete
(Excel 2007/2010 Only)

Excel 2007 introduced Formula AutoComplete, an improvement over using Ctrl+A or Ctrl+Shift+A for entering functions in formulas. When you begin to enter a function in a formula, by typing the equal sign and the first letter of the desired function, Excel displays a drop-down list of functions beginning with that letter.

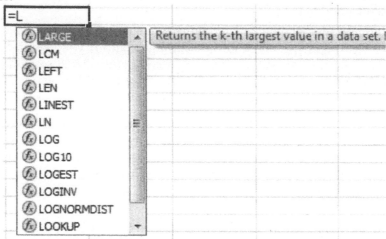

Figure 3-6. Formula AutoComplete displays a list of possible functions.

You can navigate through the list using the arrow keys, or continue to type the function name to narrow down the choices. A ScreenTip gives information about the highlighted function.

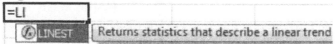

Figure 3-7 Continuing to type narrows the list of possible functions.

Once you reach the desired function name, double-click on the function name or press the Tab key to begin to enter the function. Excel enters the complete function name and the left parenthesis and displays a ScreenTip showing the syntax of the function. (Arguments in brackets are optional)

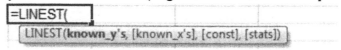

Figure 3-8. Continuing to type narrows the list of possible functions.

You can click on the function name to go to the On-line help for that function. As you enter each argument, followed by a comma, the next argument is highlighted in the ToolTip.

Excel Tip. *After entering the function name as shown in Figure 3-8, you can press Ctrl+Shift+A to enter the placeholder arguments.*

Math and Trigonometric Functions

Excel's mathematical and trigonometric functions (59 of them in Excel 2003, 60 in Excel 2007/2010) include functions that correspond to the following "scientific calculator" functions: $\sqrt{x}$, log x, ln x, Σ, e^x, π, sin x, cos x and tan x. The most useful mathematical functions are listed in Table 3-1. These functions are also listed in Appendices B and C, together with syntax, comments and examples.

Table 3-1. Selected Mathematical Functions

Function	Description
ABS	Returns the absolute value of a number
CEILING	Rounds a number to the nearest integer or to the nearest multiple of significance
EVEN	Rounds a number up to the nearest even integer
EXP	Returns e raised to the power of a given number
FACT	Returns the factorial of a number
FLOOR	Rounds a number down, toward zero
INT	Rounds a number down to the nearest integer
LN	Returns the natural logarithm of a number
LOG	Returns the logarithm of a number to a specified base
LOG10	Returns the base-10 logarithm of a number
MOD	Returns the remainder from division
MROUND	Returns a number rounded to the desired multiple
ODD	Rounds a number up to the nearest odd integer
PI	Returns the value of pi
QUOTIENT	Returns the integer portion of a division
RAND	Returns a random number between 0 and 1
ROUND	Rounds a number to a specified number of digits
ROUNDDOWN	Rounds a number down, toward zero
ROUNDUP	Rounds a number up, away from zero
SIGN	Returns the sign of a number
SQRT	Returns a positive square root
SUM	Returns the sum of arguments
SUMIF	Returns the sum of the cells specified by a given criteria
SUMIFS*	Returns the sum of the cells specified by multiple criteria
TRUNC	Truncates a number to an integer

* Excel 2007/2010 only.

As well, there are several functions, listed in Table 3-2, that return the sum of differences, products or squares of specified arrays. Although these functions operate on array arguments, they do not require that you press Ctrl+Shift+Enter.

Table 3-2. Some Mathematical Functions That Return Specialized Sums

Function	Description
SUMPRODUCT	Returns the sum of the products of corresponding array components
SUMSQ	Returns the sum of the squares of the arguments
SUMX2MY2	Returns the sum of the difference of squares of corresponding values in two arrays
SUMX2PY2	Returns the sum of the sum of squares of corresponding values in two arrays
SUMXMY2	Returns the sum of squares of differences of corresponding values in two arrays

Trigonometric Functions

Excel's trigonometric functions (SIN(angle), COS(angle), TAN(angle), etc.) require angles in radians. If the angle is in degrees, multiply it by $\pi/180$, or use the RADIANS worksheet function.

Table 3-3. Trigonometric Functions

Function	Description
COS	Returns the cosine of an angle in radians
SIN	Returns the sine of an angle in radians
TAN	Returns the tangent of an angle in radians
ACOS	Returns the arccosine of a number (the angle in radians corresponding to the cosine value)
ASIN	Returns the arcsine of a number (the angle in radians corresponding to the sine value)
ATAN	Returns the arctangent of a number (the angle in radians corresponding to the tangent value)
ATAN2	Returns the angle in radians corresponding to x- and y-coordinates
COSH	Returns the hyperbolic cosine of a number
SINH	Returns the hyperbolic sine of a number
TANH	Returns the hyperbolic tangent of a number
ACOSH	Returns the inverse hyperbolic cosine of a number
ASINH	Returns the inverse hyperbolic sine of a number
ATANH	Returns the inverse hyperbolic tangent of a number
DEGREES	Converts radians to degrees
RADIANS	Converts degrees to radians

Functions for Working with Matrices

Excel provides four functions for the manipulation of arrays or matrices: TRANSPOSE(*array*) returns the transpose of an array, MDETERM(*array*) returns the matrix determinant of an array, MINVERSE(*array*) returns the matrix inverse of an array, and MMULT(*array1*, *array2*) returns the matrix product of two arrays.

Table 3-4. Mathematical Functions for Working with Matrices

Function	Description
MDETERM	Returns the matrix determinant of an array
MINVERSE	Returns the matrix inverse of an array
MMULT	Returns the matrix product of two arrays
TRANSPOSE*	Returns the transpose of an array

* This function is listed in the Lookup & Reference category.

MINVERSE, MMULT and TRANSPOSE return an array of values. Thus you must select a range of cell of appropriate size for the result array, type the formula, and then enter the formula by pressing Ctrl+Shift+Enter.

Statistical Functions

Excel 2003 provides 80 statistical functions, including functions that return the mean, median, maximum and minimum values, average deviation, standard deviation, variance, nth quartile and rank. The most commonly used statistical functions are listed in Table 3-5.

Table 3-5. Selected Statistical Functions

Function	Description
AVEDEV	Returns the average of the absolute deviations of data points from their mean
AVERAGEIF*	Returns the average of cells specified by a given criteria
AVERAGEIFS*	Returns the average of cells specified by multiple criteria
CORREL	Returns the correlation coefficient between two data sets
COUNT	Counts how many numbers are in the list of arguments
COUNTA	Counts how many values are in the list of arguments
COUNTBLANK	Counts the number of blank cells within a range
COUNTIF	Counts the number of nonblank cells within a range that meet the given criterion
COUNTIFS*	Counts the number of nonblank cells within a range that meet multiple criteria

INTERCEPT	Returns the intercept of the linear regression line
LARGE	Returns the kth largest value in a data set
LINEST	Returns the parameters of a linear trend
MAX	Returns the maximum value in a list of arguments
MEDIAN	Returns the median of the given numbers
MIN	Returns the minimum value in a list of arguments
RANK	Returns the rank of a number in a list of numbers
RSQ	Returns the square of the Pearson product moment correlation coefficient
SLOPE	Returns the slope of the linear regression line
SMALL	Returns the kth smallest value in a data set
STDEV	Estimates standard deviation based on a sample
STDEVP	Calculates standard deviation based on the entire population
TDIST	Returns the Student's t-distribution
TINV	Returns the inverse of the Student's t-distribution
TTEST	Returns the probability associated with a Student's t-test
VAR	Estimates variance based on a sample
VARP	Calculates variance based on the entire population

* Excel 2007/2010 only.

A statistical function of considerable use for chemists is LINEST (<u>lin</u>ear <u>est</u>imation). It returns the least-squares regression parameters of the linear function that best describes a data set. LINEST is discussed in detail in Chapter 14.

Ranking a List of Number Values. The RANK worksheet function — syntax: RANK(***number,ref,order***) — returns the rank of a number in a list of numbers. However, if there are ties, the function returns the same rank number for both values that are tied.

Sorting a List of Number Values. The SMALL or LARGE worksheet functions — syntax: SMALL(***array,k***) and LARGE(***array,k***) — can be used to automatically sort a list of numbers in either ascending or descending order. The formula for ascending order, entered in cell B2 and filled down, is

=SMALL($A$2:$A$120,ROW()-1)

Logical Functions

The logical functions provided with Excel are IF, AND, OR and NOT. The latter three are almost always used in combination with the IF function.

Table 3-6. Logical Functions

Function	Description
AND	Returns TRUE if all of its arguments are TRUE
FALSE	Returns the logical value FALSE
IF	Specifies a logical test to perform
IFERROR*	Returns *value_if_error* if the argument expression evaluates to an error, otherwise returns the value of the expression
NOT	Reverses the logic of its argument
OR	Returns TRUE if any argument is TRUE
TRUE	Returns the logical value TRUE

* Excel 2007/2010 only.

The IF Function

The IF function allows you to return different results in a cell, depending on values in other cells.

The syntax of the IF function is IF(***logical_test***, *value_if_true*, *value_if_false*). *Logical_test* is an expression that evaluates to either TRUE or FALSE, e.g., B3<C3, SUM(F3:F28)<>0, etc. Since FALSE can be represented by 0 and TRUE by any non-zero value, the formula =IF(G27, *value_if_true*, *value_if_false*) tests whether G27 is nonzero or zero.

If *value_if_true* or *value_if_false* is omitted, the IF function returns TRUE or FALSE in place of the missing expression. To suppress the display of TRUE or FALSE, use a null string ("") instead of omitting the expression.

A common use of an IF function is to prevent the display of error values when values are missing or inappropriate. In Figure 3-9, the following formula is used in cell C4 to calculate the percentage change in freshman chemistry enrollment from one year to the next:

=100*(B4-B3)/B3

but if the formula is filled down to cells in which the divisor is missing, as in row 17 of Figure 3-9, the formula returns the #DIV/0! error message.

	A	B	C
1		Enrollment	
2	Year	Chem I	% Change
3	1989	363	
4	1990	328	-10
5	1991	358	9
6	1992	285	-20
7	1993	257	-10
8	1994	255	-1
9	1995	255	0
10	1996	329	29
11	1997	414	26
12	1998	481	16
13	1999	495	3
14	2000	536	8
15	2001		-100
16	2002		#DIV/0!
17	2003		#DIV/0!

Figure 3-9. A worksheet displaying error values.

	A	B	C
1		Enrollment	
2	Year	Chem I	% Change
3	1989	363	
4	1990	328	-10
5	1991	358	9
6	1992	285	-20
7	1993	257	-10
8	1994	255	-1
9	1995	255	0
10	1996	329	29
11	1997	414	26
12	1998	481	16
13	1999	495	3
14	2000	536	8
15	2001		
16	2002		
17	2003		

Figure 3-10. Display of error values suppressed by using an IF function.

If the formula is replaced by

=IF(B4<>0,100*(B4-B3)/B3,"")

the calculation is not performed for cells in which the operand in column B is missing, as shown in Figure 3-10. The following formula is equivalent.

=IF(B4,100*(B4-B3)/B3,"")

Nested IF Functions

IF functions can be nested. Most commonly a second IF function is used for *value_if_false*. Up to seven IF functions can be nested in Excel 2003, up to 64 in Excel 2007/2010. The following formula performs different operations, depending on whether C3-B3 is positive, zero or negative.

=IF(C3-B3>0, C3-B3, IF(C3-B3=0, C3, "unable to calculate"))

Nested IF functions allow you to calculate, in one cell, values that require different formulas depending on the value in one or more different cells. Otherwise you'd have to enter a different formula for each separate case and manually select the cells in which it was to be entered.

AND, OR and NOT

The AND and OR functions are similar to the comparison operators — they produce a logical result, either TRUE or FALSE, and are almost always used in conjunction with IF. AND and OR can take up to 30 arguments. AND(*logical1*, *logical2*,...) returns TRUE if all of its logical arguments are TRUE; OR(*logical1*, *logical2*,...) returns TRUE if at least one of its logical arguments is TRUE.

You can use OR to test whether a value is equal to one of the values in an array. For example, the formula

=OR(month={"Jan","Feb","Mar","Apr","May","Jun","Jul","Aug","Sep","Oct",
    "Nov","Dec"})

returns TRUE if month contains, e.g., the value Oct.

It's better to restrict use of this approach to arrays entered within formulas. If you define an array of months of the year elsewhere and use a formula such as =OR(month=array), you must remember to enter the formula as an array formula, that is, by pressing Ctrl+Shift+Enter. Otherwise the formula returns FALSE unless month is equal to Jan. You can find out much more about array formulas in Chapter 7.

Using IFERROR (Excel 2007/2010 Only)

The syntax of the IFERROR function is

IFERROR(*value, value_if_error*)

The function returns *value_if_error* if *value* evaluates to an error, or *value* if not. This is identical to IF(ISERROR(*value*), *value_if_error*, *value*).

Using COUNTIF

The COUNTIF function allows the user to count values in a range, based on a single criterion (even though Microsoft's help file uses the argument name *criteria*). The syntax of the COUNTIF function is

=COUNTIF(***range,criteria***)

and some examples of expressions using COUNTIF are

=COUNTIF($A$2:$A$120,B2)

to count the number of values in the range $A$2:$A$120 that are equal to the value in cell B2, or

=COUNTIF($A$2:$A$120,"<"&B2)

to count the number of values that are less than the value in cell B2.

Ranking a List of Number Values. The COUNTIF function can be used to return the rank of a number in a list of numbers, in the following formula entered in cell B2 and filled down:

=COUNTIF($A$2:$A$120,"<"&A2)+1

where the list is in the range$A$1:$A$120. Again, if there are ties, the formula returns the same rank number for both values that are tied, but the formula can be modified to indicate the tie values:

=COUNTIF($A$2:$A$120,"<"&A2)+1&IF(COUNTIF($A$2:$A$120,A2)-1,"(T)","")

Using SUMIF

The SUMIF function allows the user to sum values in a range, based on a single criterion (even though Microsoft's help file uses the word criteria). The syntax of the SUMIF function is

=SUMIF(***range,criteria***,*sum_range*)

The argument *range* is the range of cells containing the values to be compared to *criterion*. *Sum_range* is the range of cells containing the values to be summed; if *sum_range* is omitted, values in *range* will be summed. Figure 3-11 illustrates a use of SUMIF. A list of workers in $B$2:$B$20 produced the product values in $C$2:$C$20. Since some workers worked more than one day, we want to sum the total product produced by each worker. The formula in cell C23,

=SUMIF($B$2:$B$20,B23,$C$2:$C$20)

returns the total produced by a particular worker whose name is in cell B23.

	A	B	C
1	Day	Operator	Product
2	1	Thompson	99.1
3	2	Meitner	31.6
4	3	Meitner	11.6
5	4	Strassman	78.3
6	5	Oppenheimer	70.7
7	6	Dalton	37.7
8	7	Lawrence	13.6
9	8	Strassman	96.0
10	9	Fermi	89.4
11	10	Thompson	38.6
12	11	Meitner	68.3
13	12	Bohr	22.0
14	13	Meitner	2.2
15	14	Lawrence	53.4
16	15	Oppenheimer	33.1
17	16	Thompson	30.0
18	17	Rutherford	42.3
19	18	Fermi	97.2
20	19	Dalton	36.6
21			
22		Operator	Total
23		Dalton	74.3

Figure 3-11. Using SUMIF.

Using AVERAGEIF (Excel 2007/2010 Only)

The syntax of AVERAGEIF is identical to that of SUMIF.

Using COUNTIFS, SUMIFS and AVERAGEIFS (Excel 2007/2010 Only)

These three functions, introduced in Excel 2007, enable you to count, sum or average values using multiple criteria. The multiple criteria can be applied to values in a single column or in multiple columns. The syntax of COUNTIFS is

=COUNTIFS(*range1,criterion1*,range2,criterion2,...)

Figure 3-12 illustrates a portion of a class list, with columns containing student Name, Year Of Graduation, Major and Grade. The instructor would like to know how many students are, for example, Biology majors in the class of 97. This can be obtained using COUNTIFS.

	A	B	C	D
1	Name	YOG	Major	Grade
2	ALBERTY, Christine	96	THEATER ARTS	77.9
3	ANSARTO, Newman	97	BIOLOGY	78.3
4	BARBIERI, Joseph	94	ENV GEO SCIEN	62.2
5	BASALTA, Patrick	97	BIOLOGY	68.4
6	BLAKELY, Patrick	97	BIOLOGY	73.2
7	BLOOMSBURY, Peter	97	UNCLASSIFIED	67.7
8	BOGACZU, RobertO	97	BIOCHEMISTRY	90.8
9	BONOAN, Elwen	97	BIOLOGY	25.7
10	BOWEN, Armand	97	BIOLOGY	28.6
11	BOYLE, Seamus	97	BIOLOGY	80.3
12	BREGLEB, Keith	97	BIOLOGY	77.4
13	BRETT, Sam	97	BIOLOGY	80.3
14	BRILL, Daniel	96	UNCLASSIFIED	66.3
15	BUSSER, Susan	97	BIOLOGY	76.8
16	CAFFERTY, Christine	97	BIOLOGY	6.0
17	CALLANAN, Stephen	96	ENGLISH	66.4
18	CAMERMAN, Colleen	97	BIOLOGY	73.4
19	CAPIZZI, Jerome	97	BIOLOGY	24.9
20	CARSTAIRS, Angela	97	BIOLOGY	80.0

Figure 3-12. Portion of a class list.

The Function Arguments dialog box for COUNTIFS initially displays a single pair of *criterion_range1*, *criterion1* input boxes. As you enter values in one set, an additional set is displayed, as illustrated in Figure 3-13. (Up to 127 range/criterion pairs are allowed.)

The formula

=COUNTIFS($B$2:$B$149,97,$C$2:$C$149,"biology")

returns 60, the number of records that match the two criteria. (Note that text criteria are not case-sensitive.)

See "Using Database Functions" in Chapter 9 "Using Excel's Database Features" for another way to obtain this information.

The syntax of AVERAGEIFS is

AVERAGEIFS(*average_range, criterion_range1, criterion1,*
     criterion_range2, criterion2,...)

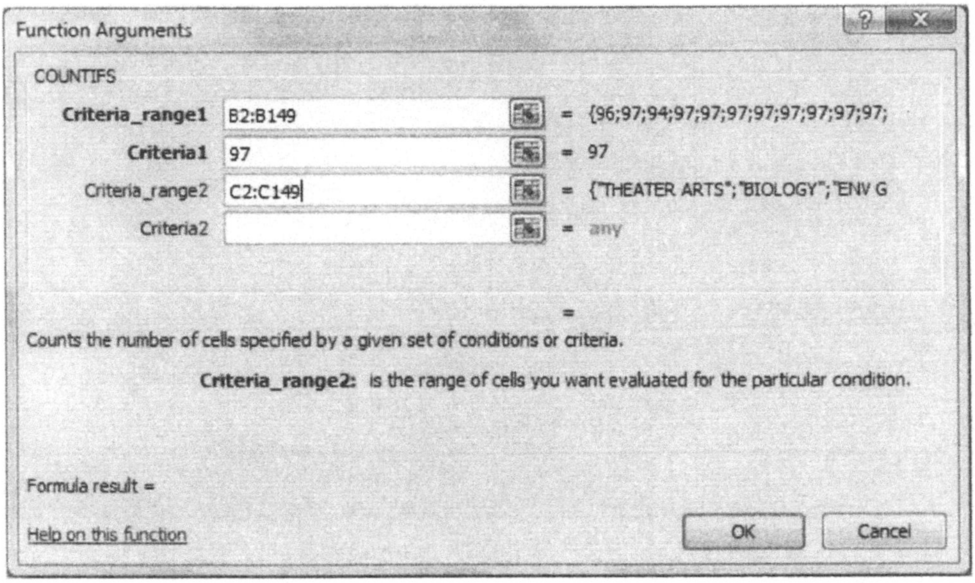

Figure 3-13. Function Arguments dialog box for COUNTIFS.

Average_range is the range of cells to be averaged. *Criterion_range1* is the range of values to be compared with *criterion1*, etc. Note the difference in order of arguments between AVERAGEIF and AVERAGEIFS. Using the data table in Figure 3-12, to return the average grade of all students who are Biology majors, you would use the expression

=AVERAGEIF($C$2:$C$149,"BIOLOGY", $D$2:$D$149).

To return the same result using AVERAGEIFS with one criterion, the expression would be

=AVERAGEIFS($D$2:$D$149, $C$2:$C$149, "BIOLOGY").

Using Conditional Formatting (Part II)

The Conditional Formatting Wizard (in the Styles group of the Home tab in Excel 2007/2010, in the **Format** menu in Excel 2003) was introduced in Chapter 1, where the "Value Is" choice in the Condition list box was discussed. The second possibility, the "Formula Is" option, requires that you enter a logical formula in the input box to the right of the Condition list box. A logical formula must evaluate to either TRUE or FALSE.

The following illustrate some examples of conditional formatting.

Color Banding. The formulas in the Condition 1 and Condition 2 input boxes of the Conditional Formatting dialog box shown in Figure 3-14 apply different background colors (cell Patterns) to dates in alternate months; in this example even months are light green and odd months light yellow, as shown in Figure

3-15.

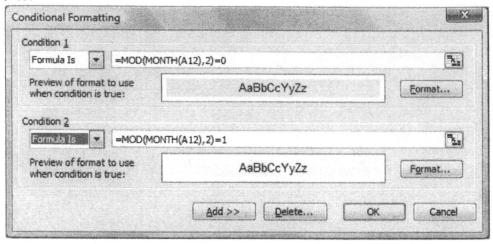

Figure 3-14. The Conditional Formatting dialog box.

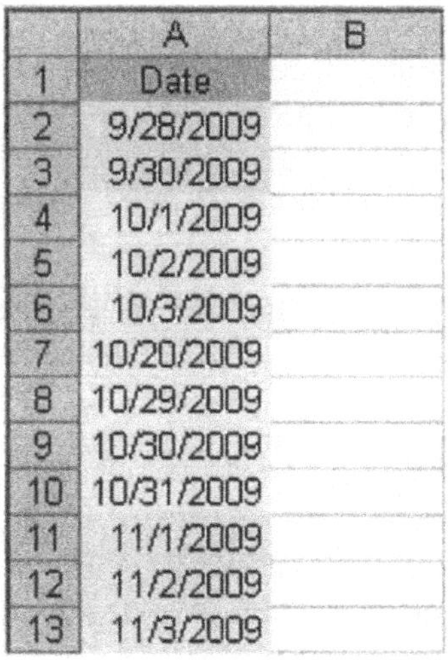

Figure 3-15. Conditional Formatting.

Flagging Invalid CAS Registry Numbers. You can enter an array formula in the input box, but you must remember to enter it by using Ctrl+Shift+Enter. For an example, see "Validating a CAS Registry Number " in Chapter 7 "Array Formulas".

Turning Conditional Formatting On and Off. You can control conditional formatting with controls that you install on the spreadsheet. For an

example, see "Using a Check Box to Enable or Disable Conditional Formatting "
in Chapter 11, "Adding Controls to a Spreadsheet".

Date & Time Functions

Excel provides a number of worksheet functions to work with dates and
times. Before we discuss these, we need to understand how Excel keeps track of
the date and time.

How Excel Keeps Track of the Date and Time

Excel records dates and times by means of a serial value. There are two
different date systems — the 1900 Date System, used by Excel for Windows,
and the 1904 Date System, used by Excel for the Macintosh. In Excel for
Windows, a date is recorded as the number of days elapsed since the base date,
January 1, 1900. The date serial number for 1/1/1900 is 1; July 1, 2010 is
represented as the serial value 40179. (In Excel for the Macintosh, the base date
is January 1, 1904 and the serial number for the start date is zero).

To see which date system is in effect, choose Excel
Options→Advanced→When Calculating This Workbook (Excel 2007/2010) or
Options... from the **Tools** menu and choose the Calculation tab (Excel 2003). If
the box for 1904 Date System is checked, the workbook was created using a
Macintosh.

If a workbook created using a Macintosh, for example, is opened using a PC,
dates entered in worksheet cells will be displayed correctly. Problems with dates
only occur if a date is copied from a workbook with one date system in effect
and pasted into a workbook that uses the other date system.

Dates can extend to the year 9999 but not prior to 1900 (or prior to 1904 in
Excel for the Macintosh).

Times are represented by the decimal part of the serial number. Thus, noon
on July 1, 2010 is represented by the date serial value 40179.5.

As you've already seen, dates and times can be entered into worksheet cells
using any one of several convenient formats: July 1 can be entered as 7-1, 7/1,
July 1, Jul 1 or 1 July, among others. All these date entries produce the date
7/1/xxxx in the formula bar (xxxx is the four-digit year) and the displayed date 1-
Jul. Excel enters the current year unless a different year is specified. If you
enter the year, either as a two-digit or four-digit number, Excel displays the date
in a different format, as e.g., 7/1/xx.

If you enter a two-digit year between 00 and 29, Excel assumes the 21st
century; thus, 7/4/18 appears in the formula bar as 7/4/2018.

Times are also recognized by Excel. If you enter 10:00 in a cell, it will be recognized as a time, and 10:00:00 AM will appear in the formula bar. Excel assumes a 24-hour clock (military time) unless you indicate differently. You can use AM/PM or am/pm designations with times. Even "2 p" can be used to enter 2:00 PM in a cell.

> ***Excel Tip.*** *Enter the current date in a worksheet by using Ctrl+ (semicolon); to enter the time use Ctrl+(colon). The date appears in the format m/d/yyyy but can be formatted differently.*

Date and Time Arithmetic

If you keep in mind that Excel stores dates and times as date serial numbers, performing date or time arithmetic is simple. For example, in a kinetics experiment you may have a table of times at which data points were recorded at irregular intervals (Figure 3-16). To analyze the data, you need the elapsed time from $t_{initial}$. Subtracting the time values from the initial value yields numbers that are decimal fractions of a day and are converted into minutes by multiplying by 24×60. The formula in cell B10 is =(A10-$A$9)*1440. Notice the use of absolute and relative references in the formula.

	A	B
8	**Date & Time**	**t, min**
9	01/15/81 10:05 AM	0
10	01/15/81 12:20 PM	135
11	01/16/81 09:10 AM	1385
12	01/19/81 09:15 AM	5710
13	01/20/81 09:20 AM	7155
14	01/23/81 09:45 AM	11500
15	01/26/81 09:20 AM	15795
16	01/28/81 09:30 AM	18685
17	02/02/81 09:10 AM	25865
18	02/05/81 09:30 AM	30205

Figure 3-16. Calculating elapsed times

Creating Date Series

In Chapter 1 we saw how to create a series of dates using AutoFill. You can create "custom" date series using formulas. For example, to create a series of weekdays, do the following: Enter dates for the first Monday through Friday of the desired series, as shown in cells A2:A6 of Figure 3-17(Left). In cell A8, enter the formula

=A2+7

and Fill Down to cell A12. This will enter the dates of weekdays of the next week, as shown in Figure 3-17(Middle). Now select cells A8:A13 and use AutoFill to fill down to produce as many weekdays as desired. A portion of the date series is shown in Figure 3-17(Right).

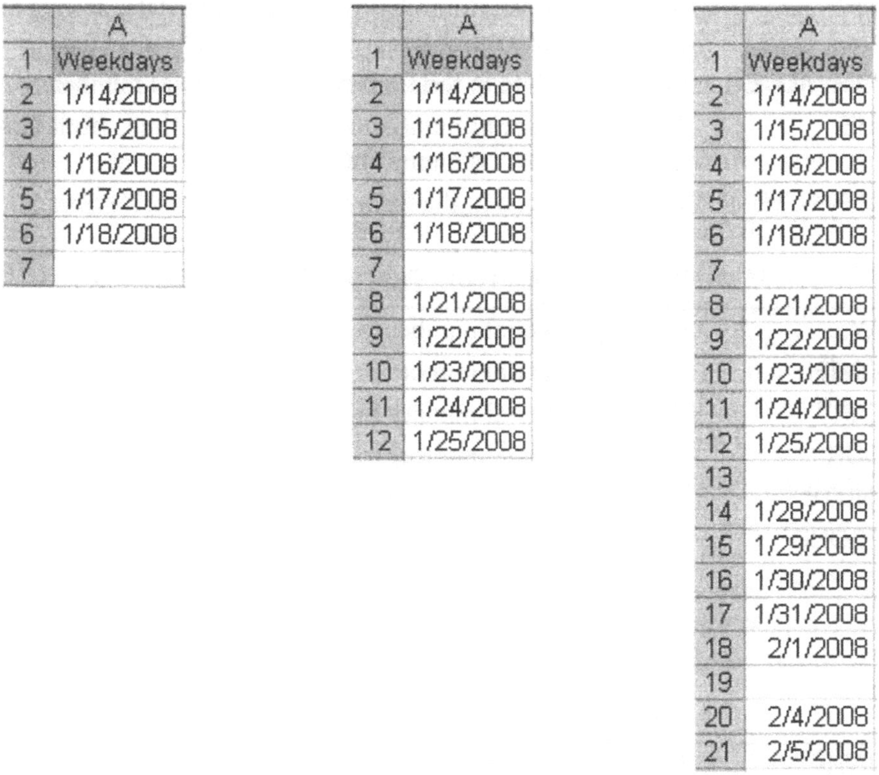

	A
1	Weekdays
2	1/14/2008
3	1/15/2008
4	1/16/2008
5	1/17/2008
6	1/18/2008
7	

	A
1	Weekdays
2	1/14/2008
3	1/15/2008
4	1/16/2008
5	1/17/2008
6	1/18/2008
7	
8	1/21/2008
9	1/22/2008
10	1/23/2008
11	1/24/2008
12	1/25/2008

	A
1	Weekdays
2	1/14/2008
3	1/15/2008
4	1/16/2008
5	1/17/2008
6	1/18/2008
7	
8	1/21/2008
9	1/22/2008
10	1/23/2008
11	1/24/2008
12	1/25/2008
13	
14	1/28/2008
15	1/29/2008
16	1/30/2008
17	1/31/2008
18	2/1/2008
19	
20	2/4/2008
21	2/5/2008

Figure 3-17. Creating a series of weekdays using date arithmetic.
(Left) Start dates entered as values. (Center) Next weekdays entered as formulas.
(Right) Date series created using AutoFill.

Using Date and Time Functions

Excel provides a number of worksheet functions that either generate a date serial number from text or values, or operate on a date serial number. Table 3-7 lists functions that create a date serial number, Table 3-8 lists functions that take a date serial number as argument, and Table 3-9 lists functions that are only available if the Analysis ToolPak is loaded (Excel 2003).

=DATE(YEAR(B2),MONTH(B2)+1,1)

Next, find the weekday of this date, using the formula (in cell B4)

=WEEKDAY(B3)

Finally, add the requisite number of days to get the first Wednesday:

=CHOOSE(B4,3,2,1,0,6,5,4)

These formulas can then be combined into a single "megaformula":

=DATE(YEAR(B2),MONTH(B2)+1,CHOOSE(WEEKDAY(DATE(YEAR
(B2),MONTH(B2)+1,1)), 3,2,1,0,6,5,4)+1)

This formula can be modified to return the first, second, third, etc., of any desired day of the week by editing the list of seven numbers. For example, to return the date of the second Thursday, the numbers are 11,10,9,8,7,13,12.

Number of Days in a Month. The following formula returns the number of days in the month specified by the date value in cell A1:

=DATE(YEAR(A1),MONTH(A1)+1,1)-DATE(YEAR(A1),MONTH(A1),1)

Text Functions

Excel provides a wide range of worksheet functions that operate on text (there are 28 listed for Excel 2003). You are already familiar with the & operator, to concatenate text or text and values. Most of Excel's text functions select or modify one or more characters within a text string. Table 3-8 lists the more useful functions.

Table 3-8. Text Functions

Function	Description
CHAR	Returns the character specified by an ASCII code
CLEAN	Removes all nonprintable characters from text
CODE	Returns the ASCII code for a character
CONCATENATE	Joins several text strings into one string
DOLLAR	Converts a number to text, using currency format
EXACT	Compares two text strings
FIND	Returns the position of one text string within another (case-sensitive)
FIXED	Formats a number as text with a specified number of decimals
LEFT	Returns a specified number of characters from a text string, beginning at the left
LEN	Returns the number of characters in a text string
LOWER	Converts text to lowercase

MID	Returns a specified number of characters from a text string, starting at a specified position
PROPER	Capitalizes the first letter in each word of a text value
REPLACE	Replaces characters within text, at a specified position
REPT	Repeats text a given number of times
RIGHT	Returns a specified number of characters from a text string, beginning at the right
SEARCH	Finds one text value within another (not case-sensitive)
SUBSTITUTE	Substitutes new text for old text in a text string
T	Converts its arguments to text
TEXT	Formats a number and converts it to text
TRIM	Removes spaces from text
UPPER	Converts text to uppercase
VALUE	Converts a text argument to a number

The LEN, LEFT, RIGHT and MID Functions

The LEN(*text*) function returns the number of characters in a text string.

The LEFT, RIGHT and MID functions are usually used to obtain a substring from a string. The LEFT(*text, num_characters*) function returns the leftmost character or characters in a text string. For example, LEFT("02167-3860",5) returns 02167. If *num_characters* is omitted, the value 1 is assumed. The RIGHT(*text, num_characters*) function is similar. If cell B7 contains a nine-digit number, then RIGHT(B7,4) returns the last four digits of the number.

The syntax of the MID function is MID(*text, start_num, num_characters*); it returns a specific number of characters from a specified position in a text string. For example, if cell A1 contains H2SO4, the expression MID(A1,3,1) returns S.

The UPPER, LOWER and PROPER Functions

Three functions change the case of a text string: the UPPER(*text*) and LOWER(*text*) functions do what their names suggest; the PROPER(*text*) function capitalizes the first letter in each word of a text string, as illustrated in column B of Figure 3-17.

The FIND, SEARCH, REPLACE, SUBSTITUTE and EXACT Functions

The REPLACE and SUBSTITUTE functions are used to modify a string.

FIND(*find_text, within_text, start_at_num*) and SEARCH(*find_text, within_text, start_at_num*) are similar. Each returns the position number of *find_text* within the text string *within_text*. FIND is case-sensitive, SEARCH is not. For example, if cell A4 contains toluene, 2-chloro-, the expression FIND(",",A4,1) returns the value 8. Unless the optional starting position is specified, the functions begin at position 1.

The following two functions are complementary: REPLACE(*text*, *start_num*, *num_characters*, *new_text*) and SUBSTITUTE(*text, old_text, new_text*, *instance_num*).

REPLACE replaces unspecified characters at a specified position within a text string. Note that, except for the inclusion of a fourth argument, *new_text*, its syntax is similar to that of the MID function.

Example: if cell A1 contains the text 2001, REPLACE(A1, 3, 2, "02") returns 2002.

SUBSTITUTE replaces specific characters within a string. For example, if cell A1 contains Et and cell B1 contains (C2H5) then SUBSTITUTE("Et3N", A1, B1) returns the text (C2H5)3N. If the optional argument *instance_num* is specified, only that instance of *old_text* will be replaced. If *instance_num* is omitted, all instances of *old_text* will be replaced.

EXACT(*text1, text2*) returns TRUE if the two strings are identical, FALSE otherwise. EXACT is case-sensitive. Simple comparison of strings is not case-sensitive. For example, the formula =("Name"="NAME") returns TRUE. Use EXACT if you want to make a case-sensitive comparison of two strings.

Example. The following formula reformats a list of names in which the original names, in column A of Figure 3-18, are in the form LAST_NAME,FIRST_NAME; the reformatted names are in column B.

=PROPER(RIGHT(A1,LEN(A1)-FIND(",",A1))&" "& LEFT(A1,FIND(",",A1)-1))

The function FIND(",",A1) returns the position of the comma in the string. The first names and/or initials are obtained using the RIGHT function; the number of characters to be returned is equal to the length of the string minus the position number of the comma. A space is concatenated, and then the last name is obtained using the LEFT function. The PROPER function is used to change the case of the string.

	A	B
1	ANTONIUS,STEPHEN J	Stephen J Antonius
2	BRUNEL,STEVEN D	Steven D Brunel
3	CARRESTO,KATHY E	Kathy E Carresto
4	LAKLIS,CLAIR L	Clair L Laklis
5	PEDROSO,BENITO A	Benito A Pedroso
6	SOUSSANE,WALID	Walid Soussane
7	WOODSTOCK,PAUL	Paul Woodstock
8	WYNDLAKE,KEVIN D	Kevin D Wyndlake
9	ZILARIO,J PATRICK	J Patrick Zilario

Figure 3-18. Manipulating text by using text functions.

The FIXED and TEXT Functions

These two functions permit you to apply number formatting to values in formulas.

FIXED(*number, decimals, no_comma_logical*) formats a number as text with a specified number of decimal places, with or without commas.

TEXT(*value, format_text*) converts a number to text and formats it, using the same number formatting symbols used the **Format** menu.

Example. When text and a number value are concatenated in a cell, the value can no longer be number-formatted by using menu commands or tool buttons, since it is no longer a number, but text. For example, if cell A1 contains today's date (e.g., July 1 2008), then the formula

="Today is "&A1

displays the result "Today is 39539", probably not the result that you intended.

The formula

="Today is "&TEXT(A1,"mm/dd/yyyy")

displays the result "Today is 07/01/2008".

The VALUE Function

Occasionally, number values will be entered in cells as text. When these text values are used in formulas, Excel can usually evaluate them as numbers and calculate the desired result. In the rare instance where Excel does not perform this conversion, VALUE(*text*) can be used to convert a text argument to a number.

The CODE and CHAR Functions

These two functions operate on or with the ASCII code value of a character. Every keyboard character is represented by an ASCII code value: For example, uppercase "A" is ASCII 65. The printable keyboard characters range from the space character, ASCII 32, to the tilde (~), ASCII 126. Some of the ASCII codes below 32 correspond to non-printing keys such as Enter (ASCII 13) or Esc (ASCII 27).

CODE(*text*) and CHAR(*number*) perform opposite functions. CODE returns the ASCII character code for a single character or the first character in a text string. CHAR returns the character corresponding to the character code. For example, CODE("a") returns 97, CHAR(36) returns $.

CHAR(10) is the line feed character, CHAR(13) is the carriage return (the ¶ symbol you see when you choose **Show ¶** from the **View** menu in Microsoft Word). Use one of these characters to insert a line break in text within a formula: CHAR(10) in Excel for Windows, CHAR(13) for the Macintosh.

Example. The formula (all in one cell, of course)

="Missing Reports as of "&TEXT(NOW(),"h AM/PM mmm dd, yyyy")
&CHAR(10)&"(X = a report that has not been received"&CHAR(10)&" or was returned for recalculation)"

in an Excel for Windows worksheet produces the text displayed in Figure 3-19. To enable the line breaks, you must choose **Cells...** in the **Format** menu, choose the Alignment tab, and check the Wrap Text box.

If you merely use Wrap Text in **Alignment** from the **Format** menu, Excel will decide where to break the text. By using CHAR(10) or CHAR(13), *you* get to decide where to break the text.

**Missing Reports as of 8 AM Dec 03, 2007
(X = a report that has not been received
or was returned for recalculation)**

Figure 3-19. Text with line breaks inserted.

Excel Tip. *The following formula will produce the correct line break character when used in either Excel for Windows or Excel for the Macintosh:*

="first text line"&CHAR(10+3*(INFO("system")="mac"))&"second line"

Lookup and Reference Functions

Excel's Lookup and Reference functions provide the ability to work with references, or to obtain values from a table based on position or value. Table 3-9 lists these functions. These are "power user" functions. Many of them are described in detail in Chapter 6, "Advanced Worksheet Formulas".

Table 3-9. Selected Lookup and Reference Functions

Function	Description
ADDRESS	Returns a reference as text to a single cell in a worksheet
AREAS	Returns the number of areas in a reference
CHOOSE	Chooses a value from a list of values
COLUMN	Returns the column number of a reference
COLUMNS	Returns the number of columns in a reference
HLOOKUP	Looks in the top row of an array for a specified value and moves down the column to return the value in a specified row
HYPERLINK	Creates a shortcut or jump that opens a document stored on a network server, an intranet, or the Internet

INDEX	Uses an index to choose a value from a reference or array
INDIRECT	Returns a reference indicated by a text value
LOOKUP	Looks up values in a vector or array
MATCH	Looks up values in a reference or array
OFFSET	Returns a reference offset from a given reference
ROW	Returns the row number of a reference
ROWS	Returns the number of rows in a reference
TRANSPOSE	Returns the transpose of an array
VLOOKUP	Looks in the first column of an array for a specified value and moves across the row to return the value in a specified column

Information Functions

Information functions include eleven IS functions that can be used to determine the data type of the value in a cell.

Table 3-10. Selected Information Functions

Function	Description
CELL	Returns information about the formatting, location, or contents of a cell
ISBLANK	Returns TRUE if the value is blank
ISERR	Returns TRUE if the value is any error value except #N/A
ISERROR	Returns TRUE if the value is any error value
ISEVEN*	Returns TRUE if the number is even
ISLOGICAL	Returns TRUE if the value is a logical value
ISNA	Returns TRUE if the value is the #N/A error value
ISNONTEXT	Returns TRUE if the value is not text
ISNUMBER	Returns TRUE if the value is a number
ISODD*	Returns TRUE if the number is odd
ISREF	Returns TRUE if the value is a reference
ISTEXT	Returns TRUE if the value is text
N	Returns a value converted to a number
NA	Returns the error value #N/A
TYPE	Returns a number indicating the data type of a value

* Only available in Excel 2003 if the Analysis ToolPak add-in is loaded.

The functions ISODD and ISEVEN are part of the Analysis ToolPak add-in. If either of these functions return the value #NAME?, then you must load the Analysis ToolPak. Alternatively, you can use the expression MOD(A4,2)=0 in place of the ISEVEN function.

Creating "Megaformulas"

As you become more experienced in constructing worksheet formulas, you will probably create more and more complicated ones. You can simplify a complicated calculation by performing it in steps, with intermediate calculations in separate cells of the worksheet. You can hide the rows or columns containing the intermediate stages of the calculation. But there are advantages to constructing a single "megaformula" in which all the intermediate calculations are combined in a single formula. You'll use less memory and recalculation of the worksheet will take less time.

You saw some examples of megaformulas earlier in this chapter. A good way to begin to construct a megaformula is to break the calculation into steps and store the results in separate cells. When the formulas are working correctly, you can combine them all in a single cell by copying and pasting. Here's an example: A list of names was imported into Excel from a word processing document. The first few entries are shown in column A of Figure 3-20. You want to create a column in Excel containing the last names, and a second column containing the first names and initial, if any.

The formulas to accomplish this are the following (the values returned by the formulas are shown in Figure 3-20): In cell B4, use the formula =LEN(A4)-LEN(SUBSTITUTE(A4," ","")) to determine the number of spaces in the text. In cell C4, use =SUBSTITUTE(A4," ","*",B4) to substitute a marker character for the last space in the name. (SUBSTITUTE accepts the optional argument *instance_number,* which specifies the instance of *find_text* that is to be substituted.) In cell D4, use =FIND("*",C4) to find the location of the marker character, which immediately precedes the last name portion of the string. In cell E4, use =RIGHT(A4,LEN(A4)-D4) to return the last name.

	A	B	C	D	E
4	Abner Coreus	1	Abner*Coreus	6	Coreus
5	Andrew G. Cosgrove	2	Andrew G.*Cosgrove	10	Cosgrove
6	Anthony Steiner	1	Anthony*Steiner	8	Steiner
7	Cheryl Ann Vigil	2	Cheryl Ann*Vigil	11	Vigil
8	Cindy A. Bronstein	2	Cindy A.*Bronstein	9	Bronstein

Figure 3-20. Portion of a worksheet to parse test into separate columns.

Finally, combine the formulas: First, **Copy** the formula in cell D4 from the formula bar (don't include the equal sign) and press the Enter button; then select cell E4; in the formula bar, select D4 in the formula and **Paste** the formula fragment. You can now delete the formula in cell D4. Repeat the process for the formulas in cells C4 and B4, pasting them into the formula in E4. The final megaformula is

```
=RIGHT(A4,LEN(A4)-FIND("*",SUBSTITUTE(A4," ","*",LEN(A4)-
    LEN(SUBSTITUTE(A4," ","")))))
```

A similar formula is used to return the first name plus initial. Finally, **Delete** columns B–D.

A formula can contain up to 8192 characters (1024 in Excel 2003), so your megaformulas can be quite complicated.

In Excel 2003, there's another limiting factor to the creation of megaformulas: A formula can contain no more than seven levels of nested functions. A nested function is a function that is located within the parentheses of another function. A simple example is

```
=SQRT(SQRT(SQRT(SQRT(SQRT(SQRT(SQRT(SQRT(1000))))))))
```

Excel color-codes the parentheses of nested functions. The colors of the left parentheses are (approximately) black, dark green, purple, brown, green, orange, pink, blue. Blue is the limit; you can't add another level of nesting once you get to blue.

If you try to exceed this limit of seven levels of nested functions, you'll get a "The formula you typed contains an error" message. Since the message doesn't say anything about too many levels of nested functions, you'll probably think that you've made an error in constructing your megaformula, and waste a lot of time trying to fix it. You'll probably never encounter this problem in Excel 2007/2010, since the maximum number of nested functions has been increased to 64.

Excel 2007 Tip. *If you open an Excel 97-2003 workbook (a workbook with .xls file extension) in Excel 2007/2010, you won't be able to exceed the limit of seven levels of nested functions. You'll have to save it in an Excel 2007 format, e.g., as an.xlsx workbook.*

Advantages and Disadvantages of Megaformulas

If you have a worksheet in which a calculation is done by means of a series of intermediate calculations, each in a separate worksheet cell, and culminating in a final result, your worksheet will calculate faster if all of the calculations are combined into a single megaformula. Practically speaking, though, you probably won't be able to detect any difference in the time required for the calculation. Yet many people still like to create megaformulas.

The downside of a megaformula is that, if you have to go back to the formula at a later date to modify it, because of its complexity you may not be able to remember exactly how it works.

Excel 2007/2010 Tip. *To make it easier to work with a long formula, you can resize the Formula Bar. Right-click anywhere inside the formula, and choose Expand Formula Bar from the shortcut menu.*

The Order in Which Excel
Performs Operations in Formulas

If several operators are combined in a single formula, Excel evaluates the formula in the following order: reference operators (colon, space, comma), negation (minus), percent (%), exponentiation (^), multiplication and division (* and /), addition and subtraction (+ and –), concatenation (&), comparison (=, <, >, <=, >=, <>).

If an expression contains operators with the same precedence (multiplication and division or addition and subtraction), Excel evaluates the operators from left to right.

Expressions enclosed in parentheses are evaluated first, no matter where they appear in the formula. Thus the expressions 5*2+3 and 5*(2+3) return different results.

Excel 2003 Formula and Function Specifications

Number of available worksheet functions	329
Maximum length of formula contents	1,024 characters
Maximum number of...	
arguments in a function	30
nested levels of functions	7
iterations	32,767
selected ranges	2,048
Earliest date allowed for calculation	January 1, 1900
Latest date allowed for calculation	December 31, 9999

Excel 2007/2010 Specifications

Number of available worksheet functions	341
Maximum length of formula contents	8,192 characters
Maximum number of...	
arguments in a function	255
nested levels of functions	64
iterations	32,767
selected ranges	2,048
Earliest date allowed for calculation	January 1, 1900
Latest date allowed for calculation	December 31, 9999

4

Excel 2007/2010 Charts

In Excel 2007, major changes were made in the creation of charts. Perhaps most importantly, the Chart Wizard with its four dialog boxes was eliminated. In its place is the Chart Tools ribbon, with three tabs: Design, Layout and Format.

Chart Types

Excel 2007/2010 provides a gallery of eleven standard chart types — Column, Line, Pie, Bar, Area, XY (Scatter), Stock, Surface, Doughnut, Bubble and Radar. (The Cylinder, Cone and Pyramid chart types of Excel 2003 are now located in the galleries of the Column or Bar chart subtypes.) Since Excel originated as a financial tool, most of the chart types are those that are useful for displaying financial and related information. Of the eleven chart types, the following seem of most use for chemists: Column, Bar, Line, Pie, XY, Radar, and Surface. These chart types will be discussed in the following sections.

All chart types except the XY chart use the x-values only for labels (called categories by Excel). The XY chart is the only one in which numeric values are used along both axes. XY charts are probably the most widely used for the display of scientific information.

Creating a Chart

There are two kinds of chart: either a separate chart sheet in a workbook, or a chart embedded in a worksheet, so that you can see both the data and the chart at the same time. An embedded chart is useful if you want to see how a curve changes as you change its parameters. As you change the values in worksheet cells, the chart will update automatically.

To create a chart, you must first select the data to be plotted, e.g., a column of x-values and one or more columns of y-values. (The data can be in rows or columns.) If the rows or columns are not adjacent, hold down the Ctrl key while selecting the separate rows or columns of data. Next, click on the Insert tab on the Ribbon. The Charts group of the Insert ribbon contains six buttons for the most popular chart types – Column, Line, Pie, Bar, Area, Scatter – plus a button for Other Charts (Figure 4-1).

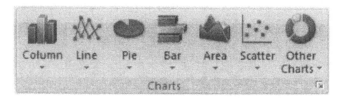

Figure 4-1. The Charts group in the Insert tab of the Ribbon.

When you click on a chart type in the Charts group, a gallery of chart subtypes will be displayed. For example, if you click on the button for a Scatter chart, a drop-down gallery is displayed (Figure 4-2), showing five Scatter chart subtypes: marker points with no connecting lines, marker points connected by straight lines, straight lines with no marker points, points connected by a smooth curve and a smooth curve with no marker points.

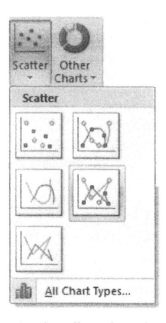

Figure 4-2. The gallery of XY chart types.

Each gallery of chart subtypes has an All Chart Types command that allows you to display a gallery of all 11 chart types (Figure 4-3).

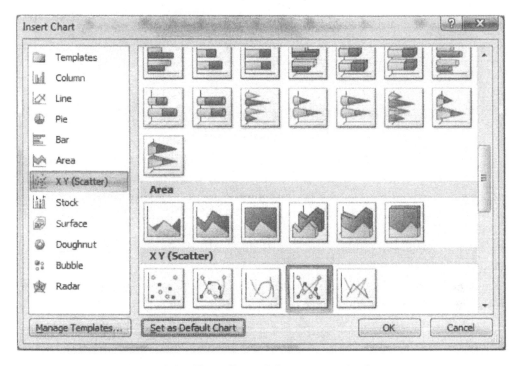

Figure 4-3. The gallery of chart types and subtypes.

In Excel 2003, charts were created by using the ChartWizard, a four-step wizard that allowed the user to choose the type of chart in Step 1, see a preview of the chart in Step 2, add or remove certain chart options in Step 3, and choose the chart location in Step 4 (the chart can be either a separate chart sheet or an embedded chart). In Excel 2007/2010, after selecting the data range, you choose the chart type from the Insert ribbon and an embedded chart is created immediately. You then add or remove chart options or change the chart location, in separate steps, using the three tabs in the Chart Tools ribbon.

Chart Tools in the Ribbon

When you create a chart, or click on a chart that has previously been created, three Chart Tools tabs will appear on the Ribbon: the Design, Layout and Format tabs. These take the place of menu commands in the Excel 2003 **Chart** menu. The Design, Layout and Format tabs shown in Figures 4-4, 4-5 and 4-6 may not look the same as shown on your screen, depending on Excel window size.

The Design tab (Figure 4-4) allows you to change a chart's type, modify the data range of the chart, or change a chart's location. The Design tab contains the commands found in the **Chart Type...**, **Data Source...** and **Chart Location...** commands in the **Chart** menu of Excel 2003 (Steps 1, 2 and 4 of the Chart Wizard).

Figure 4-4. The Design tab.

The Layout tab contains commands that in Excel 2003 were found in the **Format** menu, or in the **Chart Options...** command of the **Chart** menu (Step 3 of the Chart Wizard). This is a definite improvement, as it was never clear to me why adding gridlines was done using **Chart Options...**, adding error bars was done by using **Format**, and adding a Trendline was done by using a separate menu command altogether. In Excel 2007, all commands for adding chart elements have been collected in the Layout tab.

The Current Selection group, used to select and format chart elements, is located in the Layout ribbon.

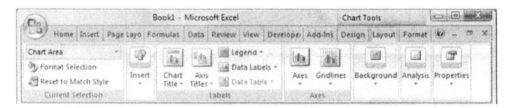

Figure 4-5. The Layout tab.

The Format Ribbon. This ribbon contains Shape and WordArt formatting options. I rarely use these features. Perhaps the most useful option on this ribbon is the Size group, which allows you to set the vertical and horizontal dimensions of a chart. The Format Selection button is located here also.

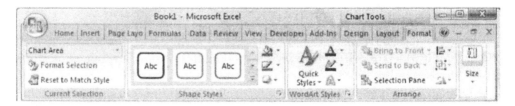

Figure 4-6. The Format tab.

The Chart Tools tabs are context-sensitive: only the tabs that are applicable for a certain action will be displayed. For example, if you select multiple charts, only the Format tab will be visible.

Activating, Resizing and Moving an Embedded Chart

To activate an embedded chart for moving, resizing or formatting, simply click on the chart. A light border will appear around the chart, as shown in Figure 4-7. To move the chart, position the mouse pointer on the border and drag to the new position. The border includes "handles" – groups of dots at the corners or sides of the chart – that can be used to resize the chart.

Figure 4-7. Chart "handles" are indicated by three dots.

To move an embedded chart, place the mouse pointer near the border of the chart and click and drag the chart to its new location. To change the size of the chart, click on one of the handles and then drag inward or outward.

To convert an embedded chart to a full chart sheet, click on the Move Chart button in the Design ribbon.

*Excel Tip   When an embedded chart is converted into a full chart sheet, the chart doesn't fill the complete chart sheet window. Unfortunately, the **Sized with Window** command in the Excel 2003 **View** menu, used to expand a chart to fill the full chart sheet window, was removed from Excel 2007.*

The SERIES Function

When you select a chart series, for example by clicking on a column in a column chart, the SERIES function or formula appears in the formula bar. (The 3-D or Surface chart does not display a series formula.). Each series in a chart has its own SERIES function. The syntax of the SERIES function is

=SERIES(*series_name*, *x_values*, **y_values**, **series_order**)

The optional argument *series_name* can be a text string, a cell reference, a named range, or omitted.

The optional argument *x_values* specifies the values used for category labels, or for *x*-values in an XY chart. The argument can be an array of numeric values or text labels enclosed in braces, a reference to a range of cells, a named range, or omitted. It can be a non-contiguous reference (separate ranges enclosed in brackets and separated by commas. If *x_values* is omitted, the integers 1,2,3 ... are used.

The required argument *y_values* specifies the values to be plotted. This argument can be an array of numeric values enclosed in braces, a reference to a range of cells, or a named range. It can be a non-contiguous reference (separate ranges enclosed in brackets and separated by commas.

The required argument *series_order* is the plotting order for the data series. It must be a whole number; it cannot be a cell reference or a name. This argument determines the automatic colors and marker symbols that are used when the chart is created. In charts that contain more than one series, it is the order in which the series names will appear in the legend; if the argument is changed manually, Excel adjusts the *series_order* of other series in the chart.

Series_order is actually the chronological order in which series are added to a chart. For example, if a chart is created with series for January, February and April, and subsequently a series for March is added, the March series will have *series_order* = 4, and the legend will show the series in that order. To put the series in the correct order, change the *series_order* for the March series to 3.

The SERIES function is only for charts; you can't use it in a worksheet formula. You can't incorporate worksheet functions or formulas directly into the SERIES function arguments, but it is possible to modify it using named formulas.

Chart Elements

A chart consists of a number of *chart elements*. For example, an XY chart can consist of some or all of the following elements: Chart Area, Plot Area, X Axis, Y Axis, Title, X Axis Title, Y Axis Title, Series 1, Series 2, Series 3, Legend, Y Axis Major Gridlines, etc. All of these chart elements can be formatted to improve the appearance of the chart.

Selecting Chart Elements

To apply or modify formatting of a chart element, you must first select that chart element. You can select a particular chart element in any one of the three ways shown below.

- Click on the chart element using the mouse. Handles (light blue circles) will indicate that the element has been selected. A particular chart element my sometimes be difficult to locate, in which case use one of the methods below.

- Use the up and down arrow keys to navigate between chart elements. Handles will indicate which element is currently selected, and the name of the currently selected chart element will appear in the Name Box, in the Current Selection group of the Format ribbon. Use the left and right arrow keys to select related chart elements within a group (e.g., Series 1 Point 1, Point 2, etc.).

- Use the Name Box drop-down button to display a list of all chart elements in the active chart. You can then click on any of the chart elements to select it.

Formatting Chart Elements

Once the desired chart element has been selected, click on the Format Selection button in the Current Selection group in the Layout ribbon or in the Format ribbon. This will display the appropriate dialog box to format the chart element. Since there are many chart types, each containing a wide range of chart elements, and since each chart element can have a wide range of formatting, it's beyond the scope of this book to cover all the formatting possibilities. Many of the most useful possibilities will be described in the following sections, but it will be up to you, the reader, to explore and experiment.

> *Excel Tip.* *In Excel 2003 you could double-click on a chart element to jump to the formatting dialog box for that chart element. That feature does not work in Excel 2007, but was added back in Excel 2010. You can't use Ctrl+Y to repeat a formatting action in either Excel 2007 or 2010.*

Creating Column or Bar Charts

A column chart compares values of elements in a group – for example, oil production figures for different areas in a given year. You can use grouped or stacked columns to show two relationships – for example, oil production figures for different areas over several years. Figure 4-8 shows an example of a chart with grouped columns, as originally created by clicking on the Column icon in the Charts group of the Insert ribbon.

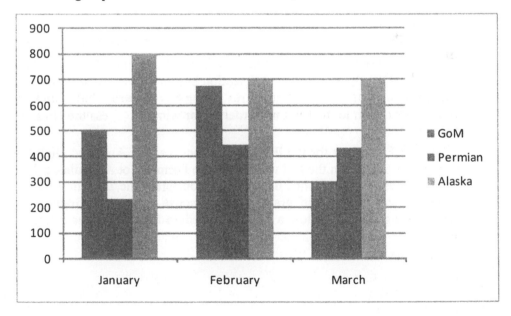

Figure 4-8. A Column chart as initially produced by Excel 2007 or 2010.

The appearance of an Excel chart can usually be significantly improved by making some formatting changes. Figure 4-9 shows the same chart after adding a title and X- and Y Axis labels, removing the border around the Chart Area, choosing a Fill Effect for the Plot Area and a Fill Effect for each of the columns, and adjusting the Overlap between columns; these formatting actions are described in detail in the following paragraphs.

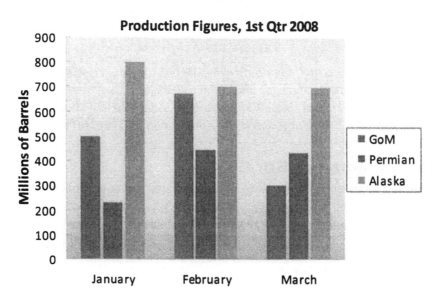

Figure 4-9. A Column chart after some formatting changes.

To remove the border around the Chart Area, or to perform other formatting on the Chart Area, click on the Format Selection button (after first selecting the Chart Area) to display the Format Chart Area dialog box (Figure 4-10). Click on the Border Color button to display the Border Color window. Press the No Line option button.

To choose a Fill Effect for the Plot Area, select the Plot Area and click the Format Selection button. In the Fill window of the Format Plot Area dialog box (Figure 4-11), press the Gradient Fill option button, press the Direction button and choose the desired type of gradient, and choose Color and Gradient to produce a pleasing effect. Follow a similar procedure for each of the series of columns.

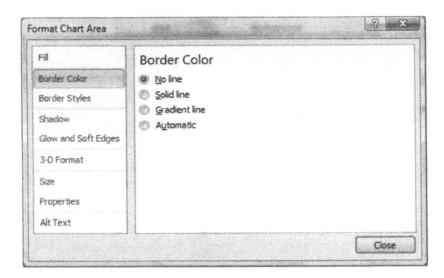

Figure 4-10. The Border Color window of the Format Chart Area dialog box.

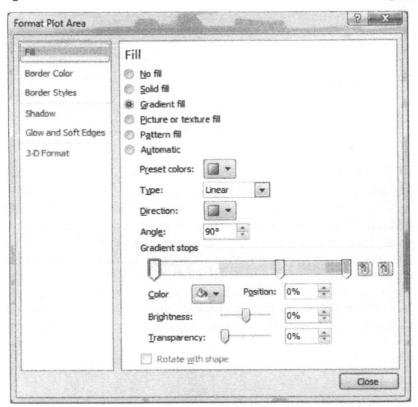

Figure 4-11. The Fill window of the Format Plot Area dialog box.

To adjust the overlap between columns, click on any one of the columns and click the Format Selection button. In the Series Options window (Figure 4-12), adjust the Series Overlap slider to produce a pleasing separation between columns.

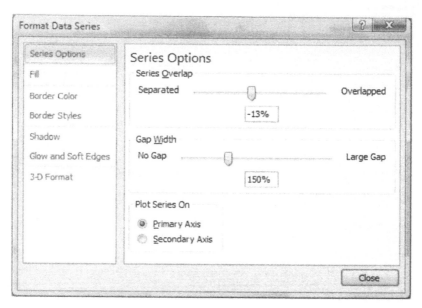

Figure 4-12. The Series Options window of the Format Data Series dialog box for a Column chart.

Creating Line Charts

Don't confuse a Line chart and an XY chart. An XY chart uses the *x*-values to determine where to plot the *y*-values, while a Line chart plots the *y*-values at equal intervals. If you select only a single range of cells, Excel will plot these values using the integers 1, 2, 3... as the category values (the *x*-values). If you select two columns or two rows of cells, that you intend to be used as the categories and the values, Excel will stubbornly plot these as two series of *y*-values and use the same series of integers 1, 2, 3... as the categories. For example, if you select the range of values in columns A and B in Figure 4-13 and choose Line chart from the Insert ribbon, Excel will produce the chart shown in Figure 4-14.

	A	B
1	X	Y
2	0.000	0.000
3	0.087	0.087
4	0.175	0.174
5	0.262	0.259
6	0.349	0.342
7	0.436	0.423
8	0.524	0.500
9	0.611	0.574

Figure 4-13. Portion of data table for a Line chart.

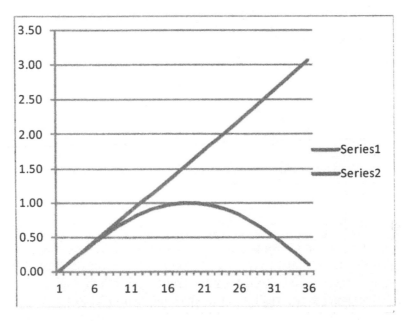

Figure 4-14. A Line chart as initially produced using the data from Figure 4-13.

To change the X Axis labels to display the i-values, remove the unwanted Series 1 and enter the range for the category labels as illustrated in Figure 4-15.

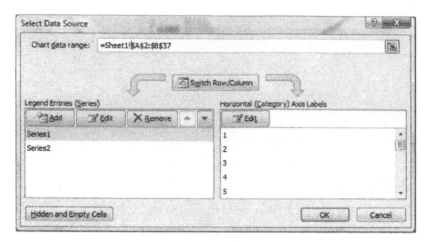

Figure 4-15. Changing the data ranges for a Line chart.

The completed chart is shown in Figure 4-16.

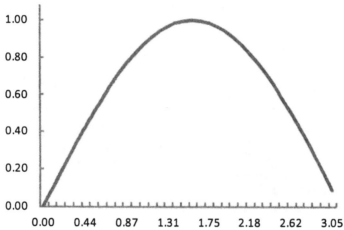

Figure 4-16. The Line chart after editing the data series.

Creating Pie Charts

A Pie chart shows the relative importance of values in a data series. There are several Pie chart subtypes, including normal or exploded Pie charts and 3-D versions. Figure 4-18 shows an example of a 3-D Exploded Pie chart.

Rather than accepting the default legend, you can add data labels to identify the segments of the pie. Click on the Layout tab; in the Labels group, click on the Data Labels drop-down button and choose More Data Label Options. In the Label Options window, uncheck the default Value option and check the Category Name option. Choose a Label Position option for a pleasing effect

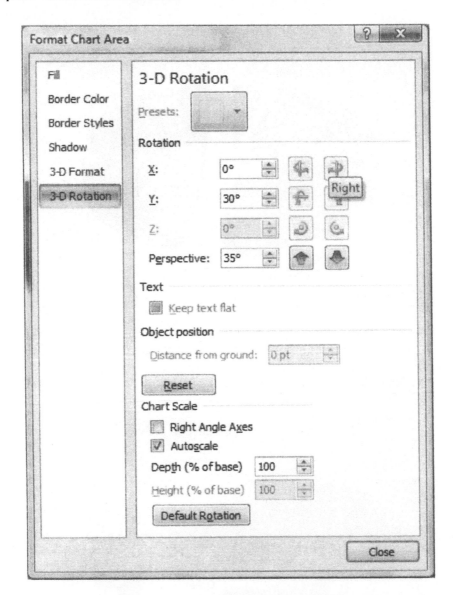

Figure 4-17. The 3-D Rotation dialog box.

You can rotate the pie for a more pleasing view. To change the perspective of the chart, select the chart area, press the Format Selection button to display the Format Chart Area dialog box, and display the 3-D Rotation window (Figure 4-17). Alternatively you can click on the 3-D Rotation button in the Background group of the Layout ribbon to display the window. The controls for rotation in the X- and Y planes and the Perspective button are active.

The completed 3-D Exploded Pie chart is shown in Figure 4-18.

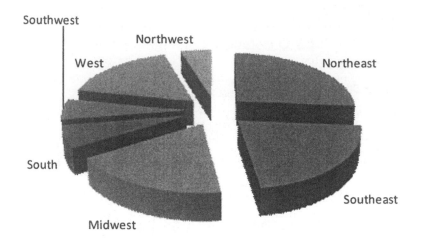

Figure 4-18. A 3-D Exploded Pie chart after some formatting changes.

A Pie chart should not contain too many categories, or should not have some values that are much smaller than the rest. However, if you can arrange the data so that all of the small values are grouped, you can create a Pie-with-Bar chart that shows the small values extracted and combined into a stacked bar, as shown in Figure 4-19.

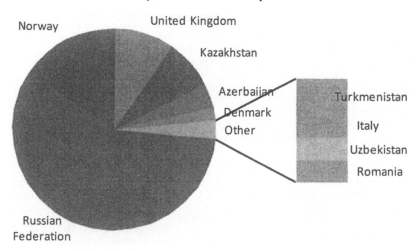

Figure 4-19. A Pie-with-Bar chart.

You can adjust the size of the bar portion of the chart, as well as the separation between the pie and the bars. Click on the pie or the bars to select the Series, click on the Format Series button, and click on Series Options in the Format Data Series dialog box to display the Series Options window (Figure 4-20). Adjust the sliders for Gap Width and Second Plot Size to obtain the desired chart appearance.

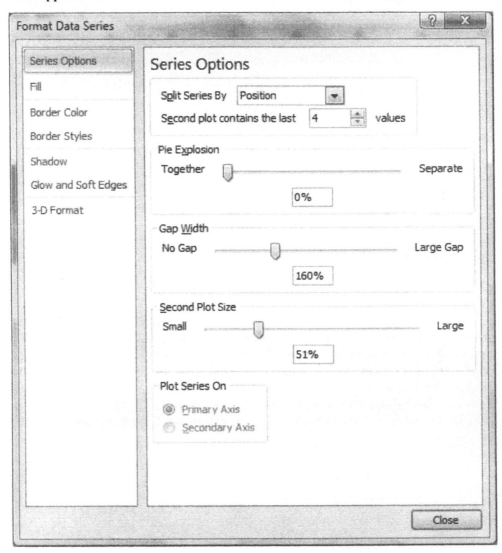

Figure 4-20. The Series Options window of the Format data Series dialog box for a Pie-with-Bar chart.

Creating Radar Charts

Although not specifically intended for this purpose, a Radar chart is a convenient way to show how a variable changes with direction – for example, over a 360° range of angles. If you want the chart to display angles from 0° to 360° in increments of 10°, create a table of values from 0° through 350°, as illustrated in Figure 4-21.

	A	B	C
1	X (radians)	Y (intensity)	X (label)
2	0	0.000	0
3	10	0.087	
4	20	0.174	
5	30	0.259	30
6	40	0.342	
7	50	0.423	
8	60	0.500	60
9	70	0.574	
10	80	0.643	
11	90	0.707	90
12	100	0.766	

Figure 4-21. Portion of a data table that shows variation of Y
with angles from 0° to 350°.

When initially created, a Radar chart is like a Line chart – that is, if you select only the range of y-values (in column B of Figure 4-21), Excel will use the integers 1,2,3... as category labels. To replace these integers with the values in column C as category labels, click on Select Data in the Data group of the Design ribbon to display the Select data Source dialog box. Press the Edit button in the Horizontal (Category) Axis Labels area (Figure 4-22) and enter a range of cells for the category labels ($C$2:$C$37 in Figure 4-21) as illustrated in Figure 4-23.

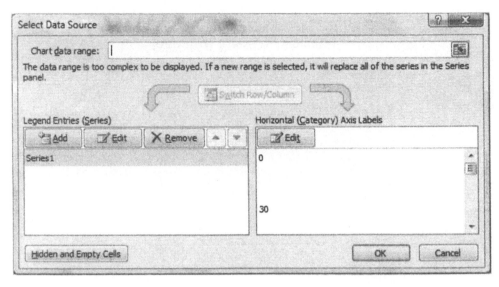

Figure 4-22. Adding a range of cells as Category labels for a Radar chart.

Figure 4-23. The Axis Labels dialog box.

Alternatively, after creating the chart from the values in column B, you can click on the data series in the chart. This will display the SERIES function for this example in the Formula bar

=SERIES(,,Sheet1!$B$2:$B$37,1)

Between the first and second commas of the above formula, enter a reference to the range of cells containing the category values, either by typing or by selecting. The formula should now look like this:

=SERIES(,Sheet1!$C$2:$C$37,Sheet1!$B$2:$B$37,1)

When you press Enter, the chart will now look like the example in Figure 4-24 (after some additional number formatting of the Y Axis).

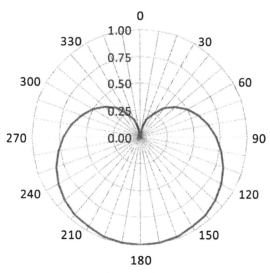

Figure 4-24. The completed Radar chart.

Creating Surface Charts (3-D Charts)

When a dependent variable z depends on the values of two independent variables x and y, you can display the relationship graphically using a 3-D chart, using Excel's built-in Surface chart format

The 3-D Surface chart type in Excel's gallery of charts is not a true 3-D chart, but rather a Line chart in two dimensions. The X- and Y axes are Categories — only the Z Axis is proportional to the data plotted. Thus if you produce a 3-D chart using Excel, you'll have to take this limitation into account. The values on the X- and Y axes must be equally spaced. Since only one Z Axis value can be charted for any pair of X- and Y Axis values, you can't produce a plot of a closed surface, such as a sphere. But this type of chart can be useful, for example, to show the effect of changing two variables on the yield of a process.

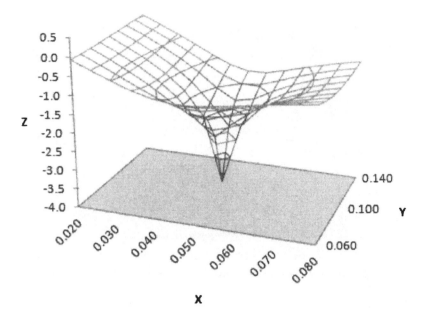

Figure 4-25. A 3-D Wireframe chart.

There are two useful format options within the sub-gallery of 3-D charts: the 3-D Surface chart and the 3-D Wireframe chart. The 3-D Surface chart uses colors to indicate areas on the surface with different ranges of Z Axis values. The 3-D Wireframe chart is identical except that colors are not used. Figure 4-25 shows an example of a 3-D Wireframe chart.

You can change the view of a 3-D chart by clicking on 3-D Rotation in the Background group in the Layout ribbon (see Figure 4-17).

Creating XY Charts

The XY Plot chart type is used often to display scientific data. In contrast to all of the other chart types produced by Excel, only the XY chart type uses the x-values to position the data points along the x direction.

Figure 4-26 shows a typical XY chart as initially produced by Excel.

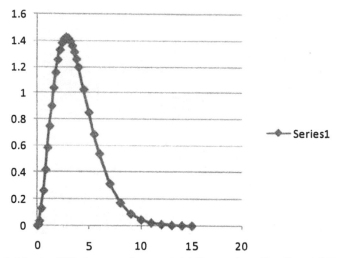

Figure 4-26. An XY Scatter chart as initially produced by Excel 2007/2010.

Figure 4-27 shows the chart of Figure 4-26 after removing the legend and gridlines, adding a chart title and axis titles, and formatting the data series and Y Axis tick mark labels.

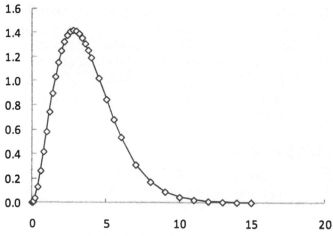

Figure 4-27. An XY chart after formatting.

Switching Between Chart Types

To change a chart from one type to another, e.g., from a Column chart to a Pie chart, click on the Design tab, click on the Change Chart Type button, and choose the desired type from the gallery. In a chart with multiple data series, you can change the chart type of a single series, to produce, for example, a combination Line-Column chart.

Formatting the Elements of an XY Chart

To format a chart element you can use the Format Selection button in the Format ribbon or the Layout ribbon, or you can use the Mini Toolbar, which is described in a following section.

Formatting Chart Elements by Using the Mini Toolbar

In Excel 2003, you can double-click on a chart element to go directly to the format dialog box for that element. This feature is not available in Excel 2007. Instead, you can use the Mini Toolbar. To display the Mini Toolbar, right-click on the chart element to be formatted. A context-sensitive toolbar/shortcut menu will be displayed. The Mini Toolbar for a Chart Title is shown in Figure 4-28. You can change the font, font size, etc., by using the Mini Toolbar buttons, or you can choose **Font...** in the shortcut menu to display the Font dialog box (Figure 4-29).

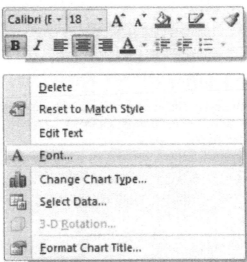

Figure 4-28. The Mini Toolbar.

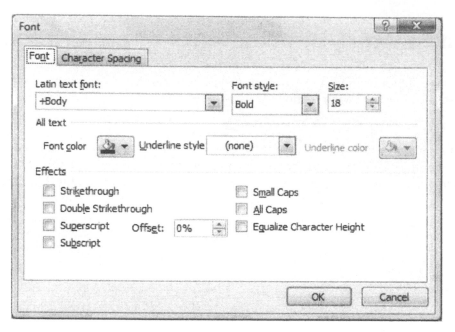

Figure 4-29. The Font dialog box accessed from the Mini Toolbar.

Formatting a Data Series

You can format the size, shape and color of the data markers and the color and thickness of the line. The Marker Options window of the Format Data Series dialog box is shown in Figure 4-30.

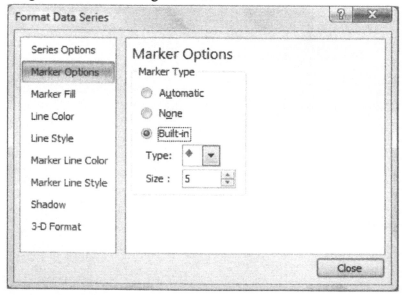

Figure 4-30. The Format Data Series dialog box showing the Marker Options screen.

I prefer a marker size of 5 (the default is 7) and a line thickness of 1 point (the default is 2.25).   You can save these preferences; see "Saving a Chart Template in Excel 2007" in Chapter 8, "Advanced Charting Techniques".

Formatting Chart Elements with Color

At first glance, the Line Color window of the Format Data Series dialog box doesn't appear to offer the possibility of changing the color of the line, and the same is true for the other dialog boxes.   The defaults color is Automatic, as shown in Figure 4-31, and there's no color palette as there is in Excel 2003.   But if you click on the Solid Line button, a drop-down color palette appears, as shown in Figure 4-32.

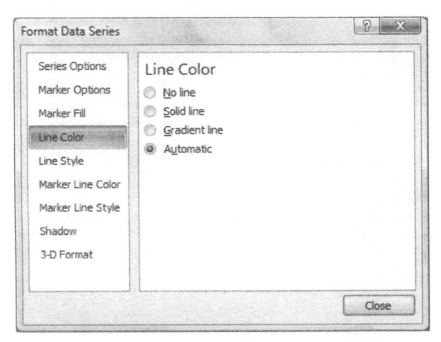

Figure 4-31. The Format Data Series (Line Color) dialog box.

The Theme Colors are the Automatic colors of Excel 2007, beginning with the fifth (medium blue) and progressing to the right (brown, olive green, violet, light blue, orange).   The next Automatic colors appear to be the ones corresponding to the Theme colors, but in the second row of the color palette.

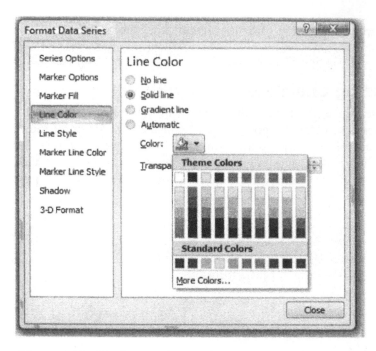

Figure 4-32. The Format Data Series (Line Color) dialog box, when the Solid Line option button is pressed.

Modifying an Axis Scale

Occasionally you will need to change the axis scales of an XY chart. Excel usually creates axis scales that include zero, but this is not suitable for every case, as in Figure 4-33, where the wavelength data ranged from 350 nm to 850 nm.

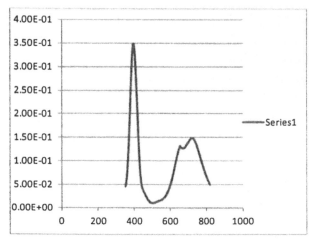

Figure 4-33. An XY chart as originally created by Excel.

To modify the X Axis scale of the chart in Figure 4-33, in the Axis Options window of the Format Axis dialog box (Figure 4-34), press the Fixed option button for the scale Minimum and enter 350 (the lowest x value) for the new value, and press OK.

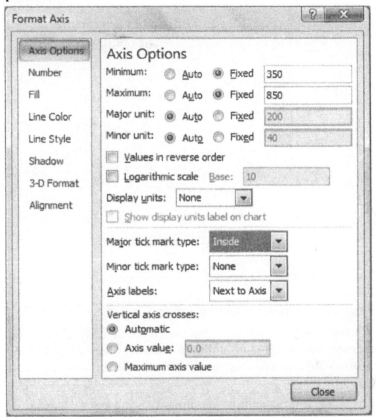

Figure 4-34. The Format Axis (Options) dialog box for axes.

One unfortunate change was made in the Minimum, Maximum, Major Unit and Minor Unit input boxes — you must click the Fixed button before you can enter a value. In Excel 2003, you can enter a value directly.

Changing the Number Format of an Axis Scale

To change the number format of the X Axis, first select the axis. Click on Format Selection in the Format ribbon. In the Number window of the Format Axis dialog box, click on the Number category and enter two decimal places.

Excel Tip. In Excel 2003, you can number-format chart axes by using number-formatting tool buttons, such as Increase Decimal *or Decrease Decimal* . *This feature is not available in Excel 2007 or Excel 2010.*

Figure 4-35 shows the same chart after some other formatting changes (tick marks, title, axis labels, etc.).

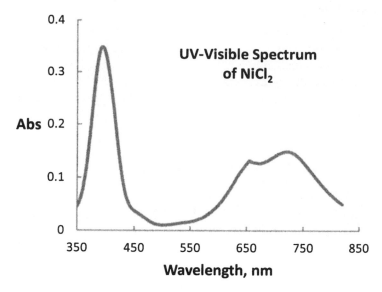

Figure 4-35. The chart in Figure 4-33 after adjustment of the scale of the X Axis.

Changing the Dimensions of a Chart

The Format ribbon contains a feature, new in Excel 2007, that makes it convenient to set the height and width of a chart. The Size group in the Format ribbon (Figure 4-36) contains two controls to set the dimensions, in inches, of an embedded chart.

You can enter a dimension in the input box and press Enter, or click on the spinner button to change the dimensions.

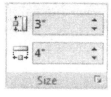

Figure 4-36. The Size group in the Format ribbon.

You can use these controls to make multiple embedded charts on a worksheet have the same dimensions. Simply make a multiple selection by holding down the Ctrl key while you click on the charts, set the dimensions, and press Enter.

When the X Values of a Category Chart Are Dates

In all chart types except XY Plots, the *x*-values are categories and are used simply as labels under the columns of a column chart (for example).   The columns are equally spaced, independent of the *x*-values, as shown in Figure 4-37. For this chart, the *x*-values on the spreadsheet were simply the text values "Jan", "Feb", etc. A chart produced from this data uses the *x*-values simply as labels, as in Figure 4-38.

	A	B
1	Jan	1
2	Feb	2
3	Mar	3
4	May	4
5	Jun	5

Figure 4-37. A data table where the X values are text.

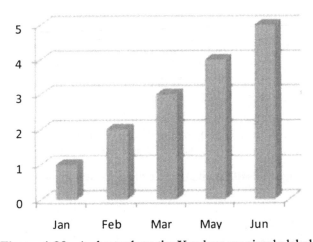

Figure 4-38. A chart where the X values are simply labels.

But if the *x*-values are actual date values, e.g., Jan 1 2008, Feb 1 2008, etc., as in Figure 4-39, the chart would look like Figure 4-40.

	A	B
1	1-Jan-08	1
2	1-Feb-08	2
3	1-Mar-08	3
4	1-May-08	4
5	1-Jun-08	5

Figure 4-39. A data table where the X values are dates.

When *x*-values are dates or times, the Scale tab in the Format Axis dialog box and the Axes tab in the Chart Options dialog box are different than they are for the ordinary column chart. In the Scale tab, shown in Figure 4-41, there are additional input boxes to specify the Base Unit, Major Unit and Minor Unit (days, months or years).

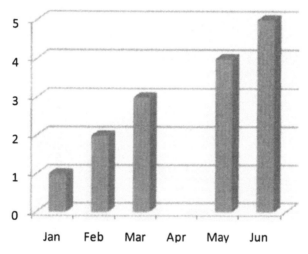

Figure 4-40. A chart where the X values are dates.

The dialog box also has Axis Type option buttons that allow you to specify Text Axis (the *x*-values are plotted as Categories), in which case the values are equally spaced along the X Axis, or as Date/time Axis, in which case the values are plotted as shown in Figure 4-40.

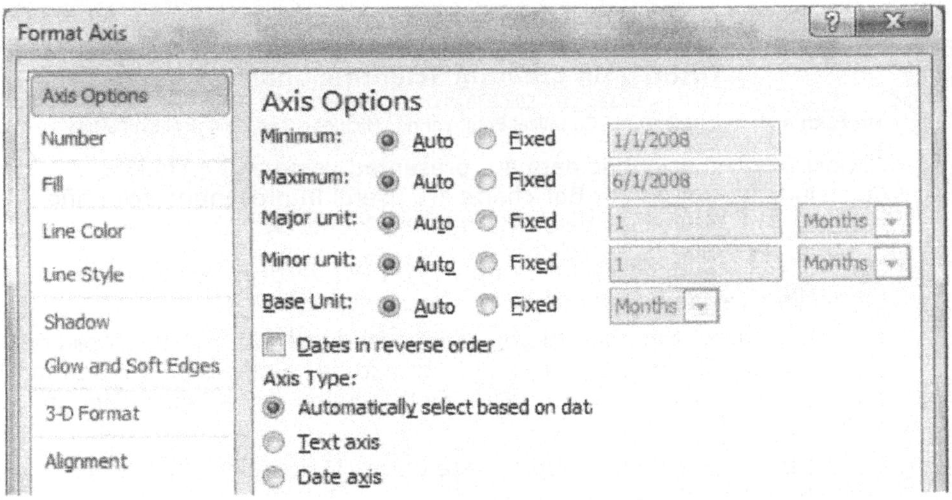

Figure 4-41. Upper portion of the Axis Options window of the Format Axis dialog box when X Values are dates.

Tutorial on Creating Scientific Charts
Based on "Guidelines on Graphing",
Dr. Faith A Morrison, Dept. of Chemical Engineering, Michigan Technological University

Almost always, scientific data are presented using an XY chart. Occasionally, Column or Bar charts are useful in presenting scientific results. The guidelines below pertain to XY charts.

1. Plot the independent variable (e.g., y) on the vertical axis and the dependent variable (e.g., x) on the horizontal axis.

2. Do not "connect the dots" of your data; use a calculated line, based on a valid scientific model, or a Trendline, instead. Show the equation for the calculated line or the Trendline on the chart, in the figure caption, or in the text.

3. Label the axes, including appropriate units. Make axis labels large enough that they are easily readable.

4. Generally, adjust the maximum and minimum of the axis scales so that there is not a lot of "wasted space" at top, bottom or sides.

5. Use standard whole numbers for the Tick Mark Labels. A series of Tick Mark Labels like 0, 5, 10... is much better than 3.3, 5.3, 7.3...

6. The numbers that label the scales of your chart should reflect the precision of your data

7. Avoid numbers with many zeros: for example, instead of Tick Mark Labels like 1000000, 2000000, 3000000..., change the axis label to "millions" and use 1, 2, 3....

8. Use a legend to label the different data series on your chart. Alternatively, you can identify the lines and symbols in a figure caption.

9. Where appropriate, data points should have some indication of associated error; error bars are most commonly used for this.

10. Do not use horizontal or vertical gridlines; they clutter the chart and make it hard to read. If you *must* use gridlines, use dashed lines, or solid lines that are lighter than the other lines in the chart.

11. Do not rely on color in a printed report; colors do not photocopy well. Use color for presentations, such as PowerPoint.

12. Make sure that the chart is large enough for your readers to see the data points and lines. When making the chart, keep the final size of the graphic in mind.

13. In the text the proper term for a chart is "Figure" and a variable is "plotted".

14. Include either a title or a figure caption but not both. Use a figure caption for a report; use a title in a presentation such as PowerPoint. A figure caption is positioned below the figure.

Excel 2007/2010 Chart Specifications

Maximum number of...

data series in one chart	255
data points in a data series for 2-D charts	32000 (Excel 2007); limited by available memory (Excel 2010)
data points in a data series for 3-D charts	4000
data points for all data series in one chart	256000
charts linked to a worksheet	Limited by available memory
worksheets referred to by a chart	255

5

Excel 2003 Charts

This chapter covers the basics of creating charts with Excel 2003. More advanced topics are covered in Chapter 8.

Chart Types

Excel 2003 provides a gallery of 14 standard chart types — bar charts, column charts, line charts and pie charts, among others. Since Excel originated as a financial tool, most of the chart types are those that are useful for displaying financial and related information — a bar chart to show sales figures for each business quarter, a line chart to show stock values each day over a one-month period, etc. All charts except the XY Scatter chart use the x-values only for labels (called categories by Excel). The XY or Scatter chart is the only one in which numeric values are used along both axes.

Of these 14 chart types, seven of them seem of most use for chemists: the Column, Bar, Line, Pie, XY Scatter, Radar, and Surface. XY Charts are probably the most widely used for the display of scientific information.

Creating a Chart

You can create a chart: either as a separate chart sheet in a workbook, or as a chart embedded in a worksheet, so that you can see both the data and the chart at the same time. An embedded chart is especially useful if you want to see how a calculated curve changes as you change its parameters. As you change the values in worksheet cells, the chart will update automatically.

Creating a Chart Using the Chart Wizard

The easiest to create a chart is to use the Chart Wizard tool button 📊 . To use the Chart Wizard, first select the data to be plotted, e.g., a column of x-values and one or more columns of y-values. (The data can be in rows or columns.) If the rows or columns are not adjacent, hold down the Ctrl key while selecting the separate rows or columns of data. Then press the Chart Wizard tool button; alternatively, you can press the Chart Wizard button first and select the data range later. The first of a series of four dialog boxes will appear. The first Chart Wizard dialog box (Figure 5-1) lets you select the desired chart format. There are tabs for two categories: Standard Types and Custom Types; Custom Types are discussed in Chapter 8, "Advanced Charting Techniques".

When you select a chart type from the Chart Type category box, the gallery of chart subtypes will be displayed on the right. For example, if you click on XY (Scatter), there are five XY Scatter chart subtypes: marker points with no connecting lines, marker points connected by straight lines, straight lines with no marker points, points connected by a smooth curve and a smooth curve with no marker points.

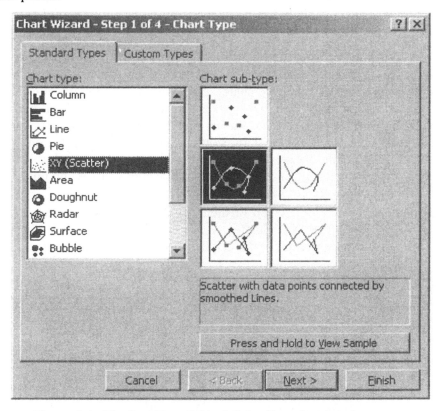

Figure 5-1. The first Chart Wizard dialog box: choosing chart type.

The Smooth Curve option will not be suitable if you are plotting data with experimental scatter. To produce a chart with a smooth calculated curve through experimental data points, see "Plotting Experimental Data Points and a Calculated Curve" in Chapter 8.

The second dialog box (Figure 5-2) displays a preview of the chart and allows you to enter or change the range of data to be plotted. The second and subsequent dialog boxes also provide the Cancel and Back buttons, in case you want to change what you've already selected.

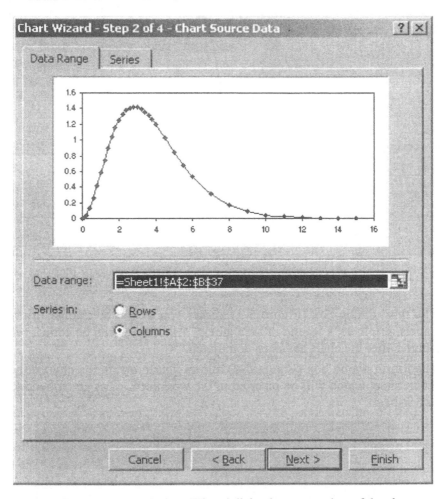

Figure 5-2. The second Chart Wizard dialog box: a preview of the chart.

series_order = 4, and the legend will show the series in that order. To put the series in the correct order, change the *series_order* for the March series to 3.

For Bubble charts only, the SERIES function includes an additional optional argument *bubble_size* that specifies the bubble sizes. This argument can be an array of numeric values enclosed in braces, a reference to a range of cells, a named range, or omitted. It can be a non-contiguous reference (separate ranges enclosed in brackets and separated by commas.

The SERIES function is only for charts; you can't use it in a worksheet formula. You can't incorporate worksheet functions or formulas directly into the SERIES function arguments, but it is possible to modify it using named formulas.

Creating Column or Bar Charts

A column chart compares values of elements in a group – for example, oil production figures for different areas in a given year. You can use grouped or stacked columns to show two relationships – for example, oil production figures for different areas over several years. Figure 5-5 shows an example of a chart with grouped columns, as originally created by the ChartWizard.

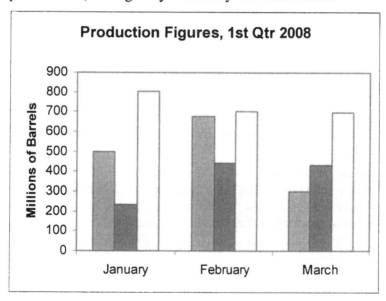

Figure 5-5. A Column chart as produced by the Chart Wizard.

The appearance of an Excel chart can usually be significantly improved by making some formatting changes. Figure 5-6 shows the same chart after adding a legend, removing the border around the Plot Area, choosing a Fill Effect for the Plot Area and a Fill Effect for each of the columns, and adjusting the Overlap between bars.

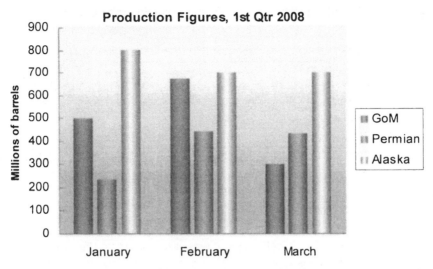

Figure 5-6. A Column chart after some formatting changes.

Creating Line Charts

Don't confuse a Line chart and an XY Scatter chart. An XY chart uses the *x*-values to determine where to plot the *y*-values, while a Line chart plots the *y*-values at equal intervals.   If you select only a single range of cells, the ChartWizard will plot these values using the integers 1, 2, 3... as the category values (the *x*-values). If you select two columns or two rows of cells that you intend to be used as the categories and the values, the ChartWizard will stubbornly plot these as two series of *y*-values and use the same series of integers 1, 2, 3... as the categories, unless you choose the Series tab in the Step 2 of 4 dialog box, enter a range of cells for the Category labels, and remove the unwanted series in the Series list box (see Figure 5-7).

Note that the points are plotted at equal intervals, independent of the values of the categories. (But see "When the X Values of a Category Chart Are Dates" later in this chapter.)

Creating Pie Charts

A Pie chart shows the relative importance of values in a data series. There are several Pie chart subtypes, including normal or exploded Pie charts and 3-D versions. Figure 5-8 shows an example of a 3-D Exploded Pie chart.

Choose 3-D View from the Chart menu, shown in Figure 5-9, to rotate a 3-D Pie chart in either the horizontal or vertical for a more pleasing view.

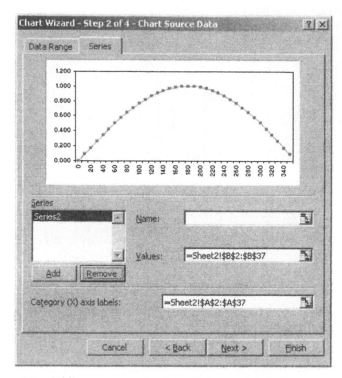

Figure 5-7. Adding a range of cells as Category labels for a Line chart.

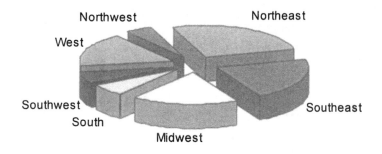

Figure 5-8. A 3-D Exploded Pie chart after some formatting changes.

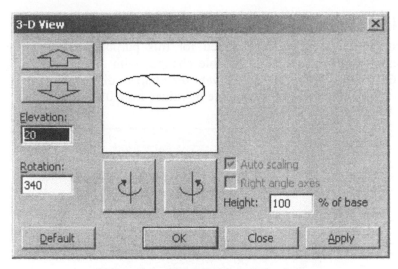

Figure 5-9. The 3-D View dialog box.

A Pie chart should not contain too many categories, or have some values that are much smaller than the rest. However, if you can arrange the data so that all of the small values are grouped, you can create a Pie-with-Bar chart that shows the small values extracted and combined into a stacked bar, as shown in Figure 5-10.

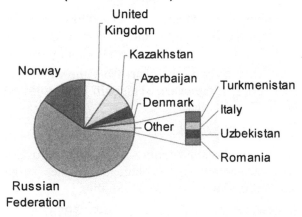

Figure 5-10. A Pie-with-Bar chart.

Creating Radar Charts

Although not specifically intended for this purpose, a Radar chart is a convenient way to show how a variable changes with direction – for example over a 360° range of angles. If you want the chart to display angles from 0° to 360° in increments of 10°, create a table of values from 0° through 350°.

	A	B	C
1	X (angle)	Y (intensity)	Y label
2	0	0.000	0
3	10	0.087	
4	20	0.174	
5	30	0.259	30
6	40	0.342	
7	50	0.423	
8	60	0.500	60
9	70	0.574	
10	80	0.643	
11	90	0.707	90
12	100	0.766	

Figure 5-11. Portion of a data table that shows variation of Y
with angles from 0° to 350°.

The ChartWizard creates a Radar chart like a Line chart – that is, if you select only the range of y-values (in column B of Figure 5-11), the ChartWizard will use the integers 1,2,3... as category labels. To replace these integers with the values in column C as category labels, choose **Source Data...** from the **Chart** menu, choose the Series tab, and enter a range of cells for the category labels ($C$2:$C$37 in Figure 5-11) as illustrated in Figure 5-12.

Alternatively, after creating the chart from the values in column B, you can click on the data series in the chart. This will display the SERIES function for this example in the Formula bar

=SERIES(,,Sheet1!$B$2:$B$37,1)

Between the first and second commas of the above formula, enter a reference to the range of cells containing the category values, either by typing or by selecting. The formula should now look like this:

=SERIES(,Sheet1!$C$2:$C$37,Sheet1!$B$2:$B$37,1)

When you press Enter, the chart will now look like the example in Figure 5-13 (after some additional number formatting of the Y Axis).

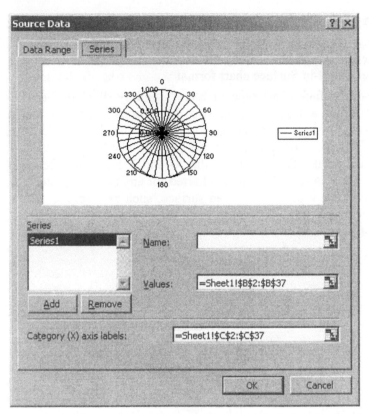

Figure 5-12. Adding a range of cells as Category labels for a Radar chart.

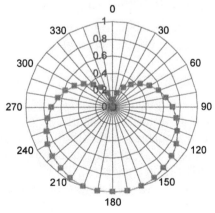

Figure 5-13. The completed Radar chart.

Creating Surface Charts (3-D Charts)

When a dependent variable z depends on the values of two independent variables x and y, you can display the relationship graphically using a 3-D chart, using Excel's built-in Surface chart format

The 3-D Surface chart type in Excel's gallery of charts is not a true 3-D chart, but rather a Line chart in two dimensions. The X- and Y axes are Categories; only the Z Axis is proportional to the data plotted. Thus if you produce a 3-D chart using Excel, you'll have to take this limitation into account and work within it. The values on the X- and Y axes must be equally spaced. Since only one Z Axis value can be charted for any pair of X- and Y Axis values, you can't produce a plot of a closed surface, such as a sphere. But this type of chart can be useful, for example, to show the effect of changing two variables on the yield of a process.

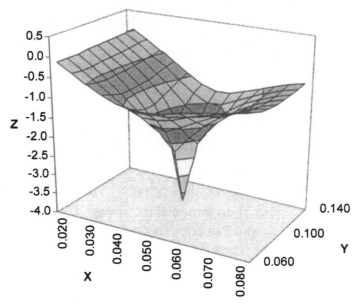

Figure 5-14. A 3-D chart.

There are two useful format options within the sub-gallery of 3-D charts: the 3-D Surface chart and the 3-D Wireframe chart. The 3-D Surface chart uses colors to indicate areas on the surface with different ranges of Z Axis values. The 3-D Wireframe chart is identical except that colors are not used. Figure 5-14 shows an example of a 3-D Surface chart.

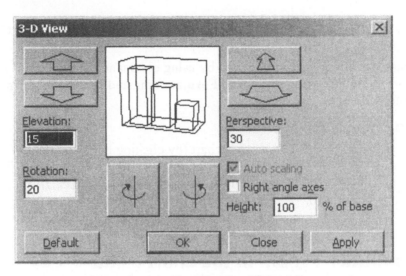

Figure 5-15. The 3-D View dialog box.

You can change the view of a 3-D chart by choosing **3-D View...** from the **Chart** menu (Figure 5-15). Use the Rotation, Elevation and Perspective buttons to view the chart from different angles. Often a more pleasing or informative chart can be produced by changing the view. To return to the original viewing angle, press the Default button.

Activating, Resizing and Moving an Embedded Chart

To activate (select) an embedded chart for moving, resizing or formatting, simply click once on the chart. When selected, an embedded chart has black "handles" at the sides and corners. To move an embedded chart, place the mouse pointer near the border of the chart and click and drag the chart to its new location. To change the size of the chart, click on one of the handles and then drag to move that edge or corner inward or outward.

To convert an embedded chart to a full chart sheet, choose **Chart Location...** from the **Chart** menu.

*Excel Tip   When an embedded chart is converted into a full chart sheet, the chart usually doesn't fill the complete chart sheet window. To expand the chart to fill the window, choose **Sized with Window** from the **View** menu.*

Formatting Charts: An Introduction

Excel scales and formats each newly created chart automatically. It does a good job, but usually there is plenty of room for improvement. Excel provides a wide range of tools for modifying a chart. A few of these are discussed in this chapter, and further details are provided in Chapter 8, "Advanced Charting Techniques".

Chart formatting is accomplished by using commands in the **Chart** menu in the Chart Menu Bar, or by using the **Format** command in the **Edit** menu in the Chart Menu Bar.

Using the Chart Menu

When you activate an embedded chart (by clicking on it) or switch to a chart sheet, the Worksheet Menu Bar is replaced by the Chart Menu Bar. The first four commands in the **Chart** menu — **Chart Type...**, **Source Data...**, **Chart Options...**, **Location...** — correspond to the four dialog boxes of the Chart Wizard. You can use these commands to modify a chart that you've already created.

Using CHART TYPE... in the Chart Menu
to Switch from One Chart Type to Another

If you created a chart as an XY Scatter plot, for example, you can change it to another chart format by choosing **Chart Type...** from the **Chart** menu and then choosing any of the other chart types. Fifteen standard types, as well as a number of more specialized custom types, such as a combination Line-Column chart, are available, plus any user-defined chart types that have been added.

Using CHART OPTIONS... in the Chart Menu
to Add Titles, Gridlines or a Legend

To add text to the chart, choose **Chart Options...** from the **Chart** menu and choose the Titles tab. The resulting dialog box, identical to Figure 5-3 but now called Chart Options, lets you add text for the chart title or the labels for the X- or Y axes. When you enter text in the Chart Title input box, for example, the title text will appear in the sample chart. Excel wraps the text if it is too long to fit on one line. In the same way you can add or remove gridlines or a legend.

Using LOCATION... in the Chart Menu
to Move or Copy an Embedded Chart
to a Separate Chart Sheet or Vice Versa

If you created an embedded chart and would like to convert it to a separate chart sheet, simply activate the chart, choose **Location...** from the **Chart** menu, press the As New Sheet option button, and then press OK. The embedded chart will be converted into a chart on a separate chart sheet. To put a copy of an embedded chart on a separate chart sheet, first make a copy of the chart and then follow the same procedure.

Formatting the Elements of an XY Chart

The **Format** menu allows you to customize any of the chart elements of an XY Scatter chart: the Plot Area, Axes, Data Series, Arrows, Text, Gridlines, etc. You can change the weight and style of lines, the color and shape of plotting symbols, the numerical ranges of axes, etc.

Excel Tip   *As you move the mouse pointer around the chart, yellow Chart Tip boxes appear, with the name of the chart element that the mousepointer is currently passing over. If you find these Chart Tips distracting, you can deactivate them by choosing Options from the Tools menu. Choose the Chart tab and un-check the Chart Tips check box.*

Selecting Chart Elements

Select a particular chart element by clicking once on it with the mouse pointer. The selected element will be indicated by the appearance of "handles"; in addition, the name of the chart element will appear in the Name Box (to the left of the formula bar). It's a good idea to look at the text in the Name Box when you choose a chart element, just to make sure that you've selected the correct chart element.

Excel Tip. *Sometimes it's difficult to select a chart element by clicking on it (for example, if two chart elements are almost superimposed). Instead of selecting with the mouse pointer, you can use the up and down arrows on the keyboard to select chart elements. This allows you to select each chart element in turn (Chart, Plot, Axis, Series 1, etc.); the name of the selected chart element is displayed in the reference area of the formula bar. By using the left and right arrows, you can select related chart elements within a group (e.g., Series 1 Point 1, Point 2, etc.).*

Formatting Chart Elements

Once the desired chart element has been selected, choose **Format** from the Chart Menu Bar. In Excel 2003 there is a single context-sensitive menu command in the **Format** menu, which appears as **Selected Axis...**, **Selected Data Series...**, etc., depending on which chart element is selected.

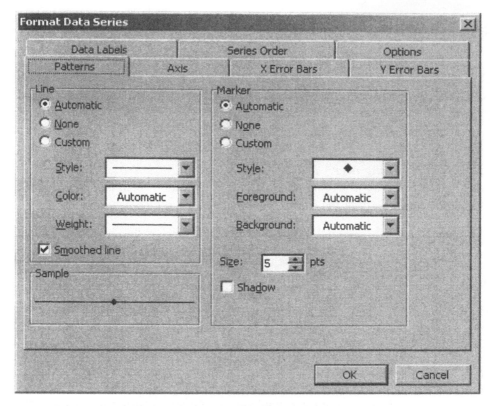

Figure 5-16. The **Patterns** dialog box for axes.

When you choose the **Format...** command, a dialog box will appear with the formatting possibilities for the selected chart element. There are different **Patterns** dialog boxes or tabs (for Chart, Plot, Series, Axis, Arrow, Gridline, Text, etc.), different **Font** dialog boxes or tabs (for Chart, Axis, Text, etc.), and so on. For example, with a Series selected, Patterns permits you to change or remove the plotting symbol or the line (Figure 5-16). With an axis selected, you can use the Patterns tab to change the style or weight of the axis, change or remove or add major and minor Tick Marks, and change or remove the Tick Labels.

Excel Tip. If you double-click on a chart element, you will go directly to the Format dialog box for that element. This feature is not available in Excel 2007 but has been re-instituted in Excel 2010.

Occasionally you will need to change the axis scales. Excel usually creates axis scales that include zero, but this is not suitable for every case, as in Figure 5-17, where the wavelength data ranged from 350 nm to 850 nm.

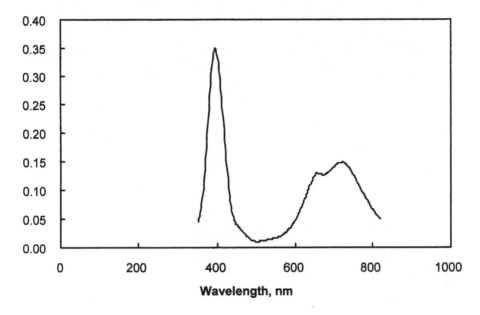

Figure 5-17. An XY chart as originally created by Excel.

With the X axis selected, the Scale tab (Figure 5-18) enables you to change the scale range, or where the Y Axis crosses the X axis, for example.   The Number tab (not shown) permits you to use the same number formats available for worksheet cells in the **Format Cells...** command to change the number formatting of the axis labels.

Excel Tip. You can also number-format chart axes by using number-formatting tool buttons, such as Increase Decimal *or Decrease Decimal* .

To modify the X axis scale of the chart in Figure 5-17, first select the axis by clicking on it. Black "handles" will appear at the ends of the axis. Then choose **Selected Axis...** from the **Format** menu and choose the Scale tab. Enter 350 (the lowest *x*-value) for the new value of Minimum and enter 850 (the largest *x*-value) for the new value of Maximum, as in Figure 5-18, and press OK.

When you change a value in any of the five input boxes in the Scale tab, the Auto check box next to the entry becomes unchecked. When Excel creates a chart, it automatically scales the axes to conform with the data but does not adjust any of the axis parameters for which Auto is unchecked.

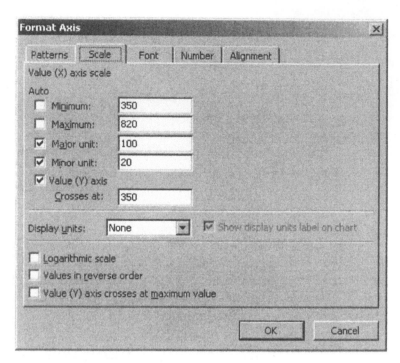

Figure 5-18. The **Scale** dialog box for axes,
with better values for maximum and minimum.

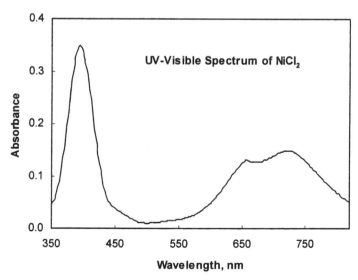

Figure 5-19. The chart in Figure 5-17 after adjustment of the scale of the X axis.

Figure 5-19 shows the same chart after some other formatting changes (tick marks, title, axis labels, etc.).

When the X Values of a Category Chart Are Dates

In all chart types except XY Scatter Plots, the x-values are categories and are used simply as labels under the columns of a column chart (for example). The columns are equally spaced, independent of the x values, as shown in Figure 5-20. For this chart, the x-values on the spreadsheet were simply the text values "Jan", "Feb", etc. A chart produced from these data uses the x-values simply as labels, as in Figure 5-21.

	A	B
1	Jan	1
2	Feb	2
3	Mar	3
4	May	5
5	Jun	6

Figure 5-20. A data table where the X Values are text.

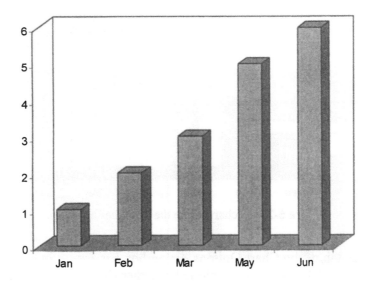

Figure 5-21. A chart where the X Values are simply labels.

But if the x-values are actual date values, e.g., Jan 1 2008, Feb 1 2008, etc., as in Figure 5-22, the chart would look like Figure 5-23.

Excel 2003 Chart Specifications

Maximum number of...

data series in one chart	255
data points in a data series for 2-D charts	32000
data points in a data series for 3-D charts	4000
data points for all data series in one chart	256000
charts linked to a worksheet	Limited by available memory
worksheets referred to by a chart	255

PART II

ADVANCED
SPREADSHEET TOPICS

6

Advanced Worksheet Formulas

This chapter shows you how to use Excel's powerful Lookup & Reference worksheet functions to construct sophisticated worksheet formulas. In addition, you'll learn how to use named references in formulas, which makes constructing complicated formulas easier and more error-free. You'll also learn techniques of formula editing and worksheet troubleshooting.

Using Names Instead of References

A name can be used in a formula to refer to a cell, a range of cells, a constant or a formula. Most often you'll use names for cell references. Names make it easier to create and to decipher complex formulas. For example, the formula

=pKa+LOG(base/acid)

is easier to understand than the formula

=$E$1+LOG(B2/C2)

Guidelines for Creating Names

The first character of a name must be a letter. Subsequent characters can be letters, numbers, the period or the underline character. Spaces are not allowed; the space character is the intersection operator (see "Operators" in Chapter 3). Excel will substitute an underline character for a space in a name. To indicate breaks between words in a name, use a period or underline character or use capitalization.

Names that look like references (e.g., A1) are not accepted. Since Excel can also use the R1C1 reference style, the letters R and C cannot be used as names.

Since the columns in Excel 2007/2010 extend to column XFD, names such as BLT1or XL4 that were valid in earlier versions of Excel will now be recognized as cell references in Excel 2007/2010. If you convert a workbook from Excel 2003 to the Excel 2007/2010 file format, you are alerted about the conflict, and an underscore character is added to the name, e.g., _BLT1.

Names are not case-sensitive; Excel will not allow you to define two names that differ only in capitalization. One you have defined a name with a particular capitalization, however, you can type it in a formula without regard to capitalization; when the formula is entered, the name will appear with the capitalization as originally defined.

Excel has a few built-in names, such as Print_Area, Print_Titles and Criteria, that you should also avoid using as range names.

Defining Names in Excel 2003

The **Name** submenu of Excel 2003's **Insert** menu (Figure 6-1) contains several commands for working with names: **Define...**, **Paste...**, **Create...**, **Apply...** and **Label...**. You will probably find **Define...** and **Create...** most useful. Use **Define...** to assign a name to a single cell or range; use **Create...** to create names for several cells or ranges simultaneously.

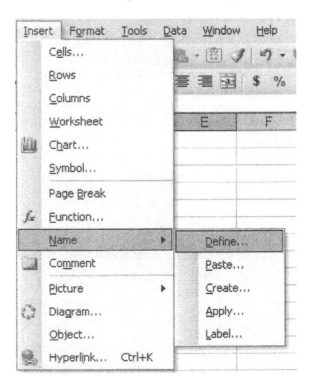

Figure 6-1. The Name submenu.

Using Define Name

To assign a name to a cell reference, first select the cell or range. Then choose **Name** from the **Insert** menu and **Define...** from the submenu to display the Define Name dialog box (Figure 6-2). Excel will propose a name in the Name box, using text from the cell immediately above or to the left of the selected cell. The absolute reference to the selected cell will appear in the Refers To box. Edit the name if desired, and press OK.

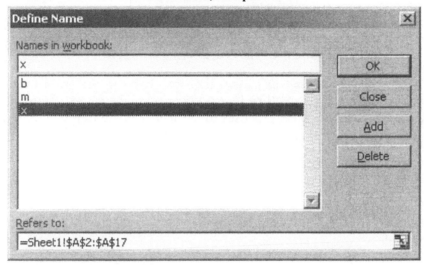

Figure 6-2. The Excel 2003 Define Name dialog box.

Using Create Names

You'll find **Create Names** very useful if you have worksheets with constants or other values arranged in a table format, as in Figure 6-3. This command allows you to assign names to ranges *en masse*.

	D	E
1	slope	3
2	intercept	5

Figure 6-3. A table of constants and names.

To use Create Names to assign names to cells, select the cells to be named and the adjacent cells containing the names. Choose **Name** from the **Insert** menu and choose **Create...** from the submenu. Excel displays the Create Names dialog box; in Figure 6-4 the Left Column check box is checked, indicating that Excel proposes to use text in the cells in the left column for names. Press the OK button. The names will be assigned to the appropriate cells or ranges.

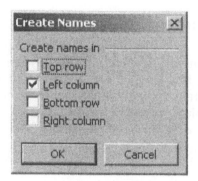

Figure 6-4. The Create Names dialog box.

Names can be enclosed in parentheses for clarity in the worksheet, as in Figure 6-5; the parentheses will be ignored in creating the names. Here the cells C4:D9 are selected and Excel proposes to use names in the Left Column. Equal signs and colons are also ignored, as in Figure 6-6.

	B	C	D
4	Initial volume, mL	(V_0)	50.00
5	Initial Concentration of aedach.3HBr, M	(CL)	0.002056
6	pH correction factor C	(C_)	-0.11
7	Concentration of titrant NaOH, M	(CB)	0.1381
8	Buret calibration factor	(calib)	0.990
9	pcKw	(pcKw)	13.78

Figure 6-5. An example of selecting cells for Create Names.

	G	H	I
4	Protonation constants:		:
5	logK1H =		10.42
6	logK2H =		9.74
7	logK3H =		8.21
8	logK4H =		5.44

Figure 6-6. Another example of selecting cells for Create Names.

If the data table is a two-dimensional one, as in Figure 6-7, cells are referenced both by row and by column. Excel proposes the row and column titles as names, as shown in the Create Names dialog box of Figure 6-8. Excel will apply the name max to the range D4:G4, the name band1 to the range D4:D6, etc. The intersection operator (the space character) can then be used to identify the named variables. For example, band3 A_0 refers to cell F5.

	B	C	D	E	F	G	H
2							
3			band1	band2	band3	band4	
4		max	29.25	22.72	18.56	11.69	
5		A$_0$	1.12	0.15	0.87	0.77	
6		s	1.60	1.54	1.38	1.5	
7							

Figure 6-7. A two-dimensional data table for Create Names.

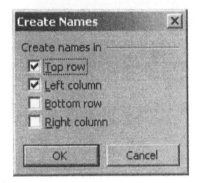

Figure 6-8. The Excel 2003 Create Names dialog box.

Using the Drop-Down Name List Box

You can also define a name by using the Name box (the cell reference area to the left of the formula bar). Simply select the cell or range on the sheet (the range will be displayed in the Name box), click the cursor in the Name box (the typing area will be highlighted), type the name and then press Enter. Excel does not propose a name based on a text label above or to the left of the selected range, as it does when you use Define Name or Create Names; you have to type in the name yourself.

You can use the Name box to view a list of names that have been assigned in the active worksheet, or to jump to a cell or range that has previously been assigned a name. Use the ▼ button to display the drop-down Name list (Figure 6-9) and select the desired name from the list; when you release the mouse button the cell or range will be selected. You can also type into the Name box the name of the cell or range that you want to select.

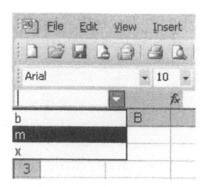

Figure 6-9. The Excel 2003 drop-down Names box.

Names Can Be Local or Global

When you assign a name to a cell or range in a workbook, by default the name is a global or workbook-level name. The name can be used in all sheets in the workbook.

If you want to create a name that is a local or worksheet-level name, that is, a name that is available only in a particular sheet, you can do so as described in the following box.

To Assign a Local or Worksheet-Level Name to a Cell or Range in Excel 2003

1. First select the cell or range. (In this example Sheet1 is the active sheet.)
2. Choose **Define...** from the **Name** submenu to display the Define Name dialog box. Excel will propose a name in the Name box.
3. Edit the name by typing e.g., Sheet1! before the name (see Figure 6-10).
4. Press OK.

If, for example, the name that you had assigned to a cell in Sheet1 was Total, the name must be edited in the Names In Workbook box to appear as Sheet1!Total. However, you can still use the name Total in your formulas in Sheet1.

A sheet-level name appears in the Define Name dialog box only when that sheet is the active sheet, with the sheet name shown on the right side of the list box, as in Figure 6-10. The name appears in the drop-down Name list box (in the formula bar) for that sheet, but does not appear when the drop-down list is selected in other sheets.

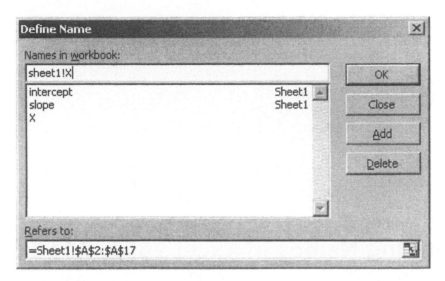

Figure 6-10. A sheet-level name in the Excel 2003 Define Name dialog box.

If you use **Move or Copy Sheet...** in the **Edit** menu to create a copy of a sheet that has defined names, the copy will have the same names but they will be worksheet-level.

If you use the same name to define both a worksheet-level and a workbook-level name, the sheet-level name takes precedence on the sheet where it is defined; only the sheet-level name appears in the list of names in the Define Name dialog box. If you delete the sheet-level name, the workbook-level name appears in the list.

You can define the same worksheet-level name for a range of sheets (usually but not necessarily referring to the same cell in each sheet). Each sheet can then have a different value for the named reference. You can define the name by following the procedure in the preceding box, once for each sheet, or you can use the method described in the following box to assign the name in all sheets at once.

**To Assign the Same Worksheet-Level Name
to the Same Cell or Range in Several Sheets**

1. Insert an extra worksheet. You will enter the name here and afterwards delete this sheet.

2. Group the worksheets, including the extra sheet (select the first sheet in the range, then hold down the Shift key and select the last sheet in the range).

3. Select the cell that you want to contain the name text, and type the name. The text will be entered in that cell in all of the grouped sheets.

4. Select both the cell or range that you want to name and the adjacent cell containing the name.

5. Use Create Names to assign the name to the cell or range in the grouped sheets. This procedure will assign a workbook-level name in the workbook but will "overwrite" it with a worksheet-level name in all of the selected sheets *except the extra sheet that you inserted.*

6. Ungroup the sheets.

7. Use Define Name to delete the workbook-level name in the extra sheet.

8. Delete the extra sheet.

Defining Names in Excel 2007/2010

In Excel 2007/2010, the ribbon commands dealing with range names are located in the Defined Names group in the Formulas tab of the Ribbon. The major change from Excel 2003 is the ease in which the scope of names can be declared global (workbook-level) or local (worksheet-level).

Using Define Name

To assign a name to a range, click on the Define Name icon to display the New Name dialog box (Figure 6-11).

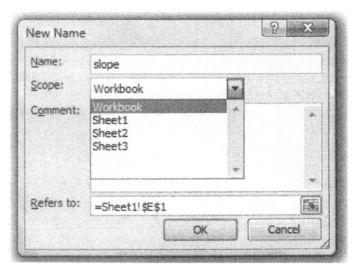

Figure 6-11. The Excel 2007/2010 New Name dialog box.

The New Name dialog box is similar to the Excel 2003 Define Name dialog box, but it has a couple of new features. Most importantly, you can more easily declare the scope of a name as either worksheet-level or workbook-level. As well, you can attach a comment to the name. Comments are displayed in the Name Manager dialog box.

Using Create Names from Selection

Using Create Names From Selection in Excel 2007/2010 is identical to using Create Names in Excel 2003. Only workbook-level names can be created.

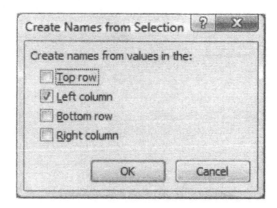

Figure 6-12. The Excel 2007/2010 Create Names From Selection dialog box.

You can use the Name box (the cell reference area to the left of the formula bar) to view a list of names for the active worksheet, or to jump to a cell or range that has previously been assigned a name. See "Using the Drop-Down Name List Box" earlier in this chapter.

Using the Name Manager

The Name Manager, a new feature in Excel 2007/2010, displays all names defined in the active workbook, with their scope, as illustrated in Figure 6-13. You can Filter (i.e., sort) names by scope or other criteria.

If you select a name and press the Edit button, the Edit Name dialog box, identical to the New Name dialog box, will be displayed. You can change Name, Refers To or Comment, but not Scope. Note that if you delete a name, the name will not be replaced by the appropriate reference.

Unlike in Excel 2003, you can make a multiple selection of names by holding down the Ctrl or Shift keys (for example, in order to delete names).

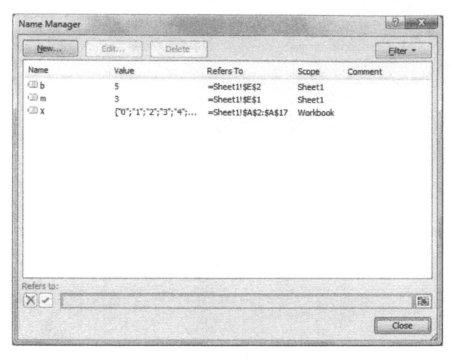

Figure 6-13. The Excel 2007/2010 Name Manager dialog box.

More about Using Names

Some additional features of the use of names in formulas, applicable in Excel 2003 and Excel 2007/2010, will be discussed in the following sections.

A Reference Using Implicit Intersection

Let's see how the range defined as x, shown in the Define Name dialog box in Figure 6-2, is used in a formula. If names were not used, the formula in cell B2 would be

=$E$1*A2+$E$2

and as you Fill Down the formula in cell B2 into cells B2:B17, the reference A2 becomes A3, A4, etc. But if names have been assigned to these cell references, the formula in cell B2 is

=m*x+b

and the formula, when Filled Down, is identical in all cells.

	A	B
1	X	Y
2	0	5
3	1	8
4	2	11
5	3	14
6	4	17
7	5	20
8	6	23
9	7	26
10	8	29
11	9	32
12	10	35
13	11	38
14	12	41
15	13	44
16	14	47
17	15	50

Figure 6-14. Implicit intersection between row 7 and the range $A$2:$A$17.

But the name x refers to a range of cells, $A$2:$A$17. Which cell in the range of x-values is to be used in the formula? Excel uses the value of x in the same row as the cell that contains the formula, as shown in Figure 6-14. This is an example of *implicit intersection*. It's similar to a reference determined by the intersection operator (see "Operators" in Chapter 3). The reference is the intersection of a horizontal range (the row occupied by the cell containing the formula) and a vertical range (the cells defined by the name x), but it's implicit because the horizontal range is only implied by the location of the formula. (You can also use an implicit intersection involving a column occupied by the cell containing a formula and a range of cells in a row.)

A Name Can Refer to a Constant or to a Formula

Instead of a reference, you can type a numeric value or a formula in the Refers To box. You can use named formulas to simplify long worksheet formulas by assigning names to parts of the formula.

In addition to simplifying complex worksheet formulas, using named formulas allows you to accomplish things in Excel that aren't otherwise possible. For examples of using named formulas, see "Returning an Array of Unique Entries in a List" in Chapter 7 or "A Drop-down List Box on a Worksheet" in Chapter 11.

Entering a Name in a Formula by Selecting

Earlier in this chapter it was recommended that you enter cell or range references in formulas by selecting, not by typing. If a name has been assigned to a cell or range, Excel will enter the name in the formula, rather than the reference, when you click on the cell or range.

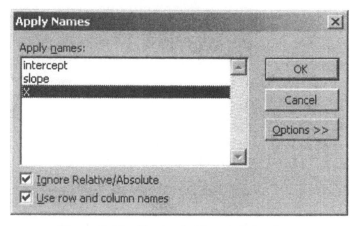

Figure 6-15. The Apply Names dialog box.

Using Apply Names

Use Apply Names (Figure 6-15) if you have created a spreadsheet with formulas using cell references and now want to replace the cell references with names. Choose **Apply Names...** (Excel 2003) or choose **Apply Names...** from the Define Name drop-down submenu, shown in Figure 6-16 (Excel 2007/2010). The names that you have assigned will be shown in the list box. The Ignore Relative/Absolute box should usually be checked. Select a name from the list and press OK.

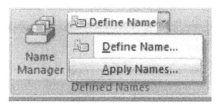

Figure 6-16. The Apply Names command is in the Define Name... drop-down menu.

Excel Tip. You can select more than one name at a time from the list box. Hold down the Ctrl key and click on the names you want to select.

Although names can be workbook-level, Apply Names only operates on the active sheet.

If you want to apply names only within a range of cells on a worksheet, select the range and then choose Apply Names.

Using Paste Name

The Use In Formula button (Excel 2007/2010) or the **Paste Name** command (Excel 2003) allows you to select a name from the list of names in the worksheet and **Paste** it into a formula as you type. There doesn't seem to be any advantage to this command over simply typing in the name, or selecting the named cell or range. Furthermore, in Excel 2003 it requires five actions – **Insert** → **Name** → **Apply...**, select the name, press OK – to insert a name into a formula using Paste Name.

But see "Using Paste List" later in this chapter for something useful in the Paste Name submenu.

Excel Tip. *Press F3 to display the Paste Name dialog box.*

Deleting Names

The Create Names dialog box lists all names that have been assigned in the workbook, even if they are no longer used or valid. If you have removed an unwanted name by deleting the cell, row or column in which it was located, the reference to that name in the Refers To box will be #REF!. Use the Delete button to delete unwanted or invalid names from the list.

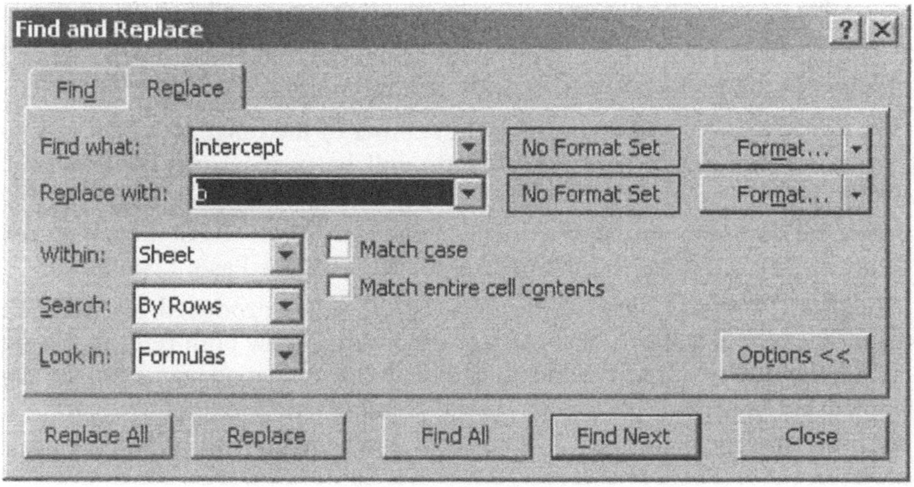

Figure 6-17. The Find and Replace dialog box.

Changing a Name

Occasionally you may want to change a defined name. This is one thing that has been improved in Excel 2007/2010.

Changing a Name in Excel 2007/2010. To change a name, click on the Name Manager icon. In the Name Manager dialog box (Figure 6-13) select the name to be changed. Press the Edit button (the button becomes active when a name is selected) to display the Edit Name dialog box, and change the name. When you change a name using the Name Manager, all instances of the old name in formulas will be changed to the new name. This is a nice feature introduced in Excel 2007/2010.

Changing a Name in Excel 2003. You can easily change all occurrences of a name in a spreadsheet by using **Replace...** from the **Edit** menu. The Replace dialog box (Figure 6-17) asks you to specify the search text and the replacement text.

Press the Options>> button to display the options part of the dialog box. When replacing a name in a formula, the "Match entire cell contents" box should not be checked. The Match Case box is not checked when the dialog box is initially displayed, but because this default option can often cause problems, it's usually a good idea to check it.

Although all occurrences of the name are changed in the worksheet, the name definition in the Define Name box is not changed, so you'll have to change it there also. Until you do so, all formulas containing the changed name will display the #NAME? error value.

Using Paste List or Paste Names...

The **Paste Names...** in Excel 2007/2010 (called **Paste List** command in Excel 2003) is useful for worksheet auditing and error tracing. It produces a list of all names used in the worksheet, with their references.

To use Paste List, select the upper-left corner of the range in which you want the list to appear; make sure that nothing will be over-written by the list. In Excel 2007/2010, click on the Use In Formula button and click on **Paste Names...** in the submenu. In Excel 2003, choose **Name...** from the **Insert** menu and choose **Paste...** from the submenu and then press the Paste List button.

Figure 6-18 shows an example of a list of names produced by **Paste List**. Inspection of the list shows that the names pKw and pcKw both refer to the same cell. Of course, it's permissible to assign more than one name to the same reference, but in general it's not good programming practice.

Sorting the list according to reference makes it easier to find duplications.

	N	O
1	CHA	=Sheet1!$C$2
2	COH	=Sheet1!$C$3
3	logK	=Sheet1!$C$5
4	pcKw	=Sheet1!$C$6
5	pH	=Sheet1!$C$9:$C$29
6	pKw	=Sheet1!$C$6
7	v	=Sheet1!$B$9:$B$29
8	V_0	=Sheet1!$C$4

Figure 6-18. Using **Paste List** to audit a worksheet.

The Label... Command (Excel 2003 Only)

You can use the column and row labels of a table to refer to a cell within the table, without assigning names. Excel will recognize the row and column labels in a table if the data table has at least one blank row and column separating it from the rest of the worksheet (more accurately, from any formulas that refer to the table).

By default, Excel does not recognize labels in formulas. To use labels in formulas, choose **Options** from the **Tools** menu, and choose the Calculation tab. Under Workbook options, check the Accept Labels In Formulas check box.

I recommend that you don't use labels, but always used defined names for cell ranges.

This feature was removed from Excel 2007.

Worksheet Functions for the "Power User"

The following sections describe worksheet functions in the Information and Lookup & Reference categories. Some of the functions in the Lookup & Reference category are among the most powerful of worksheet functions. The following sections illustrate some of the things that can be done using these functions and, in addition, show how to use Excel 4 macro functions in formulas.

Information Functions

Table 6-1 lists the functions in the information category. Most of them are IS functions. These functions are self-explanatory; for example, ISBLANK returns TRUE if the cell is empty, and FALSE otherwise.

Table 6-1. Selected Information Functions

Function	Description
CELL	Returns information about the formatting, location, or contents of a cell
ISBLANK	Returns TRUE if the cell is blank
ISERR	Returns TRUE if the value is any error value except #N/A
ISERROR	Returns TRUE if the value is any error value
ISEVEN*	Returns TRUE if the number is even
ISLOGICAL	Returns TRUE if the value is a logical value
ISNA	Returns TRUE if the value is the #N/A error value
ISNONTEXT	Returns TRUE if the value is not text
ISNUMBER	Returns TRUE if the value is a number
ISODD*	Returns TRUE if the number is odd
ISREF	Returns TRUE if the value is a reference
ISTEXT	Returns TRUE if the value is text
N	Returns a value converted to a number
NA	Returns the error value #N/A
TYPE	Returns a number indicating the data type of a value

* Only available in Excel 2003 if the Analysis ToolPak add-in is loaded.

The CELL Function

The syntax of the CELL function is CELL(**info_type**, *reference*). *Info_type* is a text value that specifies what type of information is returned. *Reference* is a reference to the cell that you want information about. Return values include the address, formatting or contents of a specified cell. See Excel's On-Line Help for more details. If *reference* is omitted, information is returned about the last cell that was changed.

If *info_type* is "filename", the function returns the full directory path to the file that contains reference, for example C:\Documents and Settings\Owner\ Desktop\[Book1.xls]Sheet1.

You can use FIND and MID to select the workbook name or sheet name. For example, the formula

=MID(CELL("filename"),FIND("]",CELL("filename"))+1,32)

returns the sheetname.

Strangely, *reference* can refer to the cell containing the formula, without generating a circular reference error message.

The N Function

The N function can be used to convert the logical values FALSE and TRUE to 0 and 1, respectively. For example, if the range $B$2:$B$67 contains dates, the formula

=$B$2:$B$67<DATE(2009,1,1)

to determine which dates fall before January 1 2009 returns the values FALSE, FALSE, TRUE, TRUE, FALSE, TRUE, FALSE, TRUE, FALSE, FALSE, ... (Only the first portion of the results are shown). To convert these logical values to 0 or 1, use the formula

=N($B$2:$B$67<DATE(2009,1,1))

Lookup Functions

Excel's Lookup & Reference category provides several functions for obtaining values from a table, based on position or value.

Table 6-2. Selected Lookup Functions

Function	Description
CHOOSE	Chooses a value from a list of values
HLOOKUP	Looks in the top row of an array for a specified value and moves down the column to return the value in a specified row
INDEX	Uses an index to choose a value from a reference or array
LOOKUP	Looks up values in a vector or array
MATCH	Looks up values in a reference or array
VLOOKUP	Looks in the first column of an array for a specified value and moves across the row to return the value in a specified column

The CHOOSE Function

Use the choose function to return a value from a list, based on an index number. The syntax of CHOOSE is CHOOSE(*index_num, value1, value2,...*). The list *value1,value2,...* can consist of up to 29 arguments, and an argument can be a range of values.

The VLOOKUP and HLOOKUP Functions

The function VLOOKUP (*lookup_value, table_array, column_index_num, range_lookup*) looks for a match in the first column of a two-dimensional array and returns a value in the relative column of *table_array* specified by *index_num,* in the row in which the match was found. The argument *lookup_value* is the value to be searched for in column 1 of the database. *Table_array* is a reference to the database; the values in the first column of *table_array* are the values to be compared with *lookup_value*. The third argument, *col_index_num*, specifies the relative column in *table_array* from which a value will be returned. The fourth (optional) argument, *range_lookup**, specifies whether you want VLOOKUP to find an exact match or an approximate match. If *range_lookup* is set to FALSE, VLOOKUP returns an exact match, or #N/A if no matching value is found. If *range_lookup* is set to TRUE, VLOOKUP returns an approximate match: the value corresponding to the table value that is equal to or less than *lookup_value*.

If the optional argument *range_lookup* is omitted, it is assumed to have the value TRUE. You can use zero as shorthand to represent FALSE, and any non-zero value to represent TRUE (I always use 0 for FALSE and 1 for TRUE).

VLOOKUP always looks in the first column of the table, and it retrieves associated information from columns to the right in the same row; you cannot use VLOOKUP to "look up" to the left. Subsequent sections will demonstrate how to perform a lookup to the left.

The function HLOOKUP (***lookup_value, table_array, row_index_num, range_lookup***) is similar to VLOOKUP, except that it "looks up" in the first row of the array and returns a value from a specified row in the same column.

The following sectiuons show how to use VLOOKUP to obtain values both from a one-way table and a two-way table.

The INDEX and MATCH Functions

The INDEX and MATCH functions are, in a sense, mirror images. Given a number value, INDEX returns the value found at that position in a range of cells; given a value to look up, MATCH returns the numerical position of that value in a range of cells.

The function INDEX (***array, row_num, column_num***, *area_num*) returns a single value from within a one- or two-dimensional range of cells, based on a specified position in the array. Non-adjacent selections are permitted; they are handled by *area_num*. See Appendix B for details.

The function MATCH (***lookup_value, array***, *match_type_num*) returns the relative position of a value in a one-dimensional array. If *match_type_num* = 1, MATCH returns the position of the largest array value that is less than or equal to *lookup_value*. The array must be in ascending order. If *match_type_num* = -1, MATCH returns the position of the smallest value that is greater than or equal to *lookup_value*. The array must be in descending order. If *match_type_num* = 0, MATCH returns the position of the first value that is equal to *lookup_value*. The array can be in any order. If no match is found, #N/A! is returned.

The MATCH and INDEX functions are also used in combination to perform interpolation; see Chapter 13.

---

* If I had named this argument, I would have called it *match_type_logical*.

Getting Values from a One-Way Table

The spreadsheet shown in Figure 6-19, containing a list of element symbols and atomic weights, is part of a one-way table (the table entries extend down to row 110). In a one-way table, the lookup values occupy a single row or column of the table.

As an example, let's imagine that we want to use the element symbol to look up a particular element in the table and obtain its atomic weight, in order to use it in a formula. The formula =VLOOKUP(symbol,$B$2:$D$110,2,FALSE) returns the atomic weight corresponding to symbol. The argument "2" in the formula specifies that once VLOOKUP finds the row containing symbol, it will return the value in column 2 (relative) of the specified database. (The formula =VLOOKUP(symbol,$B$2:$D$110,3,0) would return the electron configuration, for example.)

	A	B	C	D	E
1	Element	Symbol	At. Wt.	Elec. Config.	
2	Hydrogen	H	1.00797	1s1	1
3	Helium	He	4.0026	1s2	2
4	Lithium	Li	6.939	[He] 2s1	3
5	Beryllium	Be	9.0122	[He] 2s2	4
6	Boron	B	10.811	[He] 2s2 2p1	5
7	Carbon	C	12.01115	[He] 2s2 2p2	6
8	Nitrogen	N	14.0067	[He] 2s2 2p3	7
9	Oxygen	O	15.9994	[He] 2s2 2p4	8
10	Fluorine	F	18.9984	[He] 2s2 2p5	9
11	Neon	Ne	20.183	[He] 2s2 2p6	10
12	Sodium	Na	22.9898	[Ne] 3s1	11

Figure 6-19. Portion of a data table of the elements.

When you use VLOOKUP, you must always "look up" in the first column of the defined database and retrieve associated information from a column to the right in the same row; you cannot VLOOKUP to the left, for example. In the table shown in Figure 6-19, you cannot use VLOOKUP to return the element name corresponding to symbol. If you want to perform a lookup to the left of the lookup value, you can either construct your own lookup-type function using the MATCH and INDEX functions (see later in this chapter) or use the LOOKUP function.

Getting Values from a Two-Way Table

A two-way table has two lookup values, usually in the leftmost column and in the top row. The value to be returned from the table is the value located at the intersection of the row and column containing the two lookup values.

Once again we can use VLOOKUP to obtain the value from the table. In the preceding example, the value used for the *column_index_num* argument of VLOOKUP was the fixed value 2, because we are always going to return a value from column 2 of array. In the two-way table example shown in Figure 6-20, *column_index_num* is a variable; we must determine by means of an expression the column from which the desired value is to be returned. The MATCH expression can be used to obtain the relative column number in the table corresponding to the lookup value.

To find the value in the table for $T = 950°F$ and $P = 5000$ psia, the expression

=VLOOKUP(B44,$A$4:$P$34,MATCH(B45,$B$3:$P$3,1)+1,1)

where the T and P lookup values are entered in cells B44 and B45 respectively, returns the value 0.035 for the viscosity. The same formula, using names instead of references, is shown below:

=VLOOKUP(T,Table,MATCH(P,P_Row,1)+1,1)

	A	B	C	D	EFGHIJL	M	N	O	P
1		**Viscosity of Steam and Water, centipoise**							
2						psia			
3	T, °F	1	2	5		5000	7500	10000	12000
17	650	0.022	0.022	0.022		0.082	0.088	0.092	0.096
18	700	0.023	0.023	0.023		0.071	0.079	0.085	0.086
19	750	0.024	0.024	0.024		0.057	0.071	0.078	0.081
20	800	0.025	0.025	0.025		0.040	0.062	0.071	0.075
21	850	0.026	0.026	0.026		0.035	0.052	0.064	0.070
22	900	0.028	0.028	0.028		0.035	0.045	0.057	0.064
23	950	0.029	0.029	0.029		0.035	0.042	0.052	0.059
24	1000	0.030	0.030	0.030		0.035	0.041	0.049	0.055
25	1050	0.031	0.031	0.031		0.036	0.040	0.047	0.052
26	1100	0.032	0.032	0.032		0.037	0.040	0.045	0.050
27	1150	0.034	0.034	0.034		0.037	0.041	0.045	0.049
28	1200	0.034	0.034	0.034		0.038	0.041	0.045	0.048
29	1250	0.035	0.035	0.035		0.039	0.042	0.045	0.048
30	1300	0.037	0.037	0.037		0.040	0.043	0.045	0.048
31	1350	0.038	0.038	0.038		0.041	0.044	0.046	0.049
32	1400	0.039	0.039	0.039		0.042	0.044	0.047	0.049

Figure 6-20. Portion of a two-way data table.
Columns E – L have been partially hidden.

Creating a Custom Lookup Formula
to Obtain Values from a Table

If Excel's built-in Lookup functions do not suffice, you can perform a lookup operation by creating your own lookup formula, usually by using INDEX and MATCH.

When you use VLOOKUP, you must always "look up" in the first column of the table, and you must retrieve associated information from columns to the right in the same row; you cannot use VLOOKUP to look up to the left. If it is necessary to look to the left in a table (maybe it's not convenient or possible to rearrange the data table so as to put the columns in the proper order to use VLOOKUP), you can construct your own lookup formula using Excel's MATCH and INDEX worksheet functions. The MATCH and INDEX functions are almost mirror images of one another: MATCH looks up a value in an array and returns its numerical position, INDEX looks in an array and returns a value from a specified numerical position.

The following example illustrates how to use INDEX and MATCH to lookup to the left in a table. In the table of production figures for phosphoric acid shown in Figure 6-21, we want to find the month with the largest production.

	A	B
4	Month	Production
5	Jan	76212
6	Feb	15379
7	Mar	62220
8	Apr	83119
9	May	33872
10	Jun	80881
11	Jul	54263
12	Aug	35427
13	Sep	50361
14	Oct	71600
15	Nov	133
16	Dec	22477

Figure 6-21. A table requiring "lookup" to the left.

First, use Excel's MAX worksheet function to find the maximum value in the range of production figures. The expression

=MAX($B$5:$B$16)

returns the value 83119. Now we want to return the month value in the column to the left in the same row. We do this in two steps, as follows. First, use the MATCH function to find the position of the maximum value in the range.

The syntax of MATCH is similar to that of VLOOKUP: MATCH(**lookup_value,lookup_array**,match_type_num). If match_type_num = 0, MATCH returns the position of the first value that is equal to lookup_value. Thus the expression

=MATCH(83119,$B$5:$B$16,0)

returns 4 (the maximum value is the fourth value in the range). Finally, use the INDEX function to return the value in the same position in the array of months:

=INDEX($A$5:$A$16,4)

The specific values 83119 and 4 can now be replaced by the formulas that produced them, to yield the following "megaformula".

=INDEX($A$5:$A$16,MATCH(MAX($B$5:$B$16),$B$5:$B$16,0))

Wildcard Characters with VLOOKUP, MATCH and Some Other Functions

You can use "*" and "?" as wildcard characters in lookup_value. The question mark matches a single character, and the asterisk matches any number of characters. In order to use a wildcard character, the lookup value must be text, and the argument that controls the match type, e.g., range_lookup for the VLOOKUP function, must be set to FALSE. For example, if you are using VLOOKUP to look up in a table in which the leftmost column contains Last Name values, and the lookup value "Liv" is entered in cell A2, the formula

=VLOOKUP(A2 & "*",Sheet2!$A$2:$E$493,2,FALSE)

will return the value in column 2 of the first row in which a last name value begins with "Liv", for example, "Livingstone". The formula can only return the first instance of a match, of course. If you want to return all values in the table that match the lookup value, you'll have to create an array formula. See Chapter 7 for ideas along that line.

You can use wildcard characters in the following functions: HLOOKUP, VLOOKUP, MATCH, SEARCH, COUNTIF and SUMIF.

The LOOKUP Function

The function LOOKUP (**lookup_value, lookup_vector**, result_vector) has two syntax forms: vector and array. The vector form of LOOKUP looks in a one-row or one-column range (known as a vector) for a value and returns a value from the same position in another one-row or one-column range. The values in *lookup_vector* must be sorted in ascending order. If LOOKUP can't find *lookup_value*, it returns the largest value in *lookup_vector* that is less than or equal to *lookup_value*.

The array form of LOOKUP is similar to VLOOKUP or HLOOKUP in that it automatically looks in the first column or row of an array, but is limited to returning a value from the same position in the last column or row. Go to Excel's On-line Help for more details.

Reference Functions

Reference functions, among other things, allow you to create formulas that return a reference to a range of cells rather than to a single cell. The two functions that accomplish this are the OFFSET function and the INDIRECT function.

Table 6-3. Selected Reference Functions

Function	Description
ADDRESS	Returns a reference as text to a single cell in a worksheet
AREAS	Returns the number of areas in a reference
COLUMN	Returns the column number of a reference
COLUMNS	Returns the number of columns in a reference
INDIRECT	Returns a reference indicated by a text value
OFFSET	Returns a reference offset from a given reference
ROW	Returns the row number of a reference
ROWS	Returns the number of rows in a reference

The OFFSET Function

The OFFSET (*reference*, *rows*, *columns*, *height*, *width*) function returns a reference offset from a given reference in a one- or two-dimensional range of cells. Although OFFSET is similar to INDEX, the INDEX function returns only a single value from a one- or two-dimensional range of cells, while OFFSET can return a reference to a range of cells.

The *reference* argument can be a reference to a single cell or a range. If *reference* is a range of cells and the optional arguments *height* and *width* are omitted, then OFFSET returns a reference of the same dimensions as *reference*. To select a single cell, use the formula

=OFFSET(reference, rows, columns, 1, 1).

For a further example of using OFFSET, see "A Drop-Down List Box on a Worksheet" in Chapter 11.

The INDIRECT Function

The INDIRECT worksheet function returns a reference specified by a text string or an expression that evaluates to a text string. The cell containing the returned reference does not display the reference, but the result of the reference. INDIRECT is one of Excel's most versatile and powerful functions.

The syntax of INDIRECT is INDIRECT(***ref_text***,*a1*). *Ref_text* is a text string or a cell reference, name or expression that evaluates to a text string. For example, if cell A1 contains the value 12, the expression

=INDIRECT("B"&A1)

returns the contents of cell B12.

The optional argument *a1* specifies the reference type of *ref_text*. If *a1* is TRUE or omitted, *ref_text* is interpreted as an A1-style reference; if *a1* is FALSE, *ref_text* is interpreted as an R1C1-style reference. For example, if a range of cells 10 columns wide (columns B through K) × N rows deep is to be evaluated, and the top row of the range is always row 2, then the expression

=INDIRECT("B2:K" & N+1)

where N is a reference to a cell containing the value 100, creates a reference to the range B2:K101.

R1C1-style references are useful in the INDIRECT function in cases where the column of the reference is evaluated from an expression – it's difficult, although not impossible, to develop a formula in which a number value is converted into the letter designation of a column. For example, if the number of columns in the range is a variable W, then the expression

=INDIRECT("R2C2:R"&N+1&"C"&W+1,FALSE)

is a convenient way to create a reference to the range R2C2:R101C11.

Note that when you insert or delete rows or columns in a spreadsheet, all cell formulas that are affected are updated. However, formulas using the INDIRECT worksheet function are not updated, since the references are text.

The ADDRESS Function

The ADDRESS function is a convenient way to construct a reference for use in the INDIRECT function. The ADDRESS function returns a cell address as text, given specified row and column numbers. The syntax of the function is ADDRESS(***row_num,column_num***,*abs_num,a1,sheet_text*).

The optional argument *abs_num* specifies the type of reference to return. If *abs_num* = 1 or omitted, the function returns an absolute reference; if *abs_num* = 2, it returns a reference with absolute row and relative column; if *abs_num* =

3, it returns a reference with relative row and absolute column; if *abs_num* = 4, it returns a relative reference.

The optional argument *a1* specifies the reference type of *ref_text*. If *a1* is TRUE or omitted, ADDRESS returns an A1-style reference; if *a1* is FALSE, ADDRESS returns an R1C1-style reference. The optional argument *sheet_text* specifies the name of the worksheet to be used as the external reference. If *sheet_text* is omitted, no sheet name is used.

Unlike the INDIRECT function, which creates an address which is immediately evaluated, the ADDRESS function merely returns a text string. But you can use an ADDRESS expression as the argument of INDIRECT.

An Example Using ADDRESS. If a formula refers to a range of cells in a column, Excel uses the value in the column that is in the same row as the formula. This is referred to as "implicit intersection". (See "A Reference Using Implicit Intersection" earlier in this chapter.) Occasionally an implicit intersection will fail in a complicated formula. You can incorporate an implicit intersection expression in your formula by using ADDRESS. The following expression returns the value in the appropriate cell of the range x_values:

    =INDIRECT(ADDRESS(ROW(),COLUMN(x_values)))

Solving a Problem by Using Intentional Circular References

When a formula refers to itself, either directly or indirectly, it creates a circular reference. If a circular reference occurs, Excel issues a "Cannot resolve circular references" message and displays a zero value in the cell.

Usually circular references occur unintentionally, because the user incorrectly entered a cell reference in a formula. But occasionally a problem can be solved by intentionally creating a circular reference.

To illustrate the use of an intentional circular reference, let's perform a simple simulation: modeling heat flow in a metal plate, one edge of which is held at 100°, the other three at 0°. We subdivide the plate into a 20 × 20 matrix of cells and use the assumption that the temperature of any cell is the average of the temperatures of the four adjacent cells. Thus the formula in cell B8 is

    =(B7+A8+B9+C8)/4

When you enter the formula and press Enter, the "Cannot resolve circular references" message is displayed, and Excel displays a zero in the cell. The formula in cell B8 is filled down and filled right into the 20 × 20 range. There are a large number of circular references in this range; zero is displayed in all 400 cells, as shown in Figure 6-22.

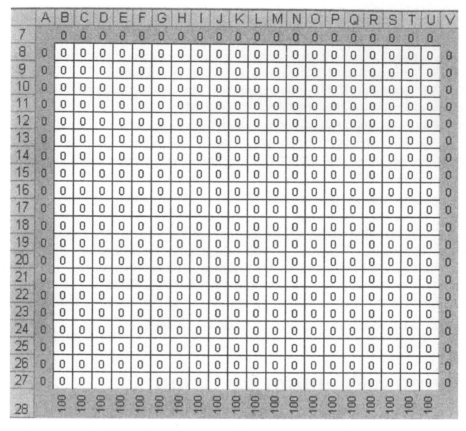

Figure 6-22. Heat flow simulation by using circular reference,
before enabling iteration.
(The cells in gray indicate the temperature of the edges of the plate)

To force Excel to evaluate the circular reference, using the results of the previous calculation cycle as start values for the next cycle, click on the Office Button → Excel Options → Formulas and look in the Calculation Options area (Excel 2007/2010), or choose **Options...**from the **Tools** menu and choose the Calculation tab (Excel 2003). Check the Iteration box. The default settings are Maximum Iterations = 100 and Maximum Change = .001; these are suitable for this example but usually a Maximum Change parameter must be entered that is suitable for the calculation at hand. When you press the OK button the circular references will be evaluated. The results of the calculations are shown in Figure 6-23.

	A	B	C	D	E	F	G	H	I	J	K	L	M	N	O	P	Q	R	S	T	U	V
7	0	0	0	0	0	0	0	0	0	0	0	0	0	0	0	0	0	0	0	0	0	
8	0	0	0	1	1	1	1	1	2	2	2	2	2	2	1	1	1	1	1	0	0	0
9	0	1	1	1	2	2	3	3	3	3	3	3	3	3	3	3	2	2	1	1	1	0
10	0	1	2	2	3	3	4	4	5	5	5	5	5	5	4	4	3	3	2	2	1	0
11	0	1	2	3	4	5	5	6	7	7	7	7	7	7	6	5	5	4	3	2	1	0
12	0	1	3	4	5	6	7	8	8	9	9	9	9	8	8	7	6	5	4	3	1	0
13	0	2	3	5	6	8	9	10	10	11	11	11	11	10	10	9	8	6	5	3	2	0
14	0	2	4	6	8	9	11	12	13	13	14	14	13	13	12	11	9	8	6	4	2	0
15	0	3	5	7	9	11	13	14	15	16	16	16	16	15	14	13	11	9	7	5	3	0
16	0	3	6	9	11	14	16	17	18	19	20	20	19	18	17	16	14	11	9	6	3	0
17	0	4	7	10	13	16	18	20	22	23	23	23	23	22	20	18	16	13	10	7	4	0
18	0	4	9	12	16	19	22	24	25	26	27	27	26	25	24	22	19	16	12	9	4	0
19	0	5	10	15	19	23	26	28	30	31	31	31	31	30	28	26	23	19	15	10	5	0
20	0	6	12	18	22	27	30	33	35	36	37	37	36	35	33	30	27	22	18	12	6	0
21	0	8	15	21	27	31	35	38	40	42	42	42	42	40	38	35	31	27	21	15	8	0
22	0	9	18	25	32	37	41	44	46	48	49	49	48	46	44	41	37	32	25	18	9	0
23	0	12	22	31	38	44	48	51	54	55	56	56	55	54	51	48	44	38	31	22	12	0
24	0	15	28	38	46	52	56	59	62	63	64	64	63	62	59	56	52	46	38	28	15	0
25	0	20	36	48	56	62	66	68	70	71	72	72	71	70	68	66	62	56	48	36	20	0
26	0	30	49	61	68	73	76	78	80	81	81	81	81	80	78	76	73	68	61	49	30	0
27	0	50	69	78	83	86	88	89	90	90	90	90	90	90	89	88	86	83	78	69	50	0
28		100	100	100	100	100	100	100	100	100	100	100	100	100	100	100	100	100	100	100	100	

Figure 6-23. Heat flow simulation by using circular reference, after enabling iteration.

This simulation merely illustrates the use of intentional circular reference calculations. Other applications could involve any calculation in which the input parameters depend on the final result: for example, the final pressure and temperature of a sample of gas after the injection of a quantity of heat energy (the heat capacity of the gas varies with temperature), or the final pH of a buffer solution (the pK_a of the buffer acid depends on the ionic strength of the solution, and the ionic strength is the result of buffer ionization).

Using Excel 4 Macro Functions in Worksheet Formulas

Excel 4 macro functions can be used in Excel formulas. Since there are some Excel 4 macro functions that provide information not available using Excel's worksheet functions, it's useful to know how to use them. Only those

Examining Formulas

When you see an error value displayed in a cell, you'll need to examine the formula. There are several things you can do to track down the error. The error value displayed can suggest a good place to begin. For example, if the error value is #NAME?, you most likely have misspelled a variable name or function, or entered a variable name that has not yet been defined.

To track down the source of an error in a lengthy, complicated formula, you can examine the value of individual references, names, functions or function arguments in one of two ways:

- Either in the formula bar or in the cell, highlight the part of the formula that you want to evaluate and press F9. The value of the selected portion of the formula will be displayed. Click on the Cancel box in the formula bar or press Esc or Undo to restore the statement; otherwise the selected portion of the formula will be permanently replaced by the numerical value.

- Select the cell you want to evaluate. Click on Evaluate Formula in the Formula group in the Formulas tab of the Ribbon (Excel 2007/2010) or choose **Formula Auditing** from the **Tools** menu, and then **Evaluate Formula** from the submenu (Excel 2003). Press the Evaluate button to evaluate the underlined part of the formula; the result will be shown in italics. Each part of a formula can be evaluated in turn. If the underlined part of the formula is a reference to another formula, you can press the Step In button to display that formula in the Evaluation box. Press Step Out to go back to the previous formula.

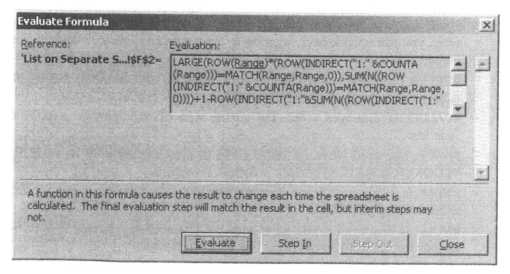

Figure 6-24. Using **Evaluate Formula...** to debug a formula.

Finding Dependent and Precedent Cells

To audit the logic of a complicated worksheet, you may want to find all cells that contain formulas that refer to a given cell (*dependent cells*) or all cells that are referred to by the formula in a given cell (*precedents*). You can search backward or forward, finding the direct precedents, then the cells that are referred to by those cells, and so on.

The Formula Auditing group in the Formula ribbon (Excel 2007/2010) or the **Formula Auditing** submenu in the **Tools** menu (Excel 2003) allows you to trace precedents, dependents or errors. Figure 6-25 illustrates a typical display when a cell is selected and **Trace Precedents** is chosen.

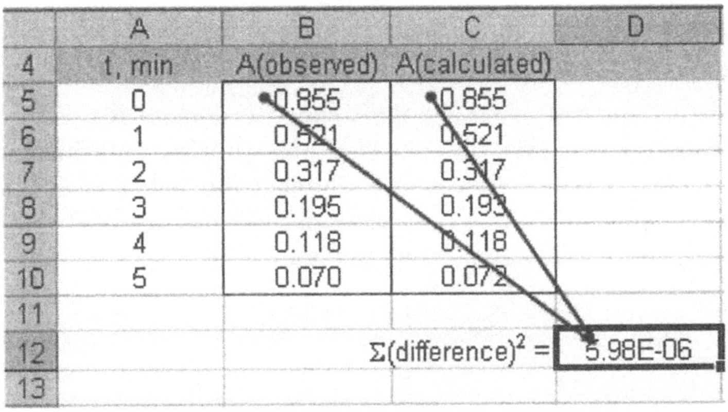

Figure 6-25. Trace Precedents display.

In Excel 2003, use the Remove All Arrows to remove all precedent or dependent arrows that you have displayed; in Excel 2007/2010 you have the option of removing all arrows, or only precedent or dependent arrows.

Repairing or Removing Links in Documents

As we saw in "Creating and Using External References" in Chapter 3, you can use an external reference in a formula to refer to a range in another worksheet or workbook. If the reference is to another workbook, it is sometimes called a link. The link is that part of an external reference, up to the exclamation mark:

```
='C:\Documents and Settings\Owner\Desktop\[Book2.xls]Sheet1'!$A$1
```

Links between documents can be broken, or a document can acquire a link when the document was not intended to have external references. The following sections describe techniques for finding, repairing or deleting links.

Re-establishing Links

Links can easily be broken if you're not careful. But first the good news: If both the dependent document and the source document are open, you can't go wrong. If you change the source reference by inserting or deleting rows or columns, or by moving the source reference, or change the sheet name, the external reference updates automatically. If you delete the source reference or the sheet containing the source reference, you'll get the #REF! error message, warning you immediately that you've made an error. If you try to move the source document to a different folder, you'll get the "Error moving file or folder" error message.

But if you've renamed or moved the source document (this requires that the source workbook is closed), or renamed the source worksheet, while the dependent document is closed, you have broken the link between source and dependent documents. The next time you open the dependent document, you'll get an error message like the one shown in Figure 6-26.

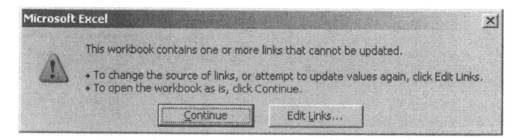

Figure 6-26. When a dependent document containing broken links is opened.

If you get a "This workbook contains links that cannot be updated" error message, there are tools to help you re-establish the links. In the message box shown in Figure 6-26, press the Edit Links... button to display the Edit Links dialog box (Figure 6-27).

Alternatively, you can display the Edit Links dialog box by choosing **Links...** from the **Edit** menu. In this case, you must press the Check Status... button to display the error message.

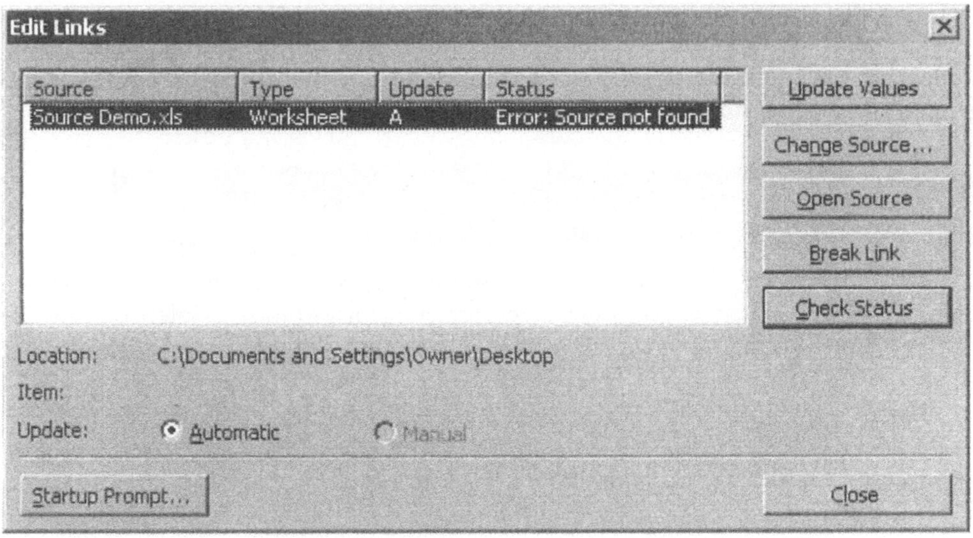

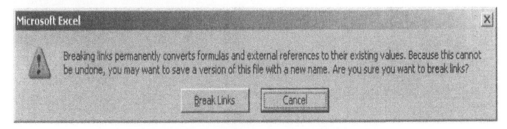

Figure 6-27. The Edit Links dialog box.

The two most likely error messages are "Error: Source not found" (the source workbook was deleted, moved or re-named) and "Error: Worksheet not found" (the source worksheet was deleted or renamed). The message "Warning: Values not updated" merely indicates that the source document is closed but the link is still correct.

If you changed the reference in a linked formula while the source document was closed, the formula will point to a different cell or range in the source sheet, and your formula will give an incorrect result. The best that you can hope for is that an error will be generated.

You can open the source workbook and re-establish the link, or press the Change Source button, navigate to the correct folder, and select it. This will change the Status to "OK".

If you choose instead to press the Break Link button, you will get the warning shown in Figure 6-28.

Figure 6-28. The Break Links warning message.

The Startup Prompt... button (bottom left corner in Figure 6-27) enables you to specifiy whether users will be alerted to the fact that the document contains links, and whether the links are updated automatically (Figure 6-29).

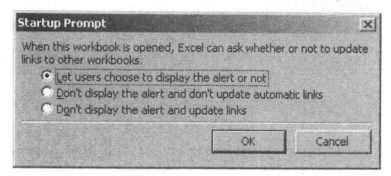

Figure 6-29. The Startup Prompt dialog box.

Finding All Links in a Workbook

There is no built-in way to find links in formulas. The best way is to click on Find & Select in the Editing group in the Home tab (Excel 2007/2010) or use **Find...** in the **Edit** menu (Excel 2003) to look for "[" or "]". Follow the steps in the box below.

To Find Links in a Workbook

1. Close all workbooks except the one you want to search.
2. Click on Find & Select in the Editing group in the Home tab (Excel 2007/2010) or click on the **Edit** menu (Excel 2003) and choose **Find...**
3. In the Find And Replace dialog box, press the Options button.
4. Enter "[" in the Find what box,
5. Select "Workbook" in the Find what box, "Formulas" in the Look In box, and press the Find All button
6. Click Find All.
7. Results will be displayed in the box at the bottom.

The above procedure does not find links in names. By far the best way to look for names with links is to use Paste Name.

7

Array Formulas

Array formulas are undoubtedly Excel's most powerful formulas. With array formulas, you can accomplish things in Excel that you can't accomplish otherwise. This chapter illustrates the point by means of some case studies.

In Excel, the terms *range* and *array* are essentially identical. They simply refer to a range of cells in a worksheet. They can be either one- or two-dimensional.

Introduction to Array Formulas

Array formulas can simplify worksheets, as illustrated in the following four examples, each of which accomplishes the same thing, namely, to calculate the sum-of-squares of residuals, where absorbance values from a first-order rate process were fitted to the equation $A_{calc} = A_0 e^{-kt}$.

In the first example, to calculate the sum of squares of the residuals, $\Sigma(A_{obsd} - A_{calc})^2$, you could use the (non-array) approach shown in the worksheet of Figure 7-1. The formula in cell D9 is

    =SUM(D2:D7)

The sum-of-squares calculation can also be done by using an array formula, shown in the second example (Figure 7-2). The formula in cell D9 is an unusual one, {=SUM((C25:C7-B2:B7)^2)}, in which a range is subtracted from a range. Normally a formula such as this would produce a #VALUE! error, but if it is entered as an array formula, by pressing Ctrl+Shift+Enter, it will be evaluated correctly. Excel indicates that the formula is an array formula by enclosing it in braces. Don't type the braces as part of the formula; they are added automatically by Excel.

	A	B	C	D
1	t, min	A(observed)	A(calculated)	(difference)^2
2	0	0.855	0.855	9.00E-10
3	1	0.521	0.521	7.40E-11
4	2	0.317	0.317	2.25E-07
5	3	0.195	0.193	2.40E-06
6	4	0.118	0.118	1.47E-08
7	5	0.070	0.072	3.34E-06
8				
9			Σ(difference)^2 =	5.98E-06

Figure 7-1. Calculating a sum of squares.

	A	B	C	D
1	t, min	A(observed)	A(calculated)	
2	0	0.855	0.855	
3	1	0.521	0.521	
4	2	0.317	0.317	
5	3	0.195	0.193	
6	4	0.118	0.118	
7	5	0.070	0.072	
8				
9			Σ(difference)^2 =	5.98E-06

Figure 7-2. Using an array formula to calculate a sum of squares.

	A	B	C	D
1	t, min	A(observed)	A(calculated)	
2	0	0.855	0.855	
3	1	0.521	0.521	
4	2	0.317	0.317	
5	3	0.195	0.193	
6	4	0.118	0.118	
7	5	0.070	0.072	
8				
9			Σ(difference)^2 =	5.98E-06

Figure 7-3. Using named arrays in an array formula
to calculate a sum of squares.

	A	B	C	D
1	t, min	A(observed)		
2	0	0.855		
3	1	0.521		
4	2	0.317		
5	3	0.195		
6	4	0.118		
7	5	0.070		
8				
9			$\Sigma(\text{difference})^2 =$	5.98E-06

Figure 7-4. Using an array formula to calculate a sum of squares.
Intermediate formulas have been eliminated.

The third example shows how using names for the A_{obsd} and A_{calc} ranges simplifies the array formula and makes it much more self-documenting. In Figure 7-3, the formula in cell D9 is

    {=SUM((A_calc-A_obsd)^2)}

The cell ranges B2:B7 and C2:C7 were previously defined as A_obsd and A_calc, respectively.

In the final example, shown in Figure 7-4, the column of A_{calc} values has been eliminated and A_calc is obtained from the formula =A_0*EXP(-k*t). All the intermediate calculations have been compressed into a single formula in cell D9, leaving only the raw data.

Array Constants

In the same way that a worksheet formula can contain a simple constant, e.g., the value 3 in the formula =3*A1+A2, an array formula can contain an *array constant*. An array constant is included in a formula by enclosing the array of values in braces. In this case you *must* type the braces. When you enter the completed array formula by pressing Ctrl+Shift+Enter, Excel automatically provides braces around the whole formula.

Within an array, values in the same row are separated by commas, and rows of values are separated by semicolons. For example, the array constant {2, 3, 4; 3, 2, −1; 4, 3, 7} represents a 3 × 3 array.

Array constants can contain number or text values, but they cannot contain references. Individual text values in an array constant must be enclosed in quotes.

An array constant is usually incorporated in a worksheet formula, but it can also be entered as a named variable in the Define Name dialog box, as shown in Figure 7-5.

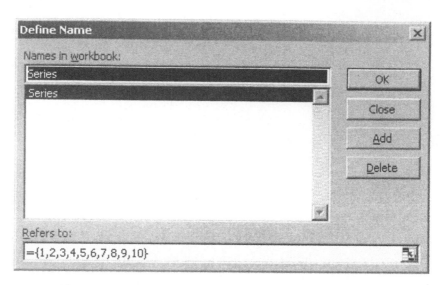

Figure 7-5. An array constant entered as a named variable.

Formulas and Functions that Return an Array Result

You can also create array formulas that *return* an array of different values in a selected range of cells. Formulas that operate on matrices, described in Chapter 13, are examples of formulas that return an array.

To use a worksheet formula that returns an array result, you must first select a suitable range of cells, with dimensions (R × C) large enough to accommodate the returned array, then type the formula in the formula bar, and finally enter the formula by pressing Ctrl+Shift+Enter. Excel will indicate that the formula is an array formula by enclosing it in braces and will enter the array formula in all the selected cells.

A few of Excel's worksheet functions are array functions; that is, they return arrays as results. One of these, the LINEST statistical function, is described in Chapter 14.

Editing or Deleting Arrays

To edit a multi-cell array formula, simply select any cell in the array. Then edit the formula in the formula bar. When you begin to edit, the braces surrounding the formula will disappear. To re-enter the edited formula, press Ctrl+Shift+Enter. The formula will be entered into all of the cells originally selected for the array.

Since a single formula is entered in all the cells of an array, you can't change or delete part of an array; if you try, you'll get a "Cannot change part of an array" message. You must delete the complete array.

Although you can't change part of an array's formulas or values, you can format individual cells. You can also **Copy** values from individual cells of an array and **Paste** them elsewhere.

You can select individual values in an array by using the INDEX function, but this is a "read-only" option. If you need to be able to change individual values in an array, you should enter the values in a range of worksheet cells, rather than as an array.

Excel Tip. *To select an entire array range (in order, for example, to **Clear** it), select any cell in the array, and then press Ctrl+/.*

Creating a "Three-Dimensional" Array on a Single Worksheet

It's possible to create an array with three dimensions by entering an array formula in each cell of a rectangular range of cells. The following example illustrates the use of a three-dimensional array to calculate an "error surface" curve such as the one shown in Figure 4-14, Surface Charts. The error-square sum, i.e., the sum of the squares of the residuals, $\Sigma(y_{obsd} - y_{calc})^2$ for a one-dimensional array of data points, was calculated for each cell of a two-dimensional array of trial values. The "best" values of the independent variables are those which produce the minimum error-square sum.

The spreadsheet shown in Figure 7-6 contains kinetic data obtained for the reaction sequence A → B → C. The concentration of the intermediate species B was measured at time intervals from 0 seconds to 100 seconds (only a portion of the data table in rows 4 and 5 is shown); the t and $[B]_{obsd}$ values were defined as named ranges t and B_obs. $[B]_{calc}$ values were obtained from the equation

$$[B] = [A]_0 \frac{k_1}{k_2 - k_1} \{e^{-k_1 t} - e^{-k_2 t}\}$$

which can be found in any standard text on chemical kinetics. The error-square sum was calculated for a range of trial values of k_1 and k_2 using an array formula.

The following array formula is entered in cell C10

{=SUM((B_obs-(C$9/($B10-C$9)*(EXP(-C$9*t)-EXP(-$B10*t))))^2)}

(note the use of mixed addressing), and then filled into the range C10:K22 by using AutoFill.

A 3-D chart of the values is shown in Figure 5-16 of Chapter 4, Excel 2003 Charts.

	A	B	C	D	E	F	G	H	I	J	K	L
1				Consecutive First Order Reactions (A->B->C)								
2				$[B]_{calc} = (k_1/(k_2-k_1))*(EXP(-k_1*t)-EXP(-k_2*t))$								
3		Experimental values										
4		t, sec	0	2	4	6	8	10	12	14	16	18
5		$[B]_{obsd}$	0.002	0.169	0.292	0.387	0.445	0.476	0.498	0.499	0.493	0.479
6												
7					Error sum = $\Sigma([B]_{calc}-[B]_{obsd})^2$							
8						k_1 trial values						
9			0.060	0.070	0.080	0.090	0.100	0.110	0.120	0.130	0.140	
10		0.020	0.833	0.829	0.839	0.860	0.889	0.922	0.958	0.995	1.034	
11		0.025	0.505	0.481	0.475	0.483	0.501	0.525	0.553	0.584	0.617	
12		0.030	0.311	0.273	0.257	0.256	0.266	0.284	0.307	0.333	0.361	
13		0.035	0.201	0.154	0.130	0.122	0.127	0.140	0.159	0.181	0.206	
14		0.040	0.144	0.090	0.060	0.048	0.049	0.058	0.074	0.093	0.116	
15		0.045	0.121	0.063	0.029	0.014	0.011	0.017	0.030	0.048	0.069	
16		0.050	0.122	0.060	0.024	0.005	0.000	0.004	0.015	0.031	0.050	
17		0.055	0.138	0.074	0.035	0.015	0.008	0.010	0.020	0.034	0.052	
18		0.060	0.164	0.099	0.058	0.037	0.028	0.029	0.037	0.050	0.067	
19		0.065	0.198	0.131	0.090	0.067	0.057	0.057	0.064	0.076	0.092	
20		0.070	0.236	0.169	0.126	0.102	0.092	0.091	0.097	0.108	0.123	
21		0.075	0.277	0.209	0.166	0.142	0.130	0.129	0.134	0.144	0.159	
22		0.080	0.319	0.252	0.209	0.184	0.172	0.169	0.174	0.184	0.197	
23												
24			At minimum:			$k_1 =$	0.10		$k_2 =$	0.05		

(left side, rows 14–18, rotated label: k_2 trial values)

Figure 7-6. Spreadsheet implementation of a three-dimensional array.

An Array Formula Example: Creating a Specialized SUMIF Formula

To illustrate the power of array formulas, let's create a formula that is similar to the SUMIF function, with one important difference: The formula sums values in the range SumRange, but uses as the criterion not whether a cell in the range TextRange *contains* a specified text value Criterion, but rather if the text in the cell *includes* the specified text.

We'll develop the formula step by step. First enter the formula

=FIND(Criterion,TextRange)

Criterion is the cell that contains the text we want to look for. TextRange is the range of cells containing the text values to look at. When you press Enter, only a single value will be displayed in the cell, but that's OK; to see the complete array of values, highlight the formula in the formula bar and press the F9 key. The array will be displayed (only the first portion of the array is shown):

{#VALUE!;#VALUE!;#VALUE!;1;#VALUE!;5;12;#VALUE!;13;1;...

The FIND function returns the numerical position of the Criterion text within each text value in TextRange; if Criterion does not occur within a string in the array, the error value #VALUE! is returned. (In our example, many of the strings in TextRange don't contain the substring Criterion.) We use the ISERROR function to convert the error values to TRUE and the number values to FALSE, and the NOT function to convert TRUE to FALSE and *vice versa*.

=NOT(ISERROR(FIND(Criterion,TextRange)))

Once again, highlight the formula and use F9 to display the result in the formula bar (only the first portion of the array is shown):

{FALSE;FALSE;FALSE;TRUE;FALSE;TRUE;TRUE;FALSE;TRUE;TRUE;...

Now we multiply the preceding expression by the array SumRange.

=NOT(ISERROR(FIND(Criterion,TextRange)))*SumRange

which produces the array of values to be summed:

{0;0;0;31.68;0;31.68;33.38;0;33.38;35.93;...

Finally, we use the SUM function to sum the values

{=SUM(NOT(ISERROR(FIND(Criterion,TextRange)))*SumRange)}

Since the formula is an array formula, you must use Ctrl+Shift+Enter to enter it.

Evaluating Polynomials or Power Series Using Array Formulas

You can evaluate series functions, for example

$$\ln(1 + x) = x - x^2/2 + x^3/3 - x^4/4 + \ldots$$

or

$$N! = 1 \times 2 \times 3 \times 4 \ldots \times N$$

in a single cell, instead of calculating individual terms in separate cells, by using an array formula. (Of course, Excel already provides the LN and FACT worksheet functions; the above are only used as examples.)

Let's begin with a simple example, the evaluation of $N!$, and evaluate it for N = 10. You could use the PRODUCT worksheet function as shown in the following example:

=PRODUCT(1,2,3,4,5,6,7,8,9,10)

Entering an explicit list of integers, as in the preceding example, is not very convenient, nor does it provide the generality needed for more complicated examples. Let's see how we can create a more general formula.

Using the ROW Function in Array Formulas

To generate an array of integers for use in array formulas, use the ROW worksheet function. When used in an array formula, the expression ROW(1:10) evaluates to the array of numbers {1;2;3;4;5;6;7;8;9;10}. Thus the formula to evaluate $N!$ could have been entered as follows:

={PRODUCT(ROW(1:10))}

In order to get the correct result from the preceding formula, you must remember to press Ctrl+Shift+Enter.

The problem with this formula is that if you insert a row above the row containing the formula, the formula becomes =PRODUCT(ROW(2:11)). You can prevent this by using the INDIRECT worksheet function, described next.

Using the INDIRECT Function in Array Formulas

The INDIRECT function creates a reference specified by a text string. In the example just referred to, the formula

={PRODUCT(ROW(INDIRECT("1:10")))}

will not change if rows are inserted.

Often, you'll want to use a variable number of terms in your array formula. You can do this by using INDIRECT. In the following example we want to evaluate $N!$, where cell A1 contains the value for N. In this example let's assume that cell A1 contains 20. We use the formula

={PRODUCT(ROW(INDIRECT("1: "&A1)))}

The text argument of the indirect worksheet function evaluates to "1:20", and the expression ROW(INDIRECT("1:"&A1)) evaluates to

{1;2;3;4;5;6;7;8;9;10;11;12;13;14;15;16;17;18;19;20}

Using Array Formulas to Work with Lists

As well as the use of array formulas themselves, you'll see that the following are invaluable when developing array formulas to operate on lists:

- use names in formulas
- logical operators

- the INDIRECT worksheet function

- the F9 key to display the value returned by a formula: to display the array produced by an expression, highlight the expression in the formula bar and press F9.

In the examples that follow, we will build up the array formulas step by step, to show how they work.

Using Multiple Criteria to Count Entries in a List

Excel's COUNTIF and SUMIF worksheet functions can return a value based only on a single criterion. For example, consider a database, the first portion of which is shown in Figure 7-7. You could use COUNTIF to return the number of students in the Chem 1 class whose major is biology, or the number whose year of graduation (YOG) is 96, but you can't use COUNTIF to find out how many students are biology majors *and* are in the class of 96. You can construct an array formula that will do this.

	A	B	C	D
1	Name	YOG	Major	Grade
2	ALBERTY, Christine	96	THEATER ARTS	77.9
3	ANSARTO, Newman	97	BIOLOGY	78.3
4	BARBIERI, Joseph	94	ENV GEO SCIEN	62.2
5	BASALTA, Patrick	97	BIOLOGY	68.4
6	BLAKELY, Patrick	97	BIOLOGY	73.2
7	BLOOMSBURY, Peter	97	UNCLASSIFIED	67.7
8	BOGACZU, RobertO	97	BIOCHEMISTRY	90.8
9	BONOAN, Elwen	97	BIOLOGY	25.7
10	BOWEN, Armand	97	BIOLOGY	28.6
11	BOYLE, Seamus	97	BIOLOGY	80.3

Figure 7-7. Using an array formula to count using multiple criteria.

A good way to construct a complicated formula such as this one is to proceed step by step, making sure that each part of the formula works properly before proceeding to the next step. The following example illustrates this procedure.

First, use Create Names to assign the names Name, YOG, Major and Grade to the values in columns A, B, C and D, respectively.

We'll begin the development of our formula by creating a formula to count the number of students whose year of graduation is 1996. Enter the formula =YOG in any worksheet cell. Only one value can be displayed in the cell, but if you highlight YOG in the formula bar and press the F9 key, the array of values will be displayed (only the first part of the array is shown):

={96;97;94;97;97;97;97;97;97;97;97;97;96;97;97;96;97;97;97;97;96...

Don't forget to press the Cancel button in the formula bar or use Undo before continuing, to revert back to the original formula.

Now, change the formula to the logical expression =(YOG=96). The parentheses are optional and are only used here for clarity. Once again, highlight the formula and use function key F9 to display the result in the formula bar (only the first portion of the array is shown):

{TRUE;FALSE;FALSE;FALSE;FALSE;FALSE;FALSE;FALSE;FALSE;...

Earlier we learned that FALSE can be represented by zero and TRUE by any non-zero value. If a logical expression is included in an arithmetic operation, FALSE becomes zero and TRUE becomes 1. Thus we can convert the array of TRUE and FALSE values to 1's and 0's by the expression =(YOG=96)*1. Again, use function key F9 to display the array of values:

{1;0;0;0;0;0;0;0;0;0;0;0;1;0;0;1;0;0;0;0;1;0;0;0;0;0;0;0;0;0;0;0;1;0;1;1;0;1;1;1;0;0;0;0;
0;0;1;0;1;1;0;0;1;0;0;0;0;0;0;0;0;0;0;0;0;0;0;1;0;0;0;0;0;0;0;0;0;0;0;1;0;0;1;0;1;0;1;0;
1;0;1;1;0;0;0;0;0;0;0;1;0;0;0;0;1;0;0;0;1;0;0;1;1;0;0;0;0;0;0;0;0;0;0;1;1;0;1;0;0;0;0;0;
0;1;0;1;1;1;1;0;0;0;0;0;0;0;0;1;0;1;0}

Instead of multiplying the logical expression by 1, you can use the N worksheet function, i.e., use the expression =N(YOG=96).

To find the number of students whose YOG = 96, we simply have to sum this array of 1's and 0's. The formula {=SUM((YOG=96)*1)} is an array formula and you must remember to press Ctrl+Shift+Enter.

By now you have probably figured out how to create a formula that handles multiple criteria. Our goal was to find how many students were biology majors in the class of 96. The array formula is

{=SUM((YOG=96)*(Major="Biology"))}

Notice that multiplying the two logical expressions has the effect of converting the result to a number, making it unnecessary to multiply by 1.

You can make the formula more general by naming two cells YOG_Criterion and Major_Criterion and changing the preceding formula to

{=SUM((YOG=YOG_Criterion)*(Major=Major_Criterion))}

Counting Common Entries in Two Lists

A similar approach can be used to find the number of entries in one list that appear in another list. In the following example, ClassList1 is the class roster of students who took Chemistry for Non-Science Majors Part I; ClassList2 is the class roster for Part II. Part I is not a prerequisite for Part II, and many students leave after taking only Part I, while other students take only Part II. How many students took both Part I and Part II?

A portion of ClassList1 is shown in Figure 7-8. Each student has a unique ID number. Thus to find the number of students in ClassList1 that appear in ClassList2, we can simply find how many ID numbers in ClassList1 (with N rows) are also found in ClassList2 (with M rows).

Once again, we begin by using the Create Names dialog box to assign the name ID1 to the range of values in column A of ClassList1 and assign the name ID2 to the range of values in column A of ClassList2.

To find how many ID values in ClassList1 are identical to ID values in ClassList2, we create a matrix of logical values of size $N \times M$, by means of the expression ID1=TRANSPOSE(ID2)), and then convert the TRUE and FALSE values to 1's and 0's by multiplying by 1, the 1's occurring when ID1=ID2.

To find the number of values that are common to the two lists, we simply need to sum the cells in the matrix. The complete formula is thus

    =SUM(1*(ID1=TRANSPOSE(ID2)))

Again, this formula is an array formula, so you must use Ctrl+Shift+Enter to enter it.

	A	B	C
1	ID Number	LAST NAME	FIRST NAME
2	08086937	Adamcik	Wendy A.
3	46457849	Adams	Daniela
4	83677323	Afayi	Tanya B.
5	16125442	Aleardo	Kathleen S.
6	89105904	Anadorelli	Jacquelyn N.
7	30219115	Apreaza	Katherine M.
8	03068210	Azmealla	Milan S.
9	54029952	Bakaymanno	Berkeley
10	49295168	Barros	Keith
11	48984878	Barros	Lisa N.
12	01624248	Bilodeau	Jonathan C.
13	72881292	Blum	Judith K.

Figure 7-8. Using an array formula to find
the number of common entries in two lists.

If the lists did not have ID numbers, a similar formula combining the last name (e.g., LN1) and first name (e.g., FN1) into a single string

    =SUM(1*(LN1&FN1=TRANSPOSE(LN2&FN2))).

will accomplish the same result.

Counting Duplicate Entries in a List

You may want to find the number of duplicate entries in a list. Figure 7-9 shows a sample list, Range1, with six entries in column A.

The following expression, used in the formula in cell B3, entered as an array formula, returns the number of duplicate entries in a list:

SUM(1*(Range1=TRANSPOSE(Range1)))-ROWS(Range1)

	A	B	C
1	Counting duplicate entries in a list with no blank entries		
2	Range1		
3	Able	There are 2 duplicate entries	
4	Baker	There are 4 unique entries.	
5	Charlie		
6	Baker		
7	Dog		
8	Able		
9			

Figure 7-9. Using array formulas to find the number of duplicate or unique entries in a list.

The expression 1*(Range1=TRANSPOSE(Range1)) creates a 2-D array of 1's and 0's, with the 1's corresponding to instances of duplicate entries. Summing this expression returns the total number of matches. Since we're comparing a list with itself, all the diagonal elements of the array will be 1's, and so we need to subtract the size of the list, which we do by using ROWS(Range1).

Counting Unique Entries in a List

Instead of counting the number of duplicates in a list, you may want to find the number of unique entries. The expression

SUM(1*(ROW(INDIRECT("1:"&COUNTA(Range1)))=MATCH(Range1,Range1,0)))

entered as an array formula in cell B4 of Figure 7-9, returns the number of unique entries in the list. Here's how it works. The expression

MATCH(Range1,Range1,0)

compares the individual values in the array Range1 (the lookup_value) to the same array (the lookup_array) and returns the relative position in the array. In this example, the array {1;2;3;2;5;1} is returned. The expression

ROW(INDIRECT("1:"&COUNTA(Range1)))

generates an array of integers, in this case {1;2;3;4;5;6}.

Comparing the two arrays returns TRUE for the unique items only, and multiplying the array of TRUE and FALSE values by 1 produces an array of 1's and 0's, in this case {1;1;1;0;1;0}. Finally, summing the array of 1's and 0's returns the number of unique items in the list. Once again, the formula is an array formula and must be entered using Ctrl+Shift+Enter.

Indicating Duplicate Entries in a List

Instead of simply returning the number of duplicate entries in a list, you probably want to know which entries are duplicates and where they are located in the list. Part of a list of addresses is shown in Figure 7-10; the complete list contains over 100 entries. The list contains many duplicates; an array formula can be used to identify them.

	A	B
1		Forward-looking
2	5625 Manzanita Ave Apt 84	
3	Rr 1	
4	722 Briarbend Dr	Duplicate in row 21
5	1639 Olympia Fields St	Duplicate in row 80
6	14333 Diplomat Dr	Duplicate in row 98
7	5000 Fawn Mdws Apt 137	
8	328 N York St	
9	12 Dodge St	Duplicate in row 13
10	15140 El Cameno Real Dr	Duplicate in row 104
11	23407 Western Ave	
12	107 Ledgewood Dr	Duplicate in row 72
13	12 Dodge St	
14	14911 Eleanor Ave	Duplicate in row 86
15	107 Ledgewood Dr	Duplicate in row 72
16	94051 Doyle Point Rd	Duplicate in row 59
17	4903 Evergreen St	Duplicate in row 31
18	3529 S Chase Ave	
19	6147 Fairway Dr	Duplicate in row 33
20	423 N Columbus St	Duplicate in row 45
21	722 Briarbend Dr	
22	17614 N 131st Dr	Duplicate in row 30
23	10023 McCartney Ln	Duplicate in row 76
24	2925 Imperial Ct	Duplicate in row 37
25	111 Ashford Dr	Duplicate in row 120
26	56 Summit Ave	Duplicate in row 95
27		Duplicate in row 84
28	5705 Northfield Rd	Duplicate in row 118
29	17614 N 131st Dr	Duplicate in row 30
30	17614 N 131st Dr	
31	4903 Evergreen St	

Figure 7-10. Using an array formula to identify duplicate entries in a list.

The following formula in cell B2 of Figure 7-10 (List is in column A, beginning in cell A2), when filled down, returns the text "Duplicate in row" N:

=IF(MAX(ROW(List)*(A2=List))=ROW(),"","Duplicate in row "
&MAX(ROW(List)*(A2=List)))

Here's how the formula works. The expression

MAX(ROW(List)*(A2=List))

returns the largest row number of a row where the list entry in cell A2 matches an entry in the list. This formula is "forward-looking"; that is, it lists only entries that are "ahead" of the examined entry, because the MAX worksheet function is used. If desired, a comparable "backward-looking" formula can be constructed.

The IF function was incorporated because if the simple expression is used, a list entry will always "find itself". Including the IF function ignores the case where the row number of the examined entry equals the row number of the element in the array.

Returning an Array of Unique Entries in a List

Instead of simply returning the number of unique entries in a list, you can create an array formula to return the array of unique values in the example in Figure 7-9.

Let's create the formula in a stepwise fashion. To simplify the formulas, we'll use Define Name to create named formulas (see "Using Create Names" in Chapter 6). These formulas will be located in the Define Name dialog box, not in worksheet cells.

The expression

=ROW(Range1).

returns the array of row numbers of the list Range1, namely {3;4;5;6;7;8}. We'll combine this with an expression we used in an earlier section in this chapter

=ROW(INDIRECT("1:"&COUNTA(Range1)))=MATCH(Range1,Range1,0)

which returns the array {TRUE;TRUE;TRUE;FALSE;TRUE;FALSE} of unique items in the list. We'll name this formula UniqLogicals.

Multiplying the two expressions

=ROW(Range1)*UniqLogicals

returns the array {3;4;5;0;7;0} containing the row numbers of the unique entries. We'll name this formula UniqRows.

The formula

=SUM(1*UniqLogicals)

returns the number of unique items; call this formula CountUniq.

To generate the series of integers {1;2;3;4} we use the formula

=ROW(INDIRECT("1:"&CountUniq)).

which we name Series.

Now we'll use the series of integers in the LARGE function to obtain the array of row numbers of the unique items. The syntax of LARGE is LARGE(***array,k***) where ***k*** is a value used to return the kth largest value in the array. Thus the formula

=LARGE(UniqRows,Series).

returns the array {7;5;4;3}; we'll name this formula UniqArrayRows.

The array is then used in the INDIRECT function to create an array of R1C1-style addresses. The INDIRECT function takes a second optional argument, a logical value that specifies the type of reference (FALSE if an R1C1-style reference). The formula is

=INDIRECT("R"&UniqArrayRows&"C"&COLUMN(Range1),0)

This expression, when filled down in a range of cells, returns the unique items in the list, as shown in Figure 7-11.

	B
39	Dog
40	Charlie
41	Baker
42	Able
43	

Figure 7-11. Using an array formula to display unique entries in a list.

The single worksheet formula (all in one cell, of course)

=INDIRECT("R"&LARGE(ROW(Range1)*(ROW(INDIRECT("1:"&COUNTA
(Range1)))=MATCH(Range1,Range1,0)),ROW(INDIRECT("1:"&SUM(1*
(ROW(INDIRECT("1:"&COUNTA(Range1)))=MATCH(Range1,Range1,0))))))
&"C"&COLUMN(Range1),0)

is equivalent to the preceding formula using names.

Sorting by Using an Array Formula

As well as abstracting values from a list, you can use an array formula to sort a list by means of a formula, rather than manually by using the **Sort...** command. This can be useful: For example, you can import raw data into Sheet1 of a workbook and have a sorted list of the data appear automatically on Sheet2.

Sorting a 1-D List of Numbers

You can use the LARGE worksheet function in an array formula to sort a list of number values automatically by means of an array formula.

For example, let's imagine that you import a list of numerical data each day and paste them into Sheet1 of a workbook. The list must then be sorted in descending order. By means of the following array formula, the list can automatically appear on Sheet2 (for example) of the workbook, sorted in descending order:

=LARGE(List,ROW()-1)

In the preceding formula, the raw data in Sheet1 was named List; the formula was entered in row 2 of Sheet2 and filled down into sufficient rows to accommodate the sorted list.

A #NUM! error value is returned if the formula is filled into more rows than are required by the raw data; the formula

=IF(ISERROR(LARGE(List,ROW()-1)),"",LARGE(List,ROW()-1))

suppresses the display of this error.

Sorting a 2-D List
Using a Column of Numbers as the Sortkey

Sorting a multi-column list using the LARGE function is only slightly more complicated, provided that there are no duplicate entries in the column on which the sort is performed. You can use the MATCH and INDEX functions to return the values in the same row as the "sortkey" value. Figure 7-12 illustrates a portion of a list of polymer research samples and some of their physical properties. The columns in this raw data table were assigned the names SampleNumber, FormulationType, etc.

The column headings were copied and pasted into row 1 of another worksheet. In this second worksheet, the table of data was sorted by means of the following formulas, using the data in column J as the "sortkey".

The formula in cell J2

=LARGE(MeltingPoint,ROW()-1)

sorts the melting point values in ascending order; ROW()-1 must be used, since the table begins in row 2. The following formula in cell A2 (and similar formulas in cells B2 through I2).

=INDEX(SampleNumber,MATCH(J2,MeltingPoint,0))

returns the appropriate values in the same row as the sortkey value.

	A	B	C	D	E	F	G	H	I	J
1	Sample Number	Formulation Type	Molecular Weight (kD)	MW Dispersity (Mw/Mn)	Specific Gravity (ASTM D792)	Refractive Index	Dielectric Constant @ 100 Hz (ASTM D150)	Rockwell Hardness (ASTM D785)	Tensile Strength PSI/1000 (ASTM D638)	Melting Point (°C)
2	91976	T	14.6	2.4	1.135	1.497	2.88	110	15.4	232
3	91977	R	14.6	2.3	1.149	1.502	2.88	110	16.9	231
4	91978	R	11.5	2.6	1.105	1.500	2.88	107	16.9	209
5	91979	R	12.0	2.8	1.117	1.503	2.94	115	14.2	208
6	91980	R	12.8	2.9	1.181	1.496	2.89	108	13.8	207
7	91981	R	12.2	2.6	1.172	1.497	2.94	111	13.9	205
8	92007	A	13.6	2.2	1.174	1.501	2.88	114	14.4	215
9	92039	U	13.6	2.1	1.184	1.500	2.92	112	15.6	212
10	92071	R	14.5	3.4	1.135	1.496	2.91	113	16.3	185
11	92103	T	15.2	3.6	1.143	1.500	2.87	109	16.7	174
12	92135	T	11.5	3.7	1.148	1.504	2.89	110	17.4	218
13	92167	T	12.0	3.4	1.139	1.498	2.94	107	15.3	210
14	92199	R	12.8	3.0	1.178	1.495	2.89	106	16.7	215
15	92231	R	12.2	2.9	1.177	1.499	2.89	113	13.1	212

Figure 7-12. A multi-column list.

	A	B	C	D	E	F	G	H	I	J
1	Sample Number	Formulation Type	Molecular Weight (kD)	MW Dispersity (Mw/Mn)	Specific Gravity (ASTM D792)	Refractive Index	Dielectric Constant @ 100 Hz (ASTM D150)	Rockwell Hardness (ASTM D785)	Tensile Strength PSI/1000 (ASTM D638)	Melting Point (°C)
2	92351	U	11.0	2.3	1.165	1.498	2.90	110	16.7	255
3	91976	T	14.6	2.4	1.135	1.497	2.88	110	15.4	232
4	91977	R	14.6	2.3	1.149	1.502	2.88	110	16.9	231
5	92135	T	11.5	3.7	1.148	1.504	2.89	110	17.4	218
6	92135	T	11.5	3.7	1.148	1.504	2.89	110	17.4	218
7	92135	T	11.5	3.7	1.148	1.504	2.89	110	17.4	218
8	92007	A	13.6	2.2	1.174	1.501	2.88	114	14.4	215
9	92007	A	13.6	2.2	1.174	1.501	2.88	114	14.4	215

Figure 7-13. Using an array formula to sort a multi-column list.

This formula works fine as long as the table does not contain duplicate values of the sortkey. However, in this example there are many sets of duplicate melting point values, the first occurring in rows 5, 6 and 7. The MATCH function then returns the first occurrence of a match; as a result, all rows having duplicate sortkeys will display the same information in the associated cells, as illustrated in Figure 7-13.

Where the raw data table contains duplicate values of the sortkey, a more complicated formula is required. The following formula in cell A2 (and similar formulas in cells B2 through I2) returns the correct values in the same row as the sortkey value. Let's examine the formulas in row 5, the first row showing duplicate values of the sortkey in column J.

{=IF(COUNTIF($J$1:J4,J5),INDEX(SampleNumber,LARGE(ROW($B$2:B 30)*(MeltingPoint=J5),COUNTIF($J$1:$J$30,J5)-COUNTIF($J$1:J4,J5))1), INDEX(SampleNumber,MATCH(J5,MeltingPoint,0)))}

Here's how this formula works. The expression

COUNTIF($J$1:J4,J5)

calculates the number of duplicate sortkey values (here MeltingPoint is the sortkey) that appear ahead of the current value; if zero, the final part of the formula is used; this is the *value_if_false* of the IF function. This expression

INDEX(SampleNumber,MATCH(J5,MeltingPoint,0))

is identical to the simple formula described earlier for the case of no duplicate sortkey values.

On the other hand, if there are duplicate sortkey values, we have to calculate which of the several rows of values to use. The expression

ROW($B$2:$B$30)*(MeltingPoint=J5)

returns the array of row numbers for which the melting point matches the sorted value in cell J5, i.e.,

{0;0;0;0;0;0;0;0;0;0;12;0;0;0;0;0;0;0;19;0;0;0;0;0;0;25;0;0;0;0;0}

The expression

COUNTIF($J$1:$J$30,J5)-COUNTIF($J$1:J4,J5)

calculates the total number of duplicates of the melting point value, minus the number of duplicates already encountered, and is thus the k value to use in the LARGE function (in this example, the expression returns 3 in row 5, 2 in row 6, 1 in row 7). Combining these two expressions returns the row number of the row that contains the appropriate information. We must subtract 1 to obtain the number to use in the INDEX function, since the array begins in row 2.

Finally, so that the formula can be filled down into many rows, where there may not be sufficient values to fill, an additional IF statement is used to suppress the #N/A error value.

{=IF(J5="","",IF(COUNTIF($J$1:J4,J5),INDEX(SampleNumber,LARGE(ROW ($B$2:$B$30)*(MeltingPoint=J5),COUNTIF($J$1:$J$30,J5)-COUNTIF ($J$1:J4,J5))-1),INDEX(SampleNumber,MATCH(J5,MeltingPoint,0)))))}

"Auto-Alphabetizing" Using an Array Formula

It's a little more complicated to sort a column of text values by using an array formula, since the LARGE and SMALL functions don't operate on text. But it is possible to do it, using the COUNTIF function to determine the number of entries that are less than a given entry. Figure 7-14 illustrates a table of names in column A that are sorted automatically by means of a formula in column D; the values in columns B and C are not required and are shown simply to make clear how the formula operates.

The array formula in cell B2,

{=COUNTIF(table,"<" &table)}

returns the number of entries in the table that are less than the value in cell A2. In order for the formula to return the correct results, you must select the range of cells B2:B11, enter the formula, and then press Ctrl+Shift+Enter.

	A	B	C	D
1	table			
2	Smith	7	4	Chi
3	Fridirici	3	6	Crowell
4	Hunnewell	5	9	France
5	Chi	0	2	Fridirici
6	Wang	9	10	Gagnon
7	Crowell	1	3	Hunnewell
8	Kearney	6	7	Kearney
9	Tubridy	8	1	Smith
10	France	2	8	Tubridy
11	Gagnon	4	5	Wang

Figure 7-14. Using an array formula to sort a range of text values.

The array formula in cell C2

{=MATCH(ROW(table)-MIN(ROW(table)),COUNTIF(table,"<" &table),0)}

returns the relative position of the numbers 0, 1, 2...9 in the array of values in column B. This array of integers is produced by the expression

ROW(table)-MIN(ROW(table))

so that the formula in cell C2 evaluates to

    {=MATCH({0;1;2;3;4;5;6;7;8;9},COUNTIF(table,"<" &table),0)}

    Finally, these expressions can be combined into the "megaformula" in cell D2

    {=INDEX(table,MATCH(ROW(table)-MIN(ROW(table)),
        COUNTIF(table,"<" &table),0))}.

    Again, you must select the range of cells D2:D11, enter the formula and then press Ctrl+Shift+Enter.

	A	B	C	D
1	LN	FN		
2	Smith	Tom	7	Chi, Bob
3	Fridirici	Dick	3	Crowell, Sam
4	Hunnewell	Harry	4	France, Albert
5	Chi	Bob	0	Fridirici, Dick
6	Wang	Albert	9	Hunnewell, Harry
7	Crowell	Sam	1	Kearney, Jane
8	Kearney	Jane	5	Smith, Edna
9	Tubridy	Tom	8	Smith, Tom
10	France	Albert	2	Tubridy, Tom
11	Smith	Edna	6	Wang, Albert

Figure 7-15. Using an array formula to sort a list of first and last names.

    A similar formula can be developed to handle values in two or more columns, for example, to sort a list of first and last names, as illustrated in Figure 7-15.

    The formula in cell D2 is

    =INDEX(LN,MATCH(ROW(LN)-2,$C$2:$C$11,0))&", "&
        INDEX(FN,MATCH(ROW(LN)-2,$C$2:$C$11,0))

    In this example I was unable to incorporate the values in column C in the megaformula.

Another Example:
Validating a CAS Registry Number

    A CAS registry number consists of three parts separated by hyphens. The first part is up to seven digits in length, the second part is two digits, and the third is a single digit checksum. For example, the CAS number for hexane is 110-54-3.

It might be useful to be able to check that a user-entered CAS number is valid. Although it's not possible to check that the CAS number corresponds to a known compound, we can determine if the checksum is correct. The checksum is calculated in the following way: the sum of the last digit times 1 plus the next digit times 2 plus the next digit times 3, etc., is computed; the checksum is the sum modulo 10 (the remainder when the checksum is divided by 10).

In Chapter 11 a CAS number is validated by checking that the string contains two hyphens, that the second part is two characters in length, and that the third part is one character in length. In the present example, by using an array formula, we will determine whether the checksum digit is correct. The steps to create the checksum formula are as follows: First convert the CAS number (in cell A2) to a string of digits by using the formula

=SUBSTITUTE(A2,"-","")

to remove the hyphens. For hexane, the formula returns 110543.

Next, create a series of integers, for this example {6;5;4;3;2;1}, by using the formula

=LEN(A2)-ROW(INDIRECT("1:"&LEN(A2)-2))-1

Now combine the two preceding expressions in the MID function to yield the formula

=MID(SUBSTITUTE(A2,"-",""),LEN(A2)-ROW(INDIRECT ("1:"&LEN(A2)-2))-1,1)

which returns the individual digits. For this example, the formula returns {"3";"4";"5";"0";"1";"1"}.

Next, multiply this series of digits by the series of integers {1;2;3;4;5} by means of the formula

=MID(SUBSTITUTE(A2,"-",""),LEN(A2)-ROW(INDIRECT("1:"&LEN(A2)-3))-
    2,1)*(ROW(INDIRECT("1:"&LEN(A2)-3)))

In this example the formula returns the array {4;10;0;4;5}.

The array formula

{=SUM(MID(SUBSTITUTE(A2,"-",""),LEN(A2)-ROW(INDIRECT
    ("1:"&LEN(A2)-3))-2,1)*(ROW(INDIRECT("1:"&LEN(A2)-3))))}

returns the sum of these products. You must enter this formula by pressing Ctrl+Shift+Enter.

The checksum is obtained by using the MOD worksheet function, in the formula

{=MOD(SUM(MID(SUBSTITUTE(A2,"-",""),LEN(A2)-ROW(INDIRECT
    ("1:"&LEN(A2)-3))-2,1)*(ROW(INDIRECT("1:"&LEN(A2)-3)))),10)}

This result of this formula can be compared with the checksum digit, by means of the formula

{=MOD(SUM(MID(SUBSTITUTE(A2,"-",""),LEN(A2)-ROW(INDIRECT
("1:"&LEN(A2)-3))-2,1)*(ROW(INDIRECT("1:"&LEN(A2)-3)))),10)
=VALUE(RIGHT(A2))}

The formula returns TRUE if the last digit of the CAS number is equal to the calculated checksum. Of course, the formula does not check that the number corresponds to an actual compound.

Excel Tip. *If you name an array formula, you don't have to remember to press Ctrl+Shift+Enter when you enter the formula in a worksheet cell. The array calculation is done "in the background".*

8

Advanced Charting Techniques

This chapter shows how to customize charts using Excel's built-in features and also shows how to create charts that go beyond the built-in capabilities of Excel. Since most scientific charts are XY scatter charts, this chapter deals mainly with that chart type.

Charts with More than One Data Series

Excel can plot several data series in the same chart. You can create a chart with multiple data series, or you can add new data series to an existing chart.

If the values in the data series are similar in magnitude, then plotting two or more sets of data is not any different from plotting one set. If the numbers are very different (for example, if one set of y-values is in the range 1–10 and the other set is in the range 0.001–0.010), you must use a chart with a secondary axis (see "Charts with Secondary Axes" later in this chapter).

Plotting Two or More Sets of Y Values in the Same Chart

If more than two columns (or rows) of data are selected for plotting, Excel uses the leftmost column or uppermost row as the independent variable (plotted on the X Axis) and the remaining rows or columns as the dependent variables (plotted on the Y Axis). Figure 8-1 illustrates a data table with one column of x data and three columns of y data to be selected for a chart. If the data series are non-adjacent, hold down the Ctrl key while you select the separated columns of data. (Excel will always use the leftmost column or uppermost row for the x-values.)

	A	B	C	D
4	t	[A]	[B]	[C]
5	0	1.0000	0.0000	0.0000
6	2	0.9048	0.0914	0.0038
7	5	0.7788	0.1996	0.0215
8	10	0.6065	0.3189	0.0745
9	15	0.4724	0.3822	0.1454
10	20	0.3679	0.4072	0.2249
11	25	0.2865	0.4069	0.3066
12	30	0.2231	0.3903	0.3865
13	40	0.1353	0.3328	0.5319
14	50	0.0821	0.2663	0.6517
15	60	0.0498	0.2047	0.7456
16	70	0.0302	0.1531	0.8167
17	80	0.0183	0.1122	0.8695
18	90	0.0111	0.0811	0.9078
19	100	0.0067	0.0579	0.9354

Figure 8-1. Spreadsheet layout for multiple Y data series.

Excel uses a different shape and color for the plotting symbols in each data series. You can change the plotting symbols or remove them if you customize the chart. Figure 8-2 illustrates the three data series from the spreadsheet of Figure 8-1.

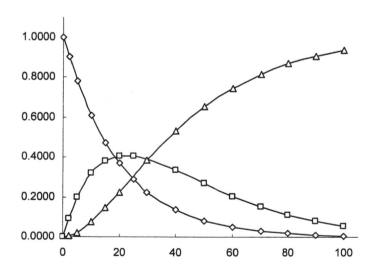

Figure 8-2. Chart with several Y data series.

Plotting Two Different Sets of X- and Y Values in the Same Chart

To plot two sets of x, y data on a single chart, use the data layout shown in Figure 8-3. A chart produced from the two data series in the spreadsheet of Figure 8-3 is shown in Figure 8-4.

	A	B	C
1	X	Y1	Y2
2	0	0	
3	1	3	
4	3	9	
5	5	15	
6	7	21	
7	9	27	
8	11	33	
9	0		0
10	2		10
11	4		20
12	6		30
13	8		40
14	10		50

Figure 8-3. Spreadsheet layout for two X and two Y data series.

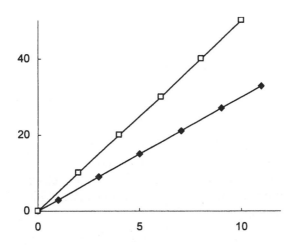

Figure 8-4. Chart with two X and two Y data series.

Alternate Data Layout
for Two Different Sets of X- and Y Values
in the Same Chart

Another way to plot two sets of x, y data on a single chart is by means of the data layout shown in Figure 8-5. To use this approach, select the x-values and $y1$-values (A2:B8) and use the Chart Wizard to create a chart. Next, select the x- and $y2$-values (A11:B16) and **Copy**, then click on the chart to activate it, then choose Home → Clipboard → Paste Special... in the Paste submenu (Excel 2007/2010) or **Paste Special...** from the **Edit** menu (Excel 2003) to display the Paste Special dialog box for charts (Figure 8-6). Check the "Add cells as: New series" and "Categories (X Values) in First Column" boxes, and then press OK.

	A	B
1	X	Y1
2	0	0
3	1	3
4	3	9
5	5	15
6	7	21
7	9	27
8	11	33
9		
10	X	Y2
11	0	0
12	2	10
13	4	20
14	6	30
15	8	40
16	10	50

Figure 8-5. Using Paste Special... for charts to add a data series.

This approach has little to recommend it for simple charts. However, the ability to **Copy** a data range from a worksheet and use **Paste Special...** to add it to a chart as a new series makes it possible to customize a chart in ways that are not built in to Excel. See "A Chart with an Additional Axis" later in this chapter for an example of this technique.

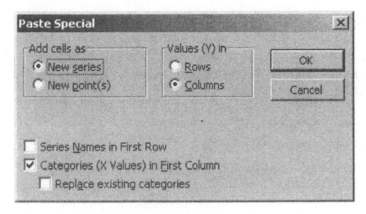

Figure 8-6. The Paste Special dialog box for charts.

Extending a Data Series or Adding a New Series

Occasionally you may want to add additional data points to a data series in a chart, or add a new series. Of course you can simply delete the chart and create a new one. But it's possible to add new data to an existing chart. You can use one of the following methods, depending on the type of chart and the layout of the data on the worksheet.

The Copy and Paste Method (For Embedded Charts or Separate Chart Sheets)

Highlight the range of data to be added to the chart and **Copy**. Switch to the chart sheet or activate an embedded chart by clicking on it. **Paste** the data. The new data will be added to the chart.

Excel usually does an excellent job of understanding how new data should be added to the chart – adding new data points to an existing series or adding a new series as required. Excel will assume that the new data are additional data points in an existing series if you select additional cells in the same rows or columns as the original x-values and y-values; if you select new rows or columns that align with the original x-values, the new data will be assumed to be new data series. Occasionally Excel will display the Paste Special dialog box, to ask you how the data should be added to the chart.

The Drag-and-Drop Method (For Embedded Charts Only)

This method is not available in Excel 2007/2010.

Highlight the range of data to be added to the chart. Using Drag-and-Drop, drag the new data to the chart. When the mouse pointer passes over the chart, the gray border indicating the selected cells will show that they will be added to the chart and a small plus sign will appear near the arrow pointer, indicating that you are adding a copy of the selected cells to the chart. Release the mouse button. The new data will be added to the chart.

The Color-Coded Ranges Method (For Embedded Charts Only)

First, click on the Chart Area (the outer border of the chart). All data series in the chart will usually be indicated on the worksheet by means of color-coded ranges (sometimes called the sizing handles): purple for x-values, blue for y-values, as shown in Figure 8-7. If you selected cells containing labels for the columns, they will be color-coded green.

To extend the range of all existing series, drag either of the handles down, as shown in Figure 8-8. To add a new Y data series, drag the y-values handle to the right.

This method can be used only for embedded charts with multiple ranges of y-values in adjacent ranges of cells. A chart with y-values in non-adjacent ranges does not display color-coded ranges when you click on the plot area.

To operate on a single data series, click on the *series*. The color-coded range of x-values and a single color-coded range of y-values will be displayed. To extend the data series, you have to drag the handle for the x-values (the purple handle) down, and repeat for the y-values. You can't add a new data series by dragging the y-values handle (the blue handle) to the right.

	A	B	C	D
1	X	Y1	Y2	
2	1	3	5	
3	3	9	15	
4	5	15	25	
5	7	21	35	
6	9	27	45	
7	11	33	55	
8	13	39	65	
9	15	45	75	
10	17	51	85	
11	19	57	95	
12	21	63	105	
13				

Figure 8-7. Chart showing color-coded ranges.

	A	B	C	D
1	X	Y1	Y2	
2	1	3	5	
3	3	9	15	
4	5	15	25	
5	7	21	35	
6	9	27	45	
7	11	33	55	
8	13	39	65	
9	15	45	75	
10	17	51	85	
11	19	57	95	
12	21	63	105	
13				

Figure 8-8. Using the color-coded ranges to extend data series in a chart.

Using SOURCE DATA...in the Chart Menu
(For Embedded Charts or Separate Chart Sheets)

Excel 2007/2010. First, switch to the chart sheet or activate an embedded chart by clicking on it. With the chart activated, click on the Select Data icon in the Data group in the Design tab to display the Select Data Source dialog box (Figure 8-9).

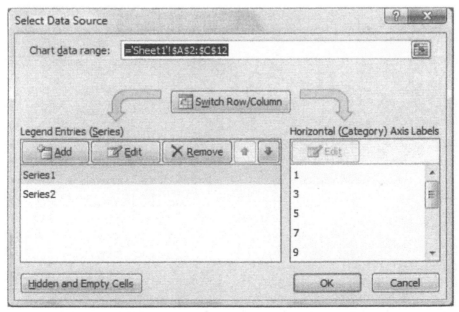

Figure 8-9. The Excel 2007/2010 Select Data Source dialog box.

Under the Legend Entries (Series) box, click the Add button to display the Edit Series dialog box (Figure 8-10). Enter the ranges of the *x*-values and *y*-values of the new or extended series, and press OK.

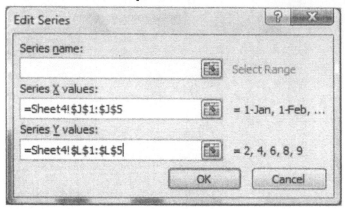

Figure 8-10. Using Edit Series to extend a data series in a chart.

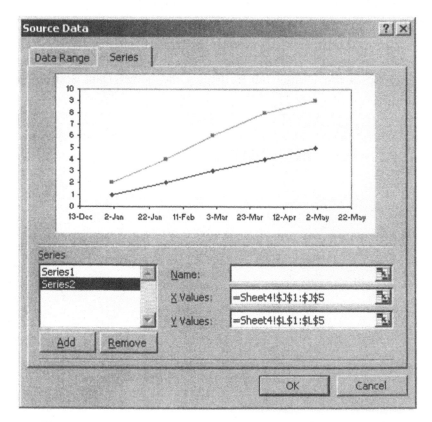

Figure 8-11. Using Source Data to add a data series in a chart.

Excel 2003. First, switch to the chart sheet or activate an embedded chart by clicking on it. Choose **Source Data...** from the **Chart** menu to display the Source Data dialog box (Figure 8-11). Choose the Data Range tab if you want to change the range of several data series all at once; choose the Series tab if you want to operate on just one series, or add a new series. To change just one data series, select Series 2 (for example); cell references to the Name (the text that will be used as the legend), the *x*-values and the *y*-values are displayed in separate text input boxes. You can enter or edit these references by typing, or by selecting. For example, use the Tab key to select the X Values input box. The worksheet will be displayed with a marquee around the *x*-values data series. Select the new range with the mouse. Tab to the Y Values input box and repeat, then press OK.

Or you can choose **Add Data...** from the **Chart** menu (Figure 8-12), and enter the range of cells for the new data.

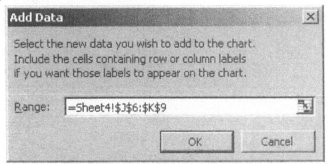

Figure 8-12. Using Add Data to extend a data series in a chart.

Editing the SERIES Function in the Formula Bar (For Embedded Charts or Separate Chart Sheets)

First, switch to the chart sheet or activate an embedded chart by clicking on it. Click on the desired data series in the chart. The definition of the data series, in the format

=SERIES(*name, x-values_ref, y-values_ref, plot_order*)

for example

=SERIES("Series2", Sheet4!$A$1:$A$11, Sheet4!$C$1:$C$11, 2)

will appear in the formula bar. Edit the references for both X- and Y Values to include the additional data points. Sometimes this is the fastest way.

Customizing Charts

The following sections illustrate some ways to improve the appearance or usefulness of a chart.

Good Charts vs. Bad Charts

Most charts produced by chemists are XY charts. Unless you choose one of the "smoothed lines" options, Excel connects plotted points by straight-line segments; thus, a chart of data that should produce a smooth curve might look like the "bad chart" shown in Figure 8-13.

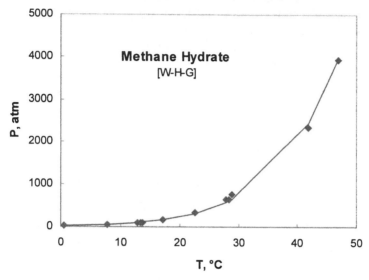

Figure 8-13. An example of a bad chart.

Plotting Experimental Data Points and a Calculated Curve

Plotting experimental data points and a smooth calculated curve is one of the most common applications of custom formatting. To do this, you need to plot two y data series — the experimental data points and a series of points to describe the calculated curve. The y_{obsd} data should be formatted as a series of symbols with no connecting line, the y_{calc} data as a line with no symbols, as in Figure 8-15. To generate a smooth calculated curve, you'll need to have the y_{calc} points fairly close together. But since having too many points can slow recalculation of a worksheet, you should try to strike a balance between the two requirements.

Of course, to plot a calculated curve you need to have an equation that fits the data. It may be the least-squares straight line (obtained from LINEST) that best fits the data, or a curve produced by an equation appropriate for the data.

Figure 8-14 illustrates a portion of a data table showing experimental data points (temperature, pressure) for a phase diagram and part of the table of calculated pressure values, where the pressure was calculated using the theoretical relationship P = A * EXP(B/(T + 273)). Since the experimental temperature data points were in the approximate range 0–50 degrees, a series of temperature values (0–50 in increments of 2) for the calculated curve was created, beginning in row 22. The values of the parameters A and B were obtained from a least-squares fit of the experimental data to the straight-line relationship $\ln Y = B/T + \ln A$.

	B	C	D
8	T	P, atm	P(calc)
9	0.5	27	
10	7.7	58	
11	12.7	97	
12	13.3	105	
13	13.5	107	
14	17.0	157	
15	22.5	335	
16	27.8	640	
17	28.4	645	
18	28.8	765	
19	41.9	2344	
20	46.9	3918	
21			
22	0		22
23	2		28
24	4		36

Figure 8-14. Spreadsheet for plotting y_{obsd} and y_{calc}.

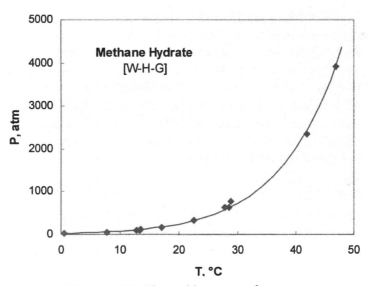

Figure 8-15. Chart with y_{obsd} and y_{calc}.

To Plot y_{obsd} and y_{calc}

1. In the same column as the experimental x-values, create a suitable range of x-values (the best way is to use AutoFill) to calculate the theoretical curve. The x-values can be appended either at the end or the beginning of the experimental x-values. Include a blank row separating the two sets of data.

2. Enter the expression for y_{calc} in a separate column and **Fill Down** to produce the calculated values.

3. Select the three data series (x, y_{obsd} and y_{calc}) and create an XY chart.

4. The chart will have the experimental and calculated values more or less superimposed. Click on the experimental data series. It may take a bit of searching and clicking to find the experimental data series; if you can't find it, you can use the arrow keys to select the appropriate data series. You'll see the name of the data series (e.g., Series 1) displayed in the Chart Elements box in the Format tab (Excel 2007/2010) or in the Name Box (Excel 2003).

5. **Excel 2007/2010**. Click on the Format Selection button to display the Format data Series dialog box. Click on Marker Options to format the marker; to change a solid symbol to an open one, click on the Solid Fill option button and choose Color = White. Then click on Line Color and press the No Line option button.

   Excel 2003. Choose **Selected Data Series...** from the **Format** menu and choose the Patterns tab. In the Patterns dialog box (see Chapter 5, Figure 5-16), choose Line = None and Marker = Automatic. If you want to change the style of data marker, select from the Style box. To change a solid symbol to an open one, choose Background = White. The symbol that you've selected is displayed in the Sample box in the lower right corner.

6. Now click on the data points for the calculated curve. Again, you'll see the name of the appropriate data series displayed. Repeat as in step 5, choosing Marker Options = None and Line Color = Automatic or Solid Line(Excel 2007/2010), or Marker = None and Line = Automatic (Excel 2003).

Charts with Secondary Axes

You've seen how Excel can plot more than one set of *y* data on a single chart. However, all the data series are plotted on a single Y Axis scale. Adding a secondary axis permits the graphing of sets of data with different X- and/or Y Axis scales. For plotting scientific data, we'll be interested in producing charts with two different Y Axis scales plotted using the same X Axis, or with two different Y Axis scales and two different X Axis scales. There may even be an occasion to create a chart with a single Y Axis and two X axes. The *secondary Y Axis* is plotted along the right side of the chart, the *secondary X Axis* along the top of the chart.

An XY Chart with a Secondary Y Axis

The data in the spreadsheet shown in Figure 8-16 has two sets of *y*-values that differ considerably in magnitude.

	A	B	C
3	X	Y1	Y2
4	0	5	0.200
5	1	8	0.125
6	2	11	0.091
7	3	14	0.071
8	4	17	0.059
9	5	20	0.050
10	6	23	0.043
11	7	26	0.038
12	8	29	0.034
13	9	32	0.031
14	10	35	0.029

Figure 8-16. Spreadsheet with two data series that differ greatly in magnitude.

If you select the range $A$4:$C$14 and create a chart of the data, you will get a chart like the one shown in Figure 8-17. The data points in the Y2 data series are essentially indistinguishable from zero. To show how the Y2 data change, we need to add a secondary Y Axis.

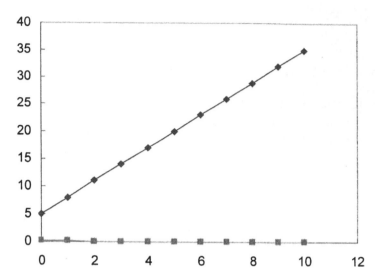

Figure 8-17. Chart created using the data in Figure 8-16.

To produce an XY scatter chart with one X Axis scale and two Y Axis scales (a primary Y Axis along the left-hand side of the chart and a secondary Y Axis along the right-hand side, as illustrated in Figure 8-19), you first create the chart using the data for both y data series and then designate one of the y data series as the series to be plotted on the secondary axis. The procedure is described in the following box.

To Create a Chart with a Secondary Y Axis
(two different Y Axis scales and the same X Axis)

1. Select all three data series to be plotted (the X Axis data series, two Y-Axis data series).

2. Create an XY chart.

3. Click on the data series whose axis you want to change.

4. **Excel 2007/2010.** Click on the Format tab. Click on Format Selection in the Current Selection group to display the Format Data Series dialog box. Click on Series Options and press the Secondary Axis option button.

    Excel 2003. Choose **Selected Data Series...** from the **Format** menu and choose the Axis tab (see Figure 8-18).

5. Press the Secondary Axis button. A preview of the combination chart will be displayed. If the chart is suitable, press the OK button.

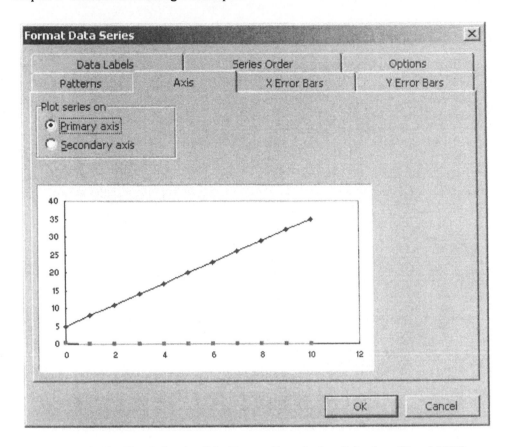

Figure 8-18. The Axis tab of the Format Data Series dialog box (Excel 2003).

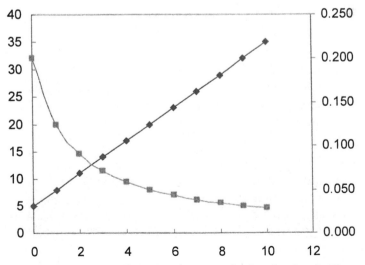

Figure 8-19. Chart with a secondary Y Axis, created using the data in Figure 8-16.

A Chart with Secondary X Axis and Y Axis

To produce an XY chart with two different X-axes and two different Y-axes, you must create a chart with one set of x and y data series, then paste the second set of data in the chart, and then designate the series to be plotted on the secondary Y Axis and the secondary X Axis, as described in the following box.

	A	B	C	D
3	X1	Y1	X2	Y2
4	0	5	0.00	0.200
5	1	8	0.01	0.125
6	2	11	0.02	0.091
7	3	14	0.03	0.071
8	4	17	0.04	0.059
9	5	20	0.05	0.050
10	6	23	0.06	0.043
11	7	26	0.07	0.038
12	8	29	0.08	0.034
13	9	32	0.09	0.031
14	10	35	0.10	0.029

Figure 8-20. Spreadsheet with two sets of data that differ greatly in magnitude.

To Create a Chart with a Secondary X Axis and a Secondary Y Axis
(two different X Axis scales and two different Y Axis scales)

1. Select the $x1$ and $y1$ data and create a chart in the usual way.

2. Select the $x2$ and $y2$ data and **Copy**.

3. Click on the chart to activate it. Choose Home → Clipboard → Paste Special… in the Paste submenu (Excel 2007/2010) or **Paste Special…** from the **Edit** menu (Excel 2003). In the Paste Special dialog box (see Figure 8-6), press buttons for New Series and X-Values In First Column, then press OK.

4. The new series will be added to the chart. If, for example, both the x- and y-values of the new series are much smaller than those of the original series, the new data will be a cluster of points near the origin.

5. Select the new series by clicking on it.

6. **Excel 2007/2010.** Click on either the Format tab or the layout tab. Click on Format Selection in the Current Selection group. Press the Secondary Axis button. (This adds the secondary Y Axis, but not the secondary X Axis.) In the Layout ribbon, click on the Axes button in the Axes group. Choose Secondary Horizontal Axis in the drop-down menu and choose Show Default Axis in the submenu.

Excel 2003. Choose **Selected Data Series...** from the **Format** menu.
Choose the Axis tab and press the Plot Series on Secondary Axis button,
then press OK. (This adds the secondary Y Axis, but not the secondary
X Axis, as shown in Figure 8-21). While the data series is still selected,
choose **Chart Options...** from the **Chart** menu and choose the Axes
tab. Check the X Axis box in the Secondary Axis category (the Y Axis
box will already be checked) then press OK.

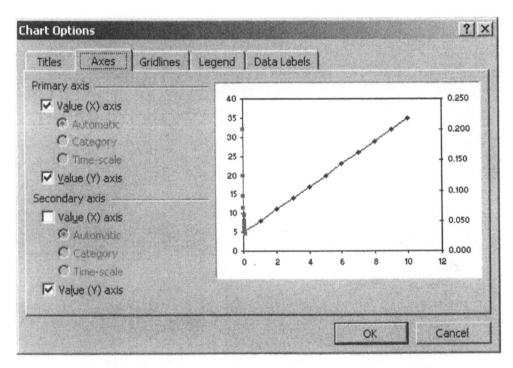

Figure 8-21. The Axes tab of the Chart Options dialog box (Excel 2003).

*Excel 2007/2010 Tip.   You can add Gridlines to either the primary or
secondary axis. This feature is not available in Excel 2003.*

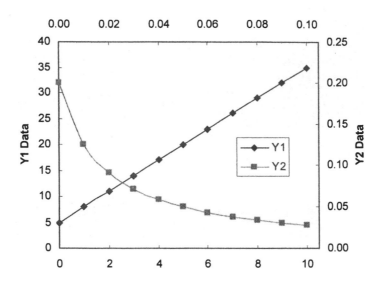

Figure 8-22. Chart created using the data in Figure 8-20
with a secondary X Axis and a secondary Y Axis.

A Column Chart with a Secondary Y-Axis

To create a column chart with a secondary Y Axis requires some additional steps. Figure 8-23 shows a table of data in which the values in the fourth series are much smaller than those in the other three series. Thus the fourth data series should be displayed using a secondary Y Axis.

	A	B	C	D	E
1		Site A	Site B	Site C	%
2	Jan	1195	1286	1395	0.15
3	Feb	1173	1227	1246	0.12
4	Mar	1196	1285	1137	0.07
5	Apr	1087	1311	1364	0.13

Figure 8-23. Spreadsheet with sets of data that differ greatly in magnitude.

If you follow the procedure in the preceding sections, the resulting chart will look like the example in Figure 8-24. The single column plotted on the secondary axis lies in front of the three columns of the primary axis.

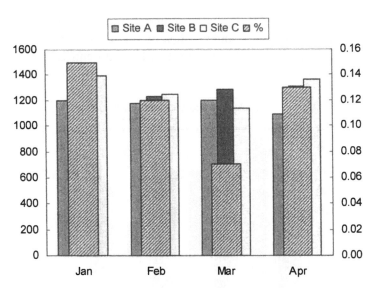

Figure 8-24. Column chart with a secondary Y Axis,
created using the data in Figure 8-23.

In order to display the secondary axis column suitably, you must insert some dummy data series. In this case, one additional column of data for the primary axis has been added, and three columns for the secondary axis, as shown in Figure 8-25. Adding dummy labels (in this example, D, E, F and G) will make it easier to modify the legend later.

	A	B	C	D	E	F	G	H	I
1		Site A	Site B	Site C	D	E	F	G	%
2	Jan	1195	1286	1395					0.15
3	Feb	1173	1227	1246					0.12
4	Mar	1196	1285	1137					0.07
5	Apr	1087	1311	1364					0.13

Figure 8-25. Modification of the spreadsheet shown in Figure 8-23.

To make the chart, select cells for all eight data series (A1$:I$5) and create a column chart. Select Series 5. For Excel 2007/2010, click on the Format tab. Click on Format Selection in the Current Selection group to display the Format Data Series dialog box. Click on Series Options and press the Secondary Axis option button. For Excel 2003, choose **Selected Data Series...** from the **Format** menu and choose the Axis tab. Repeat for Series 6, 7, and 8 ("%").

Click on the chart, and click on any of the three columns in the primary series. Choose Format Series. Choose the Options tab and adjust Overlap to, e.g., −20. This adjusts separation between columns in all three sets of columns plotted on the primary axis. Repeat with the secondary axis series.

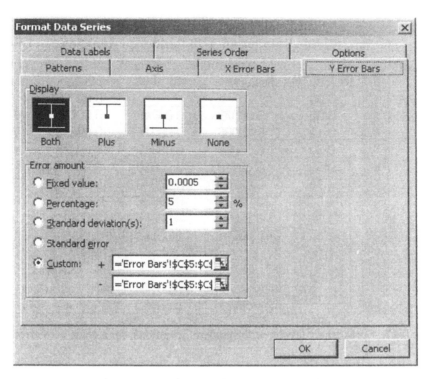

Figure 8-28. The Excel 2003 Error Bars dialog box.

Excel 2007/2010. In Excel 2007/2010, error bars are added using the Layout ribbon, not the Format ribbon.

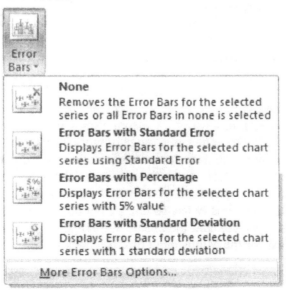

Figure 8-29. The Excel 2007/2010 Error Bars drop-down list.

Click on the series to which error bars are to be added. Click on the error bars icon in the Analysis group to display the Error Bars drop-down (Figure 8-29). To add error bars corresponding to the "Custom" category in Excel 2003 (see above), choose More Error Bars Options.

No matter which error bars option you choose, both horizontal (X Axis) and vertical (Y Axis) error bars are added automatically. To get Y Axis error bars only, for example, you have to delete the X Axis error bars

Once the X Axis error bars have been deleted, click on the Y Axis error bars, click on the Error Bars icon again, and choose More Error Bar Options again. This will display the Format Error Bars dialog box (Figure 8-30) for vertical error bars.

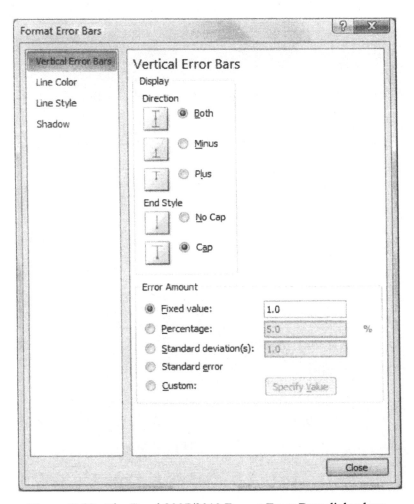

Figure 8-30. The Excel 2007/2010 Format Error Bars dialog box.

Even though you chose More Error Bar Options, Excel automatically added error bars with a fixed value of one chart division above and below each marker point. To change the error bar amount to custom a range of values on the worksheet), click the Custom option button and press the Specify Value button, to display the Custom Error Bars dialog box (Figure 8-31). Select the ranges for Positive and Negative Error Values and press OK

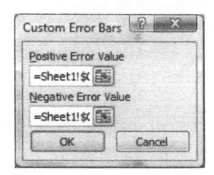

Figure 8-31. The Excel 2007/2010 Custom Error Bars dialog box.

Figure 8-32 shows the formatted chart (k_{obsd} data points and k_{calc} line) with error bars corresponding to $\pm 1 \sigma$.

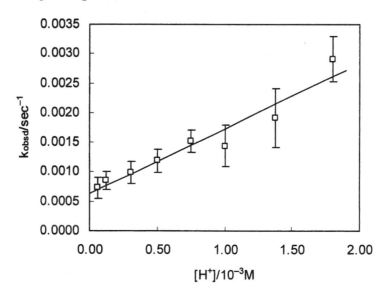

Figure 8-32. Chart with error bars.

Using Sub- or Superscripts in Chart Legends

As described in Chapter 1 and Chapter 2, you can format characters in a cell as subscript or superscript characters. In a chart, the same formatting can be applied to text characters in the chart title or the X- or Y Axis title. Unfortunately, you can't use sub- or superscripts in the legend text. Legend text in a chart is inserted in the legend box by means of Excel's code, and thus characters can't be subscripted or superscripted. The only way to provide sub- or superscripts in a legend is by using the alternate character set. Arial font provides superscript characters 0, 1 2, 3 (see the table of ASCII codes in Appendix J). Some other fonts contain both sub- and superscript characters. For example, the Arial Unicode MS font has both super- and subscript number characters 0–9 (Figure 8-33).

Unfortunately, none of the fonts that I have found provide what I would consider decent-looking characters.

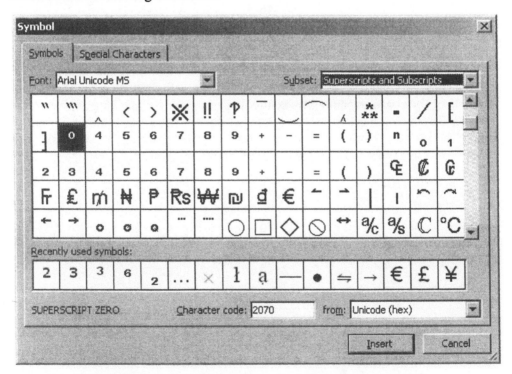

Figure 8-33. A font that provides sub- and superscript numerals.

Modifying Tick Mark Labels on Axis Scales

If an XY Scatter plot axis includes both positive and negative values, the zero tick mark label for the other axis is bisected by the axis line, as can be seen in Figure 8-34. This is not pleasing to the eye.

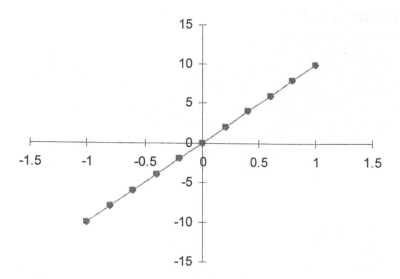

Figure 8-34. Zero tick-mark labels spoil the appearance of this chart.

You can use number formatting to eliminate the zero or other unwanted tick mark labels. Remember from Chapter 1 that the full form of a number format consists of four parts separated by semicolons: number format for positive values, number format for negative values; number format for zero; number format for text. The custom number format "0.0;-0.0;;" applied to each of the axes of the chart shown in Figure 8-34 yields the chart shown in Figure 8-35.

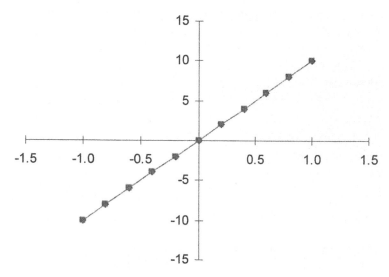

Figure 8-35. Zero tick-mark labels removed
by using a conditional number format.

Adding Data Labels to an XY Chart

Data labels are text boxes associated with each data point of a chart series. You can choose to have either the label (the X value) or the value (the Y value) displayed as the data label. Neither of these two options is too useful, but you can manually edit the Data Label text to provide more useful information.

To add data labels to a particular series in a chart, activate the chart and click on the series. Choose **Selected Data Series...** from the **Format** menu and choose the Data Labels tab. Choose either the Value option or the Label option, and press OK.

To edit the text of a particular data label text, click once on one of the labels to select all the data labels for a particular series, then a second time to select a single data label, then a third time to begin editing the text.

Manually Changing Data Labels

1. Click on any one of the data labels to select all of them. Black handles will appear.

2 To replace a data label with text, click on the data label (a box will be placed around the data label) and type new text.

or...

3 To replace a data label with a reference to a cell containing text, click on the data label (a box will be placed around the data label). Now click the cursor in the formula bar and type an equal sign, then click on the cell containing the desired text (Excel will enter a reference to the cell). Complete by pressing the Enter key or the Enter button.

You can format data labels either singly or as a group. Click once on a label to select all the data labels for a particular series, then a second time to select a single data label. The Format Data Labels dialog box has tabs for Patterns, Font, Number and Alignment. The Alignment tab allows you to position the data label either Above, Below, Center, Left or Right.

Instead of adding data labels manually, you can use the macro described in "A Sub Procedure to Apply Data Labels in a Chart" in Chapter 19.

Logarithmic Charts

In the Format Axis dialog box (in the Scale tab of Excel 2003), you will see an option to convert the axis (either the X Axis or the Y Axis) to a logarithmic scale. However, in Excel 2003 this option is very limited; the Minimum and Maximum of the scale must be powers of 10. Thus, if your y-values range from, e.g., 90 to 115, and you choose the Logarithmic Scale option, the Y Axis will extend from 1 to 1000 and your data will be an almost horizontal line in the middle of the chart. Attempting to change the Minimum and Maximum of the scale to e.g., 80 and 160 does not affect the scale of the chart.

In Excel 2007/2010 the logarithmic scale option has been somewhat improved, but is still unsatisfactory for small ranges of values.

Linking Chart Text Elements to a Worksheet

Any text element in a chart can be linked to a worksheet cell, causing the text in the worksheet to be displayed in the chart. (Some of this can be done automatically, as, for example, category labels.) But chart titles or unattached text boxes can also be linked to the worksheet. In this way titles can be generated automatically, or explanatory notes in text boxes can include information that will change as the data in the worksheet is modified.

Basically, to link a chart text element to a worksheet cell, you enter a formula as the chart text. The syntax of the formula is:

=worksheet_name!absolute_cell_reference.

Follow the procedure outlined in the following box.

The formula entered as chart text must be only a cell reference. If you want the chart text to be text concatenated with a number, e.g. to produce a title such as Half-wave Potential = -0.74 V, where the potential is a value in cell B2 of the worksheet, the complete formula ="Half-wave potential = "&B2&" V" must appear in the worksheet cell.

To Link Worksheet Text to a Chart

1. The chart must be activated.

2. With the chart as the active document, select the chart text element (Chart Title, X Axis Title, Y Axis Title or Unattached Text).

3. In the formula bar, type "=". Click on the cell in the worksheet that contains the desired text, and press Enter. The absolute reference to the cell will appear in the formula bar, and the text in the cell will appear in the chart.

To Switch Plotting Order in an XY Chart

Excel always uses the leftmost column (of those that you selected) for the *x*-values when it creates an XY chart. To change the plotting order, so that values from a column other than the leftmost one are used as the *x*-values, you must first create a chart in the normal way and then do one of the following:

Click on the desired data series in the chart. The series formula will appear in the formula bar, e.g.,

=SERIES(,Demo1!$A$1:$A$11,Demo1!$B$1:$B$11,1).

Manually edit the data series to reverse the X- and Y Values in the series, e.g.,

=SERIES(,Demo1!$B$1:$B$11,Demo1!$A$1:$A$11,1)

Excel 2007/2010: In the Design tab, click on the Switch Row/Column icon in the Data group. Occasionally, if a chart has been "fiddled with" extensively, Excel will refuse to swap the axes.

Excel 2003: Choose **Source Data...** from the **Chart** menu and choose the Series tab. The Series dialog box displays the range of x-values and y-values for a selected series. Enter the new ranges for x and y and press OK.

Getting Creative with Charts

Sometimes you may want to create a type of chart that is not "built in". With a little ingenuity you can create some useful charts. The example below illustrates how technical users can improvise to add more information to an XY plot in Excel.

A Chart with an Additional Axis

Figure 8-36 shows an Arrhenius plot, in which kinetics data is plotted in the form $\ln k$ vs. $1/T$. It is common in such plots to provide, for clarity, an additional X Axis scale showing the actual temperatures used. Often this additional X Axis scale is displayed along the bottom of the chart, sometimes it is placed at the top, as shown here. It isn't possible to provide an additional scale like this using the tools provided by the Chart Wizard, but it can be done by "hand crafting". The upper scale is a "fake" scale — the Tick Marks are a data series, the Tick Mark Labels are data labels.

Figure 8-37 shows the area of a worksheet containing the chart data, the cells producing the calculated curve, and the cells containing the data that create the upper X Axis tick marks (values in columns C and D in each case). The chart was created by selecting the experimental data points (C5:D10) and then using the Chart Wizard to create the chart. This data series (Series 1) was formatted to display marker points and no line. Next the cells containing the values for the calculated curve were selected (C15:D16), copied, and pasted on the chart using **Paste Special**, specifying Add Cells As New Series and X Values In First Column. This data series (Series 2) was formatted to display a line with no marker points.

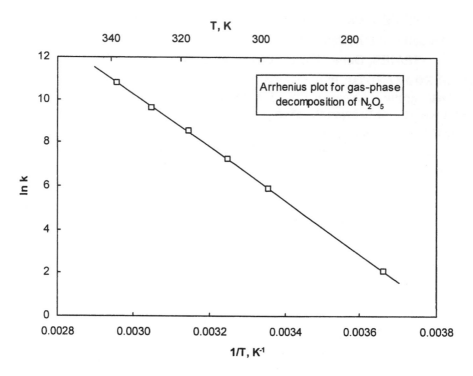

Figure 8-36. A chart with an additional axis.

The third data series in C20:D30 produces the tick marks. (To understand how these tick marks are created, remember that when Excel creates a line in an XY chart it simply connects numerical x, y coordinates with straight lines.) Since the maximum value in the Y Axis was 12, pairs of values were entered in column D (12 and 11.8, the latter having been found by trial and error to give a tick mark of suitable length). The empty rows between pairs are necessary, otherwise the tick marks would be connected to each other by a line, making a "sawtooth" pattern.

The tick marks were added to the chart by copying the table of tick mark values (C20:D30), then using **Paste Special** to add the new series (Series 3), and then formatting the series to remove the marker points and change the color of the line to black.

	A	B	C	D
3			Data	
4	T, K	k	1/T	ln k
5	338	48700	0.002959	10.793
6	328	15000	0.003049	9.616
7	318	4980	0.003145	8.513
8	308	1350	0.003247	7.208
9	298	346	0.003356	5.846
10	273	7.87	0.003663	2.063
11				
12			Calculated line	
13	(slope m =	-12376	intercept b =	47.39)
14	T, K		1/T	mx + b
15	345		0.002899	11.519
16	270		0.003704	1.555
17				
18			Tick marks for upper scale	
19	T, K		1/T	
20	280		0.003571	12
21	280		0.003571	11.8
22				
23	300		0.003333	12
24	300		0.003333	11.8
25				
26	320		0.003125	12
27	320		0.003125	11.8
28				
29	340		0.002941	12
30	340		0.002941	11.8

Figure 8-37. Worksheet layout to produce a chart with an additional axis.

Finally, the tick mark labels were added by adding data labels, either values (y-values) or labels (x-values), to this series. (It doesn't matter whether you display values or labels, since you're going to change them anyway.) Now edit the data labels individually by clicking on them and then replacing the original text with the desired text. Change the Alignment of the data labels to Above.

When this approach is applied to Series 3 to provide Tick Mark Labels, you'll get two superimposed data labels at each x value (for example, 12 and 11.8 if you chose to display values.) This means that you'll have to select one of the two labels and delete it and then select the other label and edit it. A better way to add the data labels is to add a fourth data series with a single point for each tick mark; in this example, select C20:D20,C23:D23,C26:D26,C29:D29 and **Copy**, add this as a new series, Series 4, using **Paste Special**, then format the series to set both Line and Marker to None. Then add data labels, either values (Y Values) or labels (X Values), to this series. Change the Alignment of the data labels to Above. Edit the data labels individually, replacing them with the desired text.

A Chart that Updates Automatically

If you create a chart from a data table like the one shown in Figure 8-38 and then regularly add new data, you'll find yourself extending the data series each time you add a new data point, in order to show the new data on the chart.

You can create a chart that updates automatically, so that each time you add a new line of data, the new data appears on the chart. You do this by creating three named formulas, and then use two of these formulas to modify the series function (see "The SERIES Function" in Chapter 5, "Excel 2003 Charts").

	A	B
1	Date	Sales ($)
2	3/21/2007	$105,661
3	3/22/2007	$244,332
4	3/24/2007	$272,664
5	3/25/2007	$344,765
6	3/26/2007	$386,940
7	3/28/2007	$35,090
8	3/29/2007	$16,460
9	3/30/2007	$23,010
10	3/31/2007	$47,780
11	4/1/2007	$51,920
12	4/2/2007	$38,800
13	4/3/2007	$55,060
14	4/4/2007	$98,830
15	4/5/2007	$68,050
16	4/6/2007	$273,590
17	4/7/2007	$252,520
18	4/8/2007	$292,700
19	4/9/2007	$142,000
20	4/10/2007	$157,040

Figure 8-38. Data table of a chart that updates automatically.

First, using Define Name, enter the following formula to count the number of cells containing data in column A (this requires that you have no filled cells in column A below the data, and that there are no blank rows in the data table) and assign to it the name N:

=COUNT(Sheet1!$A:$A)

Second, enter the following two formulas for the range of x-values and y-values, respectively, in the chart, and assign the names x_values and y_values:

=OFFSET(Sheet1!$A$1,1,0,Sheet1!N,1)

=OFFSET(Sheet1!$B$1,1,0,Sheet1!N,1)

Third, click on the data series in the chart that you created. The chart SERIES function will appear in the formula bar, thus:

=SERIES(Sheet1!$B$1,Sheet1! $A$2:$A$20,Sheet1! $B$2:$B$20,1)

Modify the SERIES function to read

=SERIES(Sheet1!$B$1,Sheet1!x_values,Sheet1!y_values,1)

The chart will now update automatically each time you enter new data.

Changing the Default Chart Format

Excel 2003 uses the column chart type as the default chart format, but you can change the default to any of the other built-in chart formats (maybe you'd like the XY chart type to be the default). The following sections show how to save a chart as a chart template (Excel 2007/2010) or change the Chart Wizard's default settings (Excel 2003).

Saving a Chart Template in Excel 2007/2010

In Excel 2007/2010, the Set As Default Chart button in the Change Chart Type dialog box (see Figure 4-3 in Chapter 4, "Excel 2007/2010 Charts") only sets a built-in chart type as the default chart format and does not allow you to set a custom chart format as the default. Instead, you must save the example chart as a chart template. To do this, use the steps in the following box.

To Save a Chart as a Chart Template

1. Activate the chart that you want to save as a template.
2. Click on the Design tab. In the Type group, click Save As Template to display the Save Chart Template dialog box.
3. Enter a name for the chart template, and press Save. The file will be saved as a .crtx file in the Templates folder. When you click on the Insert Chart or Change Chart Type buttons, the template will appear in the dialog box so that you can select it when you create or update a chart.

You can choose one of these chart types when you create a chart, or change the chart type of an existing chart. Unfortunately, the gallery of custom chart types does not show actual charts (unlike in Excel 2003), but only an icon indicating the type of chart, e.g., XY Scatter. As well, the name assigned to the chart is only discernable if you place the mouse pointer near the icon, as illustrated in Figure 8-39.

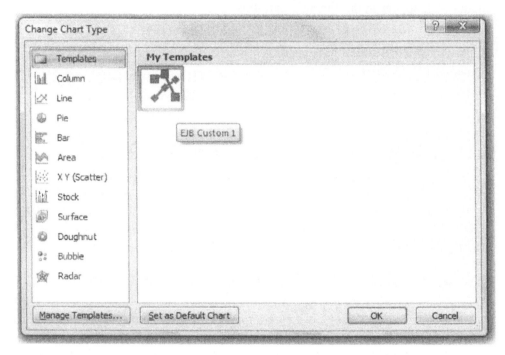

Figure 8-39. The gallery of chart templates in the Change Chart Type dialog box (Excel 2007/2010).

But you can create a chart with user-defined preferences by using the Quick Chart macro described in Chapter 19.

Changing the Default Chart Format in Excel 2003

As well as changing the default chart format from a Column chart to, for example, an XY Scatter chart, you can also save your own chart preferences as the default and thus convert the Chart Wizard from a four-step wizard to a one-step wizard. By simply pressing the Chart Wizard's Finish button in the Chart Wizard Step 1 of 4 dialog box, you can go directly to the completed chart.

Use the procedure in the following box to define a custom chart format and make that the default format. The new chart format will remain in effect through all subsequent sessions of Excel.

You can add any number of different chart formats to the list of user-defined formats and convert a chart from one format to another.

To Make a Custom Chart Format the Default Chart Format (Excel 2003)

1. Create a chart in the desired format. Usually you'll want to make the format as general as possible (do not add a title, format the axis scales, etc.), although in some instances you may want to make a chart with a specific format.

2. Activate the chart. Choose **Chart Type...** from the **Chart** menu and choose the Custom Types tab.

3. In the Select From box, press the User-defined button. A list of user-defined chart types will be displayed in the Chart Type box.

4. To add the selected chart's format to the list, press the Add button.

5. To add the selected chart's format to the list *and* make it the default chart format for all subsequent charts, press the Set As Default Chart button. In either case you'll be asked to give a name to the new custom chart type.

The procedure for deleting unwanted chart formats is shown in the box below.

To Delete a User-Defined Chart Format (Excel 2003)

1. Activate any chart.

2. Choose **Chart Type...** from the **Chart** menu and choose the Custom Types tab.

3. Press the User-Defined button in the Select From box.

4. Select the format to be deleted. The Delete button will become active.

5. Press the Delete button and press Yes to delete the format and exit.

Excel Chart Specifications (2003 and 2007/2010)

Maximum number of...

Charts linked to a worksheet	Limited by available memory
Worksheets referred to by a chart	255
Data series in one chart	255
Data points in a data series	32,000 for 2-D charts
	4,000 for 3-D charts
Data points for all data series in one chart	256,000

9

Using Excel's Database Features

This chapter demonstrates how to create and use a database on an Excel worksheet. Although database software such as Access is superior to using Excel as a database, it is sometimes useful to be able to combine the features of a database with Excel's superior calculation and data analysis features.

The Structure of a List or Database

A rectangular range of data in an Excel worksheet can be used as a *list* or *database*. A database consists of a number of *records*, each of which can contain a number of *fields*. For example, a compilation of physical properties of organic compounds, such as the one in the *CRC Handbook of Chemistry and Physics,* is a database; the row of data for a particular compound is a record and the values for the melting point, boiling point, solubility, etc., are the data fields within the record. In Excel, a list or database must be arranged in tabular form, with row or column labels; that's the only requirement.

Most fields in a database will contain values that have been entered as text, numbers or dates. A database may also contain *calculated fields*, containing values that are calculated by using Excel formulas.

Excel provides a number of menu commands and functions that permit you to use a list as a database. You can use menu commands to find records in the database that match criteria that you define, or you can use worksheet functions to extract numerical information from a database.

Creating a Database

To use Excel's database commands and functions, your list must meet the following requirements.

- The list must be in column format: Each column in the list is a field in the database, and each row in the list is a record in the database.

- The top row of the list must contain the field names, which you will use to identify the information stored in each field.

- If you leave at least one blank column and one blank row between the list and other data on the worksheet, Excel can detect and select the list automatically. If you want to use this automatic feature, however, you cannot have blank rows or columns within the list.

Defining a Database

For Excel to recognize a list as a database, simply place the mouse pointer anywhere in the list. Alternatively, you can assign the range name Database to the list. This also permits Excel to detect the database automatically. Or, you can use a reference to the desired range of cells, as illustrated in some of the examples that follow.

To assign the range name Database to the list, select the entire range of cells in the list, including the field names. Then use Formulas → Defined Names → Define Name (Excel 2007/2010) or **Insert → Name → Define** (Excel 2003) to assign the name Database to the selected range of cells. You can define only one list in a worksheet as Database; if you have more than one list in a worksheet, you will have to redefine the database range in order to switch databases.

> ***Excel Tip.*** *The name* Database *is one of Excel's built-in names. Excel recognizes the name* Database *as the reference to use in database functions. Other built-in names include* Criteria, Extract, Print_Area, Print_Titles. *Don't use any of these names as variable names except within the context of their normal use.*

Adding or Deleting Records or Fields

You can add new records to a database either by inserting new rows within the database or by entering new information below the last existing record. If you insert additional rows within the range, Excel updates the definition of the range name Database to conform to the new range. If you enter information below the last row of the defined range, you will have to redefine the range of the database by using Define Name.

If you prefer to keep the database records in the order of their entry, you can include a dummy row at the end of the database, and always insert new rows just above the dummy row. You can add new data fields (columns) in exactly the same way. Alternatively, you can use Excel's Data Form to add new records.

Updating a Database Using Data Form

Excel's Data Form is a convenient way to edit existing records or enter new data in a database. The database range must already have been defined, either by using Define Name or by selecting a cell within the block of cells that comprise the database.   Click on the Form toolbutton (Excel 2007/2010) or choose **Form**... from the **Data** menu (Excel 2003).   The Data Form dialog box will appear, with the name of the worksheet in the title bar, as in Figure 9-1.

Excel 2007/2010 Tip. The Form button is not included on the Ribbon, but you can add it to the Quick Access Toolbar.   Follow the procedure in Chapter 1 for customizing the Quick Access Toolbar.

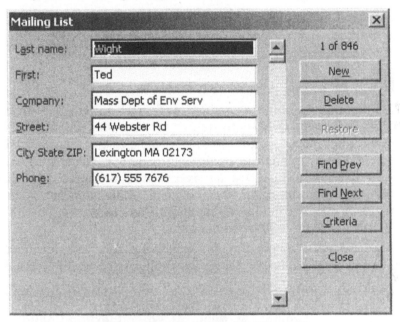

Figure 9-1. The Data Form dialog box.

Each field name in the database appears in the dialog box, along with the entries from the first record in the database. The current record and the number of records in the database are displayed in the upper right corner of the dialog box. Entries in *editable fields* display their values in a text entry box, while values that do not appear in a text box are either in calculated fields or in fields whose contents are protected.

To move from record to record, use either the Find Prev and Find Next buttons, the up and down arrow keys, or the scroll bar. The first field of each record is highlighted. Move forward from field to field within a record by using the Tab key. You can also select a field for editing by clicking the mouse pointer in it.

Edit characters in a text box in the usual way. Once a change has been made in a field, the Restore button is enabled; you can use it to restore the original data if you spot an error in the new text. To edit a single field in all records, tab forward to that field or click the mouse pointer in it; then use the up or down arrow keys to move to the same field in other records.

To add a new record, press the New button or drag the scroll button to the bottom. New Record will be displayed in the upper right corner of the dialog box. The new record will be added to the bottom of the database. Formulas of calculated fields are also included in the new record. If the new record will overwrite existing information outside the database, you'll get the "Cannot extend database" message.

The reference to Database will be updated automatically when you add or remove records.

Sorting a List

One of the most common operations performed on a list is sorting it in increasing or decreasing order with respect to the value in one of its fields.

Figure 9-2 illustrates a portion of a list of the elements, together with their symbols, atomic weights and electronic configurations. The list (which was imported into Excel by using a scanner, as described in Chapter 10) is arranged in order of increasing atomic number. It may seem redundant to include a column of atomic numbers in the list, but this column is necessary if you want to return the sorted list to its original order.

To sort the list according to any of its fields, select the complete list (be sure to select all rows and columns, or the list may become irreversibly scrambled).

	A	B	C	D	E
1	Name	Symbol	At. Num.	At. Wt.	Electron Config
2	Hydrogen	H	1	1.00797	1s1
3	Helium	He	2	4.0026	1s2
4	Lithium	Li	3	6.939	[He] 2s1
5	Beryllium	Be	4	9.0122	[He] 2s2
6	Boron	B	5	10.811	[He] 2s2 2p1
7	Carbon	C	6	12.01115	[He] 2s2 2p2
8	Nitrogen	N	7	14.0067	[He] 2s2 2p3
9	Oxygen	O	8	15.9994	[He] 2s2 2p4
10	Fluorine	F	9	18.9984	[He] 2s2 2p5
11	Neon	Ne	10	20.183	[He] 2s2 2p6
12	Sodium	Na	11	22.9898	[Ne] 3s1

Figure 9-2. A portion of a list of atomic weights and symbols of the elements.

For a large list such as this one, it's convenient to use one of the methods for selecting a block of cells described in Chapter 1: Select the first row of cells and press Ctrl+Shift+(down arrow) or place the mouse pointer on the bottom edge of the selected row, hold down the Shift key and double-click to select all cells to the bottom of the block.

Sorting a List Using Excel 2003

Once you have selected the entire list, you can use the Sort Ascending ![icon] or Sort Descending ![icon] tool buttons for simple sorting, or the **Sort...** command in the **Data** menu for more complicated sorting tasks. Choose **Sort...** from the **Data** menu to display the Sort dialog box.

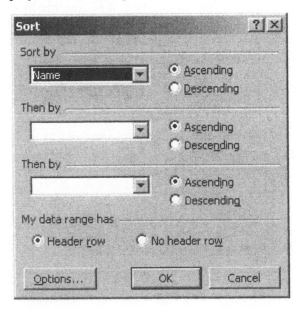

Figure 9-3. The Excel 2003 Sort dialog box.

To choose the field to sort by (sometimes called a *sortkey*), choose the field name from the Sort By list box (Figure 9-3) or type a reference in the input box; for example, to sort by atomic symbol, you can either choose symbol (field names are not case-sensitive) or enter $B$3 in the Sort By input box. Then press OK. Figure 9-4 shows a portion of the list, sorted in ascending order according to column 1. If the list is not sorted in the way that you intended, press the Undo button or choose **Undo Sort** from the **Edit** menu.

	A	B	C	D	E
1	**Name**	**Symbol**	**At. Num.**	**At. Wt.**	**Electron Config**
2	Actinium	Ac	89	[227]	[Rn] 7s2 6d1
3	Aluminum	Al	13	26.9815	[Ne] 3s2 3p1
4	Americium	Am	95	[243]	[Rn] 7s2 5f7
5	Antimony	Sb	51	121.75	[Kr] 5s2 4d10 5p3
6	Argon	Ar	18	39.948	[Ne] 3s2 3p6
7	Arsenic	As	33	74.9216	[Ar] 4s2 3d10 4p3

Figure 9-4. Portion of a list sorted according to element name.

Sorting According to More than One Field

As you can see from the Sort dialog box, it is possible to sort by up to three separate fields in Excel 2003. For example, you can sort a list of chemistry students by descending order according to year of graduation and by ascending order alphabetically within each year.

Sorting a List Using Excel 2007/2010

The toolbuttons for sorting are found in the Data tab. Just as in Excel 2003, you can use the Sort Ascending ↓ or Sort Descending ↓ tool buttons for simple sorting. For advanced sorting options, press the Sort button in the Data tab to display the Sort dialog box (Figure 9-5).

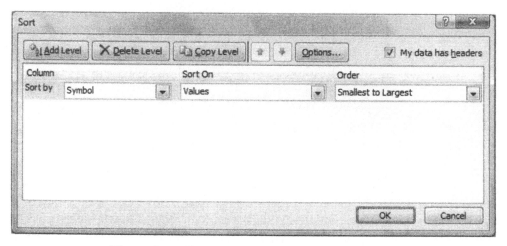

Figure 9-5. The Excel 2007/2010 Sort dialog box.

In Excel 2007/2010 you can sort with up to 64 sortkeys. Press the Add Level button to add additional sort levels. In addition to sorting by value, you can also sort by cell or font color.

Sort Options

Pressing the Options... button in the Sort dialog box displays the Sort Options dialog box and allows you to change the default sorting options.

Excel assumes that your data fields are in columns and sorts your list by rearranging rows. If you want to sort a list horizontally (i.e., to rearrange the columns of the list rather than the rows), press the Sort Left to Right button in the Orientation box (Figure 9-6).

You can choose case-sensitive sorting, in which lowercase letters follow uppercase letters in an ascending sort (AaBbCc..., not ABC...abc...). If you choose this option, the sortkeys will also be case-sensitive.

The Excel 2003 Sort Options dialog box (Figure 9-6) contains four lists for custom sorts (other than ascending or descending). For example, you can sort in the order Jan, Feb, Mar, Apr, etc. But note that if you spell the abbreviation of the ninth month as "Sept" rather than "Sep", that month won't be sorted correctly.

In Excel 2007/2010, access to the lists for custom sorts is through the Order drop-down list box; see Figure 9-5.

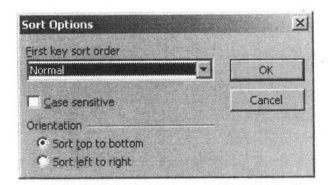

Figure 9-6. The Excel 2003 Sort Options dialog box.

You can also create a custom sort order for your own specialized application. To add a custom list in Excel 2003, choose **Options...** from the **Tools** menu and choose the Custom List tab (Figure 9-7).

You can enter the custom list values or a reference to a range of cells containing the values. Once added, the list becomes part of Excel and can be used in any workbook.

To create a custom list in Excel 2007, use Office Button→Excel Options→Top Options to display the Custom Lists dialog box, essentially identical to what is shown in Figure 9-7.

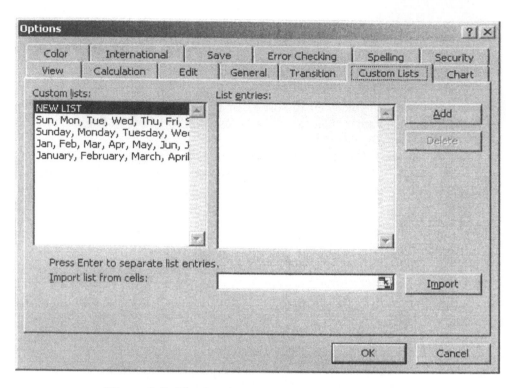

Figure 9-7. The Excel 2003 Custom Lists dialog box.

To create a custom list in Excel 2010, use File→Options→Advanced to display the Excel Options dialog box. Scroll down to the General category (near the bottom) and click on the Edit Custom Lists... button to display the Custom Lists dialog box.

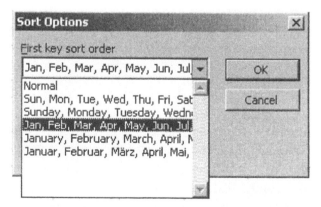

Figure 9-8. The Excel 2003 Sort Options dialog box.

Figure 9-8 shows a custom list, the months of the year in German, that was imported into Excel and is one of the available Sort Options.

Using AutoFilter
to Obtain a Subset of a List

Often you'll want to examine a subset of a list — only records for which one or more data fields match certain criteria. The process of abstracting a subset of a list is known as using a *data filter* or *filtering* the database. The Filter command makes it easy to obtain a subset of a list.

	Sample Number	Molecular Weight (kD)	MW Dispersity ("w/Mn)	Specific Gravity ("STM D792)	Refractive Index	Dielectric Constant @ 100 Hz ("STM D150)	Rockwell Hardness ("STM D785)	Tensile Strength PSI/1000 ("STM D638)	Melting Point (°C)
1									
2	91976	14.6	2.4	1.135	1.497	2.88	110	15.4	232
3	91977	14.6	2.3	1.149	1.502	2.88	110	16.9	231
4	91978	11.5	2.6	1.105	1.500	2.88	107	16.9	209
5	91979	12.0	2.8	1.117	1.503	2.94	115	14.2	208
6	91980	12.8	2.9	1.181	1.496	2.89	108	13.8	207
7	91981	12.2	2.6	1.172	1.497	2.94	111	13.9	205
8	92007	13.6	2.2	1.174	1.501	2.88	114	14.4	215

Figure 9-9. A portion of a list with AutoFilter buttons displayed.

Figure 9-9 illustrates a portion of a list of polymer research samples and some of their physical properties. To filter the list, first select the complete list; if the list is separated from the rest of the worksheet data by blank cells, you need only select any cell within the list. You can also select a partial area to be filtered. Click on the Filter button in the Data tab (Excel 2007/2010) or choose **Filter** from the **Data** menu and then choose **AutoFilter** from the submenu (Excel 2003). Excel adds drop-down arrow buttons to row 1 of the list, which ideally should contain column labels.

To use a data filter on molecular weight, for example, click the arrow button in that column (column B). Excel displays a drop-down list of all values in the column, plus (All), (Top 10) and (Custom...), as shown in Figure 9-10. To display all records that match one of the values in the selected field, you can select it from the list. For example, to display all entries with molecular weight = 12.0 kD, simply click on that value in the drop-down list box. Figure 9-11 shows the filtered list, with the two entries with molecular weight of 12.0 kD.

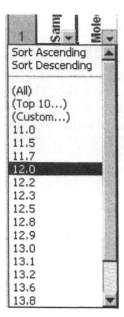

Figure 9-10. AutoFilter drop-down list box.

	A	B	C	D	E	F	G	H	I
1	Sample Number	Molecular Weight (kD)	MW Dispersity (w/Mn)	Specific Gravity STM D792)	Refractive Index	Dielectric Constant @ 100 Hz STM D 150)	Rockwell Hardness STM D785)	Tensile Strength PSI/1000 STM D638)	Melting Point (°C)
5	91979	12.0	2.8	1.117	1.503	2.94	115	14.2	208
13	92167	12.0	3.4	1.139	1.498	2.94	107	15.3	210

Figure 9-11. A list that has been filtered to display
only those entries with molecular weight of 12.0 kD.

Advanced AutoFilter in Excel 2003. To perform logical comparisons, choose Custom... from the drop-down list to display the Custom AutoFilter dialog box (Figure 9-12). To display all entries with molecular weight greater than 12.5 kD, click the comparison operator drop-down arrow and choose "is greater than". Tab over to the text box and enter 12.5, and then press the OK button. The records in the list for samples with molecular weight greater than 12 kD are displayed (Figure 9-13).

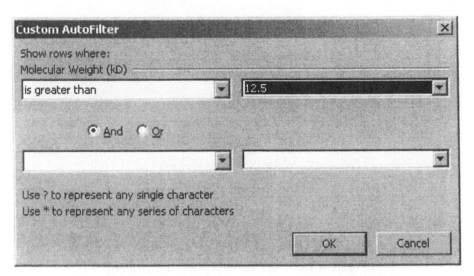

Figure 9-12. The Excel 2003 Custom AutoFilter dialog box.

	A	B	C	D	E	F	G	H	I
1	Sample Number	Molecular Weight (kD)	MW Dispersity (Mw/Mn)	Specific Gravity (ASTM D792)	Refractive Index	Dielectric Constant @ 100 Hz (ASTM D150)	Rockwell Hardness (ASTM D785)	Tensile Strength PSI/1000 (ASTM D638)	Melting Point (°C)
2	91976	14.6	2.4	1.135	1.497	2.88	110	15.4	232
3	91977	14.6	2.3	1.149	1.502	2.88	110	16.9	231
6	91980	12.8	2.9	1.181	1.496	2.89	108	13.8	207
8	92007	13.6	2.2	1.174	1.501	2.88	114	14.4	215
9	92039	13.6	2.1	1.184	1.500	2.92	112	15.6	212
10	92071	14.5	3.4	1.135	1.496	2.91	113	16.3	185
11	92103	15.2	3.6	1.143	1.500	2.87	109	16.7	174
14	92199	12.8	3.0	1.178	1.495	2.89	106	16.7	215

Figure 9-13. Portion of a list that has been filtered to display
only those records with molecular weight greater than 12.5 kD.

You can add a second data filter on data in the same field using the Custom
AutoFilter dialog box. The second filter can be either an *And filter* (e.g.,
samples with molecular weight greater than 12.5 kD and less than 13.5 kD) or an
Or filter.

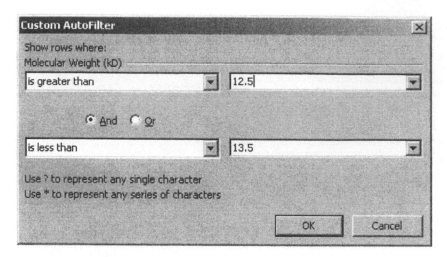

Figure 9-14. The Custom AutoFilter dialog box
with entries for an AND filter.

	Sample Number	Molecular Weight (kD)	MW Dispersity ("w/Mn)	Specific Gravity (STM D792)	Refractive Index	Dielectric Constant @ 100 Hz ("STM D 150)	Rockwell Hardness ("STM D785)	Tensile Strength PSI/1000 ("STM D638)	Melting Point (°C)
6	91980	12.8	2.9	1.181	1.496	2.89	108	13.8	207
14	92199	12.8	3.0	1.178	1.495	2.89	106	16.7	215
22	92355	13.1	1.9	1.114	1.501	2.90	108	14.7	212
23	92356	13.1	1.8	1.131	1.503	2.91	108	17.9	185
25	92358	13.0	3.0	1.188	1.500	2.88	106	13.0	218
27	92360	13.2	2.8	1.175	1.504	2.91	107	16.2	215
28	92400	12.9	2.5	1.110	1.555	2.87	111	15.1	211

Figure 9-15. A list that has been filtered to display
only those records with molecular weight between 12.5 and 13.5 kD.

To restore the complete list, simply choose **AutoFilter** from the **Data** menu again. Notice that the **AutoFilter** command is checked, indicating that the list has been filtered. Choose **AutoFilter** again to restore the complete list and remove the drop-down buttons.

Advanced AutoFilter in Excel 2007/2010.   To perform logical comparisons on the values in a column, click on the drop-down button for that column, and click on Number Filters. This will display the shortcut menu of logical comparisons. You can choose one of the logical comparisons from the submenu (Equals..., Greater Than..., etc.) or choose Custom Filter...   Either way, the Custom AutoFilter dialog box, identical to the Excel 2003 dialog box, will be displayed.

Using Multiple Data Filters

You can also add a second filter by querying data in a second field. For example, once you have filtered the list to display those records with molecular weight greater than 12.5 kD, you can display all samples with melting point greater than 215°C. Simply click on the drop-down arrow in the melting point column, choose (Custom...), enter the appropriate values in the Custom AutoFilter dialog box (Figure 9-16) and press OK.

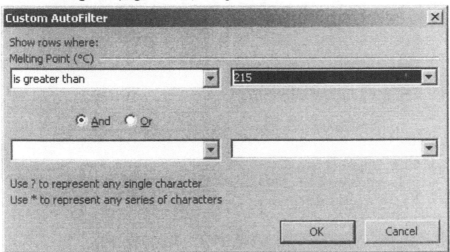

Figure 9-16. The Custom AutoFilter dialog box
with an entry to filter on a second column.

Figure 9-17 shows the subset of records having both molecular weight greater than 12.5 kD and melting point greater than 215°C.

To "undo" a particular filter, click the drop-down list button in that column and choose "(All)" from the list.

Although only three records are displayed, you can't calculate the average molecular weight of the three samples by entering the formula =AVERAGE() in a worksheet cell and selecting the displayed cells in column B that contain the molecular weight information. The range entered by selecting will be (in this example) B2:B25, and the average will not be the desired one.   The most convenient way to apply SUM, AVERAGE, STDEV, etc., to a subset is to **Copy** the desired cells, **Paste** them in a convenient location, and then perform the calculation. But see "Using Database Functions" at the end of this chapter for another way to obtain an average (for example) of a subset of a list.

	Sample Number	Molecular Weight (kD)	MW Dispersity (Mw/Mn)	Specific Gravity (ASTM D792)	Refractive Index	Dielectric Constant @ 100 Hz (ASTM D150)	Rockwell Hardness (ASTM D785)	Tensile Strength PSI/1000 (ASTM D638)	Melting Point (°C)
6	91980	12.8	2.9	1.181	1.496	2.89	108	13.8	207
14	92199	12.8	3.0	1.178	1.495	2.89	106	16.7	215
22	92355	13.1	1.9	1.114	1.501	2.90	108	14.7	212
23	92356	13.1	1.8	1.131	1.503	2.91	108	17.9	185
25	92358	13.0	3.0	1.188	1.500	2.88	106	13.0	218
27	92360	13.2	2.8	1.175	1.504	2.91	107	16.2	215
28	92400	12.9	2.5	1.110	1.555	2.87	111	15.1	211

Figure 9-17. A list that has been filtered on two fields, to display only those records with molecular weight greater than 12.5 kD and melting point greater than 215°C.

Using Advanced Filter
to Obtain a Subset of a List

You can also filter a database with multiple criteria by using Advanced Filter. Advanced Filter is more versatile than AutoFilter: It can handle more complex criteria logic, and the filtered list can be written to a separate range of the worksheet.

To use Advanced Filter, you must create a Criteria range somewhere on your worksheet, as shown later in Figures 9-19, 9-20, 9-21, 9-22 or 9-23. You can give it a range name if you wish, by using Define Name. Any name is acceptable, but Criteria is best because Excel will recognize it. You can also establish an Extract range (see "Extracting Records" later in this chapter) and give it the name Extract. If you have structured your worksheet correctly (see "Creating a Database" earlier in this chapter), you do not need to assign the name

Database to the list to be filtered; just select any cell within the list and Excel will recognize the range as a list. However, using Create Name to assign the name Database to the list is easy and will prevent Excel from misidentifying the range of cells to use.

Now click on Advanced in the Data tab (Excel 2007/2010) or choose **Advanced Filter...** from the **Filter...** submenu of the **Data** menu (Excel 2003). Excel will display the Advanced Filter dialog box (Figure 9-18). If you selected a cell within the list, the reference to the list will appear in the List Range box. If you had assigned the name Criteria to the criteria table, the reference will appear in the Criteria Range box.

You can choose to Filter The List In-Place or Copy To Another Location. Filter The List In-Place hides the rows in the list that do not meet the criteria; Copy To Another Location copies the rows that meet the criteria to another location in the worksheet or to another worksheet (see "Extracting Records" later in this chapter).

Figure 9-18. The Advanced Filter dialog box.

Defining and Using Selection Criteria

You define selection criteria by setting up a table in the worksheet with one or more field names in a row and the desired criteria below the field names. The Criteria field names must be the same as the Database field names. For example, in Figure 9-19, the range K2:K3 is the criteria table to select all records having molecular weight greater than 12.5 kD. Now select the range of cells containing the field name and the criterion and use Define Name to assign the name Criteria to the selected range of cells. At any time, only one range in a worksheet can be defined as the Criteria table.

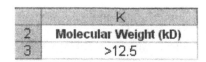

Figure 9-19. A simple criteria table.

It's good practice to copy the complete row of database field names and use it to create a Criteria table, as in Figure 9-20. You can then enter single or multiple criteria in the appropriate cells.

Although the criteria range can be located anywhere on the worksheet, do not place the criteria range below the list if you plan to add more information to the list by using the **Form** command on the **Data** menu. The new information will be added to the first row below the list, and if this row is not blank, Excel cannot add the new information.

	A	B	C	D	E	F	G	H	I
31	Sample Number	Molecular Weight (kD)	MW Dispersity (Mw/Mn)	Specific Gravity (ASTM D792)	Refractive Index	Dielectric Constant @ 100 Hz (ASTM D150)	Rockwell Hardness (ASTM D785)	Tensile Strength PSI/1000 (ASTM D638)	Melting Point (°C)
32		>12.5							

Figure 9-20. Criteria tables are best made by copying the row of database field names.

Using Multiple Criteria

You can select records on the basis of two or more criteria. Multiple criteria can be combined to produce a logical AND, a logical OR, or a combination of AND and OR.

If you enter two criteria in the same row, as in Figure 9-21, you have created an AND criterion (e.g., molecular weight >12.5 kD and melting point >215°C). If you enter two criteria in separate rows, you have created an OR criterion (e.g., molecular weight >1.5 kD or melting point <200°C) as illustrated in Figure 9-22.

If the AND criteria apply to the same field (e.g., all samples with molecular weight >12.5 kD and <13 kD), you must duplicate the field name in the Criteria range (see Figure 9-23).

Figure 9-21. A criteria table with entries for an AND filter.

Figure 9-22. A criteria table with entries for an OR filter.

Figure 9-23. A criteria table with AND logic applied to the same field.

Special Criteria for Text Entries

When you use a number value in a Criteria range, Excel returns only the records that match the field value exactly. But when you use a text value as a criterion, Excel returns all records that contain text values in the specified field that *begin* with the text value. For example, in a database of alumni information (LastName, FirstName, StreetAddress, City, State, ZIP, YearOfGraduation, etc.), using the letter N as criterion in the LastName criteria field will return all names that begin with N.

You can use wildcard characters in text criteria. The ? wildcard character represents any single character, the * character represents any number of characters. Thus, for example, the criterion *phen* will extract, from a database of chemicals, any entry that contains the character string "phen", such as phenol, 1,10-phenanthroline and benzophenone.

To extract only the records that have an exact match between a text criterion and a field value, enter the criterion in the form

="*criterion_text*"

where *criterion_text* is the text string you want to find.

Extracting Records

To extract a copy of all records that meet specified criteria, first define Database and Criteria ranges as described earlier. Then do the following to create a destination range for the extracted records: **Copy** the database field names and **Paste** them in a suitable area of the worksheet as labels for the extracted information. Select the row of field names and enough empty rows below the column labels for the extracted information; define this range as the Extract range by assigning the name Extract to the selected range of cells using Define Name.

Now choose **Advanced Filter...** from the **Filter...** submenu of the **Data** menu. Excel will display the Advanced Filter dialog box. Press the button for Copy To Another Location and, if you have assigned the name Extract, Excel will automatically enter the reference to the Extract range, as illustrated in Figure 9-24. Otherwise, enter a reference in the input box.

When you press OK, the records that meet the criteria will be copied to the Extract range (Figure 9-25). (Remember that all cells from the bottom of the Extract range to the bottom of the worksheet will be over-written.)

Figure 9-24. An extract of a database.

	Sample Number	Molecular Weight (kD)	MW Dispersity (Mw/Mn)	Specific Gravity (ASTM D792)	Refractive Index	Dielectric Constant @ 100 Hz (ASTM D150)	Rockwell Hardness (ASTM D785)	Tensile Strength PSI/1000 (ASTM D638)	Melting Point (°C)
	A	B	C	D	E	F	G	H	I
37	91976	14.6	2.4	1.135	1.497	2.88	110	15.4	232
38	91977	14.6	2.3	1.149	1.502	2.88	110	16.9	231
39	91980	12.8	2.9	1.181	1.496	2.89	108	13.8	207
40	92007	13.6	2.2	1.174	1.501	2.88	114	14.4	215
41	92039	13.6	2.1	1.184	1.500	2.92	112	15.6	212
42	92071	14.5	3.4	1.135	1.496	2.91	113	16.3	185

Figure 9-25. A portion of an extract of a database.

Excel Tip. *Be sure to select a range of empty cells as the Extract range. If you select only the field names as the Extract range, the* **Extract** *command will clear all cells below the Extract field names to the bottom of the worksheet, regardless of whether values are extracted into them, erasing any previous information. The* **Undo** *command cannot be used to reverse the action and restore the missing information.*

	A	B	C	D	E
36	Sample Number	Molecular Weight (kD)	Specific Gravity (ASTM D792)	Refractive Index	Melting Point (°C)
37	91976	14.6	1.135	1.497	232
38	91977	14.6	1.149	1.502	231
39	91980	12.8	1.181	1.496	207
40	92007	13.6	1.174	1.501	215
41	92039	13.6	1.184	1.500	212
42	92071	14.5	1.135	1.496	185

Figure 9-26. A portion of an extract containing only selected fields.

You can create an Extract range that contains only selected fields of each record. Only the fields included in the Extract range will be returned, as shown in Figure 9-26.

Using Database Functions

There are 12 *database functions* that return information about the records in a database: DAVERAGE, DCOUNT, DCOUNTA, DGET, DMAX, DMIN, DPRODUCT, DSTDEV, DSTDEVP, DSUM, DVAR and DVARP. Some of these are particularly useful and are described here. See *Microsoft Excel Worksheet Function Reference* or Excel's On-Line Help for further information.

DAVERAGE returns the average of the values in the specified field of all records that match the criteria.

DMAX or DMIN returns, respectively, the maximum or minimum value in the specified field of all records that match the criteria.

DSTDEV returns the standard deviation of the values in the specified field of all records that match the criteria.

All 12 database functions have the syntax (*database, field, criteria*). *Database* is either the reference to the database or the name assigned to it. *Field* is either the name of the field, as text, or a reference to the cell containing the filed name, or a number indicating the position of the field within the database. *Criteria* is either the reference to the criteria range or the name assigned to it.

Example. To obtain the average refractive index of all samples in the database that have molecular weight greater than 12 kD and melting point greater than 215°C, enter the database function DAVERAGE(Database, 5, Criteria). The formula returns the value 1.49948135, the average for the three samples that were extracted in the example shown in Figure 9-16.

You can use the formula =DAVERAGE(Database,"Refractive Index", Criteria) in place of the formula =DAVERAGE(Database, 5, Criteria). The field name must be identical to the field name in the database range. If the name used in the function is not identical to the name of one of the fields in the database, you'll get the #VALUE! error value.

10

Importing Data into Excel

Since you use Excel largely for the analysis of experimental data, the problem of how to get that data into Excel is crucial. Certainly you don't want to spend time transcribing data from a piece of paper into an Excel worksheet. There are several ways of transferring data to Excel. In order of decreasing preference, they are:

- directly from an instrument to an Excel worksheet
- from a data file on a CD, or from a data file received electronically, to an Excel worksheet
- from data on paper, via a scanner, to an Excel worksheet

Direct Input of Instrument Data into Excel

Most modern instruments have the capability to export data directly in Excel spreadsheet format. For those instruments lacking that capability, there are a number of software programs (some free) that are available to accept data from instruments and transfer it in real time to Excel. Most collect real-time data from an RS232 device directly into any Windows application (Excel, Access, etc.). With some programs, the user can define output strings which can be sent to an instrument to control it directly from Excel. Discussion of this and similar software is beyond the scope of this book; specific information can be obtained from the manufacturers. A Web search for "data from instrument to Excel software" will provide a number of sources.

Importing Data Files
Using the Text Import Wizard

The Text Import Wizard permits you to import data from a data file that is not in Excel format. The examples in this chapter involve files that contain numbers or dates in formats used in the United States. See Chapter 12, "Other Language Versions of Excel" for instructions on how to import data in formats used in Europe.

The Text Import Wizard can import files with two different types of file format: delimited files, or fixed width files. These file formats will be described in the following sections.

Note that the **Import External Data...** command in the **Data** menu will ultimately lead you to the Text Import Wizard dialog box, albeit by a somewhat indirect path.

Importing a Delimited File

A data file consists of a number of records. Each record can consist of a number of fields (for example, Name, Street, City, State, ZIP). A common file formats uses commas to separate fields within a record; this file format is referred to as a comma-delimited file or comma-separated values (.CSV).

Most modern instruments are controlled by a computer and can write a data file to disk. Many are interfaced to external PCs. The following example shows how to import spectrophotometric data, saved to disk as a text file, into Excel. The data file comes from a Hewlett-Packard diode-array spectrophotometer interfaced to a PC; the values were collected between 350 and 820 nm at 2 nm intervals and saved to disk as a comma-delimited text file, filename NISPEC.DAT, available on the CD that accompanies this book. Each data record consists of (wavelength, absorbance). We'd like to read the data file into Excel and get the data fields into separate columns.

To import the data, you must open the file from Excel. To ensure that the NISPEC.DAT filename is displayed in the Open dialog box, you must choose All Files in the Files Of Type list box at the bottom of the Open dialog box. The list box is shown in Figure 10-1.

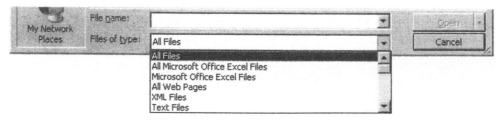

Figure 10-1. The Files Of Type list box in the Excel 2003 **Open** dialog box.

After selecting All Files from the list box, click on the NISPEC.DAT filename and press the Open button. The Text Import Wizard Step 1 of 3 dialog box (Figure 10-2) will be displayed. (The first record, sometimes referred to as the header row, usually contains information about the file. In this case the header row contains 236, the number of data points.

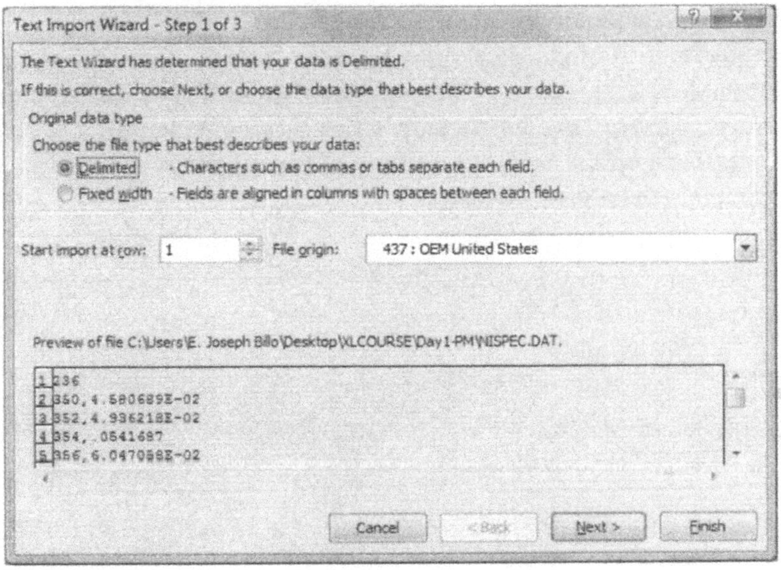

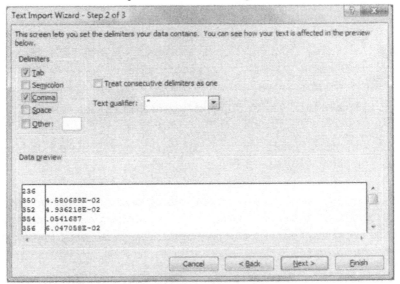

Figure 10-2. The Text Import Wizard Step 1 of 3 dialog box
with preview of data to be parsed.

Figure 10-3. The Text Import Wizard Step 2 of 3 dialog box
with preview of parsed data.

The Text Import Wizard usually recognizes whether the text file is Delimited or Fixed Width and displays the choice in the Step 1 dialog box, shown in Figure 10-2. You can override Excel's choice and manually select either the Delimited or the Fixed Width option.

If you chose Delimited in Step 1, the Step 2 of 3 dialog box allows you to select the type of delimiter and to see a preview of the parsed data (Figure 10-3).

The Step 3 of 3 dialog box (Figure 10-4) offers several options in the Column Data Format category. You can opt to import only selected columns, or apply General, Text or Date formatting. Click on a column header to select it. If you press the Date option button, a column of text values representing dates will be converted into date serial numbers.

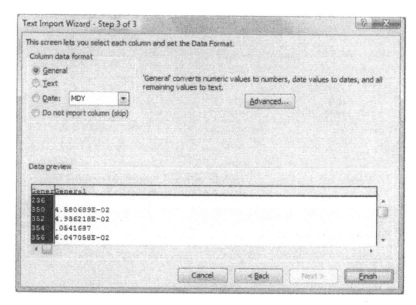

Figure 10-4. The Text Import Wizard Step 3 of 3 dialog box.

When you press Finish, a new Excel workbook will be created, containing the data. The parsed data are shown in Figure 10-5.

*Excel Tip.   The workbook is in text file format.   Be sure to **Save** it in Microsoft Excel Workbook format if you have added formulas to the worksheet and want them to be preserved; otherwise the formulas will be converted to values.*

	A	B
1	236	
2	350	4.58E-02
3	352	4.94E-02
4	354	0.054169
5	356	6.05E-02
6	358	6.84E-02
7	360	7.72E-02
8	362	9.00E-02
9	364	0.102127
10	366	0.117447
11	368	0.134659
12	370	0.153824

Figure 10-5. A portion of the text file after parsing.

Importing a Fixed-Width File

A fixed width file contains columnar data separated by spaces. If you chose Fixed Width in Step 1, the Step 2 dialog box will display a ruler on which Excel has positioned vertical bars indicating the fields. You can see an example of a fixed width file in Figure 10-13.

Using Convert Text to Columns

The Text to Columns button in the Data tab (Excel 2007/2010) or the **Text to Columns...** command in the **Data** menu (Excel 2003) allows you to parse text already present in Excel. You can parse text in a single column, either in delimited or fixed-width format, into separate columns.

	A
1	ANTONIUS,STEPHEN J
2	BRUNEL,STEVEN D
3	CARRESTO,KATHY E
4	LAKLIS,CLAIR L
5	PEDROSO,BENITO A
6	SOUSSANE,WALID
7	WOODSTOCK,PAUL
8	WYNDLAKE,KEVIN D
9	ZILARIO,J PATRICK

Figure 10-6. Text in a single column.

Parsing a Delimited File

The data in Figure 10-6, although it is already in Excel, has the same data structure as a comma-delimited text file (LastName and FirstName fields separated by a comma delimiter).

To use the Convert Text To Columns Wizard to parse the text into two separate columns, select the data in column A (the Wizard can only operate on text in a single column) and choose the Text to Columns command. This will display the Convert Text to Columns Step 1 of 3 dialog box, which looks exactly like the Text Import Wizard Step 1 of 3 dialog box that we saw earlier (Figure 10-2). Make sure that the button for Delimited is pressed, and then press the Next button to continue to the Step 2 of 3 dialog box (Figure 10-7).

Check the box for the comma delimiter. A preview of the parsed data will be shown. Press Next to continue to the Step 3 of 3 dialog box. As with the Text Import Wizard, the Convert Text To Columns Step 3 dialog box allows you to import only selected columns, or apply General, Text or Date formatting. The Step 3 dialog box contains a Destination input box (not present in the Text Import Wizard) that allows you to specify where you want the parsed data to be placed. The default option overwrites the original data column.

When you press the Finish button, the parsed data will be parsed into separate columns, as shown in Figure 10-8.

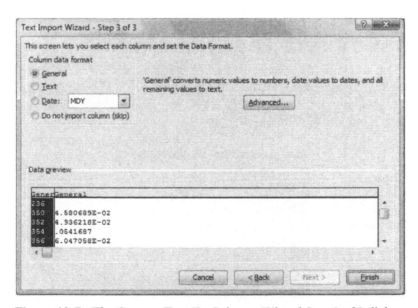

Figure 10-7. The Convert Text To Columns Wizard Step 2 of 3 dialog box, with preview of the parsed data.

	A	B
1	ANTONIUS	STEPHEN J
2	BRUNEL	STEVEN D
3	CARRESTO	KATHY E
4	LAKLIS	CLAIR L
5	PEDROSO	BENITO A
6	SOUSSANE	WALID
7	WOODSTOCK	PAUL
8	WYNDLAKE	KEVIN D
9	ZILARIO	J PATRICK

Figure 10-8. The text after parsing into two columns.

Parsing a Fixed-Width File

Figure 10-9 shows part of a data table from a scientific paper. If you **Copy** the table and **Paste** it into Excel, you'll get the result shown in Figure 10-10.

no.		log Sum$_{AA}$	log VDW$_{vol}$	clog P	$t_{R\ exp}$ [min]	$t_{R\ pred}$ [min]	Δt_R [min]
1	AA	0.70	2.1603	−0.74	3.05	4.19	1.14
2	AG	0.68	2.1047	−1.28	2.66	4.14	1.48
3	AF	1.08	2.3402	0.95	10.71	10.70	0.00
4	YL	1.18	2.4394	1.86	11.57	12.24	0.67
5	DD	0.67	2.2847	−1.98	2.63	1.43	1.19
6	ML	1.14	2.4508	−0.17	11.94	10.20	1.74
7	WW	1.35	2.5420	2.18	15.84	14.67	1.17
8	GM	0.90	2.3318	−1.90	7.30	5.74	1.56
9	GH	0.68	2.2600	−1.89	2.90	2.04	0.85
10	GL	1.02	2.2505	−0.08	9.74	9.91	0.17
11	WF	1.31	2.5058	2.41	15.49	14.44	1.05

Figure 10-9. The data table, as it appeared on the original paper.
(Data from *J. Proteome Res.* **2005**, *4*, 555)

Excel Note. When text like the above is pasted into Excel, the columns do not usually appear to be aligned, because the default font (Calibri in Excel 2007/2010, Arial in Excel 2003) is not a monospaced font, but a proportional font. In monospaced fonts, character widths are proportional to the width of the character – an "i" takes less space than a "w", whereas in monospaced fonts, each character has the same width.   (Even in proportional fonts, the number characters are monospaced, so that columns of figures line up properly.)   There are only a few monospaced fonts available in Excel; Courier, a font that looks like typewriting, is one of them. If you were to select cells A1:A13 in Figure 10-10 and change the font to Courier, you would see that the columns are perfectly lined up.

	A	B	C	D	E
1	no. peptide sequence log SumAA log VDWVol clog P				
2	1 AA 0.70 2.1603 -0.74 3.05 4.19 1.14				
3	2 AG 0.68 2.1047 -1.28 2.66 4.14 1.48				
4	3 AF 1.08 2.3402 0.95 10.71 10.70 0.00				
5	4 YL 1.18 2.4394 1.86 11.57 12.24 0.67				
6	5 DD 0.67 2.2847 -1.98 2.63 1.43 1.19				
7	6 ML 1.14 2.4508 -0.17 11.94 10.20 1.74				
8	7 WW 1.35 2.5420 2.18 15.84 14.67 1.17				
9	8 GM 0.90 2.3318 -1.90 7.30 5.74 1.56				
10	9 GH 0.68 2.2600 -1.89 2.90 2.04 0.85				
11	10 GL 1.02 2.2505 -0.08 9.74 9.91 0.17				
12	11 WF 1.31 2.5058 2.41 15.49 14.44 1.05				

Figure 10-10. Portion of the data table before parsing into columns.

To put this text in a readable form, use the Convert Text to Columns Wizard. The data to be parsed must be in a single column. Select the column of data to be parsed. Choose the Text to Columns button in the Data tab (Excel 2007/2010) or **Text to Columns...** from the **Data** menu (Excel 2003). The Text to Columns dialog boxes are essentially identical to those of the Text Import Wizard.

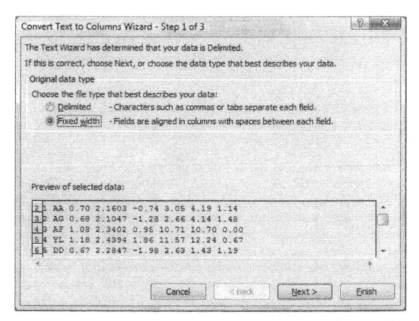

Figure 10-11. The Text to Columns Step 1 dialog box
with preview of data to be parsed.

The Step 2 of 3 dialog box shows a preview of the parsed data; Excel recognized that the data is in Fixed Width format. The Step 3 dialog box displays a ruler with break lines separating the fields. To remove a break line, double-click on it; to move a break line, drag it; to add a break line, click on the ruler.

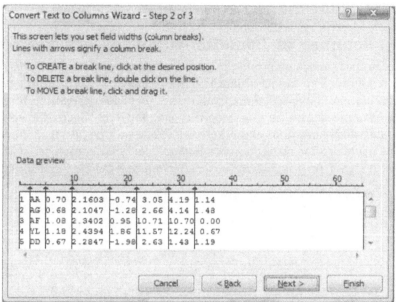

Figure 10-12. The Text to Columns Step 2 dialog box
with preview of parsed data.

As with the Text Import Wizard, the Convert Text To Columns Step 3 dialog box allows you to import only selected columns, or apply General, Text or Date formatting. The Step 3 dialog box contains a Destination input box (not present in the Text Import Wizard) that allows you to specify where you want the parsed data to be placed. The default option overwrites the original data column.

	A	B	C	D	E	F	G	H
		peptide	log					
1	no	sequence	SumAA	log VDW	clog P	tR exp	tR pred	delta tR
2	1	AA	0.70	2.1603	-0.74	3.05	4.19	1.14
3	2	AG	0.68	2.1047	-1.28	2.66	4.14	1.48
4	3	AF	1.08	2.3402	0.95	10.71	10.70	0.00
5	4	YL	1.18	2.4394	1.86	11.57	12.24	0.67
6	5	DD	0.67	2.2847	-1.98	2.63	1.43	1.19
7	6	ML	1.14	2.4508	-0.17	11.94	10.20	1.74
8	7	WW	1.35	2.5420	2.18	15.84	14.67	1.17
9	8	GM	0.90	2.3318	-1.90	7.30	5.74	1.56
10	9	GH	0.68	2.2600	-1.89	2.90	2.04	0.85

Figure 10-13. Portion of the data table, after parsing into columns.

From Hard Copy (Paper) to Excel

Occasionally you'll have data in hard copy form (from a book, report, correspondence, etc.). Unless the data set is very small, in which case it will be more efficient to enter the data manually, you'll want to use some semi-automatic way to import the data from paper copy to Excel.

Using a Scanner to Transfer Numeric Data to Excel

You can use a scanner and a software program that performs optical character recognition to create a data file from hard copy. Once imported into Excel, the data may have to be manipulated to get it into a useable form. The following example shows how to record a simple Excel macro for converting scanned data with blank lines and other undesired features into useful columnar data. See Chapter 16 for instructions on how to use the Recorder.

	A	B	C	D	E	F	G	H
1	2.51859E-4	0.0	1.21621E-4	1.42871E-2	0.0	10.0		
2								
3	0.465	0.489	0.506	0.517	0.522	0.519	0.502	0.473
4								
5	0.428	0.377	0.324	0.273	0.223	0.181	0.145	0.117
6								
7	0.093	0.074						
8								
9	2.51859E-4	0.0	1.21621E-4	2.85743E-2	0.0	10.0		
10								
11	0.569	0.581	0.576	0.555	0.529	0.496	0.459	0.416
12								
13	0.365	0.313	0.266	0.220	0.178	0.144	0.114	0.092
14								
15	0.074	0.057						
16								

Figure 10-14. Spectral data imported from a monograph (D. J. Leggett, *Computational Methods for the Determination of Formation Constants*, Plenum Press, New York, 1985).

Figure 10-14 shows scanned spectral data (absorbance values at 10 nm intervals from 500 nm to 680 nm for a metal ion at varying concentrations of a ligand) copied from a monograph and imported into Excel. Because the data in the original copy were double-spaced, the scanner put the data in alternate rows of the spreadsheet. The data consist of the following: (i) in Row 1, Row 9, etc., rows of data giving concentrations and path length, not necessary in our data table, (ii) in Row 3, Row 11, etc., rows containing eight values of absorbance measurements (for 500–570 nm) for one concentration of ligand, (iii) in Row 5,

Row 13, etc., rows containing a second eight values of absorbance (580–660 nm), and (iv) in Row 7, Row 15, etc., rows containing the final two values of absorbance (670 and 680 nm).

Of the 120 lines in data in the sheet, more than half need to be deleted. With a little bit of experience in recording macros, you can manually fix up the first lines of the data set while recording a macro and then let the macro do the rest. Chapter 16 provides detailed information on how to record a macro. Here's how to handle this example:

1. Before turning on the Recorder, select the first row of data (Row 1).

2. Choose **Macro** from the **Tools** menu and **Record New Macro...** from the **sub**menu.

3. In the Record Macro dialog box, enter Ctrl + z as the shortcut key and press OK.

3. In the Macro toolbar, press the Relative Reference button (the right button of the two on this little toolbar). The Recorder will now record all actions you perform at the keyboard or with the mouse, until you press the Stop button (the left button of the two on the toolbar).

4. Perform the following operations:

    (i) **Delete** the first two rows (the first row contains concentration information that we won't use, the second row is empty).

    (ii) **Cut** the second row of data and **Paste** it at the end of the first row.

    (iii) **Cut** the third row of data (two cells) and **Paste** at the end of the first row of data.

    (iv) Select the five empty rows below the first row of data and **Delete**.

    (v) Select the next row of data, so that the previous operations will be repeated on the correct row when automated by using the macro.

5. Press the Stop Recording button or choose **Stop Recording** from the **Macro** submenu.

Now press Ctrl+z; you should see the next rows of data rearranged into one row. Keep pressing Ctrl+z repeatedly until all data is rearranged.

The recorded macro is shown in Figure 10-15. You don't need to understand it because once it's done its work, it can be deleted. And in fact it should be deleted, because the keyboard shortcut Ctrl+z for **Undo** will be disabled as long as the workbook containing your recorded macro is open.

```
Sub Macro1()
' Macro1 Macro
' Macro recorded 12/5/2007 by Joseph Billo
' Keyboard Shortcut: Ctrl+z
    ActiveCell.Rows("1:2").EntireRow.Select
    Selection.Delete Shift:=xlUp
    ActiveCell.Offset(2, 0).Range("A1:H1").Select
    Selection.Cut
    ActiveCell.Offset(-2, 8).Range("A1").Select
    ActiveSheet.Paste
    ActiveCell.Offset(4, -8).Range("A1:B1").Select
    Selection.Cut
    ActiveCell.Offset(-4, 16).Range("A1").Select
    ActiveSheet.Paste
    ActiveCell.Offset(1, 0).Rows("1:5").EntireRow.Select
    Selection.Delete Shift:=xlUp
    ActiveCell.Rows("1:1").EntireRow.Select
End Sub
```

Figure 10-15. Recorded macro, with comments added.

To delete the macro, choose **Macro** from the **Tools** menu and **Macros...** from the submenu. Click on the macro name (Macro1 in this example) in the list of macros, and then press the Delete button.

	A	B	C	D	E	F	G	H	I	J
1	0.465	0.489	0.506	0.517	0.522	0.519	0.502	0.473	0.428	0.377
2	0.569	0.581	0.576	0.555	0.529	0.496	0.459	0.416	0.365	0.313
3	0.668	0.668	0.640	0.592	0.533	0.473	0.413	0.357	0.301	0.251

Figure 10-16. Portion of rearranged data.

A spreadsheet fragment showing some of the rearranged data is shown in Figure 10-16; Figure 10-17 is a chart of a portion of the imported data.

Using a Scanner to Transfer Graphical Data to Excel

You can use a scanner to convert graphs from strip chart recorders, published graphs, or spectra to digitized x, y data. Special software is available for this purpose. First you scan the image using a scanner. Then you import the scanned image into the digitizing program. The program converts the scanned image into an x, y ASCII file. Perform a web search for "graphical data to Excel" or "graph digitizing software" to find suitable products.

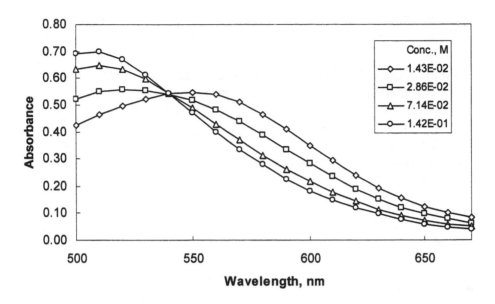

Figure 10-17. Chart created from rearranged data.

Selecting Every Nth Data Point

If you have imported a data file with a large number of data records, e.g., 8000 data points, you may wish to work with a reduced data set, say, every fiftieth point. The following sections describe three ways you can create such a list: by using AutoFill to **Fill Down** a pattern, by using Excel's Sampling tool, or by using a worksheet formula.

Using Autofill

As you saw in Chapter 1, you can use AutoFill to **Fill Down** a pattern of selected cells. You can use this feature to select every Nth value in a range.

To create a list containing, for example, every tenth data point from the spectrophotometer data shown in Figure 10-5 (the complete data set covers the range 300–820 nm with an absorbance measurement every 2 nm), use the following procedure:

- In cell C1 enter the formula =A1, and in D1 enter the formula = B1. This will copy the first line of x, y data.

- Select (highlight) the range C1:D10 (1 row of values and 9 rows of empty cells).

- Use AutoFill to fill this pattern down to the end of the data, in row 236. This will return every 10th line of x- and y-values.

- While the range is still highlighted, **Copy** the values and **Paste Special** (Values). This is necessary because we are going to **Sort** the values in the next step.

- While the range is still highlighted, choose **Sort...** from the **Data** menu and choose Ascending (Excel 2003) or click on Sort in the Sort and Filter group of the Data tab and choose Smallest to Largest. This will cause the values to rise to the top of the range. Note that this procedure succeeds only because the *x*-values are increasing monotonically. If neither the *x*-values nor the *y*-values change monotonically, add an adjacent column of integers 1, 2, 3... and use this as the sortkey.

Using the Sampling Tool

You can use Excel's Sampling tool, part of the Analysis ToolPak, to select every *N*th value from a range. The Analysis ToolPak, which provides a range of statistical tools, is accessed by clicking on Data Analysis in the Data tab (Excel 2007/2010) or by choosing **Data Analysis...** from the **Tools** menu (Excel 2003).

Since the Analysis ToolPak is an Add-In, the Data Analysis command may not be present. If the command is not present, use the procedure in the following paragraph to load the Add-In.

For Excel 2007/2010, use Office Button→Excel Options→Add-Ins→Manage Excel Add-Ins. In the list of available Add-Ins, check the box for Analysis ToolPak, and press OK. The Data Analysis button will now be available in the Analysis group in the Data tab. For Excel 2003, choose **Add-ins...** from the **Tools** menu to display the Add-Ins dialog box, check the box for Analysis ToolPak and press OK. You should now see the **Data Analysis...** command in the **Tools** menu.

Clicking the Data Analysis button or choosing **Data Analysis ...** will display the Data Analysis dialog box (Figure 10-18).

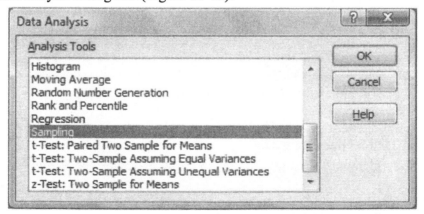

Figure 10-18. The Data Analysis dialog box.

Choose Sampling from the list of statistical tools to display the Sampling dialog box (Figure 10-19). In the Input Range box, enter the range of *x*-values. A deficiency of the Sampling tool is that it can sample in only one range at a time; if you have a column of *x*-values and a column of *y*-values, you'll have to perform the sampling operation twice.

In the Input Range box, enter the range of cells in which the sampling is to be done. The Sampling tool can perform either random or periodic sampling. Press the button for periodic sampling and enter 10 as the value for the period.

There are three output options: sending the results to a new worksheet ply, to a new workbook, or to a specified range of cells on the active sheet. For this example, we'll send the result to column C in the active sheet, so enter C2 as the start destination for the *x*-values.

Finally, repeat the whole process to sample the range of *y*-values.

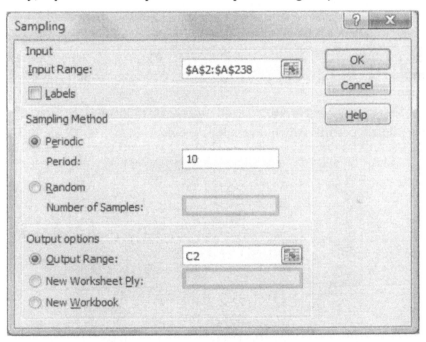

Figure 10-19. The Sampling dialog box.

Using a Worksheet Formula

Using the Sampling tool would be laborious if you needed to sample a data file with, for example, 17 columns: a column of *x*-values and 16 columns of associated *y*-values. You'd have to use the Sampling tool 17 times. In a case like this, it would be preferable to have a formula, using relative references, that could be entered once, filled down to sample the column of *x*-values, then filled right to create formulas that would sample each of the columns of *y*-values. The following formulas illustrate how to do this.

Either of the worksheet formulas described below can be used to select every N^{th} data point from a data table. Both utilize the INDEX(**array, row_num,column_num**) function to select an element from the array of data.

The first method uses names for the *x* and *y* data ranges; a separate worksheet formula must be entered for each column of data. The worksheet formula

=INDEX(XData,Nth*(ROW()-ROW(XData))+1)

when entered in cell C2 of Figure 10-5 and filled down yields the results shown in Figure 10-20. A similar formula is entered in cell D2 and filled down. XData and YData were defined as named ranges; the variable Nth contains the value 10.

The second method uses a single worksheet formula that employs a mixed reference. The formula can then be copied into multiple columns. The formula

=INDEX(A$2:A$238,Nth*(ROW()-ROW(A$2:A$238))+1)

was entered in cell C2; then **Fill Right** was used to insert it into cell D2 to yield the formula

=INDEX(B$2:B$238,Nth*(ROW()-ROW(B$2:B$238))+1)

Upon application of **Fill Down**, the values shown in Figure 10-19 are produced.

	A	B	C	D
1	236			
2	350	4.58E-02	350	0.045807
3	352	4.94E-02	370	0.153824
4	354	0.054169	390	0.34137
5	356	6.05E-02	410	0.263092
6	358	6.84E-02	430	0.091568
7	360	7.72E-02	450	0.037582
8	362	9.00E-02	470	0.023102
9	364	0.102127	490	0.011597
10	366	0.117447	510	0.01091
11	368	0.134659	530	0.01387
12	370	0.153824	550	0.016846
13	372	0.1745	570	0.023163

Figure 10-19. Using a worksheet formula to select every N^{th} data point.

11

Adding Controls to a Spreadsheet

You can create what Microsoft calls a *custom form*. A custom form is a spreadsheet that contains one or more of the controls that are normally found in a dialog box — for example, option buttons to turn options on or off, or a list box to display a list from which the user can select an item. A custom form could be the "front end" of a spreadsheet that performs simulation calculations, or searches a database, for example. Thus you can create a user-friendly document that makes it easy for users to enter information, and eliminate most input errors.

Most of the controls that you have seen in Excel's dialog boxes can be installed on a worksheet. These controls return a value to a worksheet cell (the *cell link*). You can then use this value in worksheet formulas. In addition to creating a form by adding controls, you can use hyperlinks to enable a user to easily access information, or use Data Validation to ensure that user-entered values are within specified limits.

Option Buttons, Check Boxes, List Boxes and Other Controls

The controls that can be installed on a worksheet to return a value to a worksheet cell are an option button, a check box, a list box, a drop-down list box (sometimes called a combo box), a scroll bar and a spinner. In addition, you can install a group box, which does not return a value but is used to group option buttons.

Option Button. An option button is used to select one of a group of options. When related option buttons are placed within a group box, only one button of the group can be selected. Use option buttons when only one of several possibilities is allowed at any given time.

365

Check Box. A check box is used to turn an option on or off. Use check boxes when several possibilities are allowed at any given time.

List Box. A list box displays a list of items. Several of the items in the list are visible in the box; the user scrolls up or down to display the rest of the list and then chooses an item.

Combo Box. A drop-down list box (combo box) displays a list of items. Initially only one item is visible in the window; the user presses the drop-down button to display the rest of the list and then chooses an item.

Scroll Bar. A scroll bar, with buttons and a slider, permits the user to increase or decrease a value in a worksheet cell. Maximum and minimum values can be set for the value returned by the scroll bar. A scroll bar can be installed either horizontally or vertically.

Spinner. A spinner is a button with an up arrow and a down arrow that permits the user to increase or decrease a value in a worksheet cell. Maximum and minimum values can be set for the value returned by the spinner.

How to Add a Control to a Worksheet

Adding a control to a worksheet is a three-step process, described in detail in the box on the following page: first, you display the gallery of available controls (Figure 11-1); you choose and draw the desired control; finally, you set the control properties. For example, for a list box, the control properties are (i) the range of values to be displayed in the list and (ii) the cell link, the cell to which the list box returns a value.

Each control has a title, e.g., Check Box 1, associated with it. You can select the title text and change it.

If you add Option Buttons, you will almost always place them within a group box (sometimes called a frame) and link each option button to the same cell. (The best procedure is to add the group box first, and then add the option buttons within the group box.)

 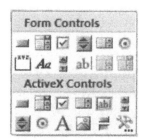

Figure 11-1. (Left) The Excel 2003 Forms toolbar, (Right) The Excel 2007/2010 Form Controls gallery

---

To Add a Control to a Worksheet

1. Click on Insert in the Controls group on the Developer tab[*] (Excel 2007/2010) or choose **Toolbars** from the **View** menu and choose **Forms** from the submenu (Excel 2003) to display the gallery of available controls. In Excel 2007/2010, Form Controls and ActiveX Controls are both displayed in a single gallery of controls; in the following sections only the Form controls will be discussed. The Excel 2003 gallery contains sixteen controls, but only nine can be used on a worksheet; the remaining ones are dimmed. The Excel 2007/2010 gallery contains twelve form controls, but again only nine can be used on a worksheet.

2. Press the button for the control you want to add. On the worksheet, click and drag to draw the control.

3. While the control is still selected (has handles around it), press the Properties button in the Controls group (Excel 2007/2010) or choose **Control...** from the **Format** menu, or right-click on the control and choose **Format Control...** from the shortcut menu (Excel 2003). Enter the control properties, described below.

4. When the control properties have been entered, click anywhere on the worksheet to un-select and activate the control. Upon activation of the control, the mouse pointer becomes the "hand'" shape when the mouse pointer is positioned over the control.

5. To select the control for moving or resizing, hold down the Ctrl key and click on the control.

When you add controls, they are numbered in the order in which you add them. For example, if you add, in turn, a frame and then add two option buttons within the frame, followed by a second frame and two more option buttons, the controls will have the titles Frame1, OptionButton2, OptionButton3, Frame4, OptionButton5, OptionButton6. Thus it's a good idea to be systematic when you add controls, so that the numbers of related controls will be in sequence.

Control Properties

You set the properties of a control by choosing **Control...** from the **Format** menu or **Format Control...** from the shortcut menu to display the Format Control dialog box for the specific control, and then choose the Control tab. Some control properties function only when the control is used with a macro. You can set these control properties, but they will have no effect on a control installed on a worksheet.

---

[*] If the Excel 2007/2010 Developer tab is not available, display it (see Chapter 5).

Table 11-1. Control Properties

Check Box   **Cell Link**: a reference to a cell that returns the state of the check box: TRUE if the box is checked, FALSE if unchecked. This logical argument can be used in a formula.

Option Button **Cell Link**: a reference to a cell that returns the number of the option button that is selected. Only one option button within a group box can be selected (pressed) at any time. The cell link returns the number of the option button within the group (e.g., 1, 2 or 3), not the number that appeared in the original title (e.g., Option Button 7, Option Button 8, Option Button 9).

List Box   **Input Range**: a reference to a range of cells whose values will be displayed in the list box.

                    Cell Link: a reference to a cell that returns a value that is the relative position of the selected item in the list. This value can then be used in a formula to return a result based on the selected item in the list.

                    Selection Type: you must use Single when a list box is installed on a worksheet.

Combo Box   **Input Range**: a reference to a range of cells whose values will be displayed in the drop-down list box.

                    Cell Link: a reference to a cell that returns the relative position of the selected item in the list.

Scroll Bar   **Minimum Value**: the minimum value that the scroll bar can return (corresponds to the top of a vertical scroll bar or the left end of a horizontal scroll bar).

                    Maximum Value: the maximum value that the scroll bar can return (corresponds to the bottom of a vertical scroll bar or the right end of a horizontal scroll bar).

                    Incremental Change: the amount the scroll slider button moves when an arrow at either end of the scroll bar is clicked. The default value is 1.

                    Page Change: the amount the scroll slider button moves when you click between the scroll button and one of the scroll arrows. The default value is 10.

                    Cell Link: a reference to a cell that returns the current value of the scroll bar. This number can be used in a formula to return a result based on the position of the scroll slider button in the scroll bar.

Spinner   The control properties for a spinner are the same as those for a scroll bar, but spinners do not have a Page Change property.

A List Box on a Worksheet

As a simple first exercise, we'll create a list box that displays a list of items. The list that will be displayed consists of the names of the months of the year. When you select a particular month from the list, the list box returns a number from 1 to 12, the relative position of the selected item in the list.

First, create a range of cells containing the list to be displayed. Fill the range A1:A12 (you can use a different worksheet than the one where you're going to install the list box) with the text January, February, etc. You can use **AutoFill** to do this.

To install the list box on a worksheet, follow the procedure outlined earlier in this chapter. First, display the Form Controls gallery (Excel 2007/2010) or the Forms toolbar (Excel 2003). Click on the List Box button (second from the right in the top row in Figure 11-1 Right) and use the mouse pointer to draw the outline of a list box. While the list box is still selected (has handles around it), right-click on the control and choose **Format Control...** from the submenu. In the Input Range box, enter the external reference to the range of cells containing the list to be displayed. In the Cell Link box, enter a cell reference, e.g., E5, then press OK.

To make the list box active, click on any cell in the worksheet. The handles around the list box will disappear.

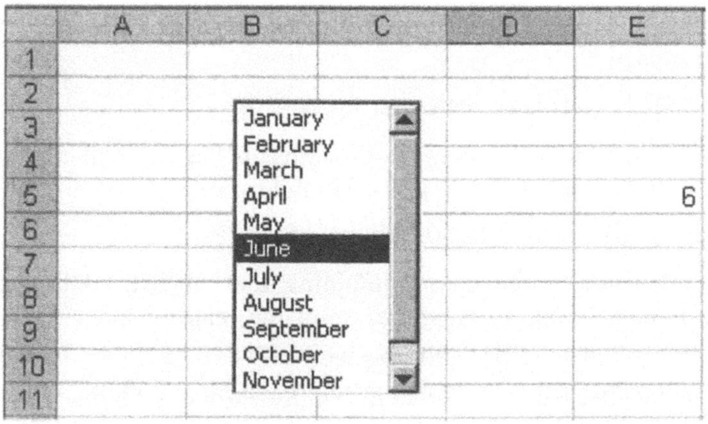

Figure 11-2. A list box installed on a worksheet. The cell link is cell E5.

The list box should now display the months of the year, as shown in Figure 11-2. When you select an item from the list, the Cell Link will return the number of the selected item. You can use the cell link value in a formula.

To select the list box for moving or resizing, click on it once while holding down the Ctrl key.

A Drop-Down List Box on a Worksheet

Figure 11-3 shows a drop-down list box that is used to display a subset of a list. This drop-down list box displays, from a list of names and telephone numbers, the subset of names that begin with a particular letter or string. The user can display all the names that begin with the letter "D", or "McD", for example. When the user chooses a name from the subset displayed in the drop-down list box, that person's telephone number is displayed in a cell.

The advantage of using the drop-down list box rather than a simple list box is that the box automatically sizes to fit the number of values displayed.

In this example, names, addresses and telephone numbers are listed in columns A, B and C of the worksheet. The range A2:A139 was assigned the name **Names**, and C2:C139 was given the name **Phones**. The list of names was sorted in alphabetical order.

Figure 11-3 shows the list box displaying the sub-list of names beginning with the letter D. To select the subset to be displayed, the user enters a letter or string in cell F2, which was assigned the name **Letter**.

The reference **SubListStart**, in cell G2, contains the formula

=MATCH(Letter&"*",Names,0)

This formula returns the relative position of the first item in the list that matches the input string. In our example, the first name that begins with D is the 34th item in the list in column A. The reference **SubListLength**, in cell H2, contains the array formula

{=SUM(1*(LEFT(Names,LEN(Letter))=Letter))}

This formula returns the number of items in the list that match the input string. These two formulas are combined in the formula

=OFFSET($A$1,SubListStart,0,SubListLength,1)

to return a reference to the cells containing the sub-list. This formula was entered as a named formula (using Define Name) and given the name **DropList**. In fact, all the formulas in this example can be entered as named formulas.

Finally, **DropList** was entered as the reference for the Input Range of the list box and cell I2 as the Cell Link.

	F	G	H	I
1	Letter	SubListStart	SubListLength	Cell Link (relative)
2	D	34	10	2
3				
4		Deckster, Meghan K ▼		
5		Daly, Josephine		
6		Deckster, Meghan K		
7		Dell, Daniel M		
8		Dickson, Martin J		
9		Dimatti, Peter A		
		Ditolla, Maria L		
10		Dominguez, Christina D		
11		Dopfer, David E		
12		Drabanti, Brad		
		Drozdowski, Christina		

Figure 11-3. A drop-down list box displaying a subset of a list. The cell link is cell I2.

The formula to return the telephone number corresponding to the name selected from the list box

=INDEX(Telephones,SubListStart+CellLink-1)

was entered in cell H6. Figure 11-4 illustrates the final result.

	F	G	H	I
1	Letter	SubListStart	SubListLength	Cell Link (relative)
2	D	34	10	2
3				
4		Deckster, Meghan K ▼		
5				
6		Telephone #:	555-5618	
7				

Figure 11-4. A drop-down list box used to return a telephone number.

Option Buttons with a Drop-Down List Box

This example illustrates the use of option buttons in combination with a drop-down list box. The database consists of a list of compounds with their CAS numbers and catalog numbers. The option buttons are used to specify whether to display the compound name, Chemical Abstract registry number (CAS number), or catalog number in the list box.

A list box was created with cell link K3. Three option buttons were positioned inside a group box (see Figure 11-5) and the labels "Option Button 1", "Option Button 2" and "Option Button 3" were changed to "Name", "CAS #" and "Catalog #", respectively. The cell link for the three buttons was cell L3. The cell links were given the names RowPointer and ColPointer, respectively.

ColPointer is used to specify whether to display the name, CAS number or catalog number in the list box. In order to do this, a named formula, BoxList, was used as the Input Range in the Format Control dialog box for the list box:

=OFFSET(listbox_table,0,ColPointer-1,table_count-1,1)

where table_count is a formula that determines the number of entries in the database, thereby allowing for expansion of the database, and listbox_table is the database, organized as described in the following paragraph.

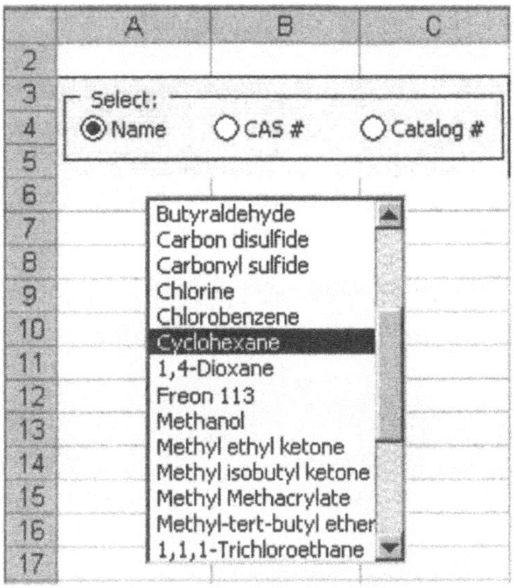

Figure 11-5. A drop-down list box used with option buttons.

The primary database is sorted by the first column, containing the compound names. When the list box displays the CAS numbers, the list needs to be sorted in that order, and similarly if the list box displays the catalog numbers. To do this, a second database, listbox_table, was created, in which each column is sorted; the list box displays values from this second database. The sorting can be done manually, and repeated if new entries are added to the database, or automatically by means of formulas similar to the ones described in Chapter 7, "Array Formulas". You can examine the formulas in the example workbook on the CD-ROM that accompanies this book.

An "output" of the list-box-plus-option-buttons combination is shown in Figure 11-6.

Name	CAS #	Catalog #
Cyclohexane	110-82-7	890346

Figure 11-6. A drop-down list box used with option buttons.

Using a Check Box to Enable or Disable Conditional Formatting

You can turn conditional formatting on or off with a check box on a spreadsheet. An example is shown on the CD that accompanies this book: For a range of values in column A, we want to highlight values that are greater than a value entered in cell $C$2. A checkbox with the label "Show Out-of-Range Values" was installed on the spreadsheet; the cell link was $E$1. Conditional formatting with the formula

=AND(A2>$C$2,$E$1=TRUE)

entered in the "Formula Is" box was applied to the range of cells in column A. When the user checks the "Show Out-of-Range Values" box, cells in the range in column A that contain values greater than the value in cell $C$2 are indicated with a red background.

How to Add a Hyperlink to a Worksheet

You can use hyperlinks in a workbook, in order to make it convenient for a user to jump to a location containing desired information. The location can be a cell or range on the same worksheet, on a different worksheet in the same workbook or in a different workbook, or it can be a bookmark in a Word document or a location on the Internet.

There are two ways to establish a hyperlink: You can either insert a hyperlink by means of a ribbon or menu command, or use the HYPERLINK worksheet function. The former is convenient for establishing a link to a fixed location, such as a cell reference. The latter allows you to create a link to a variable location.

Inserting a Hyperlink

To establish a link to a location in the same workbook, select the cell in which the hyperlink will appear. Click on Hyperlink in the Insert tab (Excel 2007/2010) or choose **Hyperlink...** from the **Insert** menu (Excel 2003) to display the Insert Hyperlink dialog box (Figure 11-7). Press the "Place in This Document" button, click on the sheet name in the "Or select a place in this document:" box, and type the cell reference for the hyperlink location (or, more conveniently, click on the name that you previously defined for the cell reference, as shown in Figure 11-7), and then press OK.

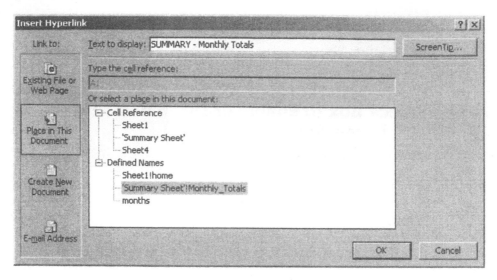

Figure 11-7. The Insert Hyperlink dialog box.

To establish a link to a specific location in another workbook, select the cell in which the hyperlink will appear, display the Insert Hyperlink dialog box, press the "Existing File or Webpage" button, and navigate to the desired workbook and select it, as illustrated in Figure 11-8. Now press the Bookmark button. Click on the sheet name in the "Or select a place in this document:" box (as in Figure 11-9) and then type the cell reference for the hyperlink location (or, more conveniently, click on the name that you previously defined for the cell reference), and then press OK.

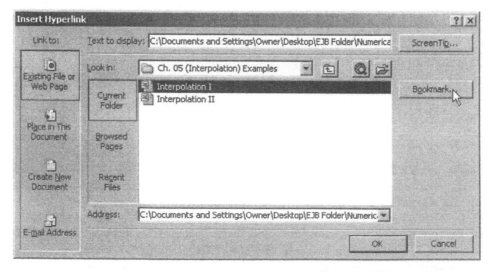

Figure 11-8. Establishing a hyperlink to a different workbook.

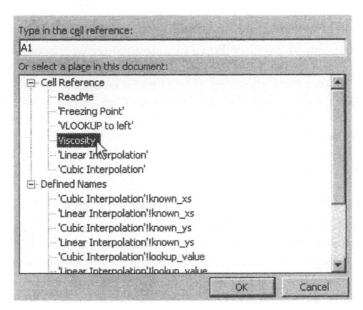

Figure 11-9. Establishing a hyperlink to a specified cell in a specified workbook.

If you don't press the Bookmark... button, and just select the workbook name and press OK, the hyperlink will merely jump to the sheet in the workbook that was the active sheet when the workbook was last saved.

Figure 11-10. Establishing a hyperlink to a web page.

A link to a web page is probably the easiest to establish: Simply press the "Existing File or Web Page" button, paste or type the URL of the desired web page in the Address box, add a suitable label in the Text To Display box, and press OK.

Using Paste as Hyperlink...

To create a link to a cell or range in a destination workbook, you can use Paste As Hyperlink in the Paste menu in the Clipboard group in the Home tab (Excel 2007/2010) or the **Paste as Hyperlink**... command in the **Edit** menu (Excel 2003). Simply **Copy** the destination cell or range, switch to the cell where you want to insert the hyperlink, and choose **Paste as Hyperlink**... (The **Paste as Hyperlink**...command is dimmed until you **Copy**.) You cannot create a hyperlink to a document that has not been saved; the command will remain dimmed even after you copy.

Using the HYPERLINK Worksheet Function

The HYPERLINK function provides much more flexibility in the creation of hyperlinks. Successfully creating a hyperlink by using the function can be problematic, though. In the following I will attempt to point out some of the pitfalls.

The syntax of the function is

=HYPERLINK(**link_location**,*friendly_name*)

where *link_location* is an expression that evaluates to the path and file name to the document to be opened, as text. The optional argument *friendly_name* is the hyperlink text that will be displayed in the cell; the hyperlink text is blue and underlined. If *friendly_name* is omitted, *link_location* is displayed.

The *link_location* argument must include the workbook name even if the link is to a cell in the same workbook. It must include the filename extension.

A few examples will illustrate the form required for the HYPERLINK arguments.

The following creates a hyperlink from a cell in Sheet1 to cell E1 of Sheet3 in the same workbook. The hyperlink text "[Book1.xls]Sheet3!$E$1" is displayed in the cell.

=HYPERLINK("[Book1.xls]Sheet3!$E$1")

The following creates a hyperlink from a cell in Book1 to cell E1 of Sheet3 in Book2, provided that both workbooks are open or both are in the same folder. The hyperlink text that is displayed is the text in cell E1.

=HYPERLINK("[Book2.xls]Sheet3!$E$1",$E$1)

The following creates a hyperlink from a cell in Book1 on the Desktop to cell E1 of Sheet3 in Book2 in a different folder, "Demo Hyperlink Folder"; Book2 does not have to be open. The hyperlink text "click here" is displayed

=HYPERLINK("'[Demo Hyperlink Folder\Book2.xls]Sheet3'!E1","click here")

In the preceding example, note the use of apostrophes, required when there are spaces in the text, and also the location of the brackets, not simply around the workbook name but around the directory path.

You need only specify enough of the directory path so that the link can be followed, but the formula

```
=HYPERLINK("'[C:\Documents and Settings\Owner\Desktop\
      Demo Hyperlink Folder\Book2.xls]Sheet3'!E1", "click here")
```

is acceptable.

Note that since the pathname is text, it does not update if the location of the hyperlinked file is changed. If you want to handle the situation in which a user might change the location of the workbook or rename it, you may be able to use a name for all or part of *link_location*. For example, the named formula

```
=MID(CELL("filename"),FIND("[",CELL("filename")),FIND("]",
      CELL("filename"))-FIND("[",CELL("filename"))+1)
```

can be used in place of the workbook name in a hyperlink formula that refers to a cell in the same workbook, so that no directory path is required. The formula updates if the user changes the name of the workbook.

As a final example, let's create a table of hyperlinks A – Z, so that a user can jump to the first entry in a database that begins with a selected letter. (You've seen these hyperlinks in many web pages, I'm sure.) In Excel, it will be more convenient to arrange the hyperlinks in a single column; part of such a table is shown in Figure 11-11.

	A
1	A
2	B
3	C
4	D
5	E
6	F

Figure 11-11. Part of a table of hyperlinks.

Cell A1 in Sheet4 contains the formula

```
=HYPERLINK("[Hyperlink  Demo.xls]Sheet5!A"&MATCH(CHAR(64+ROW())&"*",
      Sheet5!$A$2:$A$150,0)+1,CHAR(64+ROW()))
```

and the formula was filled down to row 26. Let's examine how this formula works.

The database is located on Sheet5. The leftmost column of the database consists of text entries arranged in alphabetical order, in the range $A$2:$A$150.

The expression

CHAR(64+ROW()

returns the characters "A", "B", "C", etc. in rows 1, 2, 3, respectively. This expression is used twice, once to return the lookup value for the MATCH function and once to create the expression for *friendly_name*.

The expression

MATCH(CHAR(64+ROW())&"*",Sheet4!$A$2:$A$137,0)

using the wildcard character "*", returns the position of the first entry in the database that matches the specified letter. Since the first entry in the database is in row 2, we must add one to this value, which is then used to create the hyperlink address. Thus, clicking on any letter jumps to the first entry in the table that begins with that letter.

What's the advantage of using this formula over establishing the hyperlinks manually, using **Hyperlink...** from the **Insert** menu? If you change the entries in the database, you don't have to re-do the hyperlinks.

A simple modification of the HYPERLINK formula described above allows the creation of a more compact hyperlink table, like the one shown in Figure 11-12.

	A	B	C	D	E	F
1	A	B	C	D	E	F
2	G	H	I	J	K	L
3	M	N	O	P	Q	R
4	S	T	U	V	W	X
5	Y	Z				

Figure 11-12. A table of hyperlinks with a different layout.

Using Data Validation

If you are creating a workbook in which users will enter data in order to perform calculations, you can ensure that they will be allowed to enter only valid data, by using Data Validation. You can restrict data entry in a specified cell so that only values that are less than a specified maximum value are allowed, for example.

With Data Validation, you can specify that only number values within a certain range, or only integer values, or only dates within a certain time period, or only text values, can be entered. Also, you can provide a prompt message that alerts the user when a cell with Data Validation is selected, as well as provide an error message if the user enters incorrect data.

Specifying Input Values

To specify only allowed input values for a certain cell, select the cell and click on Data Validation in the Data Tools group in the Data tab (Excel 2007/2010), or choose **Validation...** from the **Data** menu (Excel 2003). The Data Validation dialog box has three tabs: Settings, Input Message and Error Alert. In the Settings tab you can choose to allow input of one of the following data types: Whole Number, Decimal, Date, Time, Text, or a value from a specified List. If you choose Whole Number, Decimal, Date or Time, you can specify the range of allowed values: between, not between, less than, etc, as illustrated in Figure 11-13. If you choose Text, you can specify the minimum and/or maximum text length.

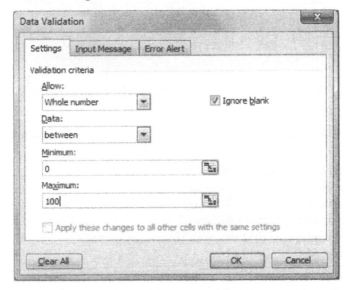

Figure 11-13. The Settings tab of the Data Validation dialog box.

As well as entering fixed values, like 0 and 100 for the minimum and maximum values in Figure 11-13, you can enter references, names or formulas. A reference must be to a cell or range on the same sheet; use a name if you want to refer to a cell or range on a different sheet. (Excel 2010 allows the user to enter a reference to a different sheet.)

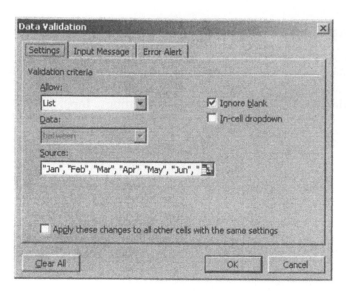

Figure 11-14. The List category in the Settings tab of the Data Validation dialog box.

If you select the List category, you can enter a list of values in the Source box, as shown in Figure 11-14. Only the values in the list are allowed entries in the cell.

It is usually more convenient to have the list of allowed items in a range of cells on the worksheet, as well as enter a reference to the range as the Source in the Data Validation dialog box. If you use a reference, the list must be on the same worksheet as the Data Validation. In order to have the list on a separate worksheet, you must assign a name to the range of cells and enter the name in the Source input box.

If the In-cell dropdown box is checked, the List category provides a feature not seen with any of the other Data Validation categories: when the user selects a cell with Data Validation, a drop-down button appears to the right of the cell and the user can choose a value to be entered in the cell, as illustrated in Figure 11-15.

If you have entered a reference to a range of cells in the Source box and wish to add additional allowed values, you must remember to insert them in the middle of the range of cells in order for Excel to update the reference. If you add the new items below the bottom of the range, the reference to the Source will not be updated. The following shows how to create a formula that allows a user to add additional items below the bottom of the list and have the Source update automatically. The procedure uses named formulas.

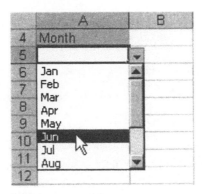

Figure 11-15. An In-cell dropdown produced by the List category of Data Validation.

In this example, the list of items is in column K, beginning in cell K2; a text label is in cell K1. Using Define Name, enter the following formula

=COUNTA($K:$K)-1

(we must subtract 1 because of the text label in cell K1). Give this formula the name ItemCount. Second, enter the formula

=OFFSET($K$2,0,0,ItemCount,1)

and give it the name ItemList. Finally, select the cell that will have Data Validation List input, and click on Data Validation in the Data Tools group in the Data tab (Excel 2007/2010) or choose **Validation...** from the **Data** menu (Excel 2003). Choose List from the Allow drop-down list, and enter =ItemList in the Source box.

Specifying an Input Message or Error Alert

You can provide an Input Message by using the Input Message tab of the Data Validation dialog box. If the user selects a cell with Data Validation, a ScreenTip box appears, like the one in Figure 11-16.

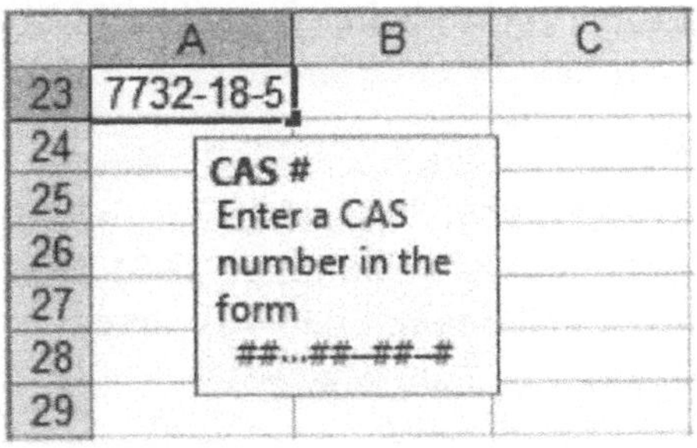

Figure 11-16. An Input Message.

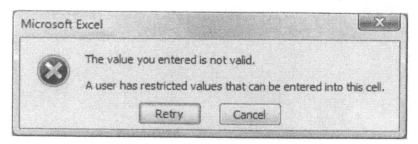

Figure 11-17. The default Error Alert message.

If the user enters a value that is outside the range restricted by the Data Validation settings, an error message is displayed. The default error message is shown in Figure 11-17.

You can create your own custom error message, like the one shown in Figure 11-18, by using the Error Alert tab.

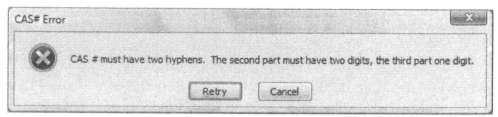

Figure 11-18. A custom Error Alert message.

Using the Custom Category

The Custom category provides greater flexibility in creating data validation restrictions. You enter a formula in the Formula box (Figure 11-19); like Conditional Formatting, the formula must evaluate to TRUE or FALSE. A few examples will indicate possibilities.

Allow certain dates. The following formula allows only dates that are at least 90 days from the current date to be entered in cell A18:

=A18>TODAY()+90

Allow only unique entries. The following formula allows only unique entries in the range $A$1:$A$100. The range $A$1:$A$100 was selected, and then the following Data Validation formula was entered.

=COUNTIF($A$1:$A$100,A1)=1

The formula

=SUM(N(($A$1:$A$100=A1))=1

is equivalent.

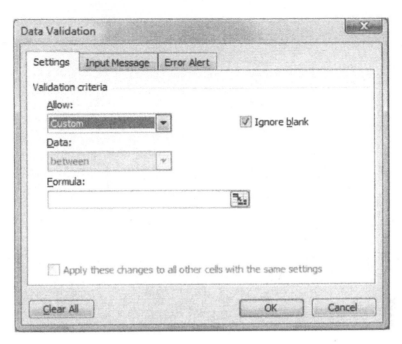

Figure 11-19. The Custom Data Validation category.

Allow entries with a specified format. Chemical Abstract registry numbers (CAS numbers) have a specific format: the number consists of three fields, separated by hyphens. The first field can be up to seven digits in length, the second contains two digits, and the third contains a single digit. The third digit is a checksum digit.

The following formula performs some simple checks to ensure that an entry (in this case in cell A14) conforms to the CAS # format.

```
=AND((LEN(A14)-LEN(SUBSTITUTE(A14,"-","")))=2),
    ((FIND("-",A14, FIND("-",A14)+1)-FIND("-",A14))=3),
    (LEN(A14)-FIND("-",A14,FIND("-",A14)+1)=1))
```

There are three logical arguments in the AND function in the preceding formula: the first determines that the entry contains two hyphens, the second determines that the second field contains two characters, and the third determines that the third field contains one character. The formula does not check the validity of the checksum digit. And of course it does not check that the number corresponds to an actual compound.

Excel Tip. I find it helpful, when creating a complicated formula like the preceding one, to create it in a worksheet cell, and then Copy and Paste to the Formula box.

12

Other Language Versions
of Excel

Excel is available in a number of language versions. This chapter shows some of the ways in which a user can handle information transmitted between Excel users in different countries, using different language versions. It will also be useful for readers of this book who are using a language version other than the U.S. version used by the author, in order to "translate" some of the U.S.-version Excel instructions into ones that apply to their version of Excel.

Covering all language versions of Excel is beyond the scope of this book; only some western European versions are discussed in this chapter. In the following, I refer to these versions as "EUR" and refer to the American version of Excel as "US".

Differences in the Display of Numbers

Numbers in Excel remain the same, no matter what the language, keyboard or Excel version. However, they are displayed differently, depending on the country and language settings.

Decimal and Thousands Separators

The United States and Canada use the period as the decimal point and the comma as the thousands separator, while most of the rest of the world uses the comma as the decimal point and the period as the thousands separator. You can change the formatting (the display) of numbers between US and EUR (or, more precisely, between what is specified by the Regional and Language Options in the Control Panel and an alternative format that you specify).

Excel 2007/2010. Click on the Office Button (Excel 2007) or the File tab (Excel 2010). Click on the Excel Options button and then click on the Advanced button. In the Editing Options category, uncheck the "Use System Separators" check box and type new separator characters.

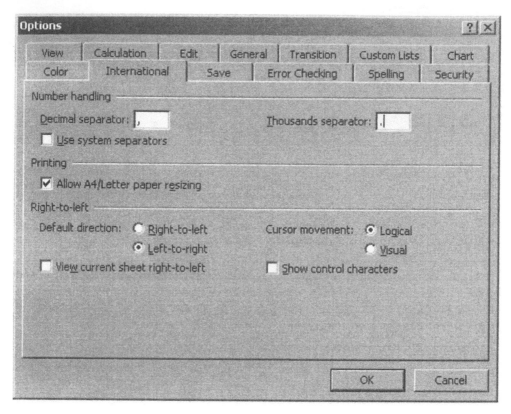

Figure 12-1. Changing decimal and thousands separator characters (Excel 2003).

Excel 2003. Choose the International tab in the **Options** command in the **Tools** menu, uncheck the "Use System Separators", and choose from the built-in options for decimal and thousands separators (Figure 12-1). When you want to use the built-in system separators again, simply check the Use System Separators check box. The previously selected decimal and thousands separators remain in the input boxes. This allows you to switch easily between display of numbers in either US or EUR format.

Argument Separators in Functions

Since the comma is used as the decimal separator in Europe, the semicolon is used as the separator between arguments in a function, e.g.,

=LINEST(*known_y's; known_x's; const; stats*)

Array Separators

In the US version of Excel, array values in the same row are separated by commas, and rows of values are separated by semicolons. For example, the array constant {1,2,3;4,5,6} represents a 2 × 3 array. However, since the comma is used in EUR versions as the decimal separator, the backslash character is used in place of the comma: {1\2\3;4\5\6}.

The characters used for decimal, thousands, function argument and array separators are summarized in Table 12-1.

Table 12-1. Separator Characters

	US	example	EUR	example
decimal point	.	10.5	,	10,5
thousands separator	,	$20,000	.	€5.000
between function arguments	,		;	
between array row elements	,	{1,2,3}	\	{1\2\3}
array rows	;	{1,2,3;4,5,6}	;	{1\2\3;4\5\6}
date	/	8/15/2010	see note*	15.8.2010

* Period, hyphen, slash or space are used, depending on country.

Differences in the Display of Dates

The format mm/dd/yy is used in the United States and Canada, while most of Europe uses dd/mm/yy. Since dates in Excel are Date Serial Numbers, independent of the Excel language version in which they are entered, a date can easily be displayed in a format suitable for almost any language and country (the *locale*).

Date Formats

When Date is selected from the list of categories in the Number tab of the Format Cells dialog box, an additional Locale list box is displayed below the list of built-in date formats. Figure 12-2 shows some of the possible date formats available when German is selected as the locale.

However, if dates are imported into Excel as text rather than as Date Serial Numbers (for example, from another program), they will have to be manipulated in order to convert from one system to the other. Some possible ways of handling dates imported as text are shown in a following section.

Date Formatting Symbols

The U.S. version of Excel uses the letters d, m and y (day, month and year) as date formatting symbols. Other languages use different letters, corresponding to the words for day, month and year in that language. For example, "year" in German is "Jahr". Date formatting symbols for some language versions of Excel are shown in Table 12-2.

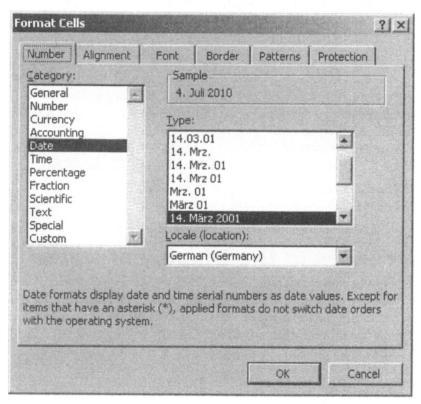

Figure 12-2. Choosing a different Locale for number formatting dates.

A custom date format created in a particular language version of Excel must use the date formatting symbols for that language.

Table 12-2. Date Formatting Symbols for Some Versions of Excel

	U.S.	French	German	Dutch	Danish	Spanish
Day, e.g., 1-31	d	j	t	d	d	d
Month, e.g., 1-12	m	m	m	m	m	m
Day, e.g., 01-31	dd	jj	tt	dd	dd	dd
Month, e.g., 01-12	mm	mm	mm	mm	mm	mm
Day, e.g., Sun-Sat	ddd	jjj	ttt	ddd	ddd	ddd
Month, e.g., Jan-Dec	mmm	mmm	mmm	mmm	mmm	mmm
Day, e.g., Sunday-Saturday	dddd	jjjj	tttt	dddd	dddd	dddd
Month, e.g., January-December	mmmm	mmmm	mmmm	mmmm	mmmm	mmmm
Year, e.g., 97	yy	aa	jj	jj	åå	aa
Year, e.g., 1997	yyyy	aaaa	jjjj	jjjj	åååå	aaaa
Hour	h	h	s	u	t	h
Minute	m	m	m	m	m	m
Second	s	s	s	s	s	s

Creating Custom Date Formats for Other Language Versions of Excel

The Locale list box appears when you choose the Date, Time or Special number formatting categories. It does not appear when you want to create a custom number format. To create a custom number format for a different language (for a different Locale), you must precede the date format by the Locale ID, enclosed in square brackets. The LCID is in hexadecimal notation. Usually numbers in hexadecimal format are represented as, for example, $040C, but Excel uses the notation $-40C for its date formatting. LCID values for some language versions are shown in Table 12-3. (There are over 200 LCIDs, for languages from Albanian to Zulu.)

Table 12-3. Locale ID Values for Some Language Versions of Excel

Language (Locale)	LCID
Czech	$-405
Danish	$-406
English (United Kingdom)	$-809
English (United States)	$-409
Finnish	$-40B
French (France)	$-40C
German (Germany)	$-407
Italian (Italy)	$-410

Thus, to create a custom number format in French while using the US version of Excel, you type, for example, [$-40C]dddd, d mmmm yyyy, to display the date in the form vendredi, 29 janvier 2010. (Note that you use the English date formatting symbols d, m and y, not the French symbols j, m and a.) With this approach you can even display dates in, for example, Hebrew (2010 ינואר 29, יום שישי) or Russian (пятница, 29 Январь 2010) in the US version of Excel.

For many European language locales, the built-in words for the days and months produced by number formatting are in lowercase. It does not appear possible to change the all-lowercase words for the days and the months to Proper case or uppercase by means of number formatting. It can be done, however, by means of a worksheet formula. If cell A1 contains the date 8/3/2010, the formula

=PROPER(TEXT(A1,"[$-40C]dddd, d mmmm yyyy"))

displays the date as Mardi, 3 Août 2010.

Importing Data

When values are imported into Excel from a text file, numbers and dates often end up as text in the spreadsheet. Text values that are in the format of the Locale of the user can often be used directly as numbers in Excel formulas, but if the values are from a different Locale, the values will usually have to be manipulated in order to convert them into numbers or dates.

Importing Values that Are Numbers

To import a text file with values in EUR format (e.g., 0,04580689 or 4,581E-02) into US Excel, use one of the following methods. Importing from US to EUR can be done in a similar way.

Method 1. Use the Text Import Wizard to import the data. In Step 3 of the Text Import Wizard, press the Advanced button to display the Advanced Text Import Settings dialog box (Figure 12-3). Choose comma as Decimal separator, period as Thousands separator. Press OK. Values are imported as numbers automatically.

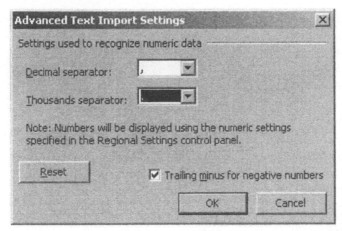

Figure 12-3. Choosing decimal and thousands separator characters.

Method 2.   Use the Text Import Wizard in the usual way to import the data. Number values are imported as text (are left-aligned). Choose Tools → Options → International tab (Excel 2003) or Excel Options → Advanced → Editing Options (Excel 2007/2010). Make sure that US separators are selected. Select the cells containing dates and use Replace All to replace commas with periods. Text values become numbers automatically.

Importing Values that Are Dates Using the Text Import Wizard

Dates imported from text files are imported as dates if they conform to a date format for the Locale. However, dates in a EUR version of Excel can be in any one of a number of formats, ranging from e.g., "29/01/2010" to "29-janv.-2010" to "ven. janv. 29 14:30 2010", that are not compatible with the US version of Excel. Use one of the methods below to convert text values to dates for the appropriate locale.

Method for Dates in the Format d/m/y. If you use the Text Import Wizard, the date 12 February 2010 in EUR format, e.g., "12/02/2010" will be imported as a date (right aligned), but incorrectly, while 29 January 2010, e.g., "29/01/2010" will be imported as text (left aligned), as in Figure 12-4.

	A
1	1/1/2010
2	1/15/2010
3	1/29/2010
4	12/2/2010
5	2/26/2010

Figure 12-4. Dates imported incorrectly using the Text Import Wizard.

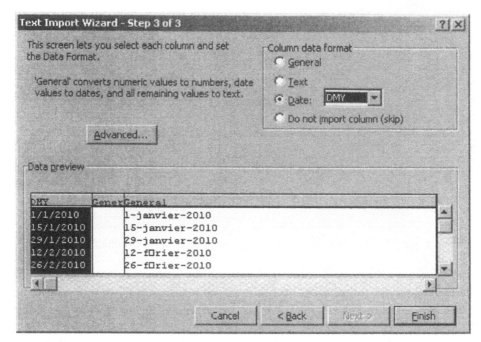

Figure 12-5. Choosing a non-US date format in the Text Import Wizard Step 3 of 3.

To avoid this problem, in the Text Import Wizard Step 3 of 3, select the column of dates by clicking on the column header, and then press the Date option button in the Column Data Format group and choose (in this example) DMY from the list of date formats, as shown in Figure 12-5.

When you press Finish, the dates will be imported correctly, as Date Serial Numbers, and formatted as dates, as shown in Figure 12-6.

	A
1	1/1/2010
2	1/15/2010
3	1/29/2010
4	12/2/2010
5	2/26/2010

Figure 12-6. Dates imported correctly by using Column Data Format.

Excel Tip. Special characters such as é will be imported incorrectly, as shown in the third column in Figure 12-5. To import special characters correctly, do the following: In the Text Import Wizard Step 1 of 3 dialog box, in the File Origin drop-down list box, choose (for example) Western European (Windows). The text imported by this method is shown in column A of Figure 12-7.

Method for Dates in Other Formats. If the dates to be imported are in a format such as "29-janvier-2010", you will have to manipulate the values. The following example shows how to handle dates in the form shown in the third column of Figure 12-5.

Use the Text Import Wizard to import the data. The date appears as text in column A of Figure 12-7. Now use the Text to Columns Wizard to parse the date into three columns (in this example, use "-" as the delimiter). The parsed data are shown in columns C, D and E in Figure 12-7. Use MATCH or VLOOKUP to return the month number 1–12 for the text values in column D. This requires that you create a lookup table with the appropriate text values, in this example the values janvier, février, mars, avril, mai, juin, juillet, août, septembre, octobre, novembre, décembre were entered in the range $L$2:$L$13. The month number in column F was returned by the formula =MATCH(D4,$L$2:$L$13,0) in cell F4. Then use the DATE worksheet function with the day, month and year numbers to return the date serial number. The formula in cell H4 is =DATE(E4,F4,C4).

	A	B	C	D	E	F	G	H
3	Original data (EUR) (as text)			Parsed data				Date (US) (Date Serial Number)
4	1-janvier-2010		1	janvier	2010	1		1/1/2010
5	15-janvier-2010		15	janvier	2010	1		1/15/2010
6	29-janvier-2010		29	janvier	2010	1		1/29/2010
7	12-février-2010		12	février	2010	2		2/12/2010
8	26-février-2010		26	février	2010	2		2/26/2010
9	12-mars-2010		12	mars	2010	3		3/12/2010
10	26-mars-2010		26	mars	2010	3		3/26/2010
11	9-avril-2010		9	avril	2010	4		4/9/2010

Figure 12-7. Converting text dates in a EUR format to Date Serial Numbers.

Worksheet Function Names in Other Languages

Different language versions of Excel use worksheet function names that are appropriate for the language and country. For example, the US worksheet function COUNT is ANZAHL in the German language version, NB in the French version, and CONTAR in the Spanish version.

Table 12-4 lists English worksheet functions and their French and German equivalents.

Table 12-4. Worksheet Functions: English, French and German

English	French	German
ABS	ABS	ABS
ACOS	ACOS	ARCCOS
ACOSH	ACOSH	ARCCOSHYP
ADDRESS	ADRESSE	ADRESSE
AND	ET	UND
AREAS	ZONES	BEREICHE
ASC	ASC	ASC
ASIN	ASIN	ARCSIN
ASINH	ASINH	ARCSINHYP
ATAN	ATAN	ARCTAN
ATAN2	ATAN2	ARCTAN2
ATANH	ATANH	ARCTANHYP
AVEDEV	ECART.MOYEN	MITTELABW
AVERAGE	MOYENNE	MITTELWERT
BETADIST	LOI.BETA	BETAVERT
BETAINV	BETA.INVERSE	BETAINV
BINOMDIST	LOI.BINOMIALE	BINOMVERT
CEILING	PLAFOND	OBERGRENZE
CELL	CELLULE	ZELLE
CHAR	CAR	ZEICHEN
CHIDIST	LOI.KHIDEUX	CHIVERT
CHIINV	KHIDEUX.INVERSE	CHIINV
CHITEST	TEST.KHIDEUX	CHITEST
CHOOSE	CHOISIR	WAHL
CLEAN	EPURAGE	SÄUBERN
CODE	CODE	CODE
COLUMN	COLONNE	SPALTE
COLUMNS	COLONNES	SPALTEN
COMBIN	COMBIN	KOMBINATIONEN
CONCATENATE	CONCATENER	VERKETTEN
CONFIDENCE	INTERVALLE.CONFIANCE	KONFIDENZ
CORREL	COEFFICIENT.CORRELATION	KORREL
COS	COS	COS
COSH	COSH	COSHYP
COUNT	NB	ANZAHL
COUNTA	NBVAL	ANZAHL2
COUNTBLANK	NB.VIDE	ANZAHLLEEREZELLEN
COUNTIF	NB.SI	ZÄHLENWENN
COVAR	COVARIANCE	KOVAR
CRITBINOM	CRITERE.LOI.BINOMIALE	KRITBINOM

DATE	DATE	DATUM
DAVERAGE	BDMOYENNE	DBMITTELWERT
DAY	JOUR	TAG
DAYS360	JOURS360	TAGE360
DB	DB	GDA2
DCOUNT	BDNB	DBANZAHL
DCOUNTA	BDNBVAL	DBANZAHL2
DDB	DDB	GDA
DEGREES	DEGRES	GRAD
DEVSQ	SOMME.CARRES.ECARTS	SUMQUADABW
DGET	BDLIRE	DBAUSZUG
DMAX	BDMAX	DBMAX
DMIN	BDMIN	DBMIN
DOLLAR	FRANC	DM
DPRODUCT	BDPRODUIT	DBPRODUKT
DSTDEV	BDECARTYPE	DBSTDABW
DSTDEVP	BDECARTYPEP	DBSTDABWN
DSUM	BDSOMME	DBSUMME
DVAR	BDVAR	DBVARIANZ
DVARP	BDVARP	DBVARIANZEN
EVEN	PAIR	GERADE
EXACT	EXACT	IDENTISCH
EXP	EXP	EXP
EXPONDIST	LOI.EXPONENTIELLE	EXPONVERT
FACT	FACT	FAKULTÄT
FALSE	FAUX	FALSCH
FDIST	LOI.F	FVERT
FIND	TROUVE	FINDEN
FINV	INVERSE.LOI.F	FINV
FISHER	FISHER	FISHER
FISHERINV	FISHER.INVERSE	FISHERINV
FIXED	CTXT	FEST
FLOOR	PLANCHER	UNTERGRENZE
FORECAST	PREVISION	SCHÄTZER
FREQUENCY	FREQUENCE	HÄUFIGKEIT
FTEST	TEST.F	FTEST
FV	VC	ZW
GAMMADIST	LOI.GAMMA	GAMMAVERT
GAMMAINV	LOI.GAMMA.INVERSE	GAMMAINV
GAMMALN	LNGAMMA	GAMMALN
GEOMEAN	MOYENNE.GEOMETRIQUE	GEOMITTEL
GROWTH	CROISSANCE	VARIATION

HARMEAN	MOYENNE.HARMONIQUE	HARMITTEL
HLOOKUP	RECHERCHEH	WVERWEIS
HOUR	HEURE	STUNDE
HYPGEOMDIST	LOI.HYPERGEOMETRIQUE	HYPGEOMVERT
IF	SI	WENN
INDEX	INDEX	INDEX
INDIRECT	INDIRECT	INDIREKT
INFO	INFO	INFO
INT	ENT	GANZZAHL
INTERCEPT	ORDONNEE.ORIGINE	ACHSENABSCHNITT
IPMT	INTPER	ZINSZ
IRR	TRI	IKV
ISBLANK	ESTVIDE	ISTLEER
ISERR	ESTERR	ISTFEHL
ISERROR	ESTERREUR	ISTFEHLER
ISLOGICAL	ESTLOGIQUE	ISTLOG
ISNA	ESTNA	ISTNV
ISNONTEXT	ESTNONTEXTE	ISTKTEXT
ISNUMBER	ESTNUM	ISTZAHL
ISPMT	ISPMT	ISPMT
ISREF	ESTREF	ISTBEZUG
ISTEXT	ESTTEXTE	ISTTEXT
KURT	KURTOSIS	KURT
LARGE	GRANDE.VALEUR	KGRÖSSTE
LEFT	GAUCHE	LINKS
LEN	NBCAR	LÄNGE
LINEST	DROITEREG	RGP
LN	LN	LN
LOG	LOG	LOG
LOG10	LOG10	LOG10
LOGEST	LOGREG	RKP
LOGINV	LOI.LOGNORMALE.INVERSE	LOGINV
LOGNORMDIST	LOI.LOGNORMALE	LOGNORMVERT
LOOKUP	RECHERCHE	VERWEIS
LOWER	MINUSCULE	KLEIN
MATCH	EQUIV	VERGLEICH
MAX	MAX	MAX
MDETERM	DETERMAT	MDET
MEDIAN	MEDIANE	MEDIAN
MID	STXT	TEIL
MIN	MIN	MIN
MINUTE	MINUTE	MINUTE

MINVERSE	INVERSEMAT	MINV
MIRR	TRIM	QIKV
MMULT	PRODUITMAT	MMULT
MOD	MOD	REST
MODE	MODE	MODALWERT
MONTH	MOIS	MONAT
N	N	N
NA	NA	NV
NEGBINOMDIST	LOI.BINOMIALE.NEG	NEGBINOMVERT
NORMDIST	LOI.NORMALE	NORMVERT
NORMINV	LOI.NORMALE.INVERSE	NORMINV
NORMSDIST	LOI.NORMALE.STANDARD	STANDNORMVERT
NORMSINV	LOI.NORMALE.STANDARD.INVERSE	STANDNORMINV
NOT	NON	NICHT
NOW	MAINTENANT	JETZT
NPER	NPM	ZZR
NPV	VAN	NBW
ODD	IMPAIR	UNGERADE
OFFSET	DECALER	BEREICH.VERSCHIEBEN
OR	OU	ODER
PEARSON	PEARSON	PEARSON
PERCENTILE	CENTILE	QUANTIL
PERCENTRANK	RANG.POURCENTAGE	QUANTILSRANG
PERMUT	PERMUTATION	VARIATIONEN
PI	PI	PI
PMT	VPM	RMZ
POISSON	LOI.POISSON	POISSON
POWER	PUISSANCE	POTENZ
PPMT	PRINCPER	KAPZ
PROB	PROBABILITE	WAHRSCHBEREICH
PRODUCT	PRODUIT	PRODUKT
PROPER	NOMPROPRE	GROSS2
PV	VA	BW
QUARTILE	QUARTILE	QUARTILE
RADIANS	RADIANS	BOGENMASS
RAND	ALEA	ZUFALLSZAHL
RANK	RANG	RANG
RATE	TAUX	ZINS
REPLACE	REMPLACER	ERSETZEN
REPT	REPT	WIEDERHOLEN
RIGHT	DROITE	RECHTS
ROMAN	ROMAIN	RÖMISCH

ROUND	ARRONDI	RUNDEN
ROUNDDOWN	ARRONDI.INF	ABRUNDEN
ROUNDUP	ARRONDI.SUP	AUFRUNDEN
ROW	LIGNE	ZEILE
ROWS	LIGNES	ZEILEN
RSQ	COEFFICIENT.DETERMINATION	BESTIMMTHEITSMASS
SEARCH	CHERCHE	SUCHEN
SECOND	SECONDE	SEKUNDE
SIGN	SIGNE	VORZEICHEN
SIN	SIN	SIN
SINH	SINH	SINHYP
SKEW	COEFFICIENT.ASYMETRIE	SCHIEFE
SLN	AMORLIN	LIA
SLOPE	PENTE	STEIGUNG
SMALL	PETITE.VALEUR	KKLEINSTE
SQRT	RACINE	WURZEL
STANDARDIZE	CENTREE.REDUITE	STANDARDISIERUNG
STDEV	ECARTYPE	STABW
STDEVP	ECARTYPEP	STABWN
STEYX	ERREUR.TYPE.XY	STFEHLERYX
SUBSTITUTE	SUBSTITUE	WECHSELN
SUBTOTAL	SOUS.TOTAL	TEILERGEBNIS
SUM	SOMME	SUMME
SUMIF	SOMME.SI	SUMMEWENN
SUMPRODUCT	SOMMEPROD	SUMMENPRODUKT
SUMSQ	SOMME.CARRES	QUADRATESUMME
SUMX2MY2	SOMME.X2MY2	SUMMEX2MY2
SUMX2PY2	SOMME.X2PY2	SUMMEX2PY2
SUMXMY2	SOMME.XMY2	SUMMEXMY2
SYD	SYD	DIA
T	T	T
TAN	TAN	TAN
TANH	TANH	TANHYP
TDIST	LOI.STUDENT	TVERT
TEXT	TEXTE	TEXT
TIME	TEMPS	ZEIT
TINV	LOI.STUDENT.INVERSE	TINV
TODAY	AUJOURDHUI	HEUTE
TRANSPOSE	TRANSPOSE	MTRANS
TREND	TENDANCE	TREND
TRIM	SUPPRESPACE	GLÄTTEN
TRIMMEAN	MOYENNE.REDUITE	GESTUTZTMITTEL

TRUE	VRAI	WAHR
TRUNC	TRONQUE	KÜRZEN
TTEST	TEST.STUDENT	TTEST
TYPE	TYPE	TYP
UPPER	MAJUSCULE	GROSS
VALUE	CNUM	WERT
VAR	VAR	VARIANZ
VARP	VAR.P	VARIANZEN
VDB	VDB	VDB
VLOOKUP	RECHERCHEV	SVERWEIS
WEEKDAY	JOURSEM	WOCHENTAG
WEIBULL	LOI.WEIBULL	WEIBULL
YEAR	ANNEE	JAHR
ZTEST	TEST.Z	GTEST

The FunctionName Translator

On the CD that accompanies this book you will find a workbook that translates worksheet function names to/from English, French, German, Danish, Spanish, Dutch, Finnish, Italian and Swedish.

PART III

SPREADSHEET MATHEMATICS

13

Mathematical Methods for Spreadsheet Calculations

This chapter describes some mathematical methods that are useful for spreadsheet calculations. For the most part, they are methods that are applicable to tables or arrays of data. The methods described range from the simple (finding the roots of a polynomial from a graph of the function) to the complex (using matrix methods to solve sets of simultaneous equations).

Interpolation

The worksheet function VLOOKUP allows you to look up a y-value for a specified value of x in a table of x- and y-values. But often it's necessary to estimate the value of y for some value of x intermediate between two adjacent data points — the process of *interpolation*. If the separation between x-values is small and the y-values do not change rapidly with x, then linear interpolation may be adequate. To calculate the value of y at a value x_i that is intermediate between x_0 and x_1, use the linear interpolation formula (equation 13-1):

$$y_i = y_0 + \frac{x_i x_0}{x_1 x_0}(y_1 y_0) \qquad (13\text{-}1)$$

Table Lookup with Linear Interpolation

You can use the MATCH and INDEX worksheet functions in a formula to perform linear interpolation in a table. The syntax of the MATCH function is MATCH(*lookup_value,lookup_array,match_type*). If *match_type* = 1, MATCH returns the position of the largest array value that is less than or equal to

403

```
Function InterpC(lookup_value, known_x's, known_y's)
'  Performs cubic interpolation, using an array of known_x's, known_y's.
'  The known_x's must be in ascending order.
'  Based on XLM code from Excel for Chemists", page 239,
'  which was based on W. J. Orvis' code.

Dim row As Integer
Dim i As Integer, j As Integer
Dim Q As Double, Y As Double

row = Application.Match(lookup_value, known_x's, 1)
   If row < 2 Then row = 2
   If row > known_x's.Count - 2 Then row = known_x's.Count - 2

For i = row - 1 To row + 2
   Q = 1
For j = row - 1 To row + 2
   If i <> j Then Q = Q * (lookup_value - known_x's(j)) / (known_x's(i) - _
      known_x's(j))
Next j
   Y = Y + Q * known_y's(i)
Next i
InterpC = Y
End Function
```

Figure 13-4. Cubic interpolation function macro.

A compact and elegant implementation of cubic interpolation in the form of an Excel 4.0 Macro Language custom function was provided by Orvis[*]. A slightly modified version, in VBA (Visual Basic for Applications) language, is provided here (Figure 13-4).   The syntax of the custom function is InterpC(*lookup_value, known_x's, known_y's*).

The cubic interpolation function can be used to produce a smooth curve through data points. Figure 13-5 illustrates interpolated values returned by the InterpC custom function, for interpolation between table values shown in Figure 13-1.

---

[*]   William J. Orvis, *Excel 4 for Scientists and Engineers*, Sybex Inc., Alameda, CA, 1993.

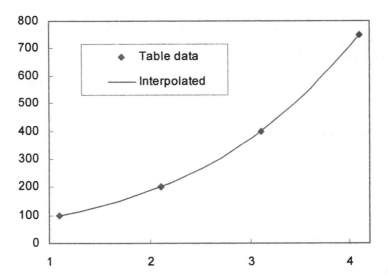

Figure 13-5. Interpolation by using a cubic interpolation function macro.

The cubic interpolation function forces the curve to pass through all the known data points. If there is any experimental scatter in the table values, the result will not be too pleasing. A better approach for data with scatter is to find the coefficients of a least-squares line through the data points, as described in Chapter 14 or 15.

Arrays, Matrices and Determinants

Spreadsheet calculations lend themselves almost automatically to the use of arrays of values. As you've seen, arrays in Excel can be either one- or two-dimensional. For the solution of many types of problem, it is convenient to manipulate an entire rectangular array of values as a unit. Such an array is termed a *matrix*. (In Excel, the terms "range", "array" and "matrix" are virtually interchangeable.)   An $m \times n$ matrix (m rows and n columns) of values is illustrated below.

$$\begin{bmatrix} a_{11} & a_{12} & \cdots & a_{1n} \\ a_{21} & a_{22} & \cdots & a_{2n} \\ \vdots & \vdots & \cdots & \vdots \\ a_{m1} & a_{m2} & \cdots & a_{mn} \end{bmatrix}$$

The values comprising the array are called *matrix elements*. Mathematical operations on matrices have their own special rules.

A *square matrix* has the same number of rows and columns. If all the elements of a square matrix are zero except those on the main diagonal (a_{11}, a_{22}, ..., a_{nn}), the matrix is termed a *diagonal matrix*. A diagonal matrix whose diagonal elements are all 1 is a *unit matrix*.

A matrix which contains a single column of m rows or a single row of n columns is called a *vector*.

A *determinant* is simply a square matrix. There is a procedure for the numerical evaluation of a determinant, so that an $N \times N$ matrix can be reduced to a single numerical value. The value of the determinant has properties that make it useful in certain tests and equations. (See, for example, "Solving Sets of Simultaneous Linear Equations" later in this chapter.)

An Introduction to Matrix Mathematics

Matrix algebra provides a powerful method for the manipulation of sets of numbers. Many mathematical operations — addition, subtraction, multiplication, division, etc. — have their counterparts in matrix algebra. Our discussion will be limited to the manipulations of square matrices. For purposes of illustration, two 3×3 matrices will be defined, namely

$$\mathbf{A} = \begin{bmatrix} a & b & c \\ d & e & f \\ g & h & i \end{bmatrix} = \begin{bmatrix} 2 & 3 & 4 \\ 3 & 2 & 1 \\ 4 & 3 & 7 \end{bmatrix}$$

and

$$\mathbf{B} = \begin{bmatrix} r & s & t \\ u & v & w \\ x & y & z \end{bmatrix} = \begin{bmatrix} 2 & 0 & 2 \\ 0 & 3 & 3 \\ 3 & 2 & 1 \end{bmatrix}$$

The following examples illustrate addition, subtraction, multiplication and division using a constant.

Addition or subtraction of a constant: $\mathbf{A} + q = \begin{bmatrix} a+q & b+q & c+q \\ d+q & e+q & f+q \\ g+q & h+q & i+q \end{bmatrix}$

Multiplication or division by a constant: $q\mathbf{A} = \begin{bmatrix} qa & qb & qc \\ qd & qe & qf \\ qg & qh & qi \end{bmatrix}$

Addition or subtraction of two matrices (both must contain the same number of rows and columns):

$$\mathbf{A} + \mathbf{B} = \begin{bmatrix} a & b & c \\ d & e & f \\ g & h & i \end{bmatrix} + \begin{bmatrix} r & s & t \\ u & v & w \\ x & y & z \end{bmatrix} = \begin{bmatrix} a+r & b+s & c+t \\ d+u & e+v & f+w \\ g+x & h+y & i+z \end{bmatrix}$$

Performing matrix algebra with Excel is very simple. Let's begin by assuming that the matrices **A** and **B** have been defined by selecting the 3R × 3C arrays of cells containing the values and naming them by using **Define Name**. To add a constant (e.g., 3) to matrix **A**, simply select a range of cells the same size as the matrix, enter the formula =A+3, and then press Ctrl+Shift+Enter. Subtraction of a constant, multiplication or division by a constant, or addition of two matrices also is performed by using standard Excel algebraic operators.

Multiplication of two matrices can be either *scalar* multiplication or *vector* multiplication. Scalar multiplication of two matrices consists of multiplying corresponding elements, i.e.,

$$\mathbf{A} \times \mathbf{B} = \begin{bmatrix} a & b & c \\ d & e & f \\ g & h & i \end{bmatrix} \times \begin{bmatrix} r & s & t \\ u & v & w \\ x & y & z \end{bmatrix} = \begin{bmatrix} a\times r & b\times s & c\times t \\ d\times u & e\times v & f\times w \\ g\times x & h\times y & i\times z \end{bmatrix}$$

Thus it's clear that both matrices must have the same dimensions $m \times n$. Scalar multiplication is commutative, that is, **A*B = B*A.**

The matrix multiplication of two matrices is somewhat more complicated:

$$\mathbf{A} \cdot \mathbf{B} = \begin{bmatrix} a & b & c \\ d & e & f \\ g & h & i \end{bmatrix} \cdot \begin{bmatrix} r & s & t \\ u & v & w \\ x & y & z \end{bmatrix} = \begin{bmatrix} ar+bu+cx & as+bv+cy & at+bw+cz \\ dr+eu+fx & ds+ev+fy & dt+ew+fz \\ gr+hu+ix & gs+hv+iy & gt+hw+iz \end{bmatrix}$$

Matrix multiplication can be accomplished easily by the use of one of Excel's worksheet functions for matrix algebra, MMULT(*matrix1, matrix2*). Matrix multiplication is not commutative, that is, **A·B ≠ B·A.** For the matrices **A** and **B** defined above, we have

$$\mathbf{A} \cdot \mathbf{B} = \begin{bmatrix} 8 & 1 & 1 \\ 9 & 8 & 1 \\ 13 & 5 & 6 \end{bmatrix} \qquad \mathbf{B} \cdot \mathbf{A} = \begin{bmatrix} 12 & 12 & 22 \\ 3 & 3 & 24 \\ 8 & 10 & 3 \end{bmatrix}$$

Matrix multiplication of two matrices is possible only if the matrices are *conformable*, that is, if the number of columns of **A** is equal to the number of rows of **B**. The opposite condition, if the number of *rows* of **A** is equal to the number of *columns* of **B**, is not equivalent. The following examples, involving multiplication of a matrix and a vector, illustrate the possibilities:

MMULT (4 × 3 matrix, 3 × 1 vector) = 4 × 1 result vector

MMULT (4 × 3 matrix, 1 × 4 vector) = #VALUE!

MMULT (1 × 4 vector, 4 × 3 matrix) = 1 × 3 result vector

The *transpose* of a matrix, indicated by a prime ($'$), is produced when rows and columns of a matrix are interchanged, i.e.,

$$\mathbf{A}' = \begin{bmatrix} a & d & g \\ b & e & h \\ c & f & i \end{bmatrix}$$

The transpose is obtained by using the worksheet function TRANSPOSE(*array*) or the Transpose option in the **Paste Special...** menu command (see "Using Paste Special to Transpose Rows and Columns" in Chapter 1).

The process of *matrix inversion* is analogous to obtaining the reciprocal of a number a. The matrix relationship that corresponds to the algebraic relationship $a \times (1/a) = 1$ is

$$\mathbf{A}\,\mathbf{A}^{-1} = \mathbf{I}$$

where $\mathbf{A}^{-1}$ is the inverse matrix and $\mathbf{I}$ is the unit matrix. The process for inverting a matrix "manually" (i.e., using pencil, paper and calculator) is complicated, but the operation can be carried out readily by using Excel's worksheet function MINVERSE(*array*). The inverse of the matrix $\mathbf{B}$ above is

$$\mathbf{B}^{-1} = \begin{bmatrix} 0.25 & 0.3333 & 0.5 \\ 0.75 & 0.6667 & 0.5 \\ 0.75 & 0.3333 & 0.5 \end{bmatrix}$$

The "pencil-and-paper" evaluation of a determinant of N rows × N columns is also complicated, but it can be done simply by using the worksheet function MDETERM(*array*). The function returns a single numerical value, not an array, and thus you do not have to use Ctrl+Shift+Enter. The value of the determinant of $\mathbf{B}$, represented by $|\mathbf{B}|$, is 12.

Numerical Differentiation

The process of finding the derivative dy/dx or slope of a function is the basis of *differential calculus*. It may be more convenient to obtain the derivative of a data set or the derivative of a worksheet formula by numeric methods, rather than by algebraic differentiation of the function.

Often a function depends on more than one variable. The *partial derivative* of the function $F(x,y,z)$, e.g., $\delta F/\delta x$, is the slope of the function with respect to x, while y and z are held constant.

First and Second Derivatives of a Data Set

The simplest method to obtain the first derivative of a function represented by a table of x, y data points is to calculate $\Delta y/\Delta x$. The first derivative or slope of the curve at a given data point x_n, y_n can be calculated using either of the following formulas:

$$\text{slope} = \frac{\Delta y}{\Delta x} = \frac{y_{n+1} - y_n}{x_{n+1} - x_n} \tag{13-2}$$

$$\text{slope} = \frac{y_n - y_{n-1}}{x_n - x_{n-1}} \tag{13-3}$$

The second derivative, d^2y/dx^2, of a data set is calculated in a similar manner, namely by calculating $\Delta(\Delta y/\Delta x)/\Delta x$.

Calculation of the first or second derivative of a data set tends to emphasize the "noise" in the data set; that is, small errors in the measurements become relatively much more important.

Points on a curve for which the first derivative is either a maximum, a minimum or zero are often of particular importance and are termed *critical points*.

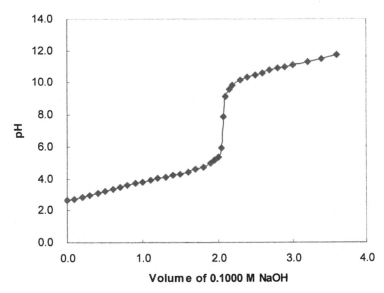

Figure 13-6. Titration curve of a weak acid.

The spreadsheet shown in Figure 13-7 uses pH titration data to illustrate the calculation of the first derivative of a data set. The raw data are in columns A and B of Figure 13-7 and are plotted in Figure 13-6.

	A	B	C	D	E	F
2	V/mL	pH	ΔV	ΔpH	V(avge)	ΔpH/ΔV
22	1.90	4.981	0.100	0.229	1.850	2.29
23	1.95	5.157	0.050	0.176	1.925	3.52
24	2.00	5.389	0.050	0.232	1.975	4.64
25	2.05	5.928	0.050	0.539	2.025	10.78
26	2.08	7.900	0.030	1.972	2.065	65.73
27	2.10	9.115	0.020	1.215	2.090	60.75
28	2.15	9.604	0.050	0.489	2.125	9.78
29	2.20	9.856	0.050	0.252	2.175	5.04
30	2.30	10.125	0.100	0.269	2.250	2.69

Figure 13-7. First derivative of titration data, near the end-point.

Since the derivative has been calculated over the finite volume $\Delta V = V_{n+1} - V_n$, the most suitable volume to use when plotting the $\Delta pH/\Delta V$ values, as shown in Figure 13-7, is

$$V_{average} = \frac{V_{n+1} + V_n}{2} \qquad (13\text{-}4)$$

The maximum in $\Delta pH/\Delta V$ indicates the location of the inflection point of the titration (Figure 13-8).

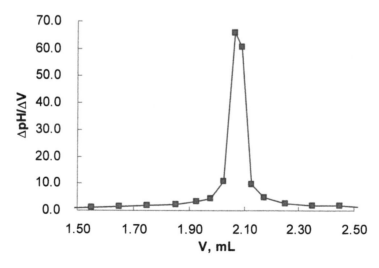

Figure 13-8. First derivative of titration data, near the end-point.

	E	F	G	H	I	J
2	V(avge)	ΔpH/ΔV	ΔV	Δ(ΔpH)	V(avge)	Δ(ΔpH)/ΔV
22	1.850	2.29	0.100	0.57	1.800	5.7
23	1.925	3.52	0.075	1.23	1.888	16.4
24	1.975	4.64	0.050	1.12	1.950	22.4
25	2.025	10.78	0.050	6.14	2.000	122.8
26	2.065	65.73	0.040	54.95	2.045	1373.8
27	2.090	60.75	0.025	-4.98	2.078	-199.3
28	2.125	9.78	0.035	-50.97	2.108	-1456.3
29	2.175	5.04	0.050	-4.74	2.150	-94.8
30	2.250	2.69	0.075	-2.35	2.213	-31.3

Figure 13-9. Second derivative of titration data, near the end-point.

The second derivative, $\Delta(\Delta pH/\Delta V)/\Delta V$, which is calculated by means of the spreadsheet shown in Figure 13-9, can be used to locate the inflection point more precisely. The second derivative passes through zero at the inflection point. Linear interpolation can be used to calculate the point at which the second derivative is zero (Figure 13-10).

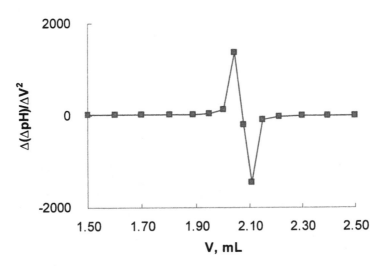

Figure 13-10. Second derivative of titration data, near the end-point.

There are more sophisticated equations for numerical differentiation. These equations use three, four or five points instead of two points to calculate the derivative. Since they usually require equal intervals between points, they are of less generality. Their main advantage is that they minimize the effect of "noise".

Derivatives of a Function

The first derivative of a formula in a worksheet cell can be obtained with a high degree of accuracy by evaluating the formula at x and at $x + \Delta x$. Since Excel carries 15 significant figures, Δx can be made very small. Under these conditions, $\Delta F/\Delta x$ is an excellent approximation to dF/dx.

The spreadsheet fragment shown in Figure 13-11 illustrates the calculation of the first derivative of a function ($F = x^3 - 3x^2 - 130x + 150$) by evaluating the function at x and at $x + \Delta x$. Here a value of Δx of 1×10^{-9} was used; alternatively, Δx could be obtained by using a worksheet formula such as =1E–9*x. For comparison, the first derivative was calculated from the expression from differential calculus: $F' = 3x^2 - 6x - 130$.

The Excel formulas in cells B12, C12, D12 and E12 are

    = t*x^3+u*x^2+v*x + w

    = t*(x+delta)^3 +u*(x+delta)^2 +v*(x+delta) + w

    =(C12-B12)/delta

    =3*t*x^2+2*u*x+v

Figure 13-12 shows a chart of the function and its first derivative.

	A	B	C	D	E
1			Numerical Differentiation		
2		(Using the equation F(x) = tx^3 + ux^2 + vx + w)			
3		(First derivative F'(x) = 3tx^2 + 2ux + v)			
4					
5	t =	1			
6	u =	-3		delta	
7	v =	-130		1.00E-09	
8	w =	150			
9				*Using differences*	*Using calculus*
10	x	F(x)	F(x+Δ)	F'(x)	F'(x)
11	-10	150	150	230.0	230.0
12	-9	348	348	167.0	167.0
13	-8	486	486	110.0	110.0
14	-7	570	570	59.0	59.0
15	-6	606	606	14.0	14.0
16	-5	600	600	-25.0	-25.0
17	-4	558	558	-58.0	-58.0
18	-3	486	486	-85.0	-85.0
19	-2	390	390	-106.0	-106.0

Figure 13-11. Calculating the first derivative of a function.

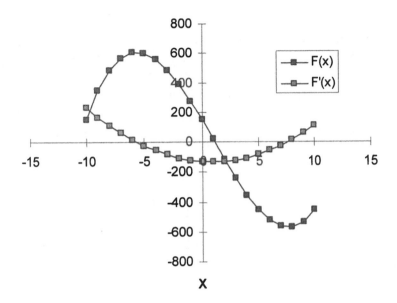

Figure 13-12. The function $F = x^3 - 3x^2 - 130x + 150$ and its first derivative.

Numerical Integration

A common use of numerical integration is to determine the area under a curve. We will describe three methods for determining the area under a curve: the rectangle method, the trapezoid method and Simpson's method. Each involves approximating the area of each portion of the curve delineated by adjacent data points; the area under the curve is the sum of these individual segments.

The simplest approach is to approximate the area by the rectangle whose height is equal to the value of one of the two data points, illustrated in Figure 13-13.

As the x increment (the interval between the data points) decreases, this rather crude approach becomes a better approximation to the area. The area under the curve bounded by the limits $x_{initial}$ and x_{final} is the sum of the individual rectangles, as given by equation 13-5.

$$\text{area} = \Sigma y_i(x_{i+1} - x_i) \tag{13-5}$$

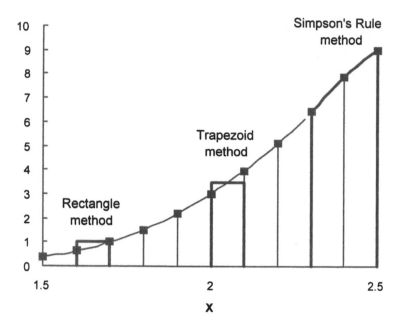

Figure 13-13. Graphical illustration of methods of calculating the area under a curve.

For a better approximation, you can use the average of the two y-values as the height of the rectangle. This is equivalent to approximating the area by a trapezoid rather than a rectangle. The area under the curve is given by equation 13-6.

$$\text{area} = \Sigma \frac{y_i + y_{i+1}}{2} (x_{i+1} - x_i) \tag{13-6}$$

Simpson's rule approximates the curvature of the function by means of a quadratic equation. To evaluate the coefficients of the quadratic requires the use of the y-values for three adjacent data points. The x-values must be equally spaced.

$$\text{area} = \Sigma \frac{y_i + 4y_{i+1} + y_{i+2}}{6} (x_{i+1} - x_i) \tag{13-7}$$

Finding the Area Under a Curve

The curve shown in Figure 13-14 is the sum of two Gaussian curves, with position and standard deviation $\mu = 90$, $\sigma = 10$ and $\mu = 130$, $\sigma = 20$, respectively. The equation used to calculate each Gaussian curve is

$$y = \frac{\exp[(x - \mu)^2 / 2\sigma^2]}{\sigma\sqrt{2\pi}} \tag{13-8}$$

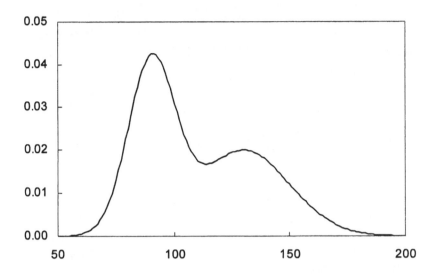

Figure 13-14. A curve that is the sum of two Gaussian curves.

The area of each Gaussian curve is equal to 1.000; thus the total area under the curve shown in Figure 13-16 is 2.000. The Excel formula used to calculate y is as follows (m is the position μ, and s is the standard deviation σ):

=EXP(-0.5*((x-curv1 m)/curv1 s)^2)/(SQRT(2*PI())*curv1 s)+EXP(-0.5*
((x-curv2 m)/curv2 s)^2)/(SQRT(2*PI())*curv2 s)

The area under the curve, between the limits $x = 50$ and $x = 200$, was calculated by using each of the preceding equations (13-5, 12-6 and 12-7): the rectangular approximation, the trapezoidal approximation and Simpson's rule, respectively. In each case a constant x increment of 10 was used. A portion of the spreadsheet is shown in Figure 13-15.

The formulas in row 9, used to calculate the area increment, are as follows:

=10*F9   (rectangular approximation)

=10*(F8+F9)/2   (trapezoidal approximation)

=10*(F8+4*F9+F10)/6   (Simpson's rule)

	E	F	G	H	I
6			Rectangular	Trapezoidal	Simpson's
7	x	Y	Approximation	Approx.	Rule
8					
9	50	0.00002	0.00020	0.00010	0.00095
10	60	0.00049	0.00487	0.00253	0.01265
11	70	0.00562	0.05621	0.03054	0.08007
12	80	0.02507	0.25073	0.15347	0.24751
13	90	0.04259	0.42594	0.33834	0.37687
14	100	0.03067	0.30673	0.36633	0.30464
15	110	0.01750	0.17498	0.24085	0.19785
16	120	0.01805	0.18046	0.17772	0.18274
17	130	0.01996	0.19960	0.19003	0.19249
18	140	0.01760	0.17603	0.18782	0.17079
19	150	0.01210	0.12099	0.14851	0.12079
20	160	0.00648	0.06476	0.09287	0.06784
21	170	0.00270	0.02700	0.04588	0.03025
22	180	0.00088	0.00876	0.01788	0.01071
23	190	0.00022	0.00222	0.00549	0.00301
24	200	0.00004	0.00044	0.00133	0.00066

Figure 13-15. Portion of a spreadsheet for calculating the area under a curve.

The area increments were summed and the area under the curve, calculated by the three methods, is shown in Figure 13-16. All three methods of calculation appear to give acceptable results in this case.

	E	F	G	H	I
6			Rectangular	Trapezoidal	Simpson's
7	x	Y	Approximation	Approx.	Rule
25					
26		Total =	1.9999	1.9997	1.9998

Figure 13-16. Area under a curve, calculated by three different methods.

Finding Roots of Equations

Sometimes a chemical problem can be reduced algebraically, by pencil and paper, to a polynomial expression for which the solution to the problem is one of the roots of the polynomial. Almost everyone remembers the quadratic formula for the roots of a quadratic equation, but finding the roots of a more complicated polynomial is more difficult. We begin by describing three methods for finding the real roots of a polynomial.

The Graphical Method

The roots of a polynomial $y = F(x)$ are the values of x that make $F(x) = 0$. One simple way to find those values is to create a spreadsheet table of x-values and corresponding y-values, then create a chart from the values. The *x-values* where $y = 0$ (where the curve crosses the x-axis) are the roots of the equation. Figure 13-17 is a graph of the function $y = x^3 - 3x^2 - 13x + 15$, for which the roots are 5, 1 and –3.

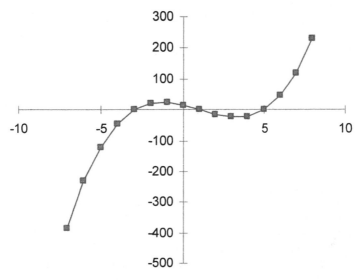

Figure 13-17. Graph of the equation $y = x^3 - 3x^2 - 13x + 15$. The roots are 5, 1 and –3.

The roots can't be read from this chart with any degree of accuracy, but it's a simple matter to create a chart of the region immediately around an intersection point and get a much more precise value. In general, though, the main use of this method is to gain an idea of the approximate value of the roots.

The Method of Successive Approximations

The graphical method can give us an approximate idea of the roots of a polynomial. To obtain a more accurate numerical result, the roots of $y = F(x)$ can be obtained by trial and error, finding the values of x that make the function y equal to zero. There are several methods, systematic rather than trial and error, that can easily be carried out on a spreadsheet; the one described in the following is sometimes referred to as the *Regula Falsi* method.

To illustrate the method, we'll find a real root of the polynomial $y = x^3 + 0.13x^2 - 0.0005x - 0.00009$. A chart of the function, using suitable values of x that show that there are three real roots of the function, is shown in Figure 13-18. There are three real roots, at approximately -0.13, -0.25 and 0.25.

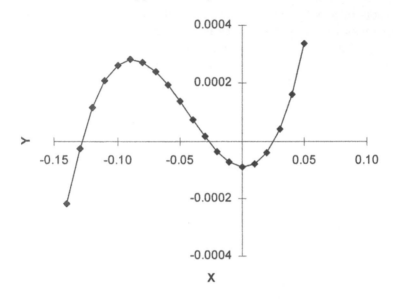

Figure 13-18. Graph of the equation $y = x^3 + 0.13x^2 - 0.0005x - 0.00009$.

We can only calculate one of the three roots at a time. In this illustration, we'll find the positive root that is near 0.025. Set up a table of x- and y-values, using x-values that span the value of the root. In Figure 13-19, I used values from 0.00 to 0.05, in increments of 0.01, and entered the formula for y in column B. The function y changes sign between $x = 0.02$ and $x = 0.03$; therefore the root must lie between these two values (the two values that span the root are indicated in italics).

	A	B
5	X	Y
6	0.00	-0.0001
7	0.01	-0.000081
8	*0.02*	*-0.00004*
9	*0.03*	*0.000039*
10	0.04	0.000162
11	0.05	0.000335

Figure 13-19. Stage 1 in solution by successive approximations. The two data points where sign change occurs are in italics.

Now create a second table, using a range of x-values between the spanning values (I used values from 0.020 to 0.030, in increments of 0.001) as shown in Figure 13-20. Once again, locate the two values that bracket $y = 0$; in this example they are $x = 0.025$ and $x = 0.026$. Repeat the process until you achieve the desired level of accuracy. You can see from Figure 13-21 that this root of the function is slightly greater than 0.0257.

	C	D
8	0.020	-0.00004
9	0.021	-3.3909E-05
10	0.022	-2.7432E-05
11	0.023	-2.0563E-05
12	0.024	-1.3296E-05
13	0.025	-5.625E-06
14	0.026	2.456E-06
15	0.027	1.0953E-05
16	0.028	1.9872E-05
17	0.029	2.9219E-05
18	0.030	0.000039

Figure 13-20. Stage 2 in solution by successive approximations. The two data points where sign change occurs are in italics.

	E	F
13	0.0250	-5.625E-06
14	0.0251	-4.8354E-06
15	0.0252	-4.0418E-06
16	0.0253	-3.244E-06
17	0.0254	-2.4421E-06
18	0.0255	-1.6361E-06
19	0.0256	-8.2598E-07
20	0.0257	-1.171E-08
21	0.0258	8.0671E-07
22	0.0259	1.6293E-06
23	0.0260	2.456E-06

Figure 13-21. Stage 3 in solution by successive approximations. The two data points where sign change occurs are in italics.

The Secant Method

Instead of "trial-and-error", it's more efficient to use a systematic method. There are several methods to find the roots of a function that are readily adaptable to spreadsheet calculation. One of these is the so-called "secant method", described in the following paragraphs.

The secant method obtains an approximation to the slope of the curve at some trial value of x, and extrapolates that slope to $y = 0$ in order to obtain a value of x that is an estimate of the root. Using this value of x as the new trial value, the process is repeated until x converges to the root. The slope of the curve is approximated by using two values of x, as illustrated in Figure 13-22.

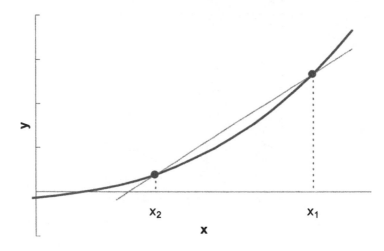

Figure 13-22. The secant method for obtaining a root of a function.

Although this may be a poor approximation to the tangent to the curve, it becomes more and more accurate as the iterations approach the root. The calculations are illustrated in Figure 13-23, applied to the function $y = x^3 + 0.13x^2 - 0.0005x - 0.00009$ shown in Figure 13-18.

The secant method requires that you enter two starting values, x_1 and x_2 (although x_2 could be calculated by means of a formula, e.g., =1.01*A7). In Figure 13-25 the two starting values are shown in bold. The formula for y was entered in cell B7 and copied into cell D7. The formulas for the slope m and for new x are

=(D7-B7)/(C7-A7)

=(E7*A7-B7)/E7

In row 8, cell A8 contains the formula =C7 and cell C8 contains the formula =F7. The formulas of the other cells in row 8 are Filled Down from the corresponding cell in row 7. Now you can select the cells in row 8 (A8:F8) and Fill Down until the x-values reach convergence, or until the y-values become sufficiently close to zero. From the last row shown in Figure 13-23 you can see that this root of the function is 0.025701434.

	A	B	C	D	E	F
5	$y = x^3 + 0.13x^2 - 0.0005x - 0.00009$					
6	x1	y1	x2	y2	m	new x
7	0.05	3.35E-04	0.0505	3.45E-04	0.02014	0.03337
8	0.0505	3.45E-04	0.03337	7.52E-05	0.01575	0.02859
9	0.033	7.52E-05	0.02859	2.54E-05	0.01044	0.02616
10	0.029	2.54E-05	0.02616	3.82E-06	0.00887	0.02573
11	0.026	3.82E-06	0.02573	2.59E-07	0.00827	0.02570
12	0.0257	2.59E-07	0.02570	3.00E-09	0.00817	0.02570
13	0.02570	3.00E-09	0.02570	2.41E-12	0.00816	0.02570
14	0.025701	2.41E-12	0.02570	2.25E-17	0.00816	0.02570
15	0.025701434	2.25E-17	0.02570	0.00E+00	0.00816	0.02570

Figure 13-23. Obtaining a root of a function by the secant method.

From Figure 13-18 it should be clear that the root calculated by this method will depend on the starting value. If 0.05 is entered in cell A7, the root 0.025701434 is returned; if -1 is entered in A7, the root -0.128437118401283 is returned; and if -0.05 is entered in A7, the root -0.027264331561808793 is returned. Thus it is up to the user to choose a starting value that will return the desired root.

Using Goal Seek...

Excel provides a built-in way to perform successive approximations, by using Data → Data Tools → What-If Analysis → Goal Seek... (Excel 2007/2010) or the **Goal Seek...** command in the **Tools** menu (Excel 2003). **Goal Seek** changes the value of a selected cell (the *changing cell*) to make the value of another cell (the *target cell*) reach a desired value.

To illustrate the use of **Goal Seek**, let's repeat finding a root of the function $y = x^3 + 0.13x^2 - 0.0005x - 0.00009$, shown in Figure 13-20. **Goal Seek** allows you to obtain the root much more easily. To use **Goal Seek**, you need a cell (the changing cell) that contains a value of x, as well as a cell (the target cell) that contains the formula that depends on the changing cell. You can use any pair of cells in Figure 13-19, 12-20 or 12-21 that contain a value of x and the corresponding function value y, or you can enter a value for x and the formula for y somewhere else in the workbook. If we want to find the most negative root of the function, enter a suitable value, such as -1, in the changing cell. Well now use **Goal Seek...** to find the value in the changing cell (I used cell A30) that makes the function (in cell B30) equal to zero.

The accuracy of the result will depend on the magnitude of the Maximum Change parameter, which determines when Goal Seek has reached convergence. To change this parameter, in Excel 2003 choose **Options...** from the **Tools** menu and then choose the Calculation tab (Figure 13-26). For Excel 2007, click on the

Office button and click on Excel Options; for Excel 2010, click on the File tab, then on the Options button. Click on Formulas to display Calculation Options.

The default value of the Maximum Change parameter is 0.001. You'll see in a moment that adjusting this parameter is critical when you are using **Goal Seek.** But for now, set the value of Maximum Change to 1E-12.

Now choose **Goal Seek** from the **Tools** menu or in the Data ribbon. As shown in Figure 13-24, enter $B$30 in the Set Cell box (the cell reference will appear there if you selected that cell before choosing **Goal Seek...**). Put the cursor in the To Value box and enter the desired value, zero. Put the cursor in the By Changing Cell box and enter $A$30 by selecting the cell or by typing. Then click OK.

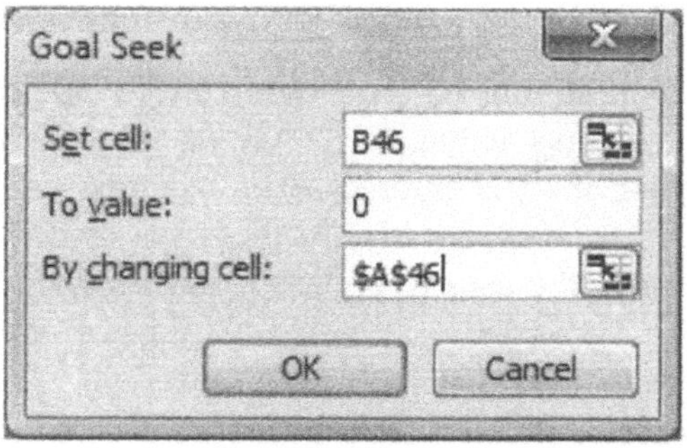

Figure 13-24. The Goal Seek dialog box.

After a few iteration cycles the Goal Seek Status dialog box (Figure 13-25) is displayed.

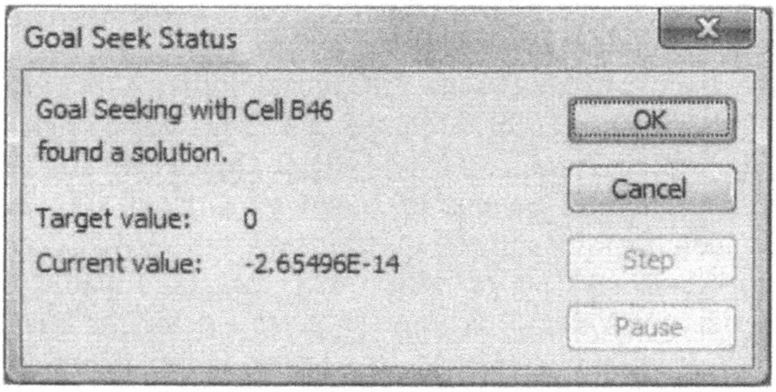

Figure 13-25. The Goal Seek Status dialog box.

Adjusting the Maximum Change parameter is critical when using **Goal Seek.** That's because Excel stops iterating when the change in the result is less than the Maximum Change parameter. Therefore the Maximum Change parameter needs to be adjusted to be much less than the value of the function. For most chemical calculations it's a good idea to set Maximum Change to 1E-12 or

1E-15 as a matter of course. Or you can set Maximum Change to zero; since the function will probably never get to be exactly zero, Goal Seek will just run for 100 iterations.

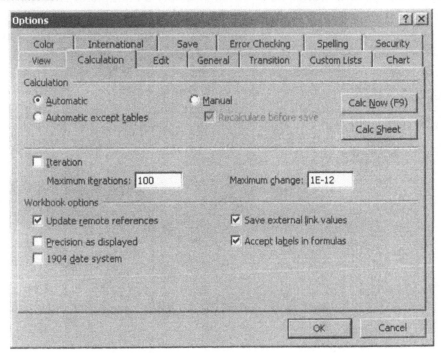

Figure 13-26. The Excel 2003 Calculation Options dialog box.

For problems requiring the variation of two or more parameters, that is, varying the values of several cells to make the value of another cell reach a desired value, you must use the Solver, which is described in detail in Chapter 15.

Solving Sets of Simultaneous Linear Equations

Sometimes a chemical system can be represented by a set of n linear equations in n unknowns, i.e.,

$$a_{11}x_1 + a_{12}x_2 + a_{13}x_3 + \cdots = c_1$$
$$a_{21}x_1 + a_{22}x_2 + a_{23}x_3 + \cdots = c_2$$
$$\cdot$$
$$a_{n1}x_1 + a_{n2}x_2 + a_{n3}x_3 + \cdots = c_n \tag{13-9}$$

where x_1 , x_2 , x_3 , ... are the experimental unknowns, c is the experimentally measured quantity, and the a_{ij} are coefficients. The equations must be linearly independent; in other words, no equation is simply a multiple of another. These equations can be represented in matrix notation by

$$\mathbf{AX} = \mathbf{C} \tag{13-10}$$

A familiar example is the spectrophotometric determination of the concentrations of a mixture of n components by absorbance measurements at n different wavelengths. The coefficients a_{ij} are the ε, the molar absorptivities of the components at different wavelengths (for simplicity, the cell path length, usually 1.00 cm, has been omitted from these equations). For example, for a mixture of three species P, Q and R, where absorbance measurements are made at λ_1, λ_2 and λ_3, the equations are

$$\varepsilon_{\lambda_1}^{P} [P] + \varepsilon_{\lambda_1}^{Q} [Q] + \varepsilon_{\lambda_1}^{R} [R] = A_{\lambda 1}$$

$$\varepsilon_{\lambda_2}^{P} [P] + \varepsilon_{\lambda_2}^{Q} [Q] + \varepsilon_{\lambda_2}^{R} [R] = A_{\lambda 2}$$

$$\varepsilon_{\lambda_3}^{P} [P] + \varepsilon_{\lambda_3}^{Q} [Q] + \varepsilon_{\lambda_3}^{R} [R] = A_{\lambda 3}$$

Thus nine coefficients are required for the determination of three unknown concentrations.

Cramer's Rule

According to Cramer's rule, a system of simultaneous linear equations has a unique solution if the determinant D of the coefficients is non-zero.

$$D = \begin{vmatrix} a_{11} & a_{12} & a_{13} \\ a_{21} & a_{22} & a_{23} \\ a_{31} & a_{32} & a_{33} \end{vmatrix}$$

Thus, for example, for the set of equations

$$2x + y - z = 0$$
$$x - y + z = 6$$
$$x + 2y + z = 3$$

the determinant is

$$\begin{vmatrix} 2 & 1 & 1 \\ 1 & 1 & 1 \\ 1 & 2 & 1 \end{vmatrix}$$

The coefficients and constants lend themselves readily to spreadsheet solution, as illustrated in Figures 12-27 and 12-28.

Using the formula =MDETERM(A2:C4), the value of the determinant is found to be –9, indicating that the system is solvable.

	A	B	C	D
1		Coefficients		Constants
2	2	1	-1	0
3	1	-1	1	6
4	1	2	1	3

Figure 13-27. Spreadsheet data for three equations in three unknowns.

	A	B	C
8	0	1	-1
9	6	-1	1
10	3	2	1

Figure 13-28. The determinant for obtaining x.

The x-values that comprise the solution of the set of equations can be calculated in the following manner: x_k is given by a quotient in which the denominator is D, and the numerator is obtained from D by replacing the kth column of coefficients by the constants $c_1, c_2, \ldots$. The unknowns are obtained readily by copying the coefficients and constants to appropriate columns in another location in the sheet. For example, to obtain x, the determinant is shown in Figure 13-28 and $x = 2$ is obtained from the formula

=MDETERM(A8:C10)/MDETERM(A2:C4)

$y = -1$ and $z = 3$ are obtained from appropriate forms of the same formula.

Solution Using Matrix Inversion

If equation 13-10 is multiplied by the inverse of **A**, we obtain the relationship

$$\mathbf{X} = \mathbf{A}^{-1}\mathbf{C} \qquad (13\text{-}11\,)$$

In other words, the solution matrix is obtained by multiplying the matrix of constants by the inverse matrix of the coefficients. To return the solution values shown in Figure 13-29, the array formula

{=MMULT(MINVERSE(A2:C4),D2:D4)}

was entered in cells E2:E4.

	A	B	C	D	E
1	Coefficients			Constants	Solution
2	2	1	-1	0	2
3	1	-1	1	6	-1
4	1	2	1	3	3

Figure 13-29. Solving a set of simultaneous equations by means of matrix methods.

Analysis of Spectra of Mixtures Using Matrix Mathematics

A common analytical problem in spectrophotometry is the analysis of a mixture of components. If the spectra of the pure components are available, the spectrum of a mixture can be analyzed to determine the concentrations of the individual components. If the mixture contains N components, then absorbance measurements at N suitable wavelengths are necessary to solve the set of N linear equations in N unknowns.

Applying Cramer's Rule to a Spectrophotometric Problem

As a simple example of the analysis of mixtures, consider an aqueous solution containing a mixture of Co^{2+}, Ni^{2+} and Cu^{2+}, to be analyzed by spectrophotometric measurements at three different wavelengths. The spectra of the individual ions and of a mixture are shown in Figures 13-30 and 13-31. The most suitable wavelengths for analysis are 394, 510 and 808 nm (determined from an examination of Figure 13-30).

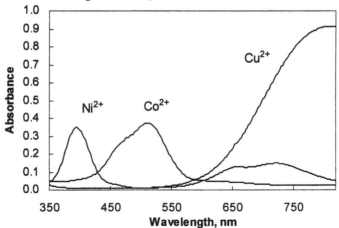

Figure 13-30. Spectra of Co^{2+}, Ni^{2+} and Cu^{2+} ions in aqueous solution (standards). (Spectrophotometric data provided by Dr. Lev Zompa.)

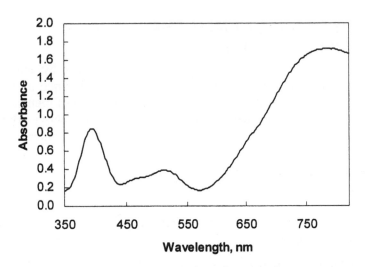

Figure 13-31. Spectrum of a mixture of Co^{2+}, Ni^{2+} and Cu^{2+} ions in aqueous solution.

The molar absorptivities of the three species at these wavelengths are shown in Figure 13-32, together with absorbance readings for a mixture of the three ions, measured in a 1.00-cm cell.

	G	H	I	J	K	L
3		*STANDARDS*				*UNKNOWN*
4		Molar Absorptivity, $M^{-1}cm^{-1}$				(Absorbance)
5	λ/nm	Co^{2+}	Ni^{2+}	Cu^{2+}		Mixture
6	394	0.995	6.868	0.189		0.845
7	510	6.450	0.215	0.199		0.388
8	808	0.469	1.179	15.053		1.696

Figure 13-32. Data table for the determination of a mixture of Co^{2+}, Ni^{2+} and Cu^{2+} ions.

	G	H	I
13	0.845	6.868	0.188
14	0.388	0.215	0.198
15	1.696	1.179	15.052

Figure 13-33. The determinant for calculating Co^{2+}.

Following the Cramer's rule procedures described above, we construct the determinant shown in Figure 13-33 to determine Co^{2+} concentration.

Using the formula

=MDETERM(O15:Q17)/MDETERM(O5:Q7)

yields the value 0.05328 M for the Co^{2+} concentration. From similar formulas, $[Ni^{2+}] = 0.1125$ M and $[Cu^{2+}] = 0.1022$ M.

Analysis of Spectra of Mixtures
Using Matrix Inversion

A set of simultaneous linear equations can also be solved by using matrices, as shown earlier in this chapter. The *solution matrix* is obtained by multiplying the matrix of constants by the inverse of the matrix of coefficients. Applying this simple solution to the spectrophotometric data used above, the inverted matrix is obtained by selecting a 3R × 3C array of cells, entering the array formula

{=MINVERSE(H6:J8)}

The inverted matrix is shown in Figure 13-34.

The solution matrix is obtained by selecting a 3R × 1C array and then entering the array formula

{=MMULT(G29:I31,L6:L8)}

The single array formula

{=MMULT(MINVERSE(H6:J8),L6:L8)}

accomplishes the same result. The solution matrix is shown in cells G38:G40 of Figure 13-35.

	G	H	I
29	-0.00452587	0.15588253	-0.00200408
30	0.14656833	-0.02249395	-0.00154432
31	-0.01134029	-0.00309449	0.06661698

Figure 13-34. The inverted matrix.

	G	H
38	0.05328	(Co^{2+})
39	0.1125	(Ni^{2+})
40	0.1022	(Cu^{2+})

Figure 13-35. The solution matrix.

Polar to Cartesian Coordinates

You may occasionally need to chart a function that involves angles. Instead of using the familiar Cartesian coordinate system (x, y and z coordinates), such functions often use the polar coordinate system, in which the coordinates are two angles, θ and ϕ, and a distance r. The two coordinate systems are related by the equations $x = r \sin \theta \cos \phi$, $y = r \sin \theta \sin \phi$, $z = r \cos \theta$. Angle θ is the angle between the vector r and the Cartesian z-axis; ϕ is the angle between the projection of r on the x,y plane and the x-axis. Since Excel's trigonometric functions only consider x- and y axes, the simplified relationships are, for angles in the $x\,y$ plane: $x = r \cos \phi$, $y = r \sin \phi$.

As an example of transformation of polar to Cartesian coordinates, we'll graph the wave function for the $d_{x^2-y^2}$ orbital in the x, y plane. The angular component of the wave function in the x, y plane is

$$\Phi = \sqrt{\frac{15}{16\pi}} \cos 2\phi \qquad (13\text{-}18)$$

and Φ can be equated to the radial vector r for the conversion of polar to Cartesian coordinates.

In the spreadsheet fragment shown in Figure 13-36, column A contains angles from 0 to 360 in 2-degree increments. Column B converts the angles to radians (required by the COS worksheet function) using the relationship =A4*PI()/180 in row 4. The formulas in cells C4, D4 and E4 are:

=SQRT(15/(16*PI()))*COS(2*B4)

=C4*COS(B4)

=C4*SIN(B4).

The chart of the x- and y-values is shown in Figure 13-37.

	A	B	C	D	E
1	Representation of Angular Wave Function of $d_{x^2-y^2}$ Orbital				
2					
3	Angle, deg	Angle, rad	d	x coord	y coord
4	0	0.0000	0.5463	0.5463	0.0000
5	2	0.0349	0.5449	0.5446	0.0190
6	4	0.0698	0.5410	0.5396	0.0377
7	6	0.1047	0.5343	0.5314	0.0559
8	8	0.1396	0.5251	0.5200	0.0731
9	10	0.1745	0.5133	0.5055	0.0891
10	12	0.2094	0.4990	0.4881	0.1038
11	14	0.2443	0.4823	0.4680	0.1167
12	16	0.2793	0.4633	0.4453	0.1277

Figure 13-36. Converting from polar to Cartesian coordinates

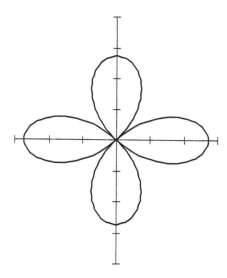

Figure 13-37. Angular wave function for $d_{x^2-y^2}$ orbital.

Significant Figures

A common question that scientists ask about Excel is the following: "Is there something in Excel to handle significant figures?" Unfortunately, there's nothing built into Excel to handle this problem: Excel's formulas calculate results with up to 15 digits, and display all of them, irrespective of how many of them are significant.

You probably remember the rules for determining the number of significant figures in the result of a calculation: For multiplication or division, the number of significant figures in the result can be no greater than the smallest number of significant figures in any of the numbers involved in the calculation; for addition or subtraction, the number of decimal places in the result can be no greater than the smallest number of decimal places in the numbers involved in the calculation. Thus the number of significant figures in the result of the following calculation should be three:

$$\frac{(0.547 - 0.033)(54.938)}{(2.36 \times 10^3)}$$

and the result should be written as 0.0120. (Excel calculates the result as 0.0119653101694915.)

There are many formulas and utilities that you can find on the Internet to return a value with a specified number of significant figures, but none that I have seen that are acceptable. They all require the user to determine the number of significant figures beforehand, and they require the use of an additional column to contain the converted value.

In my opinion, the best way to handle significant figures in Excel calculations is to simply format numbers to display three, four or five significant figures (one or two is too few and more than five is usually too many). The numbers must be in scientific number format when you do this.

But formatting a value to display, say, four significant figures does not change the underlying value. To change the underlying value, you can use Precision As Displayed in the Calculation tab of the **Options** command in the **Tools** menu (Excel 2003) or press the Office button, click on Excel Options, click on Advanced scroll down to When Calculating This Workbook, and check the box for "Set Precision as Displayed" (Excel 2007). Be aware that this converts all values (but not formulas) in the sheet to their displayed values. The warning message "Data will permanently lose accuracy." is displayed. In Chapter 19 you will find a macro that will convert only selected cells to their precision-as-displayed values.

14

Linear Regression and Curve Fitting

Excel provides several ways to find the coefficients that provide the best fit of a function to a set of data points — a process sometimes referred to as *curve fitting*. The "best fit" of the curve is considered to be found when the sum of the squares of the deviations of the data points from the calculated curve is a minimum. In the field of statistics, finding the least-squares best-fit parameters that describe a data set is known as *regression analysis*. Excel provides several ways to obtain regression coefficients; these are described in the following sections.

Linear Regression

Regression analysis is a statistical technique used to determine whether experimental variables are interdependent and to express quantitatively the relationship between them as well as the degree of correlation. For many chemical systems, the mathematical form of the equation relating the dependent and independent variables is known. In other cases, the data may be fitted by an empirical fitting function such as a power series, simply for purposes of graphing or interpolation. In any event, you must provide the form of the equation; regression analysis merely provides the coefficients.

A secondary but no less important goal is to obtain the standard deviations[*] of the regression parameters and a measure of the goodness of fit of the data to the model equation.

---

[*] In Chapters 14 and 15, the symbol σ is used for the population standard deviation (i.e., when the sample size is large) and the symbol s for the sample standard deviation (when the sample size is small).

The *method of least squares* yields the parameters that minimize the sum of squares of the residuals (the deviation of each measurement of the dependent variable from its calculated value).

$$SS_{resid} = \sum_{n=1}^{N} (y_i - y_{calc})^2 \qquad (14\text{-}1)$$

Least-Squares Fit to a Straight Line

Although it is relatively easy to draw a straight line with ruler and pencil through a series of points if they all fall on or near the line, it becomes more and more a matter of judgment if the data are scattered. The least-squares line of best fit minimizes the sum of the squares of the y deviations of individual points from the line. Regression analysis in the simplest form assumes that all deviations from the line are the result of error in the measurement of the dependent variable y.

For the least-squares straight line $y = mx + b$ through N data points, the least-squares slope and intercept are obtained from equations (14-2) and (14-3).

$$m = \frac{N\Sigma x_i y_i - \Sigma x_i \Sigma y_i}{N\Sigma x_i^2 - (\Sigma x_i)^2} \qquad (14\text{-}2)$$

$$b = \frac{\Sigma x_i^2 \Sigma y_i - \Sigma x_i \Sigma x_i y_i}{N\Sigma x_i^2 - (\Sigma x_i)^2} \qquad (14\text{-}3)$$

The *correlation coefficient*, R, is a measure of the correlation between x and y. If x and y are perfectly correlated (i.e., a perfect straight line), then $R = 1$. An R value of zero means that there is no correlation between x and y, and an R value of -1 means that there is a perfect negative correlation.

More commonly, R^2, the square of the correlation coefficient, given by equation 14-4, is used as the measure of correlation; it ranges from 0 (no correlation) to 1 (perfect correlation).

$$R^2 = \frac{(N\Sigma x_i y_i - \Sigma x_i \Sigma y_i)^2}{(N\Sigma x_i^2 - (\Sigma x_i)^2)(N\Sigma y_i^2 - (\Sigma y_i)^2)} \qquad (14\text{-}4)$$

The standard deviations of the slope m and the intercept b are given by equations 14-5 and 14-6

$$\sigma_m = \sqrt{\frac{(1 - R^2)(N\Sigma y_i^2 - (\Sigma y_i)^2)}{(N-2)(N\Sigma x_i^2 - (\Sigma x_i)^2)}} \qquad (14\text{-}5)$$

$$\sigma_b = \sqrt{\frac{(1-R^2)\{1+(\Sigma x_i)^2 /(N\Sigma x_i^2 - (\Sigma x_i)^2)\}(N\Sigma y_i^2 - (\Sigma y_i)^2)}{(N-2)N^2}} \quad (14\text{-}6)$$

Excel provides worksheet functions to calculate the least-squares slope, intercept and R^2 of the straight line $y = mx + b$. In addition, Excel provides several tools for performing multiple linear regression, to fit curves other than a straight line.

Using the SLOPE, INTERCEPT and RSQ Functions

The worksheet functions SLOPE(*known_ys*, *known_xs*) and INTERCEPT(*known_ys*, *known_xs*) return the slope m and intercept b, respectively, of the least-squares straight line through a set of data points. For example, Figure 14-1 illustrates the determination of the slope and intercept of a calibration curve from spectrophotometric data (concentration of potassium permanganate standards in column A, absorbance of the standards in column B).

The formula entered in cell B10 is

=SLOPE(B4:B8,A4:A8)

and corresponding formulas, INTERCEPT(B4:B8,A4:A8) and RSQ(B4:B8,A4:A8), were entered in cells B11 and B12 to obtain the intercept and R^2 values.

	A	B	C
	Calibration Curve		
1	**of Potassium Permanganate Solutions**		
2			
3	**C, M**	**Abs**	**A(calc)**
4	0.000E+00	0.000	0.002
5	1.029E-04	0.257	0.258
6	2.058E-04	0.518	0.513
7	3.087E-04	0.771	0.769
8	4.116E-04	1.021	1.025
9			
10	slope =	2.484E+03	
11	intercept =	0.0022	
12	$R^2 =$	0.9999	

Figure 14-1. Using SLOPE, INTERCEPT and RSQ functions.

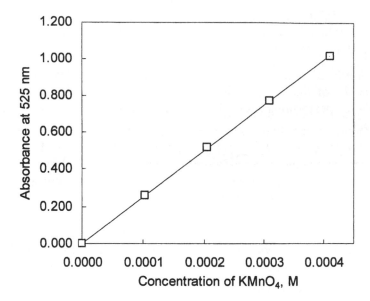

Figure 14-2. Least-squares best fit line through data points of a calibration curve.

It's a good idea to produce a chart of the data for visual inspection of the fit, as illustrated in Figure 14-2.

Column C in Figure 14-1 contains the absorbance values calculated from the slope and intercept; for example, cell C4 contains =$B$10*A4+$B$11.

Multiple Linear Regression

Linear regression is not limited to the case of finding the least-squares slope and intercept of the straight line $y = mx + b$. Multiple linear regression fits data to a model that defines y as a function of two or more independent x variables. For example, you might want to fit the yield of a biological fermentation product as a function of temperature (T), pressure of CO_2 gas (P), and fermentation time (t), for example,

$$y = a \cdot T + b \cdot P + c \cdot t + d \qquad (14\text{-}7)$$

using data from a series of fermentation runs with different conditions of temperature, pressure and time. Or the dependent variable y could be a function of several independent variables, each of which is a function of a single original independent variable, for example,

$$y = a[\text{H}^+]^3 + b[\text{H}^+]^2 + c[\text{H}^+] + d \qquad (14\text{-}8)$$

Linear regression methods can be applied to any function that is *linear in the coefficients*[*]. Many functions that produce curved *x–y* plots are linear in the coefficients, including power series such as equation 14-8 and some functions containing exponentials, such as

$$y = ae^x \qquad (14\text{-}9)$$

Although equation 14-8 is a nonlinear function (a cubic equation), it is linear in the coefficients and therefore linear regression can be used to obtain the regression coefficients *a*, *b*, *c* and *d*. Excel provides at least three ways to perform linear regression: by adding a Trendline to a chart, by using the Regression tool in the Analysis ToolPak, or by using the worksheet function LINEST. LINEST (for <u>lin</u>ear <u>est</u>imation) is the most versatile of the three, so we will begin with it.

Using LINEST to Perform Multiple Linear Regression

The worksheet function LINEST performs linear regression analysis on a set of *x,y* data points. The general form of the linear equation that can be handled by LINEST is

$$y = m_1 x_1 + m_2 x_2 + m_3 x_3 + \ldots + b \qquad (14\text{-}10)$$

LINEST returns the array of regression parameters $m_n, \ldots, m_2, m_1, b$. The syntax is LINEST(***known_ys***, *known_xs*, *const_logical*, *stats_logical*). If *const_logical* is TRUE or omitted, the regression parameters include an intercept *b*; if *const_logical* is FALSE, the fit does not include the intercept *b*. If *stats_logical* is TRUE, LINEST returns an array of regression statistics in addition to the regression coefficients $m_n, \ldots, m_1$ and *b*. The layout of the array of returned values is shown in Figure 14-3. A one-, two-, three-, four-, or five-row array may be selected. Since LINEST is an array function, you must enter the formula by pressing Ctrl+Shift+Enter.

m(n)	m(n-1)	...	m(2)	m(1)	b
std.dev(n)	std.dev(n-1)	...	std.dev(2)	std.dev(1)	std.dev(b)
R^2	std.dev(y)				
F	df				
ss(reg)	ss(resid)				

Figure 14-3. Layout of regression results and statistics returned by LINEST.

---

[*] Mathematically, a function that is linear in the coefficients is one for which *the partial derivatives of the function with respect to the coefficients do not contain coefficients*. For example, for the power series equation $y = a + bx + cx^2$, $\partial y/\partial a = 1$, $\partial y/\partial b = x$ and $\partial y/\partial c = x^2$.

Least-Squares Fit to a Straight Line
Using the LINEST Function

The worksheet function LINEST returns the parameters for multiple linear regression. As an introductory example we'll use it to find the least-squares fit to the calibration curve data of Figure 14-1. To use the LINEST function to find the slope and intercept of a least-squares straight line, select an array two columns wide and one to five rows deep. The selection is two columns wide because we are returning two regression coefficients, m and b. It can be up to five rows deep because that's the number of rows of statistical information that can be returned by LINEST. You don't need to always select five rows for the results; often three rows are sufficient, in order to display the coefficients, their standard deviations, and the R^2 value. The array of regression parameters and statistics that will be returned when a 3R × 2C array is selected is shown in Figure 14-4. LINEST returns this array of values — slope, intercept, standard deviations and R^2 — by the entry of a single formula.

m	b
std.dev. of m	std.dev. of b
R^2	SE(y)

Figure 14-4. Table of regression statistics returned by LINEST for slope and intercept of a straight line.

LINEST is an array function; to use it, with the data in Figure 14-1, you must do the following:

- Select a range two columns wide and three rows deep. For this example, the range B16:C18 was selected.
- Type the LINEST formula with its arguments: in this example =LINEST(B4:B8,A4:A8,TRUE,TRUE). You can use the following "shorthand" for the logical arguments *const* and *stats*: FALSE can be represented by 0 and TRUE by any nonzero value. For our example, you can use =LINEST(B4:B8,A4:A8,1,1).
- Enter the formula by using Ctrl+Shift+Enter.

When you "array-enter" a formula, Excel puts braces around the formula, as shown below:

{=LINEST(B4:B8,A4:A8,1,1)}

You do not type the braces; if you did, the result would not be recognized by Excel as a formula.

The regression parameters and statistics are returned, as shown in Figure 14-5. The returned values were formatted to display an appropriate number of significant figures.

	A	B	C
14	*From LINEST:*		
15		*m*	*b*
16	*parameters*	2.484E+03	0.0022
17	*std. dev.s*	1.17E+01	0.0030
18	R^2, *SE(y)*	0.99993	0.00381

Figure 14-5. Slope and intercept of a straight line, with regression statistics.

As you can see, LINEST returns a large amount of useful statistical information simply by entering a single formula: the regression coefficients, their standard deviations, the R^2 value, plus several other statistical quantities. You must, however, be familiar with the layout of regression results and statistics shown in Figure 14-3 (also shown in Excel's On-Line Help for the LINEST worksheet function) in order to know what value each cell contains.

Regression Line Without an Intercept

If the LINEST argument *const_logical* is set to FALSE, the *m* coefficient of the line of the form $y = mx$ that best fits the data is returned. Applying the LINEST function in this way to the permanganate data yields the parameters shown in Figure 14-6.

	A	B	C
21	*parameters*	2.491E+03	0
22	*std. dev.s*	6.38E+00	#N/A
23	R^2, *SE(y)*	0.99997	0.00359

Figure 14-6. Slope of a straight line through the origin, with regression statistics.

The Regression Parameters

Mathematical relationships between the regression parameters are given below (N = number of data points, k = number of independent variables, $\bar{y}$ = mean of the *y*-values):

$$df \text{ (degrees of freedom)} = N - k - 1 \qquad (14\text{-}11a)$$

if an intercept is included in the model (*const* = TRUE), or

$$df \text{ (degrees of freedom)} = N - k \qquad (14\text{-}11b)$$

if the model does not include an intercept (*const* = FALSE)

$$SS_{total} = \sum_{n=1}^{N} (\bar{y} - y_{calc})^2 \qquad (14\text{-}12)$$

$$SS_{regression} = SS_{total} - SS_{resid} \qquad\qquad (14\text{-}13)$$

$$R^2 = 1 - \frac{SS_{resid}}{SS_{total}} \qquad\qquad (14\text{-}14)$$

$$F = \frac{SS_{regression}\,/(k-1)}{SS_{resid}\,/(N-k)} \qquad\qquad (14\text{-}15)$$

$$SE(y) = \sqrt{\frac{SS_{resid}}{N-k}} \qquad\qquad (14\text{-}16)$$

The coefficient of determination, R^2 (or the correlation coefficient, R), is a measure of the goodness of fit of the data to (in this case) a straight line. If x and y are perfectly correlated (i.e., the difference between y_{obsd} and y_{calc} is zero), then $R^2 = 1$. In contrast, an R^2 value of zero means that there is no correlation between x and y. In my opinion, a value of R^2 of less than 0.9 corresponds to a rather poor fit of data to a straight line.

The $SE(y)$ parameter, the standard error of the y estimate, is sometimes referred to as the RMSD (root-mean-square deviation).

The F-statistic is used to determine whether the proposed relationship is significant (that is, whether y does in fact vary with respect to x). For many relationships observed in chemistry, a relationship will unquestionably exist. If it is necessary to determine whether the variation of y with x is statistically significant, or merely occurs by chance, you should consult a book on statistics.

The regression coefficients, the standard deviations of the coefficients, and R^2, the coefficient of determination, are the statistical parameters of interest to most chemists.

Multiple Linear Regression: An Example

In the example[*] that follows, we will use LINEST to find the correlation between the HPLC retention times (t_R) of peptides and three descriptors of the peptides: the logarithm of the sum of the retention times of the individual amino acids composing the peptide (log Sum_{AA}), the logarithm of the calculated van der Waals volume of the peptide (log V_{VDW}), and the logarithm of the calculated partition coefficient of the peptide between n-octanol and water ($c$log P). The regression equation is

$$t_R = b + m_1 \log Sum_{AA} + m_2 \log V_{VDW} + m_3\, c\log P \qquad\qquad (14\text{-}17)$$

---

[*] Tomasz Bączek, Paweł Wiczling, Michał Marszałł, Yvan Vander Heyden, and Roman Kaliszan, *J. Proteome Res.* **2005**, *4*, 555.

where b, m_1, m_2 and m_3 are the regression coefficients to be determined using LINEST.

	A	B	C
	Retention Times of Individual Amino Acids		
1	LiChrospher RP-18 column, water-acetonitrile gradient, gradient time 20 min, T = 40°C		
2	amino acid	amino acid letter code	t_R, min
3	Alanine	A	2.51
4	Arginine	R	2.69
5	Asparagine	N	2.26
6	Aspartic acid	D	2.32
7	Cysteine	C	2.63
8	Glutamic acid	E	2.53
9	Glutamine	Q	3.35
10	Glycine	G	2.29
11	Histidine	H	2.50
12	Isoleucine	I	7.70
13	Leucine	L	8.11
14	Lysine	K	2.54
15	Methionine	M	5.67
16	Phenylalanine	F	9.42
17	Proline	P	1.87
18	Serine	S	2.28
19	Threonine	T	2.44
20	Tryptophan	W	11.17
21	Tyrosine	Y	6.92
22	Valine	V	4.68

Figure 14-7. HPLC retention times of individual amino acids,
used to calculate log Sum$_{AA}$ values for peptides.

Retention times for the individual amino acids, used to calculate log Sum$_{AA}$, are shown in Figure 14-7 (only a portion of the table is shown). The van der Waals volumes and the partition coefficients were calculated by the authors using a commercially available software program.

The spreadsheet layout for calculation of the regression coefficients b, m_1, m_2 and m_3 of equation 14-17 is shown in Figure 14-8. Only a portion of the spreadsheet is shown – a total of 98 peptides were included in the analysis.

	A	B	C	D	E	F	G
1	No.	peptide sequence	log Sum$_{AA}$	log V_{VDW}	clog P	t_R (exp)	t_R (calc)
2	1	AA	0.70	2.1603	-0.74	3.05	4.07
3	2	AG	0.68	2.1047	-1.28	2.66	3.95
4	3	AF	1.08	2.3402	0.95	10.71	10.74
5	4	YL	1.18	2.4394	1.86	11.57	12.21
6	5	DD	0.67	2.2847	-1.98	2.63	1.63
7	6	ML	1.14	2.4508	-0.17	11.94	10.29
8	7	WW	1.35	2.5420	2.18	15.84	14.76
9	8	GM	0.90	2.3318	-1.90	7.30	5.80
10	9	GH	0.68	2.2600	-1.89	2.90	2.12
11	10	GL	1.02	2.2505	-0.08	9.74	9.91
12	11	WF	1.31	2.5058	2.41	15.49	14.43
13	12	GHG	0.85	2.3574	-2.63	3.06	4.19
14	13	LPQIENVKGTEDSGTT-NH2	1.72	3.1736	-9.45	11.41	10.14
15	14	VKGTEDSGTT-NH2	1.42	2.9326	-6.41	6.80	8.03
16	15	EHADLLAVVAASQKK-NH2	1.73	3.1563	-3.89	14.57	13.28
17	16	VVAASQKK-NH2	1.40	2.8874	-3.24	7.53	9.66

Figure 14-8. Portion of the spreadsheet for multiple linear regression.

The results returned by LINEST are shown in Figure 14-9. The formula

{=LINEST(F2:F99,C2:E99,1,1)}

was entered in cells $I$3:$L$7.

	I	J	K	L
2	m$_3$	m$_2$	m$_1$	b
3	0.499604	-9.89775	20.04353	11.79195
4	0.053409	1.402983	1.246636	2.2821
5	0.935621	0.957327	#N/A	#N/A
6	455.3709	94	#N/A	#N/A
7	1252.008	86.14861	#N/A	#N/A

Figure 14-9. Results returned by LINEST.

Excel Tip. *If you don't like the appearance of #N/A values in the array of values returned by LINEST, such as in Figure 14-9, you can use formatting to make the error values invisible: select the cells containing the error values, choose* **Cells...** *from the* **Format** *menu, choose the Font tab, and set the font color to white.*

Calculated t_R values for the peptides were obtained by using equation 14-17, with the regression coefficients in row 3 of Figure 14-9. Agreement between the calculated and experimental values is illustrated in Figure 14-10.

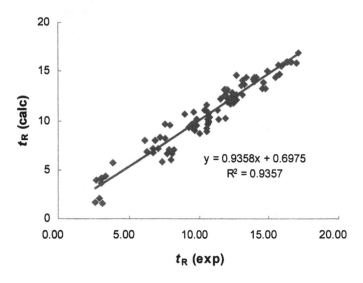

Figure 14-10. Agreement between calculated and experimental t_R values.

Fitting Data to a Power Series

It is sometimes convenient, in the absence of a suitable equation, to fit data to a power series. The equation can then be used for data interpolation. Often a power series $y = a + bx + cx^2 + dx^3$ is sufficient to fit data of moderate curvature. The lowest-order polynomial that produces a satisfactory fit should be used; if there are N data points, the highest-order polynomial that can be used is of order $(N-1)$.

The example presented in Figure 14-11 fits the solubility of oxygen in water as a function of temperature over the range 0–100°C. Columns for the T^2 and T^3 independent variables were inserted and the solubility data in column D were fitted to a cubic equation using **LINEST**.

	A	B	C	D
1	Solubility of Oxygen in Water			
2	($g\ O_2$/100 g H_2O)			
3	T, deg C	T^2	T^3	S, g/100 g
4	0	0	0	0.006945
5	5	25	125	0.006072
6	10	100	1000	0.005368
7	15	225	3375	0.004802
8	20	400	8000	0.004339
9	25	625	15625	0.003931
10	30	900	27000	0.003588
11	35	1225	42875	0.003315
12	40	1600	64000	0.003082
13	45	2025	91125	0.002858
14	50	2500	125000	0.002657
15	60	3600	216000	0.002274
16	70	4900	343000	0.001856
17	80	6400	512000	0.001381
18	90	8100	729000	0.00079
19	100	10000	1000000	0

Figure 14-11. Fitting O_2 solubility in water by a power series.
(*CRC Handbook of Chemistry and Physics, 40th edition, p. 1706.* Reproduced by
permission of Taylor and Francis Group, LLC, a division of Informa plc.)

The regression parameters returned by LINEST are shown in Figure 14-12.
All four parameters seem to be significant, since the standard deviations are, at
most, less than 5% of the parameter value.

	A	B	C	D
21	Results from LINEST			
22	-1.3E-08	2.21E-06	-0.00016	0.00685682
23	5.65E-10	8.39E-08	3.4E-06	3.6788E-05
24	0.999456	4.93E-05	#N/A	#N/A

Figure 14-12. LINEST results for the solubility of O_2 in water

Figure 14-13 shows the fit of the polynomial to the data points. Except for
the data points for 80°C and above, the deviation of the calculated line from the
data points is less than 2%.

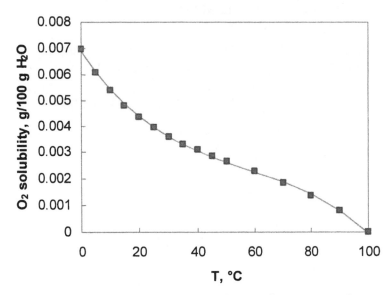

Figure 14-13. Fitting O_2 solubility in water by a power series.
The line is calculated using the coefficients in Figure 14-12.

A LINEST Shortcut for Power Series

Here's a shortcut that eliminates the need to create the columns of T^2 and T^3 in Figure 14-11. If you've read Chapter 7, "Array Formulas," and understand array constants, you'll understand how the formula

=LINEST(D4:D19,A4:A19^{1,2,3},1,1)

creates an array of the values of the independent variable W raised to the first, second and third powers. Unlike the braces that are automatically placed around an array formula when you enter it by using Ctrl+Shift+Enter, you must type the braces around the values of the array constant.

You can examine that part of the formula by highlighting A4:A19^{1,2,3} in the formula bar and pressing F9; you'll see the result displayed in the formula bar (only a portion of it is shown here):

{0,0,0;5,25,125;10,100,1000;15,225,3375;20,400,8000;...}

Note that successive array elements in a row are separated by commas, and rows of elements are separated by semicolons.

The formula, which must be entered by using Ctrl+Shift+Enter, returns the same values that are shown in Figure 14-12.

Handling Noncontiguous Ranges of *known_x's* in LINEST

One restriction in the use of LINEST is that the range of *known_x's* must be a contiguous range of cells, like $A$4:$C$13 in Figure 14-14.

	A	B	C	D
3	x_1	x_2	x_3	y
4	0.39	0.15	0.04	19.84
5	0.64	0.40	0.33	37.07
6	2.43	5.89	0.50	65.64
7	1.44	2.07	1.45	103.47
8	1.45	2.11	0.86	151.04
9	4.05	16.39	10.97	209.41
10	6.52	42.56	13.76	277.28
11	5.41	29.24	8.35	355.57
12	7.33	53.70	20.54	443.44
13	5.56	30.94	21.57	541.07

Figure 14-14. Usual data layout for multiple linear regression.

The LINEST formula used with this spreadsheet layout would be

{=LINEST(D4:D13,A4:C13,1,1)}

and the results returned by LINEST are shown in Figure 14-15.

15	m_3	m_2	m_1	b
16	17.52642	-4.54003	42.2768	17.4611
17	6.725662	5.747472	45.40945	61.16211
18	0.891522	72.46839	#N/A	#N/A

Figure 14-15. LINEST results using the data from Figure 14-14.

However, it is sometimes convenient or necessary to use columns of intermediate calculations in order to obtain the columns of independent variables. For example, in Figure 14-16, an intermediate calculation was necessary in column B in order to calculate the second independent variable *x2* in column C. A similar calculation was required in order to calculate the third independent variable *x3* in column E.

	A	B	C	D	E	F
3	x1	Intermediate Calculation	x2	Intermediate Calculation	x3	Y
4	0.39	etc.	0.15	etc.	0.04	19.84
5	0.64	etc.	0.40	etc.	0.33	37.07
6	2.43	etc.	5.89	etc.	0.50	65.64
7	1.44	etc.	2.07	etc.	1.45	103.47
8	1.45	etc.	2.11	etc.	0.86	151.04
9	4.05	etc.	16.39	etc.	10.97	209.41
10	6.52	etc.	42.56	etc.	13.76	277.28
11	5.41	etc.	29.24	etc.	8.35	355.57
12	7.33	etc.	53.70	etc.	20.54	443.44
13	5.56	etc.	30.94	etc.	21.57	541.07

Figure 14-16. Data layout for multiple linear regression,
incorporating intermediate columns of calculation.

You might think that it would be possible to enter a reference to the three non-adjacent columns of independent variables by using a LINEST expression like

{=LINEST(F4:F13,(A4:A13,C4:C13,E4:E13),1,1)}

but this formula returns a #REF! error.

One solution to the problem would be simply to copy the appropriate columns of data to a separate area of the spreadsheet and use these values in the LINEST formula.   Another solution is to use the properties of matrix multiplication and addition that are described in Chapter 13.

{=LINEST(F4:F13,A4:A13*{1,0,0}+C4:C13*{0,1,0}+E4:E13*{0,0,1},1,1)}

This formula returns the LINEST results shown in Figure 14-17

	A	B	C	D
15	m3	m2	m1	b
16	17.526417	-4.540034	42.276801	17.461098
17	6.7256623	5.7474718	45.409448	61.162108
18	0.8915221	72.468388	#N/A	#N/A

Figure 14-17. LINEST results using the data from Figure 14-16.

If you don't want to have to type in the array literals {1,0,0}, etc., you can simply create a table of the ones and zeros, like the table shown in Figure 14-18, and enter references to the appropriate ranges, as in the formula below.

{=LINEST(F4:F13,A4:A13*A23:C23+C4:C13*A24:C24+E4:E13*A25:C25,1,1)}

	A	B	C
23	1	0	0
24	0	1	0
25	0	0	1

Figure 14-18. Data table of array coefficients for use with LINEST.

How LINEST Handles Collinearity

In multiple linear regression, a case may arise where one x-variable is a linear function of other x-variables. For example, x_3 could be the sum of x_1 and x_2. In this case, x_3 has no additional predictive value; eliminating it from the model would lead to an equally accurate set of regression coefficients. This situation is called *collinearity*.

Figure 14-19, where $X_1 = 0.5X_2$, illustrates an example of collinearity.

	A	B	C	D
1	X1	X2	X3	Y
2	1	2	1	-2.577
3	2	4	4	14.941
4	3	6	9	43.453
5	4	8	16	80.699
6	5	10	25	128.958
7	6	12	36	186.789
8	7	14	49	254.818

Figure 14-19. Data table exhibiting collinearity

LINEST checks for collinearity and removes the appropriate x-variable(s) from the model. In the LINEST output, these variables will have zero values for the regression coefficient and for the corresponding standard error. The LINEST results are shown in Figure 14-20.

X3	X2	X1	const
5.014	1.397321	0	-10.447
0.02969	0.121514	0	0.424065
1	0.272118	#N/A	#N/A
362327.1	4	#N/A	#N/A
53659.22	0.296192	#N/A	#N/A

Figure 14-20. Example of results returned by LINEST when there is collinearity.

Weighted Least-Squares

If the y-values to be fitted range over several orders of magnitude, it is sometimes advisable to use a *weighting factor* in the regression. Otherwise, if the error in the measurement is proportional to the magnitude of the measurement, the residuals of the largest measurements will have an overwhelming effect on the sum of squares. The weighting factor W is applied to each data point, so that the sum of squares is calculated according to equation 14-18.

$$\text{weighted } SS_{resid} = \sum_{n=1}^{N} W_n (y_n - y_{calc})^2 \tag{14-18}$$

Some weighting functions that are often used include $W_n = 1/y_n$ and $W_n = 1/y_n^2$.

Weighting of the regression data can also be used if each y_n is an average of J_n observations. In this case the weighting factor is $W_n = J_n$.

Multiple Linear Regression Using Trendline

You can also fit a least-squares line to data points such as those shown in Figure 14-13 by adding a *trendline* to a chart. You can choose from a menu of mathematical functions—linear, logarithmic, polynomial, power, exponential— as curve-fitting functions.

To Add a Trendline in Excel 2007/2010. Select the chart by clicking on it. Click on the data series to which you want to add a trendline; the Design, Layout and Format tabs will appear on the Ribbon. On the Layout tab, press the Trendline button, and click on More Trendline Options to display the Format Trendline dialog box (Figure 14-21). The options are the same as in Excel 2003.

To change the appearance of the trendline, click on the formatting options on the left of the dialog box. (To return to the Options window, click on Trendline Options.)

To Add a Trendline in Excel 2003. Select the chart by clicking on it, and choose **Add Trendline...** from the **Chart** menu. If the chart has several data series, either select the desired data series before choosing **Add Trendline...** or choose the desired data series from the Based On Series box.

Figure 14-21. The Excel 2007/2010 Trendline Options dialog box.

In the Add Trendline dialog box, click on the Type tab (Figure 14-22) and choose the appropriate fitting function from the gallery of functional forms. (Depending on the data in the series, the exponential, power or logarithmic choices may not be available.) If you choose the polynomial form, you can select the order of the polynomial by using the spinner. If you choose 3, for example, Excel will fit a polynomial of order three (i.e., a cubic equation) to the data points. The maximum is a Trendline polynomial of order six.

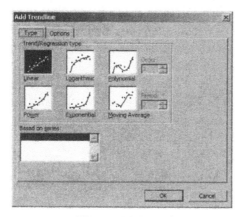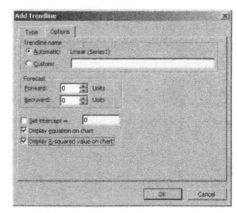

Figure 14-22. The Excel 2003 Add Trendline dialog box:
(left) the Type tab, (right) the Options tab.

Now choose the Options tab (Figure 14-22). Check the boxes for Display Equation On Chart and Display R-squared Value On Chart; then press OK. Excel displays the trendline on the chart as a heavy solid line and the least-squares equation and R^2 value as text on the chart, as shown in Figure 14-23. You can change the appearance of the trendline by clicking on the trendline and then choosing **Selected Trendline...** from the **Format** menu.

The least-squares coefficients in the Trendline equation are identical to the regression coefficients returned by LINEST (see Figure 14-12).

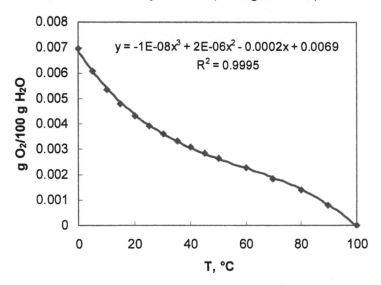

Figure 14-23. The Trendline equation shown on the chart.

There are some disadvantages of using Trendline: If you want to use the coefficients for calculations, you'll have to copy them from the chart and paste them into worksheet cells. Usually the coefficients as displayed in the chart are not precise enough for calculations, but you can number-format the text to display more figures before copying the coefficients.

Multiple Linear Regression Using the Analysis Toolpak

Linear regression can also be performed using the Add-In package called the Analysis ToolPak, which provides a series of statistical analysis tools. (As well as loading the statistical analysis tools, the Analysis ToolPak package also loads a series of additional functions called Engineering functions. If you're using Excel 2007/2010, the Engineering functions are already present in the list of worksheet functions.)

Loading the Analysis ToolPak in Excel 2007/2010. Click Data Analysis in the Analysis group on the Data tab. If the Data Analysis command is not available, you need to load the Analysis ToolPak add-in program. To load the Analysis ToolPak, press the Office button and press the Excel Options button (Excel 2007) or click on the File tab and click on Options (Excel 2010). Click on Add-Ins. Press the Manage Excel Add-Ins button to display the Add-Ins dialog box (Figure 14-24), and check the box for Analysis ToolPak.

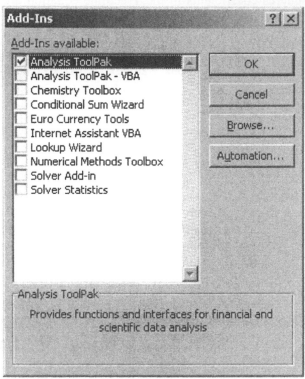

Figure 14-24. The Add-Ins dialog box.

Loading the Analysis ToolPak in Excel 2003. If you do not see the **Data Analysis...** command at the bottom of the **Tools** menu, you must load the Analysis ToolPak add-in program. Choose **Add-Ins...** from the **Tools** menu, and in the Add-Ins dialog box (Figure 14-24), check the box for Analysis ToolPak. The **Data Analysis...** command should now be present in the **Tools** menu.

Using the Regression Tool

Choose Regression from the Analysis Tools list box (Figure 14-25).

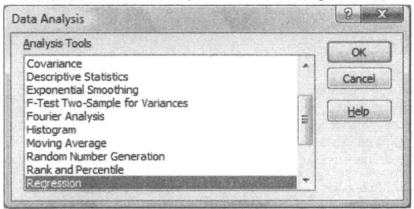

Figure 14-25. The Data Analysis dialog box.

The Regression dialog box (Figure 14-26) will prompt you to enter the range of dependent variable (*y*) values and the range of independent variable (*x*) values, as well as whether the constant is zero, whether the first cell in each range is a label, and the confidence level desired in the output summary. Then select a range for the summary table. You need select only a single cell for this range; it will be the upper left corner of the range. You can also request a table of residuals and a normal probability plot. If you select a cell or range such that the summary table will over-write cells containing values, you will get a warning message.

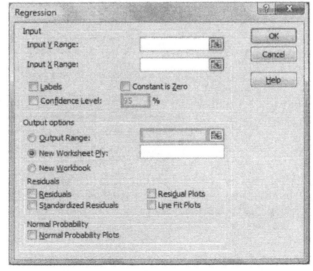

Figure 14-26. The Regression dialog box.

In contrast to the results returned by LINEST, the output is clearly labeled, and additional statistical data are provided. Regression data for the example shown in Figure 14-1 is shown in the three tables of Figure 14-27. Three tables are produced: regression statistics, analysis of variance, and regression coefficients. (The coefficients table has been broken into two parts to fit the page.)

Unlike Trendline, the values returned by the Regression tool can be used directly in calculations. But there is one disadvantage of using the Regression tool: Unlike LINEST, it is not a function. With LINEST, the returned values are dynamically linked to the original data and are updated if the input data are changed. With the Regression tool, the values are calculated from the input data and entered into worksheet cells; they do not change if you change the data.

Regression Statistics	
Multiple R	0.999966633
R Square	0.999933268
Adjusted R Square	0.999911024
Standard Error	0.003812261
Observations	5

Analysis of Variance	df	Sum of Squares	Mean Square	F	Significance F
Regression	1	0.6533136	0.6533136	44952.7706	2.3137E-07
Residual	3	4.36E-05	1.4533E-05		
Total	4	0.6533572			

	Coefficients	Standard Error	t Statistic
Intercept	0.0022	0.00295296	0.74501401
x1	2483.96501	11.715673	212.020684

P-value	Lower 95%	Upper 95%
0.49765621	-0.0071977	0.01159766
2.9688E-09	2446.68048	2521.24955

Figure 14-27. Data obtained by using Regression from the Analysis ToolPak
        top to bottom: Regression Statistics, Analysis of Variance,
        Regression Coefficients and Statistics.

Using the Regression Statistics

Regression statistics can be used to make decisions concerning the value or significance of regression coefficients — for example, whether the least-squares slopes of two lines are identical. This statistical procedure is known as *hypothesis testing*. The most common procedure is to set up a null hypothesis: for example, that the difference between the two slopes is zero. The answer — yes or no — must be stated within the context of a stated level of uncertainty. A confidence level of 95% is most commonly used; if the confidence level chosen (e.g., 99%) is too stringent, then a significant difference may be missed, while if the confidence level is too low (e.g., 50%) an insignificant difference may be accepted as real.

The *t*-test is used to test a null hypothesis. The computed *t*-statistic (calculated using one of the relationships given below) is compared with the value found in a table of *t* values corresponding to the appropriate number of degrees of freedom and at the desired confidence level. If the computed *t*-statistic exceeds the value from the table, the null hypothesis is rejected.

Testing Whether an Intercept Is Significantly Different from Zero

For the permanganate calibration curve example examined earlier in this chapter, the intercept was found to be 0.0022 absorbance units with a standard deviation of 0.0030. Here it is obvious that the postulated zero intercept lies within the confidence interval for the intercept. But consider another case, where LINEST returns $b = 0.0011$ and $\sigma_b = 0.00051$; this calibration curve also contained $N = 5$ data points. Is the intercept statistically different from zero?

To test whether the intercept b is equal to a given value b_0, compute the *t*-statistic

$$t = \frac{b - b_0}{\sigma_b} \qquad (14\text{-}19)$$

and compare it to the *t* distribution with df degrees of freedom (for a straight line, $df = N - 2$ degrees of freedom).

For the second calibration curve we obtain

$$t = \frac{0.0011 - 0}{0.00051} = 2.2 \qquad (14\text{-}20)$$

t-Table for Various Levels of Probability			
df	90%	95%	99%
1	6.31	12.70	63.70
2	2.92	4.30	9.92
3	2.35	3.18	5.84
4	2.13	2.78	4.60
5	2.02	2.57	4.03
10	1.81	2.23	3.17
15	1.75	2.13	2.95
30	1.70	2.04	2.75
60	1.67	2.00	2.66
∞	1.64	1.96	2.58

Figure 14-28. An abbreviated t-table.

If you consult a table of t values in a statistics text (a portion of such a table is shown in Figure 14-28), you will find, for 95% probability and 3 degrees of freedom, that $t = 3.18$. Since the calculated t-statistic is less than t-critical from the table, the intercept value of 0.0011 does not differ significantly from zero.

Instead of using a t-table imported from another source, you can calculate t values using Excel's TINV worksheet function TINV returns the inverse of the Student's t-distribution for the specified degrees of freedom. The syntax is TINV(*probability,degrees_freedom*), where *probability* is the probability associated with the two-tailed Student's t-distribution, and d*egrees_freedom* is the number of degrees of freedom to characterize the distribution.

The formula in cell B4 is

=TINV((100-B$3)/100,$A4)

if the values of confidence level are entered as, e.g., 90, 95, 99, or

=TINV(1-B$3,$A4)

if the values of confidence level are entered as, e.g., 0.90, 0.95, 0.9999, and formatted as percent.

A t-table generated by using the TINV function is shown in Figure 14-29.

	A	B	C	D
1	t-Table for Various Levels of Probability (calculated using TINV function)			
2		Confidence level, %		
3	df	90	95	99
4	1	6.31	12.71	63.66
5	2	2.92	4.30	9.92
6	3	2.35	3.18	5.84
7	4	2.13	2.78	4.60
8	5	2.02	2.57	4.03
9	10	1.81	2.23	3.17
10	15	1.75	2.13	2.95
11	30	1.70	2.04	2.75
12	60	1.67	2.00	2.66
13	∞	1.64	1.96	2.58

Figure 14-29. A t-table calculated using the TINV worksheet function.

Testing Whether Two Slopes Are Significantly Different

To test whether the slopes of two different straight lines are equal, use the following equation:

$$t = \frac{m_2 - m_1}{\sigma_m(pooled)} \tag{14-21}$$

The pooled estimate of the standard deviation of the slope is given as follows, provided that the number of data points is large:

$$\sigma_m(pooled) = \sqrt{\sigma_{m_1}^2 + \sigma_{m_2}^2} \tag{14-22}$$

If the numbers of data points for each line, n_1 and n_2, are not large, then you must use the following equation to calculate the pooled estimate of the standard deviation:

$$\sigma_m(pooled) =$$

$$\sqrt{\frac{(n_1 - 2)SE_{y_1}^2 + (n_2 - 2)SE_{y_2}^2}{(n_1 - 2) + (n_2 - 2)}} \sqrt{\frac{1}{(n_1 - 1)S_{x_1}^2} + \frac{1}{(n_2 - 1)S_{x_2}^2}} \tag{14-23}$$

where $S^2{}_{x1}$, the variance of the x-values in data set 1, is given by:

$$S_{x1}^2 = \frac{\Sigma x_1{}^2 - (\Sigma x_1)^2 / n_1}{(n_1 - 1)} \tag{14-24}$$

The variance can easily be obtained using the Excel function VAR. Equation 14-23 for the pooled estimate of the standard deviation is implemented, for example, by using the Excel formula

=((N.1-2)*SEY1^2+(N.2-2)*SEY2^2)/(N.1+N.2-4)*(1/((N.1-1)*var1)+1/
     ((N.2-1)*var2))

where N.1, SEY1 and var1 are the number of data points, the standard deviation of y and the variance of the x-values of data set number 1, respectively.

Testing Whether a Regression Coefficient Is Significant

In a multiple linear regression model using an empirical equation, you may want to know whether individual coefficients are significant, that is, whether they are useful in predicting the dependent variable. As in the preceding examples, if the standard deviation of the coefficient is small relative to the coefficient, the coefficient is clearly significant, while if the standard deviation of the coefficient is larger than the coefficient itself, it is clear that the coefficient is much less significant.

If you divide the coefficient by its standard deviation, the result is the t-statistic for that coefficient. The t-statistic is used to test hypotheses about the value of the coefficient. In a multiple regression model, the value of a_n/σ_n shows the relative importance of each term in the model.

Testing Whether Regression Coefficients Are Correlated

Occasionally, in a multiple linear regression model using an empirical equation, two or more independent variables are correlated among themselves, that is, one independent variable is a linear function of another independent variable. Intercorrelated variables are said to exhibit *multicollinearity*. A non-chemical example of multicollinearity might exist in the relationship between sales of a product and the demographics of the customer base. Perhaps it is found that both educational level and family income are good predictors of sales of the product, yet it is clear that these two independent variables are strongly correlated.

If two independent variables are perfectly correlated (i.e., one is an exact linear function of the other), then LINEST will return #NUM! in all cells. If they are only approximately correlated, then LINEST will return regression parameters that can be very misleading: R^2 can be very close to 1, but the coefficients may have no meaning. The standard errors of one or more of the coefficients will probably be fairly large.

The Analysis ToolPak provides a convenient method for determining the correlation between independent variables. Choose Correlation from the

Analysis Tools list box. The dialog box will prompt you to enter the range of independent variables; the dependent variables are not included. The output is a symmetrical matrix of correlation coefficients. The diagonal elements (correlation of a variable with itself) are all 1. The off-diagonal elements indicate the degree of correlation of two different independent variables. Unless a value is very close to 1.0, there is no significant correlation between independent variables.

Confidence Intervals for Slope and Intercept

The confidence limits for the slope of a straight line are given by $m \pm ts_m$, where t is obtained from the t-table for the desired confidence level and $(n - 2)$ degrees of freedom. Similarly, the confidence limits for the intercept are $b \pm ts_b$. Using the linest results for the permanganate calibration curve, shown in Figure 14-5, and using $t = 3.18$ for 95% confidence level, the 95% confidence limits for the slope are $2.484 \times 10^3 \pm (3.18)(1.17 \times 10^1)$ or $(2.484 \pm 0.037) \times 10^3$.

Confidence Limits and Prediction Limits for a Straight Line

The upper and lower *confidence limits* for y_{calc} at a particular value of x are given by equation 14-25. The confidence limits for y_{calc} are calculated by using a value of the t-statistic for a particular number of degrees of freedom and a particular probability level. Often the 90% probability level is used.

$$y_{calc} \pm t_{N-2} \sqrt{\frac{SS_{resid}}{N-2}} \sqrt{\frac{1}{N} + \frac{(x-\bar{x})^2}{(N-1)S_x^2}} \qquad (14\text{-}25)$$

The $\pm$ limits can be calculated using the Excel formula

=SQRT(SUM(d.sq)/(N-2))*SQRT(1/N+((X-X_bar)^2)/((N-1)*VAR(X)))

where d.sq is the residual squared, N is the number of observations, X_bar is the mean of the x-values, and VAR(X) returns the variance of the x-values.

The upper and lower *prediction limits* are given by the equation

$$y_{calc} \pm t_{N-2} \sqrt{\frac{SS_{resid}}{N-2}} \sqrt{1 + \frac{1}{N} + \frac{(x-\bar{x})^2}{(N-1)S_x^2}} \qquad (14\text{-}26)$$

which differs from equation 14-25 only in the extra 1 within the square root term. The prediction limits describe the confidence limits for predicting a single y-value for a particular x-value. The prediction limits contain two sources of error: the error in estimating y_{calc} and the error associated with a single measurement.

Very commonly, as in the calibration curve example, we use the measured y-value to estimate an x-value. Once the slope and intercept of a straight-line calibration curve have been established, it is easy to calculate an x-value (e.g., a concentration) from a measured y-value. The *estimation limits* of the estimated x-value are given by the equation

$$x_{meas} \pm \frac{t_{N-2}}{b} \sqrt{\frac{SS_{resid}}{N-2}} \sqrt{1 + \frac{1}{N} + \frac{(y_{meas} - \bar{y})^2}{b^2 \Sigma(x_1 - \bar{x})^2}} \qquad (14\text{-}27)$$

If y_{meas} is an average of M readings, then the equation becomes

$$x_{meas} \pm \frac{t_{N-2}}{b} \sqrt{\frac{SS_{resid}}{N-2}} \sqrt{\frac{1}{M} + \frac{1}{N} + \frac{(y_{meas} - \bar{y})^2}{b^2 \Sigma(x_1 - \bar{x})^2}} \qquad (14\text{-}28)$$

15

Nonlinear Regression Using the Solver

In this chapter you'll learn how to use the Solver, Excel's powerful optimization package, to perform nonlinear least-squares curve fitting. Users who are familiar with the Solver will see a number of changes in the Excel 2010 version of the Solver.

Nonlinear Regression

The function

$$y = F(x_1, x_2, ..., a_0, a_1, a_2, ...) \qquad (15\text{-}1)$$

where y is the dependent variable, $x_1, x_2, ...$ are the independent variables and $a_0, a_1, a_2, ...$ are coefficients, can be either *linear* or *nonlinear* in the coefficients. The word "linear" in linear regression does not mean that the function is a straight line, but that the function is linear in the coefficients — that is, the partial derivatives with respect to each coefficient are not functions of other coefficients.

An example of a function that is linear in the coefficients is

$$y = a_0 + a_1x + a_2x^2 + a_3x^3 \qquad (15\text{-}2)$$

and it should be clear from reading Chapter 14 that LINEST can be used to obtain the regression coefficients for a data set that can be described by such an equation. However, if the function is one such as

$$y = \exp(a_0 + a_1x) \qquad (15\text{-}3)$$

or

$$y = \exp(-a_1x) - \exp(-a_2x) \qquad (15\text{-}4)$$

then it is not linear in the coefficients, since the equation cannot be rearranged to obtain an expression containing separate a_iZ_i terms. Some nonlinear equations can be transformed into a linear form. Equation 15-3, for example, can be transformed by taking the base-*e* logarithm of each side, to yield the linear equation

$$\ln y = a_0 + a_1 x \qquad (15\text{-}5)$$

while others, such as equation 15-4, cannot be converted into a linear form and are said to be *intrinsically nonlinear*. LINEST can't be used on an equation of this type; we must use a different method.

Using the Solver to Perform Nonlinear Least-Squares Curve Fitting

There are many published computer programs and commercial software packages for nonlinear regression analysis, but you can perform nonlinear regression very easily by using the Solver. When applied to the same data set, the Solver gives essentially the same results as commercial software packages.

Using the Solver for Optimization

The Solver is an optimization package that finds a maximum, minimum or specified value of a *target cell* by varying the values in one or several *changing cells*. It accomplishes this by means of an iterative process, beginning with trial values of the coefficients. The value of each coefficient is changed by a suitable small increment, the new value of the function is calculated, and the change in the value of the function is used to calculate improved values for each of the coefficients. The process is repeated until the desired result is obtained. The Solver uses gradient methods or the simplex method to find the optimum set of coefficients.

With the Solver you can apply constraints to the solution. For example, you can specify that a coefficient must be greater than or equal to zero or that a coefficient must be an integer. Solutions to chemical problems will rarely use the integer option, and although the ability to apply constraints to a solution may be tempting, it can sometimes lead to an incorrect solution.

The Solver is an Add-In, a separate software package. To save memory, it may not automatically be loaded whenever you start Excel. If the Solver Add-in has already been loaded, you will see the **Solver...** command in the Data tab in Excel 2007 or in the Excel 2003 **Tools** menu. If Solver is not loaded, follow the appropriate procedure in the following sections.

Loading the Solver in Excel 2007/2010. Press the Office button (Excel 2007), or click on the File tab (Excel 2010). Press the Excel Options button to display the Excel Options window. Click on Add-Ins. In the "View and manage Microsoft Office add-ins" window, press the Go… button to the right of "Manage Excel Options" to display the list of available add-ins. Check the box for the Solver Add-In, and press OK.

Loading the Solver in Excel 2003. Choose the **Add-Ins…** command in the **Tools** menu. In the Add-Ins dialog box, check the box for Solver Add-In, and press OK

The Solver is a product of Frontline Systems Inc. (P.O. Box 4288, Incline Village, NV 89450, www.solver.com).

Changes to the Solver in Excel 2010

Some changes were made to the Solver in Excel 2010. The earlier version of the Solver provided two solving methods: (a) the Simplex method for linear problems and (b) the GRG (Generalized Reduced Gradient) method for nonlinear problems. In Excel 2010, a third method, the Evolutionary method, has been added. The Evolutionary method uses genetic algorithms to find its solutions. The Evolutionary method may be useful when the model is not smooth nonlinear, that is, when it involves spreadsheet functions such as IF or VLOOKUP. To perform multiple nonlinear least-squares regression, you should choose the GRG Nonlinear method.

The Classic Menus utility loads the "old" version of the Solver. To use this enhanced version of the Solver with Classic Menus, you must use the Add-ins command to unload Classic Menus, load the Excel 2010 Solver, then reload Classic Menus.

Using the Solver for Least-Squares Curve Fitting

To use the Solver to perform multiple nonlinear least-squares curve fitting, follow the procedure outlined in the accompanying box.

**To Use the Solver to Perform
Nonlinear Least-Squares Curve Fitting**

1. Start with a worksheet containing the data (independent variables x_1, x_2, … and dependent variable y_{obsd}) to be fitted.

2. Add a column containing y_{calc}-values, calculated by means of an appropriate formula and involving the x-values and one or more coefficients to be varied.

3. Add a column to calculate the square of the residual ($y_{obsd} - y_{calc}$) for each data point.

4. Calculate the sum of squares of the residuals.

5. Use **Solver…** to minimize the sum of squares of residuals (the target cell) by changing the coefficients of the function (the changing cells).

The target cell and the changing cells must be on the active sheet. Your model can involve external references to values in other worksheets or workbooks, however.

Since the Solver operates by a search routine, it will find a solution most rapidly and efficiently if the initial estimates that you provide are close to the final values. Conversely, it may not be able to find a solution if the initial estimates are far from the final values. Good estimates become more important as the model becomes more complex.

To ensure that the Solver has found a *global minimum* rather than a *local minimum*, it's a good idea to obtain a solution using different sets of initial estimates. You can do this manually, using starting points selected based on your knowledge of the problem. Or, if you are using Excel 2010, you can choose the Multistart option in the GRG tab of the Solver Options dialog box. The Multistart method automatically runs the GRG method from a number of randomly chosen starting points. In my opinion, running the Solver manually with a number of sets of well-chosen starting values is to be preferred over the Multistart method.

The least-squares regression coefficients that are returned by the Solver may be slightly different, depending on the starting values that you provide.

Using the Solver: An Example

The following example illustrates the ease with which the Solver can be used to perform nonlinear least-squares curve fitting. Here we analyze kinetics data (absorbance vs. time) from a biphasic reaction involving two consecutive first-order reactions ($A \rightarrow B \rightarrow C$) to obtain two rate constants and the molar absorptivity of the intermediate species B.

The equations for the concentrations of the species A, B and C in a reaction sequence of two consecutive first-order reactions can be found in almost any kinetics text. The expression for $[B]_t$ is

$$[B]_t = [A]_0 \frac{k_1}{k_2 - k_1} \left(e^{-k_1 t} - e^{-k_2 t} \right) \tag{15-6}$$

where $[A]_0$ is the initial concentration of the reactant species and k_1 and k_2 are the rate constants. This is an example of an intrinsically nonlinear equation.

For the example that follows, a stopped-flow spectrophotometer was used to obtain kinetics data[*] for the reaction of a nickel(II) complex NiL_2^{2+} of a substituted bidentate diamine ligand (L = 2,3-dimethyl-2,3-diaminobutane) with cyanide ion. The reaction is biphasic; one diamine ligand is replaced by two cyanide ligands, and then the second diamine ligand is replaced:

---

[*] J. C. Pleskowicz and E. J. Billo, *Inorg. Chim. Acta* **1985**, *99*, 149.

$$NiL_2^{2+} + 2CN^- \rightarrow NiL(CN)_2 + L \qquad\qquad (15\text{-}7)$$

$$NiL(CN)_2 + 2CN^- \rightarrow Ni(CN)_4^{2-} + L \qquad\qquad (15\text{-}8)$$

The species $NiL(CN)_2$, which is formed and then decays during the reaction, absorbs at 243 nm, where NiL_2^{2+}, $Ni(CN)_4^{2-}$, CN^- or L do not absorb appreciably. The stopped-flow data (measured manually from an oscilloscope trace used to collect the data) are shown in columns A and B of Figure 15-1.

	A	B	C	D	E
1	Consecutive First-Order Reactions (A → B → C)				
2	$[B]_t = [A_0] \cdot k_1 \cdot (EXP(-k_2 \cdot t) - EXP(-k_1 \cdot t))/(k_1 - k_2)$				
3	Only B absorbs at the monitoring wavelength (243 nm). Path length 0.4 cm				
4	Absorbance = molar absorptivity × path length × concentration $A = \varepsilon_B \times 0.4 \times [B]$				
5	Conc (mol/L)	4.00E-05		k_1 (sec^{-1})	**1.000**
6	ε_A (cm L/mol)	0		k_2 (sec^{-1})	**0.200**
7	ε_B (cm L/mol)	**3.00E+03**			
8	ε_C (cm L/mol)	0			
9	t, sec	A(obsd)	[B]	A(calc)	∂^2
10	0.0	0.0000	0.00E+00	0.0000	0.0E+00
11	0.2	0.0047	7.10E-06	0.0085	1.5E-05
12	0.6	0.0129	1.69E-05	0.0203	5.5E-05
13	1.0	0.0163	2.25E-05	0.0271	1.2E-04
14	1.4	0.0188	2.55E-05	0.0306	1.4E-04
15	1.8	0.0201	2.66E-05	0.0319	1.4E-04
16	2.2	0.0208	2.67E-05	0.0320	1.3E-04
17	2.6	0.0208	2.60E-05	0.0312	1.1E-04
18	3.0	0.0205	2.50E-05	0.0299	8.9E-05
19	4.0	0.0178	2.16E-05	0.0259	6.5E-05
20	5.0	0.0149	1.81E-05	0.0217	4.6E-05
21	6.0	0.0118	1.49E-05	0.0179	3.7E-05
22	7.0	0.0090	1.23E-05	0.0147	3.3E-05
23	8.0	0.0070	1.01E-05	0.0121	2.6E-05
24	9.0	0.0052	8.26E-06	0.0099	2.2E-05
25	10.0	0.0038	6.76E-06	0.0081	1.9E-05
26				$\Sigma\delta^2 =$	**1.03E-03**
27					(target cell)

Figure 15-1. The spreadsheet before optimization of coefficients by the Solver. The target cell and changing cells are shown in bold.

In the expression for $[B]_t$ (equation 15-6), $[A]_0$ is the initial concentration of the reactant species NiL_2^{2+}, and k_1 and k_2 are the rate constants for reactions in equations 14-7 and 14-8. We use $[B]_t$ together with Beer's law to write

$$A_{obsd} = \varepsilon b[B]_t \tag{15-9}$$

where A_{obsd} is the measured absorbance, ε is the molar absorptivity of the intermediate species $NiL(CN)_2$, and b is the path length of the stopped-flow cell. The parameters ε, k_1 and k_2 are the regression parameters that we want to obtain by nonlinear least-squares curve fitting.

In Figure 15-1, columns C and D contain the concentration of the intermediate species B and the absorbance. The formulas in cells C10 and D10 are (the optical path length was 0.4 cm)

=C_A*k_1*(EXP(-k_2*t)-EXP(-k_1*t))/(k_1-k_2)

=0.4*E_B*C10

In column E the squares of the residuals are calculated. The sum of squares is in cell E26.

The changing cells (B7, E6, E7) and the target cell (E26) are shown in bold. Initial values, estimated from the data, were 3000, 1 and 0.2, respectively.

To use the Solver, click on Solver in the Analysis group in the Data ribbon (Excel 2007/2010) or choose **Solver...** from the **Tools** menu (Excel 2003). You may see the message "Set Target Cell must be a single cell on the active sheet". Press OK to continue. The Solver Parameters dialog will then be displayed: The Excel 2003/Excel 2007 dialog box is shown in Figure 15-2, the Excel 2010 dialog box in Figure 15-3.

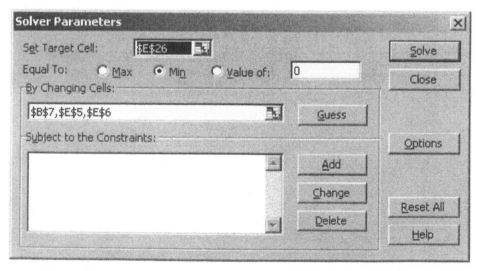

Figure 15-2. The Solver Parameters dialog box (Excel 2007 or 2003).

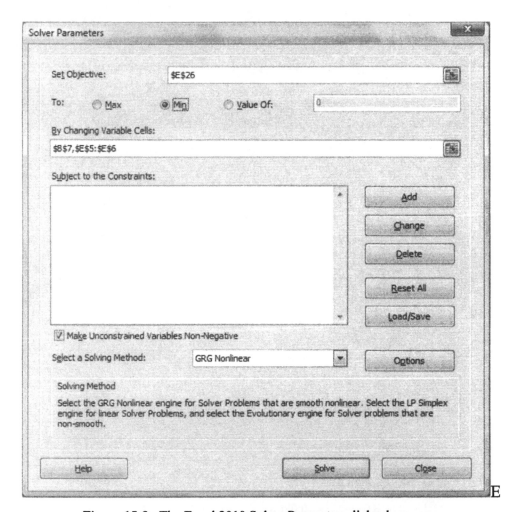

Figure 15-3. The Excel 2010 Solver Parameters dialog box.

There is one important change in the Excel 2010 version of the Solver Parameters dialog box. The user can now choose one of three solving methods: Simplex LP (used for linear problems), GRG Nonlinear (used for smooth nonlinear problems), and Evolutionary (which be used for any type of problem, even if not linear or smooth nonlinear). The GRG method will probably be the method of choice for most scientific problems.

To solve the problem displayed in Figure 15-1:

1. In the Set Cell box, type E26, or select cell E26 with the mouse. (If you selected E26 before running the Solver, $E$26 will appear in the Set Cell box.)

2. You want to minimize the sum of squares, so press the Min button.

3. Select cells B7 and E6:E7 so that they appear in the By Changing Cells box.

4. Press the Options button, check "Use Automatic Scaling" and press OK. The reason for choosing this option will be explained in a subsequent section.

5. Do not enter any constraints.

6. Press the Solve button.

	A	B	C	D	E
1	Consecutive First-Order Reactions (A → B → C)				
2	$[B]_t = [A_0] \cdot k_1 \cdot (EXP(-k_2 \cdot t) - EXP(-k_1 \cdot t))/(k_1 - k_2)$				
3	Only B absorbs at the monitoring wavelength (243 nm). Path length 0.4 cm				
4	Absorbance = molar absorptivity × path length × concentration $A = \varepsilon_B \times 0.4 \times [B]$				
5	Conc (mol/L)	4.00E-05		k_1 (sec^{-1})	0.639
6	ε_A (cm L/mol)	0		k_2 (sec^{-1})	0.285
7	ε_B (cm L/mol)	2.53E+03			
8	ε_C (cm L/mol)	0			
9	t, sec	A(obsd)	[B]	A(calc)	∂^2
10	0.0	0.0000	0.00E+00	0.0000	0.0E+00
11	0.2	0.0047	4.66E-06	0.0047	8.5E-11
12	0.6	0.0129	1.16E-05	0.0118	1.3E-06
13	1.0	0.0163	1.62E-05	0.0164	2.6E-09
14	1.4	0.0188	1.89E-05	0.0191	1.0E-07
15	1.8	0.0201	2.04E-05	0.0206	2.2E-07
16	2.2	0.0208	2.09E-05	0.0211	7.6E-08
17	2.6	0.0208	2.07E-05	0.0209	1.2E-08
18	3.0	0.0205	2.01E-05	0.0203	4.6E-08
19	4.0	0.0178	1.75E-05	0.0177	2.1E-08
20	5.0	0.0149	1.44E-05	0.0145	1.3E-07
21	6.0	0.0118	1.15E-05	0.0116	3.9E-08
22	7.0	0.0090	8.98E-06	0.0091	5.7E-09
23	8.0	0.0070	6.94E-06	0.0070	8.6E-11
24	9.0	0.0052	5.31E-06	0.0054	2.8E-08
25	10.0	0.0038	4.04E-06	0.0041	8.2E-08
26				$\Sigma \delta^2 =$	2.06E-06
27					(target cell)

Figure 15-4. Regression coefficients (in bold) returned by the Solver.

As the problem is set up and solved, messages will appear in the status bar at the bottom of the screen. The value of the target cell is displayed in the status bar after every iteration. In a few seconds or less, the Solver finds a solution and displays the Solver Results dialog box. You have the option of accepting the Solver's solution or restoring the original values. Press the Accept Solver Solution button. The spreadsheet will be displayed with the final values of the target cells (Figure 15-4).

Excel Tip. Don't introduce constraints (e.g., to force a constant to be greater than or equal to zero) if you're using the Solver to obtain the least-squares best fit. The solution will not be the "global minimum" of the error-square sum, and the regression coefficients may be seriously in error.

There are some additional controls in the Solver Parameters dialog box:

- The Add..., Change... and Delete buttons are used to apply constraints to the model. Since the use of constraints is to be avoided, these buttons are not of much interest.

- Pressing the Guess button will enter references to all cells that arc precedents of the target cell. In the example above, pressing the Guess button enters the cell references $A$10:$B$25,$E$7,$B$5,$E$6:$E$7 in the By Changing Cells box. I don't recommend this.

- The current Solver model is automatically saved with the worksheet. The Reset All button permits you to "erase" the current model and begin again.

The fit of the curve to the data points is shown in Figure 15-5.

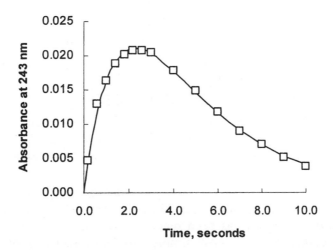

Figure 15-5. Curve calculated from the coefficients obtained by the Solver.

Excel Tip.   Discontinuous functions in your Solver model can cause problems. They can be either discontinuous mathematical functions such as TAN, which has a discontinuity at $\pi/2$, or worksheet functions that are inherently "discontinuous", such as IF, ABS, INT, ROUND, CHOOSE, LOOKUP, HLOOKUP or VLOOKUP. The Evolutionary solving method in Excel 2010 may be able to handle this type of problem.

Comparison with a Commercial Nonlinear Least-Squares Package

In Figure 15-6, the Solver results are compared with the results obtained by using a commercial nonlinear regression analysis program. It is clear that the Solver provides results that are essentially identical to those from the commercial software package. The slight differences (ca. 0.001%) arise from the fact that the coefficients are found by a search method; the final" values will differ, depending on the convergence criteria used in each program.

A major disadvantage of the Solver is that it does not provide the standard deviations of the coefficients. This problem is addressed later in this chapter (see "Statistics of Nonlinear Regression").

Comparison of results from the Solver and from a commercial software package		
	Solver	*NLLSQ*
k_1	0.63874	0.63878
k_2	0.28540	0.28539
ε	2526.50	2526.39
$\Sigma\delta^2$	2.0620E-06	2.0620E-06

Figure 15-6. Comparison of regression coefficients.

Solver Options (Excel 2007 or 2003)

The Options button in the Solver Parameters dialog box displays the Solver Options dialog box (Figure 15-7) and allows you to control the way Solver attempts to reach a solution. The default values of the options are shown in Figure 15-7.

The Max Time and Iterations parameters determine when the Solver will return a solution or halt. If either Max Time (100 seconds) or Iterations (100) is exceeded before a solution has been reached, the Solver will pause and ask if you want to continue. For most simple problems, these limits will not be exceeded. In any event, you don't need to adjust Max Time or Iterations, since if either parameter is exceeded, the Solver will pause and issue a "Continue anyway?" message.

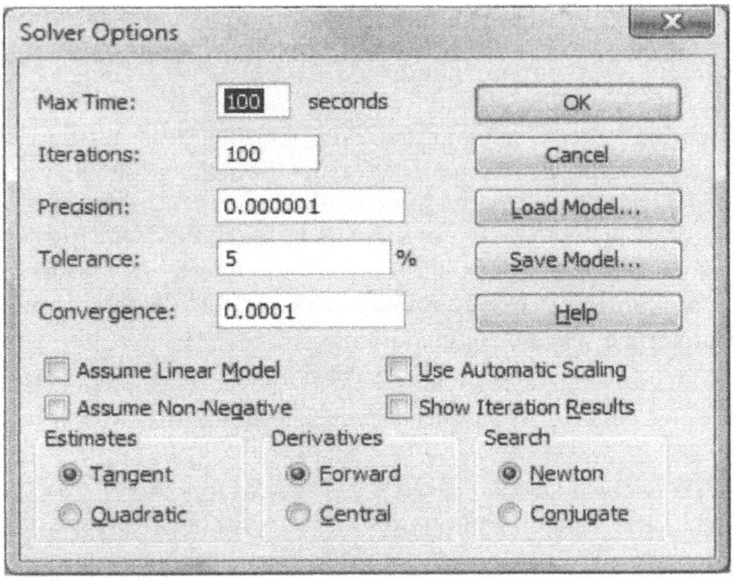

Figure 15-7. The Solver Options dialog box (Excel 2003 or 2007).

Both the Precision and Tolerance options apply only to problems with constraints. The Precision parameter determines the amount by which a constraint can be violated. The Tolerance parameter is similar to the Precision parameter, but applies only to problems with integer solutions. Since adding constraints to a model that involves minimization of the error-square sum is not recommended, neither the Precision nor the Tolerance parameter is of use in nonlinear regression analysis.

The Convergence parameter was introduced in Excel 97; in earlier versions it was fixed and could not be changed by the user. Unlike Goal Seek's Maximum Change parameter in the Calculations Options dialog box (see Chapter 13, Figure 13-26), which is an absolute convergence limit, the Solver's Convergence parameter is relative; the Solver will stop iterating when the *relative* change in the target cell value is less than the number in the Convergence box for the last five iterations. Thus you don't have to scale the convergence limit to fit the problem, as is required with the Maximum Change parameter in the Calculations Options dialog box.

If the function is linear, checking the Assume Linear Model box will speed up the solution process.

If the Show Iteration Results box is checked, the Solver will pause and display the result after each iteration.

The Use Automatic Scaling option should be checked if there are large differences in magnitude between the Changing Cells, as in the previous example, or between the Target Cell and the Changing Cells.

The Estimates, Derivatives and Search parameters can be changed to optimize the solution process. The Search parameter specifies which gradient search method to use: The Newton method requires more memory but fewer iterations, whereas the Conjugate method requires less memory but more iterations. The Derivatives parameter specifies how the gradients for the search are calculated: The Central derivatives method requires more calculations but may be helpful if the Solver reports that it is unable to find a solution. The Estimates parameter determines the method by which new estimates of the coefficients are obtained from previous values; the Quadratic method may improve results if the system is highly nonlinear.

The current Solver model is automatically saved with the worksheet. The Save Model and Load Model buttons permit you to save multiple Solver models.

Solver Options (Excel 2010)

The Excel 2010 Solver Options dialog box contains three tabs. The All Methods tab, shown in Figure 15-8, contains the controls concerned with limits and constraints, as well as the Use Automatic Scaling and Show Iteration Results check boxes. The GRG Nonlinear tab (Figure 15-9) allows the user to change

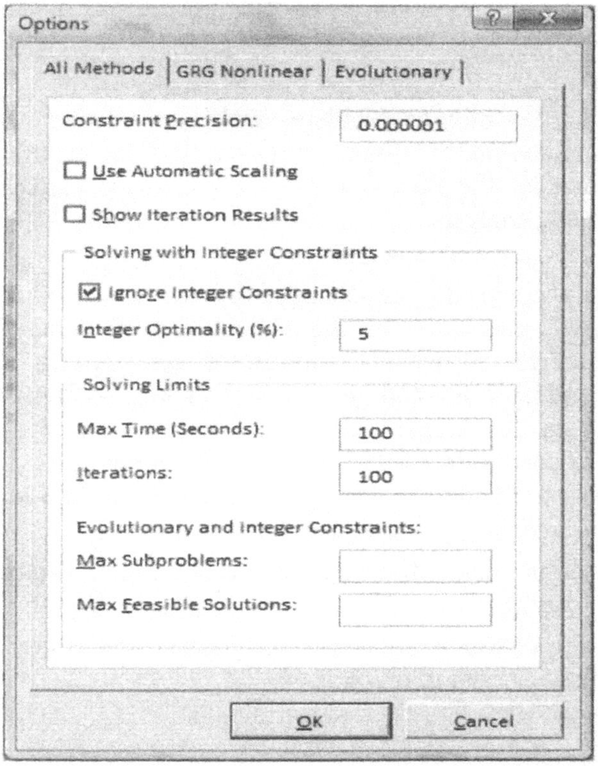

Figure 15-8. The All Methods tab of the Solver Options dialog box (Excel 2010).

the convergence limit or to specify Forward or Central Derivatives. The default is Forward Derivatives; for some problems for which the Solver seems to be unable to find a solution, choosing Central Derivatives instead of Forward Derivatives will permit the Solver to converge to a set of reasonable values.

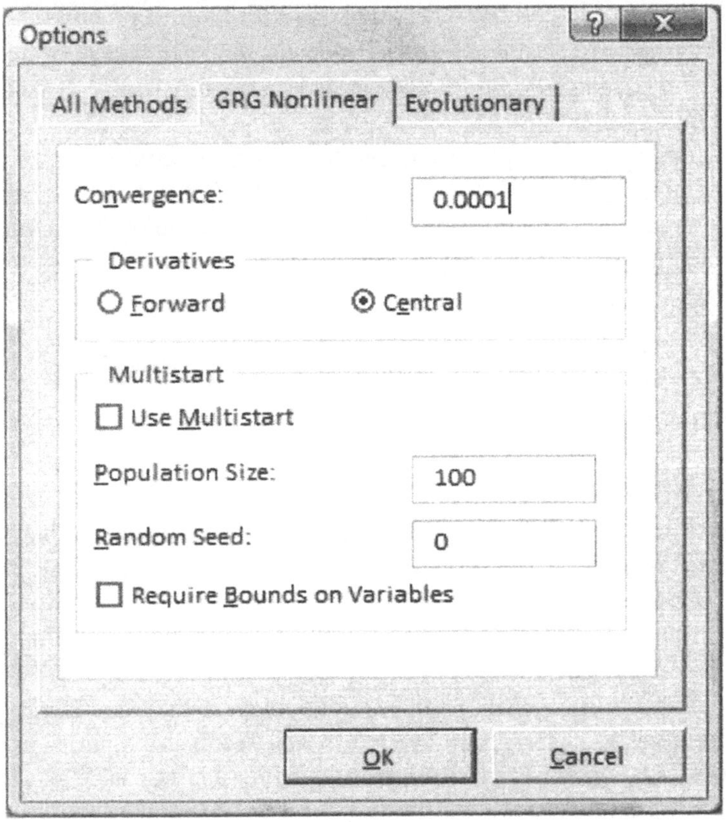

Figure 15-9. The GRG Linear tab of the Solver Options dialog box (Excel 2010).

Excel Tip. If, after you've made changes to a worksheet, a Solver model that had previously converged to a reasonable solution refuses to converge, and all attempts to find the problem have failed, use the Reset All button to erase the current model. Then re-enter references to the Target Cell and the Changing Cells. This may solve the problem.

The "Use Automatic Scaling" Option Is Important for Many Chemical Problems

For some models the Solver may refuse to converge satisfactorily. This may occur when there is a large difference in magnitude between changing cells or between changing cells and the target cell. The Solver may "refuse" to vary one or more changing cells or vary them by only an insignificant amount. You can usually overcome this problem by checking the Use Automatic Scaling option in the Solver Options dialog box.

In the preceding example, in order to obtain a satisfactory result it was necessary to use Automatic Scaling, because the initial values of the three parameters to vary were 3000, 1 and 0.2. (Try solving the problem without using Automatic Scaling to see what happens.)

Some Additional Solver Examples

The following sections illustrate some additional uses of the Solver.

Deconvolution of Spectra

The resolution of a complex absorption spectrum into individual absorption bands may be necessary if information about the position, height or width of individual bands is required. There are a number of computer programs designed for the deconvolution of spectra, but you can do a reasonable job with Excel.

The procedures described next were developed for the deconvolution of electronic absorption spectra (UV–visible spectra) but are equally applicable to the deconvolution of infrared, Raman or NMR spectra. UV–visible spectra differ from vibrational spectra in that the number of bands is much smaller and the bandwidths are much wider. Band shape may also be different. UV–visible spectra are also usually recorded under conditions of high resolution and high signal-to-noise ratio. Spectra from older instruments usually require manual digitization from a spectrum on chart paper at, e.g., 10-nm intervals. With the widespread use of computer-controlled instruments, it is a simple matter to obtain a file of spectral data at, e.g., 1-nm intervals. In fact, it may be necessary to reduce the size of the data set to speed up calculations.

Mathematical Functions for Spectral Bands. A symmetrical spectral band can be described by three parameters: position (wavelength or frequency corresponding to the absorption maximum), intensity (absorbance or molar absorptivity at the band maximum) and width (usually the bandwidth at half-height). The band shape functions most commonly used for deconvolution are the Gaussian function and the Lorentzian function[*].   Both are symmetrical functions.   UV–visible spectra generally have a Gaussian band shape.   The Lorentzian function is useful for the simulation of NMR spectra.   The log-normal band function has been applied to unsymmetrical spectral band shapes.

Many spectral bands can be closely approximated by a Gaussian line shape when the independent variable v is in energy units, e.g., cm^{-1}. The absorbance A at a wavenumber v is given by equation 15-10, where A_{max} is the band maximum, v_{max} is the wavenumber of the band maximum and Δv is the half-width.

$$A = A_{max} \exp\left[-(4 \ln 2)\frac{v - v_{max})^2}{\Delta v^2} \right] \tag{15-10}$$

The corresponding equation for a Lorentzian line shape is given by

$$A = \frac{A_{max}}{1 + 4\dfrac{(v - v_{max})^2}{\Delta v^2}} \tag{15-11}$$

For unsymmetrical bands, the equation for the log-normal line shape is

$$A = A_{max} \exp\left\{ -\frac{\ln 2}{(\ln \rho)^2}\left[\ln\left(\frac{v - v_{max}}{\Delta v}\frac{\rho^2 - 1}{\rho} + 1 \right) \right]^2 \right\} \tag{15-12}$$

for the region $v \geq v_{max} - (\Delta v \, \rho / (\rho^2 - 1))$ and $A = 0$ elsewhere.   The asymmetry parameter ρ is given by

$$\rho = \frac{v_R - v_{max}}{v_{max} - v_L} \tag{15-13}$$

where v_R and v_L are the right and left portions of the half-width.

A simpler form of the Gaussian band shape, where σ is simply treated as an adjustable parameter, is given in equation 15-14.   This is the equation that will be used in the following treatment.   By using an embedded chart to compare calculated and experimental data, you can fairly easily find a set of A_{max}, v_{max} and σ values that approximate the band shape, to use as initial guesses for the deconvolution procedure outlined in the box on the following page.

---

[*]   P. Pelikán, M. Č eppan and M. Liš ka, *Applications of Numerical Methods in Molecular Spectroscopy*, CRC Press, Boca Raton FL, 1993.

$$A = A_{max} \exp\left[-\frac{(\upsilon - \upsilon_{max})^2}{\sigma^2} \right]$$
                                                                                                    (15-14)

Deconvolution of a Spectrum: An Example. The spreadsheet shown in Figure 15-10 illustrates the deconvolution of the UV–visible spectrum of a nickel(II) complex. Four bands are apparent in the spectrum, one a weak shoulder lying between relatively intense bands at approximately 350 and 550 nm. The fourth band appears only as the tail of a fairly intense band lying at longer wavelengths.

The formulas in cells C10 (converting wavelength λ to wavenumber ν) and D10 (calculating the Gaussian band profile of band 1) are

=10000/A10

=band1 A_0*EXP(-(((C10-band1 max)/band1 s)^2)/2)

The Solver was used to vary the values in cells D4:F6 and G4:G5 to make cell I7 a minimum. Because the data did not permit a complete resolution of band 4, cell G6, the bandwidth parameter for band 4, was held constant at the reasonable value of 1.5. The results are shown on the spreadsheet. The resolved spectrum (solid line), with the four bands (broken lines), is shown in Figure 15-11.

Tackling a Complicated Spectrum. For a complicated spectrum, it may be helpful to operate on a reduced-size data set. Many spectrometers record absorbance readings at 1-nm intervals; a complete UV–visible spectrum (200–700 nm) contains 500 data points. If the spectrum contains eight bands, you're performing calculations on more than 4000 cells. Start with a data set consisting of every 10th data point, for example. After you have achieved a reasonably good fit to this data set, use these values as initial parameters for the complete data set.

It may be necessary to first minimize portions of the spectrum separately.

	A	B	C	D	E	F	G	H	I
1				\multicolumn{4}{c}{**$[Ni(2,3,2\text{-tet})]^{2+}$ spectrum**}			
2				\multicolumn{4}{c}{(b = 5 cm, c = 0.0252 M, pH = 10.6)}			
3				band1	band2	band3	band4		
4			ν_{max}	29.25	22.72	18.56	11.69		
5			A_{max}	1.12	0.15	0.87	0.77		
6			σ	1.60	1.54	1.38	1.5		$\Sigma(\delta^2)$
7			λ_{max}	342	440	539	855		0.0148
9	λ,nm	Abs	ν,cm^{-1}	band1	band2	band3	band4	Sum	δ^2
10	300	0.173	33.33	0.043	0.000	0.000	0.000	0.043	
11	310	0.274	32.26	0.191	0.000	0.000	0.000	0.191	0.007
12	320	0.514	31.25	0.512	0.000	0.000	0.000	0.512	0.000
13	325	0.694	30.77	0.714	0.000	0.000	0.000	0.714	0.000
14	330	0.871	30.30	0.903	0.000	0.000	0.000	0.903	0.001
15	335	1.026	29.85	1.046	0.000	0.000	0.000	1.046	0.000
16	340	1.126	29.41	1.118	0.000	0.000	0.000	1.118	0.000
17	345	1.141	28.99	1.109	0.000	0.000	0.000	1.109	0.001
18	350	1.036	28.57	1.028	0.000	0.000	0.000	1.028	0.000
19	355	0.908	28.17	0.896	0.000	0.000	0.000	0.896	0.000
20	360	0.737	27.78	0.737	0.001	0.000	0.000	0.738	0.000
21	370	0.406	27.03	0.429	0.003	0.000	0.000	0.432	0.001

Figure 15-10. Deconvolution of a UV–visible spectrum.

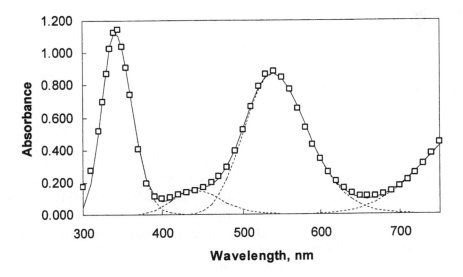

Figure 15-11. Deconvoluted spectrum.

Deconvolution of a Spectrum

1. Start with a table of wavelength, absorbance data pairs.

2. You may wish to use wavenumber rather than wavelength, in which case, create a column of wavenumbers.

3. Determine the number of bands necessary to describe the spectrum, either by inspection or by using the first derivative of the spectrum, $\Delta A/\Delta x$. ($\Delta A/\Delta x = 0$ indicates a band maximum.)

4. Estimate the half-width of the bands by using one or more bands not overlapped by other bands. As first approximation, use this value for all bands in the spectrum

4. Set up a table of ν_{max}, A_{max} and σ for each band.

5. Calculate the band profile for each contributor.

6. Sum the individual band contributions.

7. Calculate the sum of squares of the residuals $(A_{obsd} - A_{calc})^2$.

8. Create an embedded chart, plotting A_{obsd} and A_{calc}.

9. Initial manual adjustment of the parameters is helpful if the spectrum is complicated (more than three or four bands, especially if they are overlapped strongly).

10. Use the Solver to minimize sum of squares of residuals by varying (ultimately) the $3N$ parameters for the N bands in the spectrum.

Determination of Binding Constants by NMR Measurements

Nuclear magnetic resonance spectroscopy is a powerful tool for the determination of structural information of complexes in solution. It can also be used for the examination of solution equilibria.

The appearance of the NMR spectrum of a complex will depend on whether chemical exchange of the ligand between the free state and the bound state is slow or fast. If exchange is slow relative to the "NMR time scale" (i.e., lifetimes of minutes or longer), the spectrum of both species, the bound ligand and the free ligand, will be observable. If the system is truly at equilibrium, then the concentration of each of the species can be obtained from the NMR spectrum, and the equilibrium constant can be determined. This case is rarely encountered.

If the system is at the *fast exchange limit* (i.e., lifetimes of milliseconds or less), the magnetic environment experienced by a proton will be averaged over the environments of the free and bound states, and an NMR singlet will be observed, at a frequency that is the weighted average of the time spent in the two states. This leads to equation 15-15 for the position of the NMR singlet under conditions of fast exchange, where the α's are the fractions of the total ligand in the free and bound states.

$$\delta_{obsd} = \alpha_{free}\,\delta_{free} + \alpha_{bound}\,\delta_{bound} \qquad (15\text{-}15)$$

Calculations. Only systems undergoing fast exchange, and only 1:1 binding, will be discussed here. Depending on the magnitude of the binding constant, the following situations may be observed:

Case I: Both δ_{free} and δ_{bound} can be measured independently. The calculations in this case are identical to those for the spectrophotometric determination of the pK_a of an indicator and will not be discussed further.

Case II: Only δ_{free} can be measured independently. There are two unknowns to be determined, K and δ_{bound}, and these can be obtained as the slope and intercept of a linear transformation of the data.

Case III: Neither δ_{free} or δ_{bound} can be measured independently. A curve-fitting approach is necessary, using the Solver.

Monomer–Dimer Equilibrium.   As part of a study of host–guest complexation, the hydrogen-bonded dimerization of the substituted urea N-phenyl-N'-(2-pyridyl)urea (U) in 1:1 CH_2Cl_2/toluene was studied. The chemical shift of the high-field urea proton is especially sensitive to concentration (see Figure 15-14).

The variation in chemical shift was analyzed assuming dimer formation. The equations are as follows:

dimerization: $\qquad\qquad\qquad 2\,U \leftrightharpoons U_2 \qquad\qquad K = [U_2]\,/\,[U]^2 \quad (15\text{-}16)$

mass balance: $\qquad\qquad [U]_T = [U] + 2\,[U_2] \qquad\qquad\qquad\qquad (15\text{-}17)$

chemical shift: $\qquad\qquad \delta_{obsd} = \alpha_1\,\delta_1 + \alpha_2\,\delta_2 \qquad\qquad\qquad\quad (15\text{-}18)$

where $\alpha_1 = [U]/[U]_T$ and $\alpha_2 = 2\,[U_2]/[U]_T$ are the fractions of U_T in the monomeric and dimeric form, respectively.

Equations 14-16, 14-17 and 14-18 yield the following expressions for the concentration of free $[U]$ and the chemical shift:

$$[U] = \frac{\sqrt{8K[U]_T + 1} - 1}{4K} \qquad (15\text{-}19)$$

$$\delta_{calc} = \frac{[U]\delta_1 + 2K[U]^2\delta_2}{[U]_T} \qquad (15\text{-}20)$$

The Solver can be used to minimize the sum of squares of residuals, $\Sigma(\delta_{obsd} - \delta_{calc})^2$, in order to find the best values of K, δ_1 and δ_2. The results are shown in Figures 14-12 and 14-13.

The Spreadsheet. Figure 15-12 shows a portion of the spreadsheet and illustrates the layout used for generating the theoretical curve. The experimental data are in A11:A18 and B11:B18. Column C contains the expression for free [U]

=(SQRT(8*K*A11+1)-1)/(4*K)

and column D contains the expression for the calculated chemical shift:

=(C11*delta1+2*K*(C11^2)*delta2)/A11

Below the data section lies an extensive table (not shown), used to obtain the smooth calculated curve, in rows 20 to 118.

	A	B	C	D	E
1	Dimerization of N-phenyl-N'-(2-pyridyl)urea (U)				
2	(Chemical shift δ of high field urea proton)				
3		K =	50.0		
4		δ_1 =	7.000		
5		d_2 =	10.000		
6	Equations used:				
7	free [U]	=(SQRT(8*K*A11+1)-1)/(4*K)			
8	δ_{calc}	=(C11*delta1+2*K*(C11^2)*delta2)/A11			
9					
10	[U]total	δ(obsd)	[U] free	δ(calc)	resid^2
11	0.0010	7.074	0.0009	7.252	3.2E-02
12	0.0020	7.298	0.0017	7.438	2.0E-02
13	0.0050	7.718	0.0037	7.804	7.4E-03
14	0.0070	7.929	0.0047	7.966	1.3E-03
15	0.0100	8.128	0.0062	8.146	3.2E-04
16	0.0300	8.736	0.0130	8.697	1.5E-03
17	0.0800	9.358	0.0237	9.110	6.1E-02
18	0.1000	9.503	0.0270	9.190	9.8E-02
19				$\Sigma\delta^2$ =	2.2E-01

Figure 15-12. The spreadsheet before using the Solver, with initial estimates of K, δ_1 and δ_2. (Data provided by Dr. Steve Bell.)

The Solver can be used to minimize the sum of squares of residuals (in cell E19), the changing cells being C3 (K), C4 (delta1) and C5 (delta2). The results are shown in Figure 15-13.

	A	B	C	D	E
1	Dimerization of N-phenyl-N'-(2-pyridyl)urea (U)				
2	(Chemical shift δ of high field urea proton)				
3		K =	39.1		
4		δ_1 =	6.840		
5		d_2 =	10.588		
6	Equations used:				
7	free [U]	=(SQRT(8*K*A11+1)-1)/(4*K)			
8	δ_{calc}	=(C11*delta1+2*K*(C11^2)*delta2)/A1'			
9					
10	[U]total	δ(obsd)	[U] free	δ(calc)	resid^2
11	0.0010	7.074	0.0009	7.094	4.2E-04
12	0.0020	7.298	0.0018	7.293	2.6E-05
13	0.0050	7.718	0.0038	7.706	1.4E-04
14	0.0070	7.929	0.0050	7.897	1.0E-03
15	0.0100	8.128	0.0066	8.115	1.6E-04
16	0.0300	8.736	0.0142	8.813	5.9E-03
17	0.0800	9.358	0.0262	9.359	1.6E-06
18	0.1000	9.503	0.0299	9.466	1.4E-03
19				$\Sigma\delta^2$ =	9.0E-03

Figure 15-13. The spreadsheet after refinement of the constants.

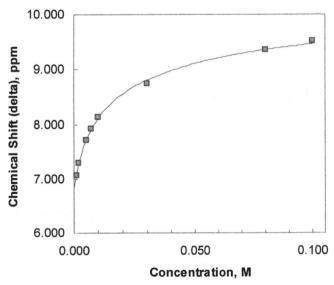

Figure 15-14. The fit of the chemical shift data. The curve is generated by using equations and constants found in the text and in Figure 15-13.

To plot the theoretical binding curve, generate δ_{calc} for a range of [U], as follows. In cell A20 type 0.000, in cell A21 type 0.001, and then use AutoFill to generate values from 0.000 to 0.100 in steps of 0.001. **Copy** the expression for [U] and **Paste** in cell C20. **Fill Down** to end of table. Transfer the expression for δ_{calc} from any cell in row D to cell F20 (copy the formula from the formula bar to transfer correctly). **Fill Down** to end of table.

Other Examples: NIST Datasets

The NIST (National Institute of Standards and Technology) website provides a number of datasets for testing nonlinear regression. The data sets are ranked as lower, average and higher levels of difficulty. The datasets can be found at http://www.itl.nist.gov/div898/strd/nls/nls_main.shtml or go to www.nist.gov and search for "regression data sets".

I have analyzed several of the NIST datasets with the Solver; they are included on the CD-ROM that accompanies this book.

Statistics of Nonlinear Regression

As you've seen, the Solver finds the set of least-squares regression coefficients very quickly and efficiently. However, it does not provide the standard deviations of the coefficients. Without these, the Solver's solution is essentially useless. The following illustrates how to obtain the standard deviations of the regression coefficients after obtaining the coefficients by using the Solver.

The standard deviation of the regression coefficient a_i is given by[*]

$$\sigma_i = \sqrt{\mathbf{P}_{ii}^{-1}}\ \text{SE}(y) \qquad (15\text{-}21)$$

where $\mathbf{P}_{ii}^{-1}$ is the ith diagonal element of the inverse of the $\mathbf{P}_{ij}$ matrix,

$$\mathbf{P}_{ij} = \sum_{n=1}^{N} \frac{\delta F_n}{\delta a_i} \frac{\delta F_n}{\delta a_j} \qquad (15\text{-}22)$$

$\delta F_n/\delta a_i$ is the partial derivative of the function with respect to a_i evaluated at x_n and

---

[*]   K. J. Johnson, *Numerical Methods in Chemistry*, Marcel Dekker, New York, 1980, p. 278.

$$SE(y) = \sqrt{\frac{SS_{resid}}{N-k}} \tag{15-23}$$

The quantities SS_{resid}, N and k are as defined in Chapter 14.

The $\delta F/\delta a_i$ terms can be calculated for each data point by numerical differentiation. The term a_i is varied by a small amount from its optimized value while the other a_j terms are held constant. The differential $\delta F/\delta a_i \approx \Delta F/\Delta a_i = (F_{new} - F_{opt})/(a_{new} - a_{opt})$ is calculated for each data point. Since Excel carries 15 significant figures, the change in a_i can be made very small, so that $\delta F/\delta a_i = \Delta F/\Delta a_i$. This process is repeated for each of the k regression coefficients. Then the cross-products $(\delta F/\delta a_i)(\delta F/\delta a_j)$ are computed for each of the N data points and the $\Sigma(\delta F/\delta a_i)(\delta F/\delta a_j)$ terms are obtained. The $\mathbf{P_{ij}}$ matrix of $\Sigma(\delta F/\delta a_i)(\delta F/\delta a_j)$ terms is constructed and inverted. The terms on the main diagonal of the inverse matrix are then used with equation 15-21 to calculate the standard deviations of the coefficients. This method may be applied to either linear or nonlinear systems.

A Macro to Provide Regression Statistics

The calculation described in the preceding section can be performed on a spreadsheet, but it is cumbersome to apply. The macro Solver Statistics.xls on the CD-ROM that accompanies this book was written to perform the calculations. It returns the standard deviations of the coefficients, the correlation coefficient and the $SE(y)$ or RMSD; it can be applied to linear or nonlinear regression.

To calculate the standard deviations of regression coefficients obtained by using the Solver, the macro uses the approach outlined in the preceding section. The $SS_{residuals}$, $SS_{regression}$ and $SE(y)$ are calculated from the known y's and calculated y's. The partial differentials $\delta y/\delta a_i$ for each of the k regression coefficients are calculated for all N data points by a procedure similar to that used in "Derivatives of a Function" in Chapter 13. A table of products of partial differentials is created and used to create a $k \times k$ matrix, and the matrix is inverted. Then equation 15-21 is used to calculate the standard deviations, using the diagonal elements of the inverted matrix. Finally, the standard deviations, the correlation coefficient and $SE(y)$ are returned to the source worksheet.

Using the Solver Statistics Macro

The Solver Statistics.xls macro is an Auto_Open macro; when you **Open** the document, the macro will run automatically. The macro hides the Solver Statistics.xls workbook; to examine it, choose **Unhide** in the View ribbon (Excel 2007) or from the **Window** menu (Excel 2003).

Or, to make the use of the Solver Statistics macro more convenient, use Save As to save a copy of Solver Statistics.xls as an Add-In macro; save it in the Add-Ins folder. Then you can load or unload the Add-In by using the **Add-Ins...** command.

If you're using Excel 2007/2010, the **Solver Statistics...** command appears in the Add-Ins tab. If you're using Excel 2003, a new menu command, **Solver Statistics...**, will be installed directly under the **Solver...** command in the **Tools** menu. If the Solver Add-In has not been loaded, the **Solver Statistics...** command will be at the bottom of the menu. The command will remain in the menu until you exit from Excel, if you unload the add-in using the Add-Ins command, or if you **Unhide** Solver Statistics and **Close** it.

When you choose the **Solver Statistics...** command, a sequence of four dialog boxes will be displayed, and you will be asked to select four cell ranges: (i) the y_{obsd} data, (ii) the y_{calc} data, (iii) the Solver coefficients, and (iv) a $3R \times nC$ range of cells to receive the statistical parameters. The Step 1 dialog box is shown in Figure 15-15. The y-values can be in row or column format. The Solver coefficients (Step 3 of 4 dialog box) can be in non-adjacent cells.

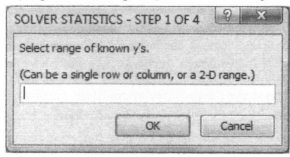

Figure 15-15. Solver Statistics.xls Step 1 of 4 dialog box.

The Solver Statistics macro performs extensive error checking of the input values. One of the several error messages is shown in Figure 15-16.

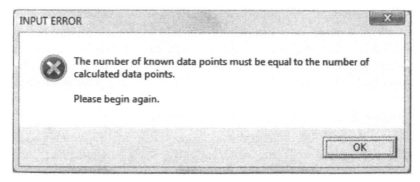

Figure 15-16. One of the possible error messages returned by Solver Statistics.

The array of values returned is in a format similar to that returned by LINEST: The regression coefficients are in the first row, the standard errors of the coefficients are in the second row, and the R^2 and SE(y) or RMSD parameter are in the third row, as shown in Figure 15-17.

coeff(1)	coeff(2)	...	coeff(N)
std.dev(1)	std.dev(2)	...	std.dev(N)
R^2	std.dev(fit)		

Figure 15-17. Layout of regression parameters and statistics returned by the Solver Statistics macro.

The values *coeff(1)* to *coeff(n)* are not calculated by the macro; they are echoed simply to indicate which standard deviation is associated with which coefficient (since the Solver coefficients can be in non-adjacent cells).

If the Solver Statistics.xls macro is used with the kinetics data of Figure 15-4, the regression parameters shown in Figure 15-18 are returned.

0.638743	0.285404	2526.495
0.047725	0.019109	138.5363
0.997222	0.000398	

Figure 15-18. Regression parameters returned by the Solver Statistics macro

An Additional Benefit from Using the Solver Statistics Macro

There is an additional major advantage in using the Solver Add-in and the Solver Statistics.xls macro, even for functions that are linear or can be rearranged to a linear form. For example, if the function is

$$y = \frac{abx}{1 + bx} \tag{15-24}$$

it can be re-cast as a linear function by taking the reciprocal of each side and rearranging to give

$$\frac{1}{y} = \frac{1}{abx} + \frac{1}{a} \tag{15-25}$$

Plotting $1/y$ vs $1/x$ yields a straight line with slope $1/ab$ and intercept $1/a$. LINEST can be used to provide the regression coefficients $1/ab$ and $1/a$, and their associated standard deviations. The coefficients a and b can be obtained from the regression coefficients (a = 1/intercept, b = intercept/slope). However, relationships dealing with the propagation of error must be used to calculate the standard deviations of a and b from the standard deviations of $1/a$ and $1/ab$. In contrast, when the Solver is used, the expression does not need to be rearranged, y_{calc} is calculated directly from equation 15-24, the Solver returns the coefficients a and b, and Solver Statistics returns the standard deviations of a and b.

PART IV

EXCEL'S
VISUAL BASIC
FOR APPLICATIONS

16

Visual Basic for Applications: An Introduction

In early spreadsheet programs, a macro was simply a string of keystrokes that could be recorded, saved and repeated to automate a simple keyboard operation. In Microsoft Excel, macros are written in a complete programming language that provides the capability to perform iterative calculations, or to take different actions based on the results of logic functions.

Macro programming was introduced in Excel 4.0. A macro written in Excel 4.0 Macro Language (XLM) is a series of statements or commands on an Excel macro sheet, which looks much like an Excel worksheet, with rows and columns. Macro statements in XLM can use any of Excel's worksheet functions, but in addition there are over 400 macro functions that can be used. Macros written in XLM can still be run in Excel 2010, but Excel 2010 is the last version that will support Excel 4.0 macros.

Beginning with Excel 5/95, Microsoft introduced a new macro language for Excel — Microsoft Visual Basic for Applications, or VBA. The basic VBA programming language, albeit with variations to accommodate the particular application, is used in Excel, Word, Access and PowerPoint. Unfortunately, Visual Basic macros are not supported in Excel 2008 for the Macintosh.

The following eight chapters cover programming and applications of VBA.

The Visual Basic Editor

VBA macros are located on *module sheets*. In Excel 5/95, module sheets were sheets in a workbook, just like worksheets and chart sheets; you could click on a sheet tab and view the VBA code. But beginning with Excel 97, the VBA programming environment became much more sophisticated.   It is more professional and has more features, but it's also much more confusing for the beginner. As far as programming in VBA is concerned, there is no appreciable difference between Excel 2003 and Excel 2007/2010.

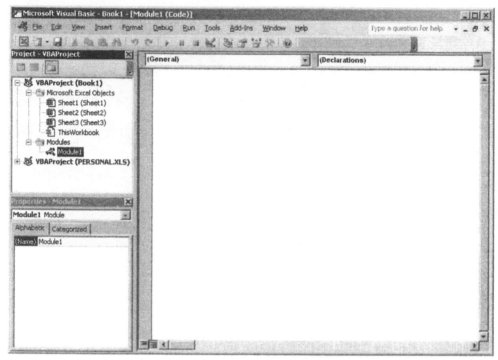

Figure 16-1. The Visual Basic Editor window.

To access the Visual Basic Editor in Excel 2007/2010, click on the Developer tab and click on the Visual Basic button in the Code section. (If the Developer tab is not present on the Ribbon, do the following: If you are using Excel 2010, click the File tab, click on Options, click on Customize Ribbon to display the "Customize the Ribbon" dialog box. In the Main Tabs list on the right, check the box for the Developer tab. If you are using Excel 2007, press the Office button, click on Excel Options, and in the Popular Options dialog box, check the box for "Show Developer tab in the Ribbon".) In Excel 2003, choose **Macro** from the **Tools** menu and then **Visual Basic Editor** from the submenu.

Excel Tip. You can toggle between the VBA code window and the active worksheet by pressing Alt+F11.

The Visual Basic Editor screen contains three important windows: the Code window, the Project Explorer window, and the Properties window. Procedures are viewed or typed in the Code window, which corresponds to a module sheet in Excel 95. Use the Project window (upper left window in Figure 16-1) to navigate among the available modules in open workbooks. If the Project window is not visible, choose **Project Explorer** from the **View** menu, or click on the Project Explorer tool button ⬚ to display it.

In the Project Explorer window you will see a hierarchy "tree" with a node for each open workbook. In the example illustrated in Figure 16-2, there are three open workbooks: Book1, Keyboard shortcuts.xla, and PERSONAL.XLS. The node for Book1 has a node (a folder icon) labeled Microsoft Excel Objects; click on the folder icon to display the nodes it contains — an icon for each sheet in the workbook and an additional one labeled ThisWorkbook. If you double-click on any one of these nodes, you will display the code sheet for it, but these code sheets are for a special type of procedure called an automatic procedure (see Chapter 23).

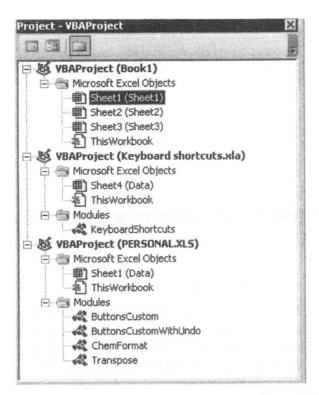

Figure 16-2. The Visual Basic Editor's Project Explorer window.

Sub or **Function** procedures must be created on a module sheet. To insert a module sheet, choose **Module** from the **Insert** menu on the **Visual Basic** menubar. A folder icon labeled Modules will be inserted; if you click on this icon, the node for Module1 will be displayed.

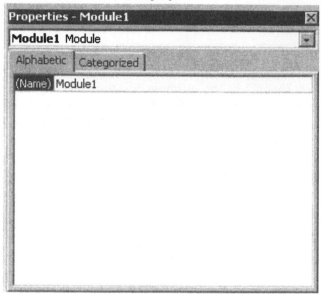

Figure 16-3. The Visual Basic Editor's Properties window.

To change the name of the module from Module1 to a more descriptive one, choose **Properties Window** from the **View** menu, or click on the Properties Window tool button. In the Properties Window (Figure 16-4), highlight the name Module1 and type the new name.

Visual Basic Procedures and Modules

VBA macros are usually referred to as *procedures*. A single module sheet can contain many procedures.

There Are Two Kinds of Procedure: Sub Procedures and Function Procedures

You can create two different kinds of procedures: **Sub** procedures, often called command macros, and **Function** procedures, called function macros, custom function macros, or user-defined functions. Although these procedures can use many of the same VBA commands, they are distinctly different.

Sub procedures can automate any Excel action. For example, a command macro might be used to create a report by inserting a new worksheet, copying selected ranges of cells from other worksheets and pasting them into the new worksheet, formatting the data in the new worksheet, adding headings, and printing the new worksheet. Command macros are usually "run" by clicking on Macros in the Code section of the Developer tab (Excel 2007/2010) or by choosing **Macro** from the **Tools** menu and **Macros...** from the submenu (Excel 2003). They can also be run by means of an assigned shortcut key, by being called from another macro, or in several other ways.

Function procedures augment Excel's library of built-in functions. A function macro is used in a worksheet in the same way as, for example, the SQRT function. It is entered in a worksheet cell, performs a calculation, and returns a result to the cell in which it is located. For example, a custom function macro named ALPHA can be used to calculate α_j, the fraction of an acid–base species in one of its protonated forms H_jX at a particular pH. The function takes three *arguments*: the pH of the solution, the range of pK_a values of the weak acid, and the coefficient j. This function, useful in constructing distribution diagrams, titration curves, and so on, is described in Chapter 20.

Both kinds of macro can incorporate decision-making, branching, looping, subroutines and many other aspects of programming languages.

The Structure of a Sub Procedure

The structure of a **Sub** procedure is shown in Figure 16-4. The procedure begins with the keyword **Sub** and ends with **End Sub**. It has a ProcedureName, a unique identifier that you assign to it. (The name can be a long one, since you never have to type it.) A **Sub** procedure has the possibility of using one or more arguments (see "Using Subroutines" in Chapter 17), but for now we will create **Sub** procedures that do not take arguments. Empty parentheses are still required, even if a **Sub** procedure uses no arguments.

```
Sub ProcedureName(Argument1, ...)
    VBA statements
End Sub
```

Figure 16-4. Structure of a **Sub** procedure.

The Structure of a Function Procedure

The structure of a **Function** procedure is shown in Figure 16-5. The procedure begins with the keyword **Function** and ends with **End Function**. It has a *FunctionName*, a unique identifier that you assign to it. (The name should be long enough to indicate the purpose of the function, but not too long, since you will probably be typing it in your worksheet formulas.) A **Function** procedure usually takes one or more arguments; the names of the arguments should also be descriptive. Empty parentheses are still required, even in the rare case that a **Function** procedure takes no arguments.

```
Function FunctionName(Argument1, ...)
    VBA statements
    FunctionName = result
End Function
```

Figure 16-5. Structure of a user-defined function.

The function's *return statement* directs the procedure to return the result to the *caller* (usually the cell in which the function was entered). The return statement consists of an assignment statement in which the name of the function is equated to a value, e.g.,

```
FunctionName = result
```

Using the Recorder
to Create a Simple Sub Procedure

Excel provides the Recorder, a useful tool for creating command macros. When you turn on the Recorder, all subsequent menu and keyboard actions will be recorded until you Stop Recording. The Recorder is sufficient for creating simple macros, but you can't use it to incorporate logic, branching or looping. After using the Recorder to create some simple macros, you'll probably view it as simply a tool to create fragments of macro code for incorporation into more complex macros. Macros that involve only the use of menu or keyboard commands can be created using the Recorder.

The Recorder creates Visual Basic commands. You don't have to know anything about Visual Basic to record a command macro in Visual Basic. This provides a good way to gain some initial familiarity with Visual Basic.

To illustrate the use of the Recorder, we'll record the action of applying scientific number formatting to a number in a cell. Before preparing to record the macro, select a cell in a worksheet and enter a number. This will be the number that we're going to format.

Recording a Simple Macro Using Excel 2007/2010

First, make sure the Developer tab is visible on the Ribbon. If the Developer tab is not visible, do the following: Click the File tab, click Options, and then click the Customize Ribbon category. In the Main Tabs list, check the Developer box, and then click OK.

Click on the Developer tab and click on Record Macro in the Code section. You can also begin recording by clicking on the Record New Macro... toolbutton, which is located in the Status Bar, as shown in Figure 16-6.

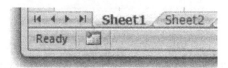

Figure 16-6. The Excel 2007/2010 Record New Macro tool button.

Clicking on Record New Macro... displays the Record Macro dialog box (Figure 16-7).

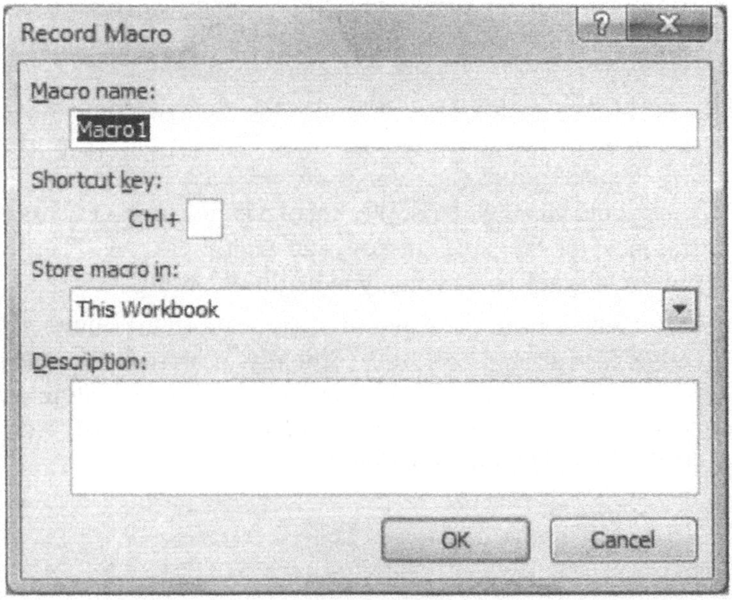

Figure 16-7. The Record Macro dialog box.

The Record Macro dialog box displays the default name that Excel has assigned to this macro: Macro1, Macro2, etc. Change the name in the Macro Name box to ScientificFormat (no spaces are allowed in a name). The "Store Macro In" box should display This Workbook; if not, choose This Workbook. Enter "e" in the box for the shortcut key, then press OK.

When you begin recording, the Record Macro command in the Developer tab will be replaced by Stop Recording. Now click on the Home tab; in the Number section, click on the drop-down button to display the list of number formatting possibilities, and click on Scientific. Finally, click on the Developer tab and click on Stop Recording. You can also stop recording by pressing the Stop Recording button, shown in Figure 16-8.

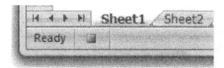

Figure 16-8. The Excel 2007/2010 Stop Recording tool button.

Recording a Simple Macro Using Excel 2003

As in the previous example, we'll record the action of applying scientific number formatting to a number in a cell.

First, select a cell in a worksheet and enter a number. Choose **Macro** from the **Tools** menu, then **Record New Macro...** from the submenu, and press OK. The Record Macro dialog box (Figure 16-7) will be displayed.

The Record Macro dialog box displays the default name that Excel has assigned to this macro: Macro1, Macro2, etc. Change the name in the Macro Name box to ScientificFormat (no spaces are allowed in a name). The "Store Macro In" box should display This Workbook; if not, choose This Workbook. Enter "e" in the box for the shortcut key, and then press OK. The Macro Stop toolbar will appear (Figure 16-9) when you begin recording.

Now choose **Cells...** from the **Format** menu, choose the Number tab, choose Scientific number format, and press OK. Finally, press the Macro Stop button. .If the Macro Stop toolbar doesn't appear, you can always stop recording by using the **Tools** menu: In the Macro submenu the **Record New Macro...** command will be replaced by **Stop Recording**.

Figure 16-9. The Excel 2003 Macro Stop toolbar.

Examining the Macro Code. To examine the macro that you have just recorded, click on Visual Basic in the Code section of the Developer tab (Excel 2007/2010), or choose **Macro** from the **Tools** menu and **Visual Basic Editor** from the submenu (Excel 2003). Click on the node for the module in the active workbook. This will display the code module sheet containing the Visual Basic code. The macro should look like the example shown in Figure 16-10.

```
Sub ScientificFormat()
' Macro1 Macro
' Macro recorded 7/12/2008 by E. J. Billo
    Selection.NumberFormat = "0.00E+00"
End Sub
```

Figure 16-10. Macro for scientific number-formatting, recorded in VBA.

This macro consists of a single line of code. You'll learn about Visual Basic code in the chapters that follow.

To run the macro, enter a number in a cell, select the cell, and then press the shortcut key combination that you designated when you recorded the macro. The number should be displayed in the cell in scientific format.

> *Excel Tip. If you have trouble locating the code module containing your macro, here's what to do "when all else fails": Click on Macros in the Code section of the Developer tab (Excel 2007/2010), or choose **Macro** from the **Tools** menu and **Macros...** from the submenu (Excel 2003). Highlight the name of the macro in the Macro list box, and press the Edit button. This will display the code module sheet containing the Visual Basic code.*

The Personal Macro Workbook

The Record Macro dialog box allows you to choose where the recorded macro will be stored. There are three possibilities in the "Store Macro In" list box: This Workbook, New Workbook and Personal Macro Workbook. The Personal Macro Workbook (PERSONAL.XLS) is a workbook that is automatically opened when you start Excel. Since only macros in open workbooks are available for use, the Personal Macro Workbook is the ideal location for macros that you want to have available all the time. In addition, macros that are saved in the Personal Macro Workbook do not produce the "This workbook contains macros" message each time you start Excel and the Personal Macro Workbook is opened.

Normally the Personal Macro Workbook is hidden; to view it, click on the View tab and click on Unhide in the Window section (Excel 2007/2010), or choose **Unhide...** from the **Window** menu (Excel 2003). If you don't yet have a Personal Macro Workbook, you can create one by recording a macro as described earlier, choosing Personal Macro Workbook from the "Store Macro In" list box.

Running a Sub Procedure

In the preceding example, the macro was run by using a shortcut key. There are a number of other ways to run a macro. One of the most common is to use the Macro dialog box. Again, enter a number in a cell, select the cell, then click on Macros in the Code group of the Developer tab (Excel 2007/2010), or choose **Macro** from the **Tools** menu and **Macros...** from the submenu (Excel 2003). The Macro dialog box will be displayed (Figure 16-11). This dialog box lists all macros in open workbooks. To run the macro, select it from the list and then press the Run button.

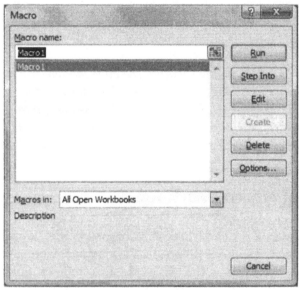

Figure 16-11. The Macro dialog box.

Here are some (but not all) of the ways to run a command macro:

- from the Macro dialog box, as described in this chapter.

- by means of a shortcut key, as described in this chapter.

- by means of a button on a worksheet, as described in Chapter 11.

- as a subroutine called by another macro, as described in Chapter 17.

- by means of a custom menu command, as described in Chapter 21.

- by means of a custom button on a toolbar, as described in Chapter 22.

- by means of an automatic procedure or event-handler procedure, as described in Chapter 23.

Assigning a Shortcut Key to a Sub Procedure

If you didn't assign a shortcut key to the macro when you recorded it, but would like to do so "after the fact", click on Macros in the Code group of the Developer tab (Excel 2007/2010), or choose **Macro** from the **Tools** menu and **Macros...** from the submenu (Excel 2003). Highlight the name of the macro in the Macro Name list box, and press the Options... button. You can now enter a letter for the shortcut key: Ctrl+(key) or Shift+Ctrl+(key).

Creating a Simple Custom Function

As a first example of a **Function** procedure, we'll create a custom function to convert temperatures in degrees Fahrenheit to degrees Celsius.

Function procedures can't be recorded; you must type them on a module sheet. You can have several macros on the same module sheet, so if you recorded the ScientificFormat macro earlier in this chapter, you can type this custom function procedure on the same module sheet. If you do not have a module sheet available, insert one by choosing **Module** from the **Insert** menu on the **Visual Basic** menubar.

Type the macro as shown in Figure 16-12. DegF is the argument passed by the function from the worksheet to the module (the Fahrenheit temperature); the single line of VBA code evaluates the Celsius temperature and returns the result to the caller (in this case, the worksheet cell in which the function is entered).

```
Function FtoC(DegF)
   FtoC = (DegF – 32) * 5 / 9
End Function
```

Figure 16-12. Fahrenheit to Celsius custom function.

Using a Function Macro

A custom function is used in a worksheet formula in exactly the same way as any of Excel's built-in functions. You can enter it in a formula by typing it, or by using Insert Function. The workbook containing the custom function must be open.

To use Insert Function, follow the procedure described in Chapter 3. Select the worksheet cell or the point in a worksheet formula where you want to enter the function (e.g., cell B2 in Figure 16-13). Click on the Formula tab and click on the Function Wizard button in the Function Library group to display the Insert Function dialog box (Excel 2007/2010), or choose **Function...** from the **Insert** menu or press the Function Wizard toolbutton f_x to display the Insert

Function dialog box (Excel 2003), or. Scroll through the Function Category list and select the User Defined category. The FtoC function will appear in the Paste Function list box. When you press OK, the Function Arguments dialog box will be displayed. Enter the argument, or click on the cell containing the argument to enter the reference (cell A2 in Figure 16-13), and then press the OK button.

	A	B
1	T, °F	T. °C
2	212	100

Figure 16-13. Result returned by the FtoC custom function.

You can also type the function name, with or without the opening parenthesis, and then press Ctrl+A or Ctrl+Shift+A, as described in Chapter 3. The function placeholder argument will be displayed, highlighted so that you can enter a value or reference (Figure 16-14).

	A	B
1	T, °F	T. °C
2	212	100
3	32	=FtoC(DegF)

Figure 16-14. Entering a custom function by using Ctrl+Shift+A.

Unfortunately, if you're entering the custom function in a different workbook than the one that contains the custom function, the function name must be entered as an external reference, e.g. PERSONAL.XLS!FtoC. This can make typing the function rather cumbersome, and it means that you'll probably enter the function by using Excel's Paste Function. But if the workbook containing the custom function is saved as an Add-In workbook, you can use the function in any workbook. See "Creating Add-In Function Macros" in Chapter 20.

How Do I Save a Macro?

A macro is part of a workbook, just like a worksheet or a chart. To save the macro, you simply save the workbook.

17

Programming with VBA

This chapter provides an overview of macro programming using VBA. If you are familiar with programming in other computer languages, many of the programming techniques described in this chapter will be familiar.

Creating Visual Basic Code

VBA has a wide range of commands, functions and methods that can be used to create custom applications, and this tends to make VBA confusing for the beginner (or even the non-beginner). This chapter covers the basics that you will need to get started.

Entering VBA Code

As you type your VBA code, Excel checks each line for errors. A line that contains one or more errors will be displayed in red. Variables usually appear in black. Other colors are also used: Comments (see later) are usually green and VBA keywords (**Function**, **For,** etc.) usually appear in blue.

You can enter numbers in E format, but they will automatically be converted to floating point. You can't enter numbers as percentages; the percent symbol has another meaning in VBA (see "Type Declaration Characters" later in this chapter).

If you type a long VBA expression, it will not wrap to the next line but will simply disappear off the screen. You will have to insert a *line-continuation character* (a space followed by the underscore character followed by Enter) to cause a line break in a line of VBA code, as in the following example:

```
ReturnValu = InputBox("Enter validation code number", _
"Validation of this copy of SOLVER.STATS")
```

The line-continuation character can't be used within a string.

Several VBA statements can be combined in one line by separating them with colons. For example, the procedure in Figure 17-1 can be replaced by the more compact one in Figure 17-2 or even by the one in Figure 17-3.

```
Sub MultiBeeper()
For J = 1 To 10
Beep
Next J
End Sub
```

Figure 17-1. A simple VBA **Sub** procedure.

```
Sub MultiBeeper()
For J = 1 To 10: Beep: Next J
End Sub
```

Figure 17-2. A **Sub** procedure with several statements combined.

```
Sub MultiBeeper() : For J = 1 To 10: Beep: Next J : End Sub
```

Figure 17-3. A **Sub** procedure in one line.

When you type VBA code in a module, it's good programming practice to use Tab to indent related lines for easier reading.

Components of Visual Basic Statements

VBA code consists of *statements*. Statements are constructed by using VBA commands, operators, variables, functions, objects, properties or methods. (VBA Help refers to keywords such as **Beep, Do** or **Exit** as statements, but here they'll be referred to as commands, and we'll use "statement" in a general way to refer to a line of VBA code.)

Operators

VBA operators include the arithmetic operators (+, -, *, /, ^), the text concatenation operator (&), and the comparison operators (=, <, >, <=, >=, <>).

Excel Tip. Be sure to leave a space on either side of the concatenation operator; otherwise it will be mistaken for a type-declaration character (see "Type Declaration Characters" later in this chapter).

Variables and Arguments

Variables are the names you create to indicate the storage locations of values or references. Arguments are values that are passed from a worksheet to a **Function** procedure, for example.

The value of a variable is determined by an *assignment statement*. An assignment statement assigns the result of an expression to a variable or object; the form of an assignment statement is

variable = expression

In an Excel worksheet, you have to use Define Name or Create Name to assign a name to a variable, but variable names in VBA are automatically assigned as you type the VBA code in a module. There are just a few rules for naming variables or arguments:

- The first character must be a letter.
- A name cannot contain a space or a period.
- The type-declaration characters (%, $, #, !, &) cannot be embedded in a name.
- You shouldn't use any of the VBA reserved words, such as Function, Range or Value.

You should make variable names as descriptive as possible, but avoid overly long names which are tedious to type. You can use the underscore character to indicate a space between words (e.g., formula_string). Don't use a period to indicate a space, since VBA uses the period character in a specific way, described below. The most common form for variable names uses upper- and lowercase letters (e.g., FormulaString).

Objects, Properties and Methods

VBA is an *object-oriented* programming language. *Objects* in Microsoft Excel are the familiar components of Excel, such as a worksheet, a chart, a toolbar or a range. Objects have *properties* and *methods* associated with them. Objects are the nouns of the language, properties are the adjectives, and methods are the verbs.

Objects

Some examples of VBA objects are the **Workbook** object, the **Worksheet** object, the **Chart** object and the **Range** object. A complete list of objects in Microsoft Excel can be found in Excel's On-line Help. You can also use the Object Browser to see the complete list of objects. To display the Object Browser dialog box, choose **Object Browser** from the **View** menu.

You can also refer to *collections* of objects. A collection is a group of objects of the same kind. A collection has the plural form of the object's name (e.g., **Worksheets**). **Worksheets** refers to all worksheets in a particular workbook.   To refer to a particular worksheet, use, for example, **Worksheets**("Spectrum1").

Figure 17-6 shows an abbreviated hierarchical list of the most useful objects.

Application	MenuBar	Menu	MenuItem	
	ToolBar	ToolbarButton		
	Workbook	Chart	Axis	AxisTitle
				GridLines
				TickLabels
			etc.	
		Worksheet	PageSetup	
			Range	Areas
				etc.

Figure 17-6. Partial list of objects, arranged in hierarchical order.

There is a hierarchy of objects. You specify the object by specifying its location in a hierarchy, separated by periods, e.g.,

Workbooks("Deconvolution").**Worksheets**("Spectrum1").**Range**("E5")

If Deconvolution is the active workbook, you can omit reference to the workbook and simply use the reference

Worksheets("Spectrum1").**Range**("E5")

If Spectrum1 is the active sheet, you can omit reference to the workbook and worksheet and simply use the reference **Range**("E5").

Some Useful Objects

The objects you probably will use most often in the beginning are the **Workbook** object, the **Worksheet** object and the **Range** object. Note that there is no cell object, only the **Range** object. In later chapters you'll learn about the **Dialog**, **MenuBar**, **ToolBar** or **Chart** objects and the objects they in turn contain.

"Objects" that Are Really Properties

Although **ActiveCell** and **Selection** are properties, not objects, you can treat them like objects. (**ActiveCell** is a property of the **Application** object, or the **ActiveWindow** property of the **Application** object.) The **Application** object has the following properties that you can treat just as though they were objects: the **ActiveWindow**, **ActiveWorkbook**, **ActiveSheet**, **ActiveCell**, **Selection** and **ThisWorkbook** properties. Since there is only one **Application** object, you can omit the reference to **Application** and simply use **ActiveCell**.

You Can Define Your Own Objects

VBA allows you to equate a variable to an object, but the variable does not automatically become an object. If you then attempt to use the variable in an expression that requires an object, you'll get an "Object required" error message. You must use the **Set** command to define a variable as an object, as in the following example.

Set RangeToUse **= Worksheets(**"Sheet3"**).Range(**"A1:A10"**)**

You can then use, for example, the ColumnWidth property of the RangeToUse object.

Making a Reference to a Cell or a Range

One of the most important things you'll need to master is making a reference to a cell or range of cells. In brief, you'll need to be able to send values from a worksheet to a module sheet so that you can perform operations on the worksheet data, and you'll need to be able to send results back from the module sheet to the worksheet.

You can reference a particular cell or range within a worksheet in a number of ways: by using the **Selection** or **ActiveCell** keywords or by using the **Range** keyword or the **Cells** keyword .

Don't use the **Select** keyword unless you actually need to select cells in a worksheet. For example, to copy a range of cells, you could use the statements shown in Figure 17-4, and in fact this is exactly the code you would generate using the Recorder. But you can do the same thing much more efficiently by using the code shown in Figure 17-5.

```
Range("D1:D20").Select
Selection.Copy
```

Figure 17-4. VBA code fragment by the Recorder.

```
Range("D1:D20").Copy
```

Figure 17-5. A more efficient way to accomplish the same thing, without selecting cells.

Making a Reference to a Selected Cell or Range

You may wish to create a macro that operates on the contents of a user-selected cell or range. To do this you can use either the **ActiveCell** keyword or the **Selection** keyword.

Note the difference between the active cell and the selection: **Selection** can be a range of cells or a single cell; **ActiveCell** (always a single cell) refers to the cell in the upper-left corner of a range if applied to a range of cells.

Making a Reference to a Cell
Other than the Active Cell

Instead of a macro that operates on the contents of the active cell or a selection, you may want to create a macro that copies values from specified rows and columns in a worksheet, independent of where the cursor has been "parked" by the user. To accomplish this, you can use either the **Range** keyword or the **Cells** keyword to make a reference to a cell or range. The syntax of the **Cells** keyword is **Cells(*row, column*)**. The first two of the following expressions refer to cell B3, the next two to the range B3:E27.

Range("B3")

Cells(3,2)

Range("B3:E27")

Range(Cells(3, 2), Cells(27, 5))

The preceding examples are "absolute" references, since they always refer to B3 or B3:E27. You can also use what could be called a "computed" reference, in which the reference depends on the value of a variable. The **Cells** keyword is conveniently used in this way. For example, the expression

Cells(J,2)

allows you to select any cell in column B, depending on the value assigned to the variable J. The **Range** keyword can be used in a similar way by using the concatenation operator, e.g.,

Range("B" & J)

See the Tutorial on the following page for a variety of code expressions to refer to a cell or range.

Tutorial on Ways to Refer to a Cell or Range

1. Use the **Range** keyword with a reference:
   Range("B1:D10"**)**
2. Use the **Cells** keyword with row and column numbers:
   Cells(15, 5**)**
3. Use the **Range** keyword with a range name:
   Range("ref1"**)**

   The range name ref1 was assigned using **Insert → Name → Define**. Benefit: the reference updates if the spreadsheet is modified.
4. Use the **Cells** keyword with variables:
   Cells(RowNum, ColNum)
5. Use the **Range** keyword with a variable:
   Range(ref2**)**

   The variable ref2 is text.
6. Use the **Range** keyword with ampersand:
   TopRow = 2: BtmRow = 12
   Range("F" & TopRow & ":G" & BtmRow)

   The **Range** argument evaluates to "F2:G12")
7. Use the **Range** keyword with two **Cells** expressions:
   Range(Cells(1, 1**)**, **Cells(**5, 5**))**

   This expression refers to the range A1:E5.
8. Use the **Range** keyword with **Cells**(index):
   Range("A5:A12"**).Cells(**3**)**

   This expression refers to cell A7; it provides a way to select individual cells within a specified range.
9. Use the **Range** keyword with **Offset**:
   Range("A1:A12"**).Offset(**3, 1**)**

   This expression refers to the range B4:B15.
10. Use the **Range** keyword with **Offset** and **Resize**:
    Range("A1:A12"**).Offset(**3, 1**).Resize(**1, 1**)**

    This expression refers to cell B4.
11. Use the **Union** keyword:
    Union(Cells(1, 1**)**, **Cells(**2, 2**)**, **Cells(**3, 3**))**

    This expression refers to cells A1, B2, C3.
12. Use the **Intersect** keyword:
    Intersect(Range("A5:H5"**)**, **Range(**"E1:E10"**))**

    This expression refers to cell E5.

References Using the Union or Intersect Method

VBA can create references by using methods that are the equivalents of the worksheet union operator or intersection operator described in Chapter 3. The **Union** method creates a reference that includes multiple selections, e.g., A1,B5 or G3:L3,G5:L5. The syntax of the **Union** method is **Union**(*range1, range2, …*). The **Intersect** method creates a reference that is common to two references (e.g., F4:F6 E5:I5). The syntax of the **Intersect** method is **Intersect**(*range1, range2*). Both *range1* and *range2* must be **Range** objects.

Getting Values from a Worksheet

To transfer values from worksheet cells to a procedure, use a reference to a worksheet range in an assignment statement, like one of the following.

variable1 = **ActiveCell.Value**

QZ = **Worksheets**("Sheet1").**Range**("A9").**Value**

MyVal = **Range**("A" & x) .**Value**

CellVal = **Cells**(StartRow+x,StartCol) .**Value**

The **Formula** property can be used to obtain the formula in a cell, as text, rather than its value.

Sending Values to a Worksheet

To send values from a module sheet back to a worksheet, simply use an assignment statement like one of the following. Numbers, text, the contents of a VBA variable, and even a worksheet formula can be entered into worksheet cells or ranges.

Cells(1, 2).**Value** = 5

Range("E1").**Value** = "Jan.-Mar."

Worksheets("Sheet1").**Range**("A1").**Value** = variable2

Cells(1, 3).**Formula** = "=sum(F1:F10)"

Properties

Objects have *properties* that can be set or read. Some properties of the **Range** object are **ColumnWidth, NumberFormat, Font** and **Value**. A property is connected to the object it modifies by a period. For example,

CelFmt = **Range**("E5").**NumberFormat**

returns the number format of cell E5 and assigns it to the variable CelFmt, and

Range("E5"). **NumberFormat** = "0.000"

sets the number format of cell E5.

Properties can also modify properties of an object: for example,

Range("A5:E5").**Interior.ColorIndex** = 8

Properties of the Range Object

The large number of properties of each object accounts for most of the difficulty in learning VBA. For example, there are 95 properties of the **Range** object in Excel 2007 (up from 93 in Excel 2003):

AddIndent, Address, AddressLocal, AllowEdit, Application, Areas, Borders, Cells, Characters, Column, Columns, ColumnWidth, Comment, Count, CountLarge, Creator, CurrentArray, CurrentRegion, Dependents, DirectDependents, DirectPrecedents, End, EntireColumn, EntireRow, Errors, Font, FormatConditions, Formula, FormulaArray, FormulaHidden, FormulaLocal, FormulaR1C1, FormulaR1C1Local, HasArray, HasFormula, Height, Hidden, HorizontalAlignment, Hyperlinks, ID, IndentLevel, Interior, Item, Left, ListHeaderRows, ListObject, LocationInTable, Locked, MDX, MergeArea, MergeCells, Name, Next, NumberFormat, NumberFormatLocal, Offset, Orientation, OutlineLevel, PageBreak, Parent, Phonetic, Phonetics, PivotCell, PivotField, PivotItem, PivotTable, Precedents, PrefixCharacter, Previous, QueryTable, Range, ReadingOrder, Resize, Row, RowHeight, Rows, ServerActions, ShowDetail, ShrinkToFit, SmartTags, SoundNote, Style, Summary, Text, Top, UseStandardHeight, UseStandardWidth, Validation, Value, Value2, VerticalAlignment, Width, Worksheet, WrapText, XPath

Some Useful Properties

Some of the most useful properties of the **Range** object are listed in Table 17-1. Some properties, such as **Column** or **Count**, are read-only.

Table 17-1. Some Useful VBA Properties

Column	Returns a number corresponding to the first column in the range.
ColumnWidth	Returns or sets the width of all columns in the range.
Count	Returns the number of items in the range.
Font	Returns or sets the font of the range.
Formula	Returns or sets the formula.
Name	Returns or sets the name of the range.
NumberFormat	Returns or sets the format code for the range.
Row	Returns a number corresponding to the first row in the range.
RowHeight	Returns or sets the height of all rows in the range.
Text	Returns or sets the text displayed by the cell.
Value	Returns or sets the contents of the cell or range.

Methods

Objects also have *methods*. Methods can operate on an object or on a property of an object. Some methods that can be applied to the **Range** object are the **Copy** method, the **Cut** method, the **FillDown** method or the **Sort** method. For example,

Range("A1:E1").Clear

clears the formulas and formatting in the range A1:E1.

Some Useful Methods

VBA Help lists 267 methods. Many of them correspond to familiar menu commands. For example, **Copy**, **Cut**, **Clear** and **Sort** can be performed on a range of cells. Some useful VBA methods are listed in Table 17-2.

Table 17-2. Some Useful VBA Methods

Activate	Activates an object (sheet, etc.).
Clear	Clears an entire range.
Close	Closes an object.
Copy	Copies an object to a specified range or to the Clipboard.
Cut	Cuts an object to a specified range or to the Clipboard.
FillDown	Copies the cell(s) in the top row into the rest of the range.
Select	Selects an object.

Two Ways to Specify Arguments of Methods

VBA methods usually take one or more arguments. The **Sort** method, for example, takes 10 arguments. The syntax of the **Sort** method is

*object.***Sort***(key1, order1, key2, order2, key3, order3, header, orderCustom, matchCase, orientation)*

Object is required; all arguments are optional.

You can specify the arguments of a method in two ways. One way is to list the arguments in order as they are specified in the preceding syntax, i.e.,

Selection.Sort Range("A2", 1)

In the preceding example, only the arguments **key1** and **order1** were specified; the remaining arguments are optional and were not required.

You must include commas to indicate omitted arguments, as in the following example:

Set known_Ys = **Application.InputBox** _
("Select the range of Y values", "STEP 1 OF 2", , , , , , 8)

The second way is to use the name of the argument as it appears in the preceding syntax, with the := operator, to specify the value of the argument. The arguments can appear in any order, as in the following (all in one line of code, of course):

Selection.Sort Key1:=**Range**("A2"), Order1:=xlAscending, _
Key2:=**Range**("B2"), Order2:=xlAscending, Key3:=**Range**("C2"), _
Order3:=xlDescending, Header :=xlGuess, OrderCustom:=1, _
MatchCase:=**False**, Orientation:=xlTopToBottom

Arguments with or without Parentheses

As well as performing an action, methods create a return value. The return value can be either **True** or **False**. Even the **ChartWizard** method creates a return value: **True** if the chart was created successfully, **False** if the method failed. Usually you aren't interested in the return value.

An example of a method that creates a return value that can be used is the **CheckSpelling** method. The **CheckSpelling** method has the following syntax:

Application.CheckSpelling(word)

If you use this method, you'll need the return value (either **True** or **False**) to determine whether the word is spelled correctly.

If you want to use the return value of a method, you must enclose the arguments of the method in parentheses, as if it were a function. If the arguments are not enclosed in parentheses, the return value will not be available for use. Put another way, the expression

result = **Application.CheckSpelling(ActiveCell.Value)**

does not produce a syntax error, while the expression

result = **Application.CheckSpelling ActiveCell.Value**

does give a syntax error.

VBA Functions

The functions available in VBA are similar to the functions available in Excel itself. There are 108 VBA functions listed in VBA Help. Table 17-3 lists some of the more useful ones for numerical calculations.

Table 17-3. Some Useful VBA Functions

Abs	Returns the absolute value of a number.
Asc	Returns the character code of a character.
Chr	Returns the character corresponding to a code.
Exp	Returns *e* raised to a power.
Fix	Returns the integer part of a number (truncates).
Int	Returns the integer part of a number (rounds down).
IsArray	Returns **True** if the variable is an array.
IsNull	Returns **True** if the expression is null (i.e., contains no valid data).
IsNumeric	Returns **True** if the expression can be evaluated to a number.
LBound	Returns the lower limit of an array dimension.
LCase	Converts a string into lowercase letters.
Left	Returns the leftmost characters of a string.
Len	Returns the length (number of characters) in a string.
Log	Returns the natural (base-*e*) logarithm of a number.
Mid	Returns a specified number of characters from a string.
Right	Returns the rightmost characters of a string.
RTrim	Returns a string without trailing spaces.
Sqr	Returns the square root of a number.
Str	Converts a number to a string.
UBound	Returns the upper limit of an array dimension.
UCase	Converts a string into uppercase letters.

Most of these functions take one or more arguments; refer to VBA Help for information about the arguments.

Using Worksheet Functions with VBA

Many useful worksheet functions do not have a VBA equivalent. To use an Excel worksheet function within VBA, use the syntax

Application.*WorksheetFunction*

and supply arguments for the function just as you would in a worksheet. For example, to use the SUBSTITUTE function in VBA, use the code

Application.Substitute(*text,old_text,new_text,instance_num*)

e.g.,

Application. Substitute (Range("A1")," ","",1)

VBA Commands

Commands in VBA are similar to commands in BASIC or FORTRAN. Table 17-4 lists some of the most useful VBA commands.

Table 17-4. Some Useful VBA Commands

Beep	Makes a "beep" sound.
Dim	Declares an array and allocates storage for it.
Do...Loop	Delineates a block of statements to be repeated.
Else	Optional part of **If...Then** structure.
ElseIf	Optional part of **If...Then** structure.
End	Terminates a procedure.
End If	Terminates block of statements begun by **If**.
Exit	Exits a **Do..., For..., Function... or Sub...** structure.
For Each...Next	Delineates a block of statements to be repeated.
For...Next	Delineates a block of statements to be repeated.
Function	Marks the beginning of a **Function** procedure.
GoSub...Return	Delineates a subroutine.
GoTo	Unconditional branch.
If...Then...Else	Delineates a block of conditional statements.
On...GoSub	Branch to one of several specified subroutines.
On...GoTo	Branch to one of several specified lines.
Select Case	Executes one of several blocks of statements.
Set	Assigns an object reference to a variable or property.
Stop	Stops execution.
Sub	Marks the beginning of a **Sub** procedure.
Until	Optional part of **Do...Loop** structure.
While	Optional part of **Do...Loop** structure.
With...EndWith	Delineates a block of statements to be executed on a single object.

VBA Data Types

VBA uses a range of different data types. Table 17-5 lists the built-in data types.

The Variant Data Type

The **Variant** data type is the default data type in VBA. Like Excel itself, the **Variant** data type handles and interconverts between many different kinds of data: integer, floating point, string, etc. The **Variant** data type automatically chooses the most compact representation.

Unless you declare a variable's type, VBA will use the **Variant** type. You can save memory space if your procedure deals only with integers, for example, by declaring the variable type.

Declaring Variables

VBA uses the **Variant** data type as the default data type for variables and arguments. The **Variant** data type permits Excel to switch between floating-point, integer and string variables as required.

To force a particular variable or argument to take a specified data type, use the **Dim** statement, e.g.,

Dim ChemFormula **As String**

Specifying the Data Type of an Argument

You can specify the data type of an argument passed to a **Function** proceduret. For example, if the **Function** procedure MolWt has two arguments, formula (a string) and decimals (an integer), then the statement

Function MolWt (formula **As String**, decimals **As Integer**)

declares the type of each variable. If an argument of an incorrect type is supplied to the function, a #VALUE! error message will be displayed.

Specifying the Data Type Returned by a Function Procedure

You can also specify the data type of the return value. If none is specified, the **Variant** data type will be returned. In the example of the preceding section, MolWt returns a floating-point result. If you want to specify double precision floating-point, use an additional **As Type** expression in the statement, e.g.,

Function MolWt (formula **As String**, decimals **As Integer**) **As Double**

Table 17-5. VBA's Built-in Data Types

Data type	Storage required	Range of values
Logical	2 bytes	**True** or **False**
Integer	2 bytes	–32,768 to 32,767
Long integer	4 bytes	–2,147,483,648 to 2,147,483,647
Single precision	4 bytes	–3.402823E38 to –1.401298E–45 for negative values; 1.401298E–45 to 3.402823E38 for positive values
Double precision	8 bytes	–1.79769313486232E308 to 4.94065645841247E–324 for negative values; 4.94065645841247E–324 to 1.79769313486232E308 for positive values
Currency	8 bytes	–922,337,203,685,477.5808 to 922,337,203,685,477.5807
Date	8 bytes	
Object	4 bytes	any Object reference
String	1 byte/character	
Variant	16 bytes + 1 byte/character	any numeric value up to the range of a **Double** or any text

String Data Types

Strings can be stored either as *variable-length strings* (the default data type) or as *fixed-length strings*. To declare a string variable as fixed length, use the statement in a **Dim** statement (more about the **Dim** statement in Chapter 18).

String * *length*

For example, the following statement sets aside storage for a two-dimensional array of names and addresses, containing fixed-length strings of 32 characters:

Dim AddressList(500, 4) **As String** * 32

If a string of length less than 32 characters is assigned to the array AddressList, trailing spaces are added to fill out the string length. If a string of more than 32 characters is assigned to AddressList, the string will be truncated.

The Boolean (Logical) Data Type

A logical variable can have only the value **True** or **False**. The keywords **True** and **False** are often implied. For example, the expressions

If (j > N) = **True Then** ALPHA = 0

and

If (j > N) **Then** ALPHA = 0

are equivalent.

You can use other data types as Boolean variables. When a variable is used in a logical expression, zero is converted to **False** while any non-zero value is converted to **True**. Thus the expression

If j **Then** *expression*

tests for a non-zero value of the variable **j**.

When Boolean variables are converted to other data types, **False** becomes zero but **True** is converted to −1.

Type Declaration Characters

You can also declare a variable's type by appending a type-declaration character to the variable name, a technique from older versions of BASIC. The type-declaration characters include % for integer variables and $ for string variables.

Program Control

If you are familiar with computer languages such as BASIC or FORTRAN, you will find yourself quite comfortable with most of the material in this section.

Decision-Making (Branching)

VBA supports **If...Then** and **If...Then...Else** structures, very similar to the Excel worksheet function IF.

The **If...Then** statement can be on a single line:

If (x = j) **Then** numerator = 10 ^ (logbeta - pH * x)

This is the so-called Simple If statement. If multiple commands are to be executed following a logical expression, use the Block If statement, as in Figure 17-7.

```
If (pKa_logical = False) Then
    logbeta = logbeta + pKs_or_logKs(x)
    denom = denom + 10 ^ (logbeta - pH * x)
    etc.
End If
```

Figure 17-7. Example of VBA **If...End If** structure.

In addition to the Simple If or the Block If structures, there is also the **If...Then...Else** structure:

If LogicalExpression **Then** statement **Else** statement

In addition to **If...Then...Else**, there is also **If...Then... ElseIf... End If**, as illustrated in Figure 17-8.

```
If LogicalExpression1 Then
    statements
ElseIf LogicalExpression2 Then
    statements
ElseIf LogicalExpression3 Then
    statements
    etc.
End If
```

Figure 17-8. The VBA **If...ElseIf...End If** structure.

The **Select Case** statement provides an efficient alternative to the series of **ElseIf** *condition* statements.   The syntax of the **Select Case** statement is illustrated in Figure 17-9.

TestExpression is evaluated and used to direct program flow to the appropriate **Case**.   ExpressionListN can be a single value, a list of values separated by commas, or a range of values specified by the expression lowervalue **To** uppervalue. The comparison operators must be used with the **Is** keyword, in an expression like **Case Is** > 5. (If you forget to include the **Is**, it will be inserted automatically.)

The optional **Case Else** statement is executed if TestExpression doesn't match any of the values in any of ExpressionListN. It is good programming practice to always include a **Case Else**, to handle unforeseen circumstances.

The example shown in Figure 17-10 illustrates the use of **Select Case** to calculate the pK_a value of a polyprotic acid. Since data at or near the equivalence points cause large calculation errors, the pK_a is calculated only for NBar values in the range $0.2 - 0.8$, $1.2 - 1.8$, $2.2 - 2.8$ or $3.2 - 3.8$. The expression used to calculate the pK_a from the NBar parameter depends on the number of protons bound, i.e., on the value of NBar. The **Select Case** statement is used to direct program flow to the appropriate expression.

Note that a range of values is indicated by using the **To** keyword.

```
Select Case TestExpression
    Case ExpressionList1
        statements
    Case ExpressionList2
        statements
    Case ExpressionList3
        statements
    Case Else
        statements
End Select
```

Figure 17-9. The VBA **Select Case** structure.

Logical Operators

You are already familiar with the **And** and **Or** operators, but VBA provides in addition the **Xor** (exclusive or) operator. The operators have the following syntax:

expression1 **And** expression2   **True** if both expressions are **True**.

expression1 **Or** expression2   **True** if either expression is **True**.

expression1 **Xor** expression2   **True** if one expression is **True**, the other **False**.

```
NBar = (ZP * CR + CA + COH - CH - CNa) / CR

Select Case NBar
Case 3.2 To 3.8
    pK = pH + Application.Log((NBar - 3) / (4 - NBar))
Case 2.2 To 2.8
    pK = pH + Application.Log((NBar - 2) / (3 - NBar))
Case 1.2 To 1.8
    pK = pH + Application.Log((NBar - 1) / (2 - NBar))
Case 0.2 To 0.8
    pK = pH + Application.Log((NBar) / (1 - NBar))
Case Else
    pK = ""
End Select
```

Figure 17-10. An example of the **Select Case** structure.

The preceding expressions must evaluate to **True** or **False**; that is, they must be logical expressions. The logical operators are almost always used in combination with **If** statements.

More than one **And** or **Or** can be combined in a single statement. For example,

If Char = " " **Or** Char = "*" **Or** Char = "," **Or** Char = "(" **Or** Char = "/" **Then**...

evaluates to **True** if any one of the logical expressions is **True**.

Parentheses are often necessary to control the logic of the expression. For example, each of the expressions

> **If** (expression1 **And** expression2) **Or** expression3 **Then**...

> **If** expression1 **And** (expression2 **Or** expression3) **Then**...

has eight different possible combinations of expression1, expression2 and expression3; two of them give different outcomes, depending on which expression is used.

Looping

The loop structures in VBA are similar to those available in other programming languages.

For...Next Loops

The syntax of the **For...Next** loop is given in Figure 17-11.

```
For Counter = Start To End Step Increment
     statements
Next Counter
```

Figure 17-11. The VBA **For...Next** structure.

Step Increment in the **For** statement is optional. If it is omitted, Increment is set equal to 1. Increment can be negative.

For Each...Next Loops

The **For Each...Next** loop structure is similar to the **For...Next** loop structure, except that it executes the statements within the loop for each object within a group of objects. Figure 17-12 illustrates the syntax of the statement.

```
For Each Element In Group
     statements
Next Element
```

Figure 17-12. The VBA **For Each...In...Next** structure.

The **For..Each...Next** loop returns an object variable in each pass through the loop. You can access or use all of the properties or methods that apply to Element. For example, in a loop such as

> **For Each** cel **In Selection**

the variable cel is an object that has all the properties of a cell (a **Range** object): **Value**, **Formula**, **NumberFormat**, etc.

Do While... Loop

The **Do While...Loop** is used when you don't know beforehand how many times the loop will need to be executed. The syntax is shown in Figure 17-13.

```
Do While LogicalExpression
    statements
Loop
```

Figure 17-13. The VBA **Do While...Loop** structure.

```
Do
    statements
Loop While LogicalExpression
```

Figure 17-14. Alternate form of **Do While...Loop** structure.

An alternate format of this type of loop places **While** *LogicalExpression* at the end of the loop, as exemplified in Figure 17-14.

Note that this form of the **Do While** structure executes the loop at least once.

Exiting from a Loop or from a Procedure

Often you use a loop structure to search through an array of values or collection of objects, looking for a certain value or property. Once you find a match, you don't need to cycle through the rest of the loops. You can exit from the loop using the **Exit For** (from a **For...Next** loop or **For Each...Next** loop) or **Exit Do** (from a **Do While...** loop). The **Exit** statement will normally be located within an **If** statement. For example,

If CellContents.**Value** <= 0 **Then Exit For**

Use the **Exit Sub** or **Exit Function** to exit from a procedure. Again, the **Exit** statement will normally be located within an **If** statement.

Exit statements can appear as many times as needed within a procedure.

Subroutines

A subroutine is a procedure that is called by another VBA program. It's good programming practice to break up a complicated task into simpler tasks and write subroutines to do each task. The separate subroutines are called by a main program.

There are several ways to execute a subroutine within a main program. The most common is by using the **Call** command, as illustrated in Figure 17-15. MainProgram calls the subroutine Task1, which requires arguments.

```
Sub MainProgram()
etc.
Call Task1(argument1,argument2)
etc
End Sub

Sub Task1(ArgName1,ArgName2)
etc
End Sub
```

Figure 17-15. A main program that calls a subroutine.

Note that the variable names of the arguments in the calling statement and in the subroutine do not have to be the same, but only listed in the same order.

Figure 17-26 shows an example of a custom function that calls a subroutine.

```
Function RK1(XAddress, YAddress, X, Y, H, FormulaText)
Dim T1 As Double, T2 As Double, T3 As Double, T4 As Double
Dim result As Double

Call eval(XAddress, YAddress, X, Y, FormulaText, result)
T1 = result * H
Call eval(XAddress, YAddress, X + H / 2, Y + T1 / 2, FormulaText, result)
T2 = result * H
Call eval(XAddress, YAddress, X + H / 2, Y + T2 / 2, FormulaText, result)
T3 = result * H
Call eval(XAddress, YAddress, X + H, Y + T3, FormulaText, result)
T4 = result * H
RK1 = Y + (T1 + 2 * T2 + 2 * T3 + T4) / 6
End Function
'+++++++++++++++++++++++++++++++++++++++++++++++++++++++++++++
Sub eval(XAddress, YAddress, XArg, YArg, FormulaText, result)
Dim T As String
T = FormulaText
T = Application.Substitute(T, XAddress, XArg)
T = Application.Substitute(T, YAddress, YArg)
result = Evaluate(T)
End Sub
```

Figure 17-16. The custom function RK1 calls the subroutine eval.

Scoping a Subroutine

A subroutine can be **Public** or **Private**. Public subroutines can be called by any subroutine in any module. The default for any **Sub** procedure is **Public**. A **Private** subroutine can be called only by other subroutines in the same module.

To declare the subroutine Task3 as a private subroutine, use the statement

Private Sub Task3()

One good use of the **Private** statement is to prevent a **Sub** procedure without arguments (a main program) from being listed in the Macro Run dialog box, if you don't want users to be able to run the procedure or even to see the name of the procedure in the list.

Interactive Macros

VBA provides two built-in dialog boxes for display of messages or for input: **MsgBox** and **InputBox**.

Msgbox

The **MsgBox** dialog box allows you to display a message, such as "Please wait..." or "Access denied". The box can display one of four message icons, and there are many possibilities in the number of buttons, and their captions, that can be displayed.

The syntax of the **MsgBox** function is

MsgBox (*prompt_text, buttons, title_text, helpfile, context*)

where *prompt_text* is the message displayed within the box, *buttons* specifies the buttons to be displayed, and *title_text* is the title to be displayed in the Title Bar of the box. For information about *helpfile* and *context*, refer to *Microsoft Excel Visual Basic Reference*.

For example, the VBA expression

MsgBox "You entered " & incr & "." & **Chr**(13) & **Chr**(13) & _
"That value is too large." & **Chr**(13) & **Chr**(13) & "Please try again.", 48

produces the message box shown in Figure 17-17.

Figure 17-17. A **Msgbox** display.

The value of *buttons* determines the type of message icon and the number and type of response buttons; it also determines which button is the default button. The possible values are listed in Table 17-6. The values 0 – 5 specify the number and type of buttons, values 0 or 16 – 64 specify the type of message icon, and values 0, 256, 512 specify which button is the default button. You can add together one number from each group to form a value for *buttons*. For example, to specify a dialog box with a Warning Query icon, with Yes, No and Cancel buttons, and with the No button as default, use the value 32 + 3 + 256 = 291 as the *buttons* argument.

Table 17-6. Values for the *buttons* Argument of **MsgBox**

buttons	**Description**
0	Display OK button only.
1	Display OK and Cancel buttons.
2	Display Abort, Retry and Ignore buttons.
3	Display Yes, No and Cancel buttons.
4	Display Yes and No buttons.
5	Display Retry and Cancel buttons.
0	No icon
16	Display Critical Message icon.
32	Display Warning Query icon.
48	Display Warning Message icon.
64	Display Information Message icon.
0	First button is default.
256	Second button is default.
512	Third button is default.

The four possible message icons are shown in Figure 17-18.

Figure 17-18. MsgBox icons (left to right):
Critical Message, Warning Query, Warning Message, Information Message.

Msgbox Return Values

MsgBox returns a value that indicates which button was pressed. This allows you to take different actions, depending on whether the user pressed the Yes, No or Cancel buttons, for example. To get the return value of the Message Box, use an expression like the following (parentheses are required):

ReturnValu = MsgBox (*prompt_text, buttons, title_text, helpfile, context*)

The return values of the buttons are as follows: OK, 1; Cancel, 2; Abort, 3; Retry, 4; Ignore, 5; Yes, 6; No, 7.

InputBox

The **InputBox** allows you to pause a macro and request input from the user. There are both an **InputBox** function and an **InputBox** method.

The syntax of the **InputBox** function is

InputBox(*prompt_text, title_text, default, x_position, y_position, helpfile, context*)

where *prompt_text* and *title_text* are as in **MsgBox**. *Default* is the expression displayed in the input box, as a string. The horizontal distance of the left edge of the box from the left edge of the screen, and the vertical distance of the top edge from the top of the screen are specified by *x_position* and *y_position*, respectively. For information about *helpfile* and *context*, refer to *Microsoft Excel Visual Basic Reference*.

If the user presses the OK button or the Return key, the **InputBox** function returns as a value whatever is in the text box. If the Cancel button is pressed, the function returns a null string. The following example produces the input box shown in Figure 17-19.

ReturnValu = **InputBox**("Please enter a value now.", "INPUT BOX DEMO")

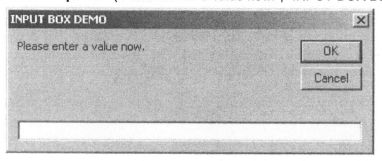

Figure 17-19. An **InputBox** display.

The syntax of the **InputBox** method is

Object.InputBox(*prompt_text, title_text, default, x_position, y_position, helpfile, context, type_num*)

The differences between the **InputBox** function and the **InputBox** method are the following: (i) *default* can be any data type and (ii) the additional argument *type_num* specifies the data type of the return value. The values of *type_num* and the corresponding data types are listed in Table 17-7. Values of *type_num* can be added together. For example, to specify an input dialog box that accepts number or string values as input, use the value 1 + 2 = 3 for *type_num*.

type_num	Data type
Table 17-7. InputBox Data Type Values	
0	Formula
1	Number
2	String
4	Logical
8	Reference (as a Range object)
16	Error value
64	Array

Using a Dialog Box to Open a Workbook

In addition to the VBA **MsgBox** and **InputBox**, there are other built-in dialog box possibilities. To illustrate, we'll look at possible ways to open a workbook from within a macro.

As an example, you may have created a macro that uses information from another workbook (the source workbook). There are several possible ways to access the source workbook, which are listed in order of increasing utility.

Case 1. The source workbook must be open before the user runs the macro. A typical line of code in a procedure using this approach might be

RI = **Workbooks**("Polymers")**.Worksheets**("PhysProperties")**.Range**("E2:E700")

If the source is not open, the procedure fails.

Case 2. The source workbook does not need to be open, but must be in the same folder as the workbook containing the macro. The source workbook is opened by the procedure, using the **Open** method. The filename of the source workbook is known and is coded in the procedure. If the source is moved to a different directory, the procedure fails.

```
Sub OpenDemo1()
Workbooks.Open Filename:="Polymers.xls"
End Sub
```

Figure 17-20. Simple procedure to open a workbook.

Case 3. The source workbook does not need to be open, and does not need to be in the same folder as the workbook containing the macro, but must be in a known location. The source workbook is opened by the procedure, using the **Open** method. The filename of the source workbook is known and is coded in the procedure. The source must be in a known location, which is coded in the procedure; if the source is moved to a different directory, the procedure fails.

```
Sub OpenDemo1()
Workbooks.Open Filename:="C:\Documents and Settings\Owner _
    \Desktop\Polymers.xls"
End Sub
```

Figure 17-21. Simple procedure to open a workbook.

Case 4. The source workbook can be any one of a number of workbooks; the filename is input by the user, using the **InputBox** method. The source workbook does not need to be open, but does need to be in the same folder as the workbook containing the macro, or in a known location. The source workbook is opened by the procedure, using the **Open** method.

```
Sub OpenDemo3()
FileToOpen = Application.InputBox("Enter name of workbook to open",_
    "OPEN DEMO #3")
Workbooks.Open FileToOpen
End Sub
```

Figure 17-22. Simple procedure to open a workbook.

Using the GetOpenFilename Method

The preceding approaches are not too satisfactory. What we would like to do is to display the **Open** dialog box, allow the user to navigate through directories, locate the desired source workbook, and open it. VBA provides two possible ways to display the **Open** dialog box. One, using the **GetOpenFilename** method, a VBA method specifically to display the **Open** dialog box, is discussed in the paragraphs immediately following. The other, using the **Dialogs** object, is a general way to display, at least in theory, any of Excel's dialog boxes. It will be discussed in the following section.

The **GetOpenFilename** method displays the **Open** dialog box. The user can navigate to a different directory, select a file, and press the Open button. But the **GetOpenFilename** method doesn't open the document, it merely captures the filename and directory path. You then use the filename in the **Open** method.

The syntax of the **GetOpenFilename** method is

Application.GetOpenFilename(*FileFilter, FilterIndex, Title, ButtonText,* _
    MultiSelect)

The optional *FileFilter* argument is a string specifying the file type(s) that will be displayed in the **Open...** dialog box. For example, you can choose to display only text files.

The optional *Title* argument allows you to specify a custom title for the dialog box. The default is "Open".

See Excel's On-line Help for further information about the arguments.

```
Sub OpenDemo4()
FileToOpen = Application.GetOpenFilename(Title:="Open Demo", _
    FileFilter:="Text Files (*.txt), *.txt")
If FileToOpen <> False Then
Workbooks.Open FileToOpen
MsgBox "The file that was opened is " & FileToOpen & Chr(13) & _
    Chr(13) & "Now continue with the procedure"
End If
End Sub
```

Figure 17-23. Using **GetOpenFileName** to open a workbook.

There is a parallel **GetSaveAsFilename** method that displays the **Save As...** dialog box and returns the filename and directory path.

Using Excel's Built-In Dialog Boxes

You can use the **Dialogs** collection object to return one of Excel's built-in dialog boxes. The syntax is

Application.Dialogs(*xlDialogConstant*)

where *xlDialogConstant* is a number or enumerated constant to return one of Excel's built-in dialog boxes. Each enumerated constant is formed from the prefix "*xlDialog*" followed by the name of the dialog box. For example, to return the **Open** dialog box, the constant is *xlDialogOpen*. You can also use the expression **Application.Dialogs**(1) to display the **Open** dialog box. To obtain a complete list of dialog box constants, search for "dialog box constants" in Excel VBA Help.

There is only one method that operates on the **Dialog** object: the **Show** method. The procedure in Figure 17-24 illustrates the code to display the **Open** dialog box.

```
Sub OpenDemo5()
Application.Dialogs(xlDialogOpen).Show
OpenedFile = ActiveWorkbook.Name
MsgBox "The file that was opened is " & OpenedFile & Chr(13) & Chr(13) _
& "Now continue with the procedure"
End Sub
```

Figure 17-24. Using **Dialogs** to open a workbook.

A few of the more useful dialog box constants are shown in Table 17-8.

Table 17-8. Some Built-In Dialog Box Constants

xlDialogConstant	Number
xlDialogFileDelete	6
xlDialogOpen	1
xlDialogPageSetup	7
xlDialogPrint	8
xlDialogPrinterSetup	9
xlDialogPrintPreview	222
xlDialogSaveAs	5
xlDialogSort	39
xlDialogWorkbookNew	302
xlDialogZoom	256

Although, for Excel 2007, the VBA expression **Application.Dialogs.Count** returns the surprising large value 1101, there are only about 125 that are useful in Excel 2003 or 2007/2010. Some of the others in the 1101 are dialog boxes from earlier versions of Excel. For many numeric values of *xlDialogConstant*, the expression **Application.Dialogs(***number***)** does not display a dialog box.

Other Ways to Display Built-In Dialog Boxes

In addition to the **Show** method, there are some other ways to display Excel's built-in dialog boxes.

Using the Execute Method. To display an Excel 2003 dialog box, you can use the **Execute** method, e.g.,

```
CommandBars("Worksheet Menu Bar").Controls("Edit") _
.Controls("Go To...").Execute
```

To specify the menu or the menu command you can either use the menu text or a number. For example, the Go To command is command 16 in the Excel 2003 **Edit** menu, so the following line of code displays the Go To dialog box:

```
CommandBars("Worksheet Menu Bar").Controls("Edit") _
.Controls(16).Execute
```

It's safer to use the menu text, since menus can be customized, as we saw in Chapter 20.

For some dialog boxes, this method may not provide complete functionality. For example, when the Go To dialog box is displayed using this method, the Special button is inactive.

Using SendKeys with Shortcut Keys. If a dialog box has an associated shortcut key, you can use the **SendKeys** method (see the VBA On-line Help for a description of the **SendKeys** method). For example, to display the Format Cells dialog box, for which the shortcut key is Ctrl+1, use the expression

```
Application.SendKeys "^1"
```

This method is applicable to Excel 2003 and Excel 2007/2010. Only a limited number of dialog boxes have associated shortcut keys, though.

Using SendKeys with Access Keys Every menu command can be accessed with Access keys, sometimes called hotkeys. To use an access key at the keyboard, simply hold down the Alt key and press the underlined letter in the menu that contains the command you want, followed by the underlined letter in the command that you want. You can use **SendKeys** to access dialog boxes in this way. For example, to display the Print dialog box, for which the hotkey sequence is Alt+F,P, use the expression

   Application.SendKeys "%fp"

This method can be used to display any dialog box in Excel 2003 or Excel 2007/2010.

Using Excel 4 Macro Commands

We saw in Chapter 6 that Excel 4 macro commands can be used in Excel formulas. They can also be used within VBA procedures. Since there are some Excel 4 macro commands that provide information not available in VBA, it's useful to know how to use them. And a few Excel 4 macro commands execute faster than the corresponding VBA code. (Use the Recorder to create a Page Setup procedure, run it, and you'll see what I mean.)

Appendix G lists some useful Excel 4 macro functions.

Use the expression **Application.ExecuteExcel4Macro(**_string_**)** to run an Excel 4.0 macro function and return the result of the function. The argument _string_ is the Excel 4 macro language function (without the equal sign), enclosed in quotes. For example, the Excel 4 macro function GET.DOCUMENT(50) returns the total number of pages that would be printed based on current settings. As far as I can determine, VBA does not provide this information.

I use the GET.DOCUMENT(50) function in a simple macro to warn me, when I press the Print toolbutton, if a large number of pages are going to be printed. The macro is shown in the following listing, and an example of the message box that is produced is shown in Figure 17-26. In Chapter 19 I describe how to assign this macro to the Print toolbutton.

```
Sub PrintWarning()

Dim NumPages As Integer
Dim ButtonValue As Integer
Dim RtnValu As Integer

NumPages = Application.ExecuteExcel4Macro("get.document(50)")
ButtonValue = 52

If NumPages = 1 Then
    ActiveWindow.SelectedSheets.PrintOut Copies:=1, Collate:=True
    Exit Sub
End If
If NumPages > 1 Then ButtonValue = ButtonValue + 256
RtnValu = MsgBox("You are about to print " & NumPages & " pages." _
    & Chr(13) & Chr(13) & "Continue anyway?", ButtonValue, _
    "PRINTER WARNING")
If RtnValu = 7 Then Exit Sub
ActiveWindow.SelectedSheets.PrintOut Copies:=1, Collate:=True
End Sub
```

Figure 17-25. The PrintWarning VBA code.

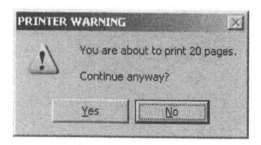

Figure 17-26. Example of the PrintWarning display.

Testing and Debugging

When an error occurs during execution of a procedure, VBA will stop execution and display a run-time error message. There are a large number (over 50) of these run-time error messages. The following shows a few of the more common ones.

Subscript out of range	Attempted to access an element of an array outside its specified dimensions.
Property or method not found	Object does not have the specified property or method.
Argument not optional	A required argument was not provided.

The line of code in which the error occurred will be highlighted, usually in yellow (see Figure 17-27). As a rule, after you have corrected the error in your VBA code, the line will still be highlighted, and you won't be able to run the macro. Press F5 to run the procedure.

Tracing Execution

When your program produces an error during execution, or executes but doesn't produce the correct answer, it is often helpful to execute the code one statement at a time and examine the values of selected variables during execution. If your procedure contains logical constructions (**If** or **Select Case**, for example), simply stepping through code will allow you to verify the logic.

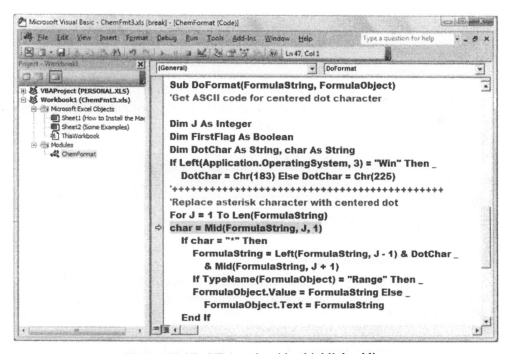

Figure 17-27. VBA code with a highlighted line.

Stepping Through Code

To step through the code of a **Sub** procedure, follow the steps in the following box. There are two ways you can begin the process.

To Step Through VBA Code

1. In Excel 2007/2010, click on the Developer tab of the Ribbon, and click on the Macro button to display the Macro dialog box. In Excel 2003, choose **Macro** from the **Tools** menu and **Macros...** from the submenu.

2. Select the macro in the Macro Name list box and press the Step Into button. This will display the code module containing the procedure. The first line of the procedure will be highlighted, usually in yellow (see Figure 17-27).

3. Choose **Toolbars** from the **View** menu and **Debug** from the submenu to display the Debug toolbar (Figure 17-28).

or...

1. Activate the Visual Basic Editor. The VBA code module must be visible.

2. Choose **Toolbars** from the **View** menu and **Debug** from the submenu to display the Debug toolbar (Figure 17-28).

As you step through the code, the next statement to be executed is highlighted, as shown in Figure 17-27. Use the Step Into toolbutton ⬚ or press F8 to step through the procedure. Press F5 to run the macro from the current line.

Figure 17-28. The VBA Debug toolbar.

Adding a Breakpoint

A breakpoint allows you to halt execution at a specified line of code, rather than having to step through the code from the beginning. There are several ways to add a breakpoint:

- Opposite the line of code where you want to set the breakpoint, click in the gray bar on the left side of the VBA module sheet. The line of code will be highlighted (usually in red-brown) and a breakpoint indicator, a large dot of the same color, will be placed in the margin (see Figure 17-29).

- Place the cursor in the line of code where you want to set a breakpoint.

  Press the Toggle Breakpoint button 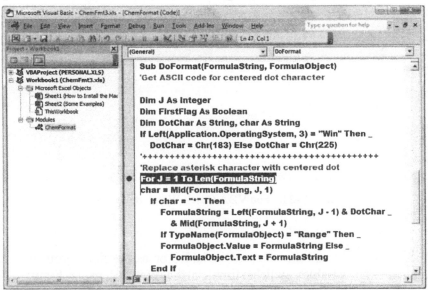 on the Debug toolbar.

- Insert a **Stop** statement in the VBA code. See Appendix E for details.

- Enter a break expression in the Add Watch dialog box (see "Examining the Values of Variables" later in this chapter).

When you run the macro, the code will execute until the breakpoint is reached, at which point execution will stop. You can now step through the code one statement at a time or examine the values of selected variables.

Since you can't "run" a **Function** procedure, the *only* way to step through a **Function** procedure is to add a breakpoint, and then recalculate the formula containing the custom function.

To remove a breakpoint, click on the breakpoint indicator, or place the cursor on the highlighted line and press the Toggle Breakpoint button, or delete a **Stop** statement.

Examining the Values of Variables
Using the Mouse Pointer

You can examine the values of selected variables while in Break Mode. You enter Break Mode if your procedure generated a run-time error and halted, or your procedure reached a line with a breakpoint or a **Stop** statement.

Figure 17-29. VBA code with a breakpoint.

Once in Break Mode, to see the current value of a variable, highlight the variable by double-clicking on it, or simply "hover" the cursor over the variable.

The current value of the variable will be displayed in a yellow "InfoBox" next to the cursor, as illustrated in Figure 17-30.

$$M = \boxed{(Y2 - Y1) / (X2 - X1)}$$
$$\boxed{(Y2 - Y1) / (X2 - X1) = 1.0000000302571E\text{-}08}$$

Figure 17-30. Displaying the value of a variable while in break mode.

Examining the Values of Variables
Using the Watch Pane

You can also display the values of selected variables in the Watch Pane. There are several ways to add variables or expressions to be displayed in the Watch Pane:

- Highlight the variable or expression and then choose **Quick Watch...**

- Press the Quick Watch button 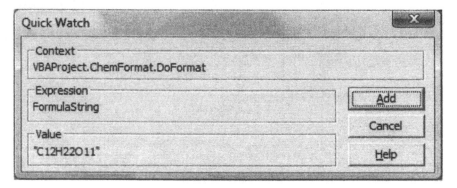 on the Debug toolbar, to display the Quick Watch dialog box (Figure 17-31).

- Highlight the variable or expression and then choose **Add Watch...** from the **Debug** menu to display the Add Watch dialog box (Figure 17-32).

- Press Shift + F9.

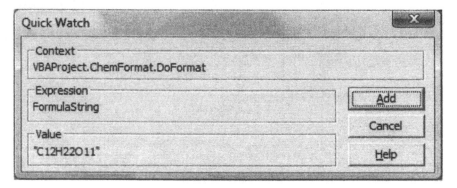

Figure 17-31. The VBA Quick Watch dialog box.

To see the values of the selected variables or expressions, you must be in Step mode. The variables will be listed in the Watches pane (Figure 17-33), which is usually located below the Code window. The current values of the variables will be displayed as you step through the code.

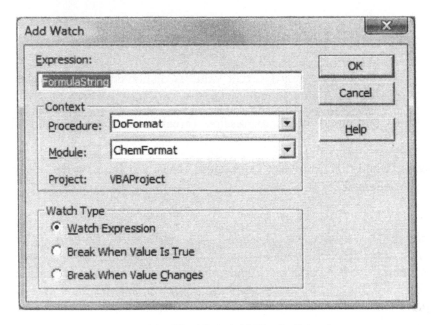

Figure 17-32. The VBA Add Watch dialog box.

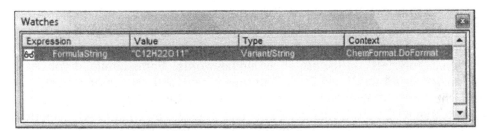

Figure 17-33. The VBA Watches pane.

To remove a variable or expression from the Watches pane, select it in the Watches pane, choose **Edit Watch** from the **Debug** menu, and then press Delete.

Watch expressions are not saved with your code.

Using Conditional Watch

A conditional watch expression causes VBA to enter break mode only when a variable changes in value or when an expression evaluates to **True**.

To establish conditional watch expressions, choose **Add Watch** from the **Debug** menu and press the appropriate Watch Type button (see Figure 17-32). There are three possibilities, which are indicated by different icons in the Watches pane:

    Watch expression (current value is displayed in the Watches pane when VBA enters break mode)

    Conditional break expression (break occurs when expression is True)

    Conditional break expression (break occurs when expression changes)

Excel Tip. You can adjust the widths of the Expression, Value and Context columns in the Watches pane by placing the mouse pointer on the separator bar to the left of the Value or Context header; the pointer will change to the ✛ pointer shape, and you can drag the separator bar to adjust the column width.

Using VBA On-Line Help

To see information about a particular VBA keyword, you can either use VBA's **Help** menu or use the Object Browser. The latter is a more direct route to information about a specific keyword, but the former sometimes provides some additional "tutorials" on the use of certain keywords. For example, if you search for "range" using the Excel 2003 Help menu, you will see, among many other links, the links "Referring to Cells by Using the Range Object", "Referring to Named Ranges", "Referring to Cells and Ranges by Using A1 Notation", "Working with 3-D Ranges" and "Referring to Multiple Ranges".

Using VBA Help (Excel 2007/2010)

Click on **Microsoft Visual Basic for Applications Help** in the **Help** menu, or simply press F1.

To see information about, for example, the Workbook object, enter "workbook object" in the Search box and click on the Search button. The Help screen displays search results for "workbook object". Click on "Workbook Object" in the list of results to display a screen describing syntax, comments and examples.

At the bottom of the page, you can click on the link "Workbook Object members" to see links to the complete list of Methods, Properties and Events.

There are 77 Methods and 96 Properties of the Range object, for example, listed in Excel 2010.

Using VBA Help (Excel 2003)

Click on **Microsoft Visual Basic Help** in the **Help** menu, enter a term in the Search box and press Enter. You can also highlight a term in existing code and press F1 in order to see the Help file for that keyword.

To see information about, for example, the Workbook object, enter "workbook object" in the Search box and click on the Search button to display search results for "workbook object". Click on the "Workbook Object" link in the list of results to display syntax, comments and examples.

Depending on the object selected, "Properties", "Methods" or "Events" links will appear the top of the page. You can click on the link to see a complete list

Using the Object Browser

Another way to get to the Help screen for an Object, Property or Method is to use VBA's Object Browser. Choose **Object Browser** from the **View** menu, or simply press F2 to display the Object Browser screen. Choose "Excel" in the Project/Library list box at the top left. (It should say "<all libraries>" at this initial point.) The "Classes" list box will now display a list of Excel's objects (objects are indicated by 🗐). You can scroll through the list and select an object, whereupon the list of Properties and Methods for the selected object will appear in the list on the right. In Figure 17-34 the Range object has been selected and the Methods (indicated by a ➾) and Properties (indicated by 🖼) are shown.

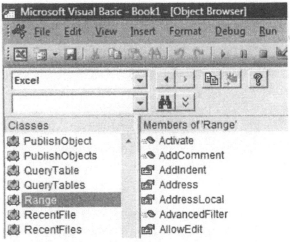

Figure 17-34. The VBA Object Browser window.

To sort the list of Properties and Methods, right-click on the grey header (containing, in this case, "Members of Range") and click on Group Members in the shortcut menu, as illustrated in Figure 17-35.

To display the Help file for a Property or Method, select it and then press the Help button (the [?] button above the list) to display the Help file for the selected keyword, as illustrated in Figure 17-36.

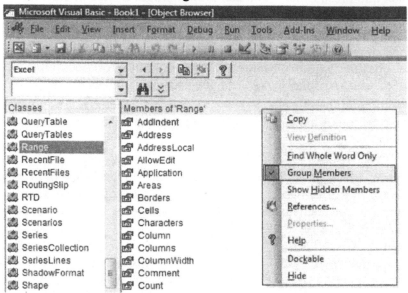

Figure 17-35. The VBA Object Browser window, with Properties and Methods grouped.

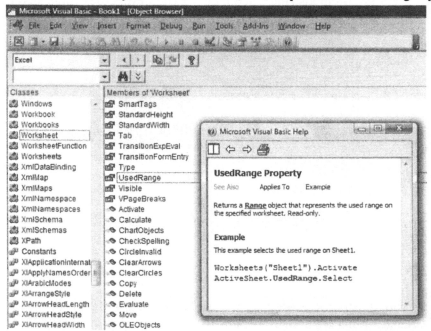

Figure 17-36. A help file displayed by using the Object Browser.

To display the list of VBA functions, choose VBA from the Project/Library list box. You can then select functions by category: DateTime, Information, Math, Strings, etc. Figure 17-37 illustrates the list of VBA Math functions.

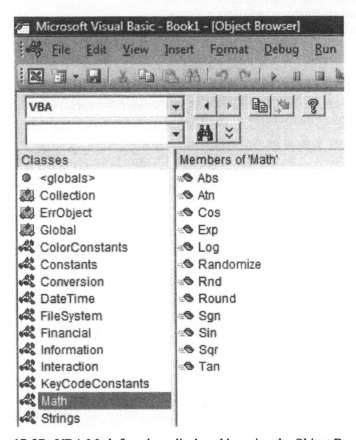

Figure 17-37. VBA Math functions displayed by using the Object Browser.

Some Ways to Improve
the Performance (i.e., Speed)
of Your VBA Procedures

- Declare variable types with **Dim** statements.
- . Avoid using **Select** if possible
- Turn off screen updating and automatic recalculation, as shown in Figure 17-38.

```
'Save current state of some settings, then turn off to enable code to run
faster
SavedScreenUpdateState = Application.ScreenUpdating
SavedCalcState = Application.Calculation
Application.ScreenUpdating = False
Application.Calculation = xlCalculationManual

'Your VBA code

'Restore state to user's settings
Application.ScreenUpdating = SavedScreenUpdateState
Application.Calculation = SavedCalcState
```

Figure 17-38. VBA code to turn off screen updating.

18

Working with Arrays
in VBA

Many scientists make extensive use of arrays in their calculations. Because arrays in VBA can be confusing, this chapter provides detailed coverage of this important topic.

Visual Basic Arrays

If you're familiar with other programming languages, you are probably familiar with the concept of an array. An array is a collection of related variables denoted by a single name, such as Sample. You can then specify any element in the array by using an index number: Sample(1), Sample(7), etc.

Dimensioning an Array

The **Dim** statement is used to declare the size of an array and also the number of dimensions. Since we'll be working with Excel, one-and two-dimensional arrays will be commonplace, and occasionally, even a three-dimensional array might be useful. It's considered good programming practice to put the **Dim** statements at the beginning of the procedure.

To create a 2-D array with dimensions of 500 rows by 2 columns, use the statement

> **Dim** Spectrum (500, 2)

Since multidimensional arrays such as the one above can use up significant amounts of memory, it's important to define the data type of the variable. The complete syntax of the **Dim** statement is

> **Dim** *VariableName* (*Lower* **To** *Upper*) **As** *Type*

The optional *Lower* **To** can be omitted. *Type* can be **Integer**, **Single**, **Double**, **Variant**, etc. (See the complete list of data types in "VBA Data Types" in Chapter 17.) If an array is to hold values of different data types, e.g., "Bromobenzene" and 1.495, then the data type must be **Variant**.

Several variables can be dimensioned in a single **Dim** statement, but there must be a separate **As** Type for each variable.

Unless specified otherwise, VBA arrays begin with an index of 0. Thus the statement

> **Dim** Sample(100)

establishes array storage for 101 elements, Sample(0) through Sample(100). Since worksheet ranges, worksheet functions and worksheet arrays use (or assume) a lower array index of one, I strongly recommend that you always set the lower array index of all arrays to 1.

There are two ways to specify the lower bound of an array in the **Dim** statement. The **Option Base 1** statement specifies that all arrays in a module begin with a lower index of 1. The **Option Base 1** statement is used at the module level: that is, it must appear in a module sheet ahead of any procedures.

Alternatively, you can use the statement

> **Dim** Sample (1 **To** 100)

if you want to specify the lower array index of only a particular array.

Use the Name of the Array Variable to Specify the Whole Array

You can refer to the complete array by using the array variable name in your code. You can include the parentheses or omit them.

Dynamic Arrays

If you don't know what array size you will need to handle a particular problem, you can create a *dynamic array*. This will allow you to declare a variable as an array but set its size later. Dimension the array using the **Dim** command, using empty parentheses; use the **ReDim** command later to specify the array size, e.g.,

```
Dim MeanX(), MeanY()
   :
'Get number of cells to use in calculation
Ncells = XValues.Count
ReDim MeanX(Ncells), MeanY(Ncells)
```

You can also use the **ReDim** command to change the number of dimensions of an array.

The **ReDim** command can appear more than once in a procedure.

Preserving Values in Dynamic Arrays

You can also use the **ReDim** statement to resize an existing array during execution of the code. For example, your procedure initially may need an array of 5 × 2000 elements, but later, during execution on a particular data set, a much smaller number of array elements may be needed. You can free up memory by resizing the array during execution. However, the **ReDim** statement re-initializes the array (numbers are set to zero, strings to null, objects to **Nothing**, for example). You can preserve the values in an existing array by using the **Preserve** keyword, e.g.,

> **Dim MeanX(), MeanY()**
>
> :
>
> **ReDim Preserve MeanX(Ncells / 2), MeanY(Ncells / 2)**

But, there's a limitation. Only the upper bound of the last (i.e., outermost) dimension of a multidimensional array can be changed.

If you use **Preserve**, you can't use the **ReDim** command to change the number of dimensions of an array.

Using Arrays in Sub Procedures: Passing Values from Worksheet to VBA Module

There are at least two ways to get values from a worksheet into a VBA array in a **Sub** procedure. You can either set up a loop to read the value of each worksheet cell and write the value to the appropriate element of an array that you had previously dimensioned, or you can assign the VBA array to a worksheet range. These two methods are described in the following sections.

Using a Loop to Transfer Values from a Worksheet to a VBA Array

The **Sub** procedure shown in Figure 18-1 reads values from worksheet cells and writes them to elements of a VBA array by means of a loop. Either the **Cells** keyword or the **Range** keyword can be used; in the example that follows, the **Range** keyword is used, with the concatenation operator, to access the appropriate cell in the range. In this example the values to be copied are in column A, beginning in row 2. To illustrate an additional point, the values extend down to a row that is determined by the VBA code (a piece of code that you may find useful later).

The code to obtain the row number of the last-used row in the block of cells in column A was obtained by using the Recorder to record the VBA code corresponding to Ctrl + Shift + (down arrow).

A Range Specified in a Sub Procedure Becomes an Array Variable

If a variable in a VBA **Sub** procedure is set equal to a range of cells in a worksheet, that variable can be used as an array. The variable can be equated to a range reference or to a name that refers to a reference.

> TestArray = **Range**("A2:A10")

No **Dim** statement is necessary; it will give a run-time error if used.

A one-row or one-column reference becomes a one-dimensional array; a rectangular range becomes a two-dimensional array of dimensions *array(rows, columns)*.

The lower index of these arrays is always 1. Although arrays created *within* VBA by using a **Dim** statement have a lower array index of zero unless specified otherwise (by means of the **Option Base 1** statement, for example), when you transfer a range of cells from a worksheet to VBA, an array is created with lower array index of 1.

```
Sub ArrayDemo1()
'Reads values from column A, beginning in row 2, into a VBA array.

Dim J As Integer
Dim LastRow As Integer, NCells As Integer
Dim TestArray()

LastRow = Range("A2").End(xlDown).Row
NCells = LastRow - 1
ReDim TestArray(NCells)
For J = 2 To LastRow
TestArray(J - 1) = Range("A" & J)
Next J
MsgBox "First array element = " & TestArray(1) & Chr(13) & _
"Last array element = " & TestArray(NCells)
End Sub
```

Figure 18-1. Reading values into a VBA array.

The **Set** command is used to create an object variable in VBA. The code in Figure 18-2 creates an array of "cell" objects in VBA. You can access any of the properties of the array elements, as shown in Figure 18-2

```
Sub ArrayDemo2()
' Uses the range A1:A12 as an array

Set TestArray = Range("A2:A12")
MsgBox "Fifth array element = " & TestArray(5)
End Sub
```

Figure 18-2. Another way to create a VBA array.

Some Worksheet Functions Used Within VBA Create an Array

If you use a worksheet function within VBA that returns an array, the lower array index will be 1. Such worksheet functions include: LINEST, TRANSPOSE, MINVERSE, MMULT. Another function that returns an array is the VBA function **Caller** when used with a menu command or toolbutton.

Using Arrays in Sub Procedures: Passing Values from VBA Module to Worksheet

There are at least two ways to send values from a VBA array to a worksheet. You can either use a loop to write the value of each array element to a worksheet cell, or you can assign the value of the VBA array to a worksheet range.

Using a Loop to Transfer Values from a VBA Array to a Worksheet

To write the value of each array element to a worksheet cell, you can use either the **Cells** keyword or the **Range** keyword. In the example shown in Figure 18-3, the **Cells** keyword is used.

This method is straightforward, although sometimes not as convenient as the method in the following section.

```
Sub ArrayDemo3()
'Demo to illustrate writing array values to a sheet by means of a loop.

Dim J As Integer
Dim TestArray(13)

'Puts the numbers 11, 12, 13...22  in a VBA array.
For J = 1 To 12: TestArray(J) = 10 +J: Next
'Then writes the array elements to cells B2:B13.
For J = 1 To 12
Cells(J, 2).Value = TestArray(J)
Next J
End Sub
```

Figure 18-3. Passing values from a VBA array to a worksheet.

Equating a VBA Variable to a Worksheet Range

In the example shown in Figure 18-4, a 2-D range on a worksheet becomes an array in VBA. Then the array elements are written back to the worksheet with a single line of code.

This is much more convenient than the loop method, where two nested loops would be required to read or write a 2-D range. However, a problem arises when you use this method with a 1-D range, as described next.

```
Sub ArrayDemo4()
'Demo to illustrate writing array values to a sheet by writing the array.

TestArray = ActiveSheet.Range("A2:B22")
Range("D2:E22") = TestArray
End Sub
```

Figure 18-4. Another way to write values from a VBA array to a worksheet.

A One-Dimensional Array Assigned to a Worksheet Range Can Cause Problems

Arrays can cause some confusion when you write the array back to a worksheet by assigning the value of the array to a worksheet range.

VBA considers a one-dimensional array to have the elements of the array in a row. This can cause problems when you select a range of cells in a column and assign an array to it, as in the **Sub** procedure shown in Figure 18-5.

```
Option Base 1
Sub ArrayDemo5()
'Illustrates that 1-D VBA arrays have elements in a row, not a column.
'Was expected to put the numbers 11, 12, 13... in E1:E10.
'But instead writes element(1) in all cells.

Dim J As Integer
Dim TestArray(10)

'Puts the numbers 11, 12, 13... in a VBA array.
For J = 1 To 10: TestArray(J) = 10 + J: Next
'Then writes the array elements to cells E1:E10.
Range("E1:E10").Value = TestArray
End Sub
```

Figure 18-5. From a VBA array to a worksheet: the row-column problem.

If you run the preceding **Sub** procedure, you will find that cells E1 through E10 will all contain 11, the first element of the array. That's because you have tried to put a "horizontal" array of values into a "vertical" range of cells. (And even more confusing, if you had not specified **Option Base 1** in your module, the empty zeroth element of the array would have been written to all the cells, and it would have appeared that the procedure had failed to do anything.)

However, if you write the array to a row of cells instead of a column, e.g.,

> **Range**("E1:N1").**Value** = TestArray

then each cell of the range will receive the correct array value.

There are at least three ways to "work around" this problem caused by "horizontal" and "vertical" arrays. One way is to use a loop to write the elements of the array to individual worksheet cells in a column.

A second way is to specify both the row and the column dimensions of the array, so as to make it an array in a column, as illustrated in the **Sub** procedures shown in Figure 18-6 and 18-7.

```
Sub ArrayDemo6()
'Second method to "work around" the row-column problem:
'specify the row and column dimensions.

'Puts the numbers 11, 12, 13... in a VBA array.
Dim TestArray(10, 1)
For J = 1 To 10: TestArray(J, 1) = 10 + J: Next
'Then writes the array elements to cells E1:E10.
Range("E1:E10").Value = TestArray
End Sub
```

Figure 18-6. A second way to "work around" the row–column problem.

If you don't use the **Set** keyword when you assign a range to a variable, you can avoid the "row-column" problem by specifying both row and column array indices, as in Figure 18-7.

```
Sub ArrayDemo7()
' Use the range A2:A13 as an array without the Set keyword
' Specify both row and column indices of the array.

TestArray = Range("A2:A13")
MsgBox "Fifth array element = " & TestArray(5, 1)
End Sub
```

Figure 18-7. A second way to "work around" the row–column problem.

A third way is to use the TRANSPOSE worksheet function (Figure 18-8):

```
Sub ArrayDemo8()
'Another method to "work around" the row-column problem:
'use Transpose.  Note that Transpose creates a 1-base array.

Dim J As Integer
Dim TestArray(10)
'Puts the numbers 11, 12, 13... in a VBA array.
For J = 1 To 10: TestArray(J) = 10 + J: Next
NewArray = Application.Transpose(TestArray)
Range("E1:E10").Value = NewArray
End Sub
```

Figure 18-8. A third way to "work around" the row–column problem.

Using Arrays in Function Procedures: From Worksheet To Module

You can create **Function** procedures that use arrays as arguments, or return an array as a result.

A Range Passed to a Function Procedure Automatically Becomes an Array

If a range argument is passed in a function macro, the range can be used as an array in the VBA procedure. No **Dim** statement is necessary. Thus the expression

Function Deming(XValues, YValues)

passes the worksheet ranges XValues and YValues to the VBA procedure where they can be used as arrays.

A one-row or one-column reference becomes a one-dimensional array; a rectangular range becomes a two-dimensional array of dimensions *array(rows, columns)*.

The **Function** procedure in Figure 18-9 is identical to Excel's INDEX worksheet function: it passes a range and a number as arguments and returns the specified element of the array. The name of the array in the procedure must be the name of the placeholder argument of the function.

```
Option Base 1
Function ArrayDemo9(range, num)
'Shows the use of a range of cells as an argument
'Returns the element of the array specified by num.

ArrayDemo9 = range(num)
End Function
```

Figure 18-9. Passing an array as an argument in a **Function** procedure.

Passing an Indefinite Number of Arguments Using the ParamArray Keyword

Occasionally a **Function** procedure needs to accept an indefinite number of arguments. The SUM worksheet function is an example of such a function; its syntax is =SUM(number1,number2,...). To allow a **Function** procedure to accept an indefinite number of arguments, use the **ParamArray** keyword in the argument list of the function, as in the following expression

Function ConcatenateSpecial(**ParamArray** String1())

Only one argument can follow the **ParamArray** keyword, and it must be the last one in the function's list of arguments. The argument declared by the **ParamArray** keyword is an array of **Variant** elements. Empty parentheses are required.

Interestingly, even if you use the **Option Base 1** statement, the lower bound of the array is zero.

Elements in the array of arguments passed using the **ParamArray** keyword can themselves be arrays.

Using Arrays in Function Procedures: Returning an Array of Values as a Result

There are several ways to enable a **Function** procedure to return an array of values. The most obvious is to assemble the values in an array and return the array. The procedure shown in Figure 18-10 illustrates a function that returns an array of two values. The function requires no arguments. The user must select a range of two cells, enter the function and press Ctrl + Shift + Enter.

```
Option Base 1
Function ArrayDemo10()
'Shows how to return an array of results
'The returned array is in a row of cells

Dim results(2) As Double
valu1 = 5: valu2 = 0.12345
results(1) = valu1
results(2) = valu2
ArrayDemo10 = results
End Function
```

Figure 18-10. A **Function** procedure that returns an array.

To use this function the user must select a horizontal range of cells, enter the function, and press Ctrl+Shift+Enter.

To return a two-dimensional array of results, simply create a two-dimensional array within the VBA procedure, populate the array with the desired values, and return the array, as shown in Figure 18-11.

```
Option Base 1
Function ArrayDemo11()
'Shows one way to return a 2-D array of values:
'Create an actual array by using the Dim statement.

Dim results(2, 2) As Variant
Dim valu1 As String, valu2 As String
Dim valu3 As Double, valu4 As Double
valu1 = "a": valu2 = "b"
valu3 = 5: valu4 = 0.12345
results(1, 1) = valu1
results(1, 2) = valu2
results(2, 1) = valu3
results(2, 2) = valu4
ArrayDemo11 = results
End Function
```

Figure 18-11. Returning a 2-D array of values.

Again, you must select an appropriate range of cells to contain the result, enter the function, and press Ctrl + Shift + Enter.

To summarize, two things can cause problems with VBA arrays:

- By default, VBA arrays begin with a lower array index of zero. Be sure to use **Option Base 1**.

- One-dimensional VBA arrays have values in a row. This can cause a problem if you write the array to a range of cells in a column.

PART V

SOME
APPLICATIONS
OF VBA

19

Command Macros

A command macro (a VBA **Sub** procedure) can automate any sequence of actions that can be performed by the use of menu commands or keystrokes. Many simple but useful command macros can be created entirely by using the Recorder, as described in Chapter 16. But command macros can also carry out much more complicated actions. In this chapter we'll look at some examples of creating more advanced **Sub** procedures.

Creating Advanced Macros in VBA

Chapter 17 focused on the basic tools needed to create macros to automate chemical worksheet calculations: how to transfer values from a sheet to a VBA module, how to perform calculations within a VBA module, how to perform logical branching and iterative looping, and how to send values back from a VBA module to a worksheet. In this chapter we'll use these tools to create some useful **Sub** procedures.

A Sub Procedure
to Format Text as a Chemical Formula

You'll find this macro useful if you label rows or columns in your worksheets with chemical formulas, such as "$CH_3CH=CH_2$" or "moles of H_3PO_4". Formatting a text label that represents a chemical formula can require extensive subscripting of number characters. Each subscripting action requires the following actions: Select the character(s), choose **Format** → choose **Cells...** → click on Font tab → click on Subscript check box → press OK. The initial version of this macro simply subscripts every character that is a number symbol. Once we've created the basic macro, we'll add some enhancements.

There are four steps in the basic version of this procedure: (i) Obtain the contents of the active cell, (ii) set up a loop so that we can examine each character in turn, (iii) test to see if the character is a number, and (iv) subscript the character.

The VBA keywords required for the first three steps were described in Chapter 17. We'll use the **ActiveCell** keyword to obtain the string, use **For... Next** to create a loop and the **Mid** Function to access each character within the loop and use an **If** statement to test each character. The only bit of code we'll need that wasn't described in Chapter 17 is the code to subscript a character. You could spend some time looking in the VBA On-line Help, but a much easier way to obtain the correct code is to use the Recorder. Simply turn on the Recorder, perform the action of subscripting a character, then go to the Visual Basic Editor and examine the recorded code. You'll find that the expression to subscript specified characters in a text string in the active cell uses the **Characters(*start, length*)** method to specify the characters:

object.Characters(start, length).Font.Subscript = True

Simply **Copy** the code fragment and **Paste** it into your macro. It will have to be modified a bit, as shown in Figure 19-1.

The complete macro is shown in Figure 19-1. Right now you'll have to use the Macro Run dialog box to run the macro, which isn't very convenient. You can assign a shortcut key to the macro, as described in Chapter 16. In Chapter 23 you'll learn how to create a custom tool button and assign the macro to it.

```
Sub ChemicalFormatDemo1()
Dim J As Integer
Dim char As String
For J = 1 To Len(ActiveCell)
    char = Mid(ActiveCell, J, 1)
    If Asc(char) >= 48 And Asc(char) <= 57 Then ActiveCell. _
        Characters(Start:=J, Length:=1).Font.Subscript = True
Next J
End Sub
```

Figure 19-1. A simple **Sub** procedure to format text as a chemical formula.

Adding Enhancements
to the Chemical Format Macro

The simple macro in Figure 19-1 formats the text in a single cell. We'd like our macro to be able to format text in a single cell or in a range of cells. To do this, we simply need to add an outer **For..Each...Next** loop. We'll add the lines of code **For Each** cel **In Selection** and **Next** cel at the beginning and the end, respectively, of the original macro, and change three instances of **ActiveCell** to cel. Note that in the **For Each** cel **In Selection** loop shown in Figure 19-2, the variable cel is an object variable; thus we can use the code

cel.**Characters**(Start:=J, Length:=1).**Font**.**Subscript** = **True**

Secondly, the macro subscripts every number character in a string. We'd like our macro to be able to handle formulas of hydrates (e.g., $CuSO_4 \cdot 5H_2O$) or other text strings containing numbers that should not be subscripted, such as "$CaSO_4 \cdot 1/2H_2O$" or "H_3PO_4 (85%)". The **Sub** procedure in Figure 19-2 includes code that handles these situations, by using a logical variable FirstFlag. Number characters are subscripted only when FirstFlag = **True**, and FirstFlag is set to **False** when any of several special characters are encountered while looping through the string.

Some examples of text formatted by using the ChemicalFormatDemo2 macro are shown in Figure 19-3.

```
Sub ChemicalFormatDemo2()
Dim J As Integer
Dim char As String
Dim FirstFlag As Boolean
Dim cel As Object
For Each cel In Selection
FirstFlag = True
   For J = 1 To Len(cel)
      char = Mid(cel, J, 1)
      If IsNumeric(char) Or char = "." Then
         If FirstFlag = True Then GoTo EndLoop
         If char = "." Then GoTo EndLoop
         cel.Characters(Start:=J, Length:=1).Font.Subscript = True
      Else
         FirstFlag = False
         If char = " " Or char = "*" Or Asc(char) = 165 Or char = "," _
            Or char = "(" Or char = "/" Then FirstFlag = True
      End If
EndLoop:
   Next J
Next cel
End Sub
```

Figure 19-2. The ChemicalFormat macro with some additional features added.

	A
1	$C_{12}H_{22}O_{11}$
2	$CuSO_4 \cdot 5H_2O$
3	$CaSO_4 \cdot 1/2H_2O$
4	H_3PO_4 (85%)
5	$H_2SO_4 + 2NaOH = Na_2SO_4 + 2H_2O$
6	$Fe_{0.95}O$

Figure 19-3. Some examples of formatting with the ChemicalFormat macro.

Adding More Enhancements

Finally, we'd like our macro to be able to format text in a worksheet cell or range, in a chart title, or in a textbox. As well, we don't want the macro to crash if we attempt to run it when no sheet is active, for example. The **Sub** procedure shown in Figure 19-4 uses the VBA keywords **TypeName(*ActiveSheet*)** and **TypeName(*Selection*)** to determine that the correct kind of sheet is active and that an appropriate selection has been made, before attempting to format the selection.

```
Sub ChemicalFormat()
'Formats text, e.g., H2SO4, as a chemical formula (subscripts numbers).
'Operates on a cell, a range of cells, text in a chart, or a textbox.
Dim cel As Object
'MAKE SURE WE ARE ON A WORKSHEET OR CHART SHEET
If TypeName(ActiveSheet) <> "Worksheet" And _
    TypeName(ActiveSheet) <> "Chart" Then Exit Sub
'GO TO CORRECT CODE FOR SELECTION TO BE FORMATTED
Select Case TypeName(Selection)
    Case "Range"    'FORMAT TEXT IN A CELL OR RANGE
        For Each cel In Selection
            Call DoFormat(cel.Value, cel)
        Next cel
        Exit Sub
    Case "AxisTitle"    'FORMAT TEXT IN A CHART
        Call DoFormat(Selection.Characters.Text, Selection)
        Exit Sub
    Case "ChartTitle"  'FORMAT TEXT IN A CHART
        Call DoFormat(Selection.Characters.Text, Selection)
        Exit Sub
    Case "TextBox"  'FORMAT TEXT IN A TEXT BOX
        Call DoFormat(Selection.Characters.Text, Selection)
        Exit Sub
End Select
End Sub
```

Figure 19-4. The ChemicalFormat macro with some additional features added.

```
Sub DoFormat(FormulaString, FormulaObject)
Dim J As Integer
Dim char As String
Dim FirstFlag As Boolean
FirstFlag = True
For J = 1 To Len(FormulaString)
  char = Mid(FormulaString, J, 1)
  If IsNumeric(char) Or char = "." Then
    If FirstFlag = True Then GoTo EndLoop
    If char = "." Then GoTo EndLoop
    FormulaObject.Characters(Start:=J, Length:=1)_
        .Font.Subscript = True
  Else
    FirstFlag = False
    If char = " " Or char = "*" Or Asc(char) = 165 Or char = "," _
        Or char = "(" Or char = "/"  Then FirstFlag = True
  End If
EndLoop:
Next J
End Sub
```

Figure 19-5. The DoFormat subroutine of the ChemicalFormat macro.

To develop the code to handle text in a chart or in a text box, I used a one-line **Sub** procedure containing either

Msgbox TypeName(*ActiveSheet*)

or

Msgbox TypeName(*Selection*)

to display the keywords associated with selected chart elements or a text box. Again, this was faster and more convenient than looking in reference books or using the On-line Help. Once the correct keywords had been found, it was a relatively simple matter to modify ChemicalFormat2 to handle text in different environments.

The inner-loop code in Figure 19-2 was converted into a general subroutine (now called DoFormat) to examine the text, find the number characters, and format them. The main program simply ensures that a worksheet or chart is active, determines the kind of text to be formatted, and then calls the subroutine. The subroutine call passes two arguments, a simple variable containing the text to be examined (FormulaString) and an object variable containing the text to be subscripted (FormulaObject).

Since several different kinds of text were to be formatted, a **Select Case** construction was used. As it turned out, the same syntax is used to subscript text in a chart or in a text box. Nevertheless, the original **Select Case** structure was left in place.

A Sub Procedure
to Apply "Precision as Displayed"
to a Selected Range of Cells

A cell might contain the value 0.0119653101694915 but be number-formatted to display 0.01197. Checking the box for Precision As Displayed in Office Button → Excel Options → Advanced → When Calculating This Workbook (Excel 2007/2010) or in the Calculation tab of the **Options** command in the **Tools** menu (Excel 2003) will convert the cell value to the displayed value. Although you might fear that formulas are converted to values, that's not the case; formulas remain formulas.

Precision As Displayed converts *all* values on a sheet to the number of figures displayed. But sometimes we'd like to convert only a selected range of cells. This macro applies "Precision as Displayed" to a selected range of cells on a worksheet.

Since there doesn't seem to be a VBA property that returns the as-displayed value of a cell, we'll use the worksheet function TEXT(*value, format*) in a VBA macro. The TEXT function formats a value according to a format-code string and returns it as text. The simple macro based on this is shown in Figure 19-6. To use it, select a range of cells, format them to display the desired number of figures, and run the macro.

```
Sub PrecisionAsDisplayedDemo()
Dim valu As Double
Dim fmt As String

For Each cel In Selection
   valu = cel.Value
   fmt = cel.NumberFormat
   cel.Value = Application.Text(valu, fmt)
Next cel
End Sub
```

Figure 19-6. The PrecisionAsDisplayedDemo macro.

Once we've determined that the macro operates as desired, we can add some enhancements. I added two message boxes. The first one warns the user that the change is irreversible. The second one warns the user that cells containing formulas have been selected; it also allows the user to opt to convert formulas to values-as-displayed.

```
Sub PrecisionADisplayed()
Dim ButtonValu As Integer
Dim valu As Double
Dim fmt As String

ButtonValu = MsgBox("Selected data will permanently lose accuracy." _
    & Chr(13) & Chr(13) & "OK to continue?", 49, _
    "PRECISION AS DISPLAYED")
If ButtonValu = 2 Then Exit Sub

For Each cel In Selection
  If cel.HasFormula = True Then
    ButtonValu = MsgBox("Selected range contains one or more formulas." _
    & Chr(13) & Chr(13) & "OK to convert formulas to values?", 49, _
    "PRECISION AS DISPLAYED")
    If ButtonValu = 2 Then Exit Sub Else Exit For
  End If
Next cel

For Each cel In Selection
  valu = cel.Value
  fmt = cel.NumberFormat
  cel.Value = Application.Text(valu, fmt)
Next cel
End Sub
```

Figure 19-7. The final PrecisionAsDisplayed macro.

A Sub Procedure
to Apply Data Labels in a Chart

As described in Chapter 4, the Data Labels tab in the Format Data Series dialog box allows you to add data labels to an XY chart. But only the *x*-values or the *y*-values can be used as labels. You'd probably like to use some other text as data labels. For example, in Figure 19-8, the *x*-values and *y*-values for a chart are in columns B and C. We would like to use the values in column A as data labels. You can do this manually, by adding the data labels, and then manually editing each one to enter the text you want, but a macro will make the task much easier.

	A	B	C
1	peptide	clog P	t_R (exp)
2	AA	-0.74	3.05
3	AG	-1.28	2.66
4	AF	0.95	10.71
5	YL	1.86	11.57
6	DD	-1.98	2.63
7	ML	-0.17	11.94
8	WW	2.18	15.84
9	GM	-1.90	7.30
10	GH	-1.89	2.90
11	GL	-0.08	9.74
12	WF	2.41	15.49

Figure 19-8. Data for a chart. The text values to be used as labels are in column A.

In the following section we will develop a **Sub** procedure to assign text values in a range of cells as the data labels for an XY chart. Here's how we'll begin the process of creating the macro: we'll use the Recorder to obtain the code for adding data labels, selecting a data label and entering text. We'll use this code as a framework on which to construct our macro.

```
Sub Macro1()
ActiveSheet.ChartObjects("Chart 1").Activate
ActiveChart.SeriesCollection(1).Select
ActiveChart.SeriesCollection(1).ApplyDataLabels _
Type:=xlDataLabelsShowValue, AutoText:=True, LegendKey:=False
ActiveChart.SeriesCollection(1).DataLabels.Select
ActiveChart.SeriesCollection(1).Points(4).DataLabel.Select
Selection.Characters.Text = "abc"
Selection.AutoScaleFont = False
With Selection.Characters(Start:=1, Length:=3).Font
        .Name = "Geneva"
        .FontStyle = "Regular"
        .Size = 10
        .Strikethrough = False
        .Superscript = False
        .Subscript = False
        .OutlineFont = False
        .Shadow = False
        .Underline = xlUnderlineStyleNone
        .ColorIndex = xlAutomatic
    End With
    ActiveChart.ChartArea.Select
End Sub
```

Figure 19-9. Macro1 as recorded.

The code shown in Figure 19-9 was obtained when the Recorder was used to record the following actions: selecting a chart, selecting a data series, choosing **Selected Data Series** from the **Format** menu and applying data labels, selecting a data label, and typing the text "abc" in place of the original label. In this **Sub** procedure we have almost all the code we need to create the initial version of our macro.

In the first version of our macro, the chart must be the active document, and the user must have selected the data series to which the data labels are to be attached before running the macro. In a later version we can allow the user to select the data series while the macro is running.

The four lines of code from Macro1 that we'll use (after modifying them) are as follows:

```
ActiveChart.SeriesCollection(1).ApplyDataLabels _
Type:=xlDataLabelsShowValue, AutoText:=True, LegendKey:=False

ActiveChart.SeriesCollection(1).DataLabels.Select

ActiveChart.SeriesCollection(1).Points(4).DataLabel.Select

Selection.Characters.Text = "abc"
```

To complete our macro, we'll need to add some code (i) so that the user can specify a range of cells that contain the labels, (ii) to set up a loop so that we can loop through all the data points, and (iii) to change **SeriesCollection(1)** to the general case.

To specify the range of cells to use as data labels, we'll use the **InputBox** method. We'll define the range as an object variable, so that we can use a **For...Each...Next** loop. For a code example, see "You Can Define Your Own Objects" in Chapter 17. To convert **SeriesCollection(1)** into the general case, remember that a specific item in a collection can be referred to by its index number or by its name. We'll use the code

```
SeriesName = Selection.Name
```

and

```
ActiveChart.SeriesCollection(SeriesName)
```

to specify the selected data series.

After the macro was completed, some testing revealed that in most cases, the code containing the **Select** keyword could be eliminated. The completed macro is shown in Figure 19-10.

```
Sub Labeler()
Dim SeriesName As String
Dim LabelRange As Object, cel As Object
Dim J As Integer

' Get the name of the chart data series.
SeriesName = Selection.Name

' Input the range of cells to be used as labels.
Set LabelRange = Application.InputBox("Select the range of cells
containing the labels", "CHART LABELER STEP 1 OF 1", , , , , , 8)

' Apply the standard data labels.
ActiveChart.SeriesCollection(SeriesName).ApplyDataLabels _
Type:=xlDataLabelsShowValue, AutoText:=True, LegendKey:=False

' Set up a loop to replace data labels with new ones.
J = 1
For Each cel In LabelRange
    ActiveChart.SeriesCollection(SeriesName).Points(J). _
DataLabel. Characters.Text = cel.Value
    J = J + 1
Next cel
End Sub
```

Figure 19-10. The completed ChartLabeler macro.

There's more work to be done on this macro, since if the user has not selected a data series, the macro fails. If you wish, you can modify the macro by adding code similar to the first lines in the procedure shown in Figure 19-4.

A more sophisticated version of the macro (Chart Labeler 2.xls) can be found on the CD-ROM that accompanies this book. The macro installs a new menu command, **Add Data Labels...**, in the Add-Ins tab (Excel 2007/2010) or in the Chart menu (Excel 2003), as shown in Figure 19-11.

The macro makes use of a custom dialog box to pass user-selected options to the macro. The dialog box is shown in Figure 19-12. The user enters a range of cells on the worksheet to be used as data labels, such as $A$2:$A$12 in Figure 19-10. The label position (Above, Below, Centered, Left, Right) can be specified, and the labels can be either (a) references linked to the worksheet or (b) simple text. Only in the latter case can the data labels be formatted.

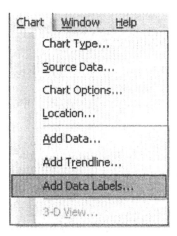

Figure 19-11. New menu command installed by the Chart Labeler 2 macro.

The chart with data labels added is shown in Figure 19-13. The position of several of the data labels was adjusted manually to avoid overlapping text.

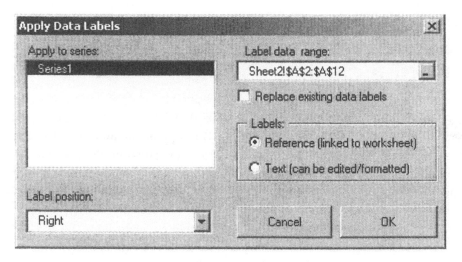

Figure 19-12. The Chart Labeler 2 dialog box.

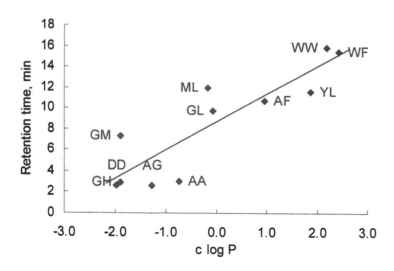

Figure 19-13. A chart with data labels.

Quick Chart:
Create Excel 2007/2010 Charts
with One Click

If you've read Chapters 4 and 5 on Excel 2007/2010 and Excel 2003 charts, you probably know that although it's possible to change the default chart type in Excel 2003, and even set your own chart preferences as the default, thus making the Chart Wizard a one-step wizard, it's not possible to do that in Excel 2007/2010. In Excel 2007/2010, you can save a particular chart as a template, but to create a new chart using that template requires 6 clicks (Insert → Other Charts → All Chart Types… → Templates → select desired template → OK).

The following simple macro allows you to create a chart with one toolbutton click. You can incorporate your own chart preferences in the code.

The VBA code for the Quick Chart procedure is shown in Figure 19-14. I provided a bit of error-checking at the beginning of the procedure, in case the user selected something other than a worksheet, or selected a range of empty cells. The Recorder was used to provide most of the code that creates the chart.

I have shown two ways to incorporate chart preferences in the procedure. In the first method, shown in the code following "MODIFY THE CHART USING THESE RECORDED PREFERENCE", I used the recorder to obtain the code for modifying the appropriate chart elements. My preferences include tick marks inside, no legend, smaller marker size and thinner line. You'll have to use the Recorder if you want to modify or add preferences. This method has the capability to handle all the chart series in the chart. In the second method, the single line of code following "OR USE PREFERENCES FROM A SAVED TEMPLATE", I used the Recorder to obtain the code for using preferences from a chart template that I had created previously. The code providing the chart template is "hard-wired"; if you want to change templates, or if you change the location or name of the template file, you'll have to use the Recorder to get the directory path to the template.

The Quick Chart procedure on the CD-ROM that accompanies this book uses only the first method.

```vba
Private Sub QuickChartCMD()
Dim NumBlanks As Integer
Dim ChartRange As String
Dim cel As Object, ser As Object

'   MAKE SURE WE ARE ON A WORKSHEET
If TypeName(ActiveSheet) <> "Worksheet" Then
    MsgBox "A worksheet must be the active sheet.", 16,_
        "QUICK CHART ERROR MESSAGE"
    Exit Sub
End If

'   MAKE SURE WE HAVE NOT SELECTED EMPTY CELLS
If TypeName(Selection) <> "Range" Then GoTo ExitError
For Each cel In Selection
    If cel.Formula = "" Then NumBlanks = NumBlanks + 1
Next
If NumBlanks = Selection.Count Then GoTo ExitError

'   CREATE THE CHART
    ChartRange = Selection.Address(True, True, 1, True)
    ActiveSheet.Shapes.AddChart.Select
    ActiveChart.SetSourceData Source:=Range(ChartRange)

'   MODIFY THE CHART USING THESE RECORDED PREFERENCES
For Each ser In ActiveChart.SeriesCollection
    With ser.Border
        .ColorIndex = xlAutomatic
        .Weight = xlThin
        .LineStyle = xlContinuous
    End With
    With ser
```

```
            .MarkerBackgroundColorIndex = xlAutomatic
            .MarkerForegroundColorIndex = xlAutomatic
            .MarkerStyle = xlAutomatic
            .Smooth = True
            .MarkerSize = 5
            .Shadow = False
        End With
    Next ser
    ActiveChart.Axes(xlValue).MajorGridlines.Delete
    ActiveChart.Axes(xlValue).MajorTickMark = xlInside
    ActiveChart.Axes(xlCategory).MajorTickMark = xlInside
    ActiveChart.Legend.Delete

    '  OR USE PREFERENCES FROM A SAVED TEMPLATE
     ActiveChart.ApplyChartTemplate _
        ("C:\Users\E. Joseph Billo\AppData\Roaming\Microsoft\Templates
          \Charts\EJB Custom 1.crtx")
    Exit Sub
    ExitError:
        MsgBox "You must select cells to plot.", 16, _
              "QUICK CHART ERROR MESSAGE"
    End Sub
```

Figure 19-14. The Quick Chart macro.

This procedure can be assigned to a custom button. The VBA code to install the custom toolbutton and put the ChartWizard image on it is on the CD-ROM that accompanies this book.

20

Custom Functions

Chapter 16 provided an introduction to **Sub** procedures and **Function** procedures. By now it should be clear that a **Sub** procedure (a command macro) is a computer program that you "run"; it can perform actions such as formatting, opening or closing documents, etc. A **Function** procedure (a user-defined function) is a computer program that calculates a value and returns it to the cell in which it is typed. User-defined functions can add convenience and functionality to your spreadsheet calculations.

Some Additional Tools for Creating Custom Functions

The following sections illustrate some features that can be incorporated into advanced custom functions.

Arrays as Arguments or as Return Values

Chapter 17 illustrated how a custom function can take an array as argument or return an array as result. Briefly, if the user enters a range of cells as an argument in a custom function, that argument can be treated as an array in VBA. If the custom function returns an array of values, the user must select an appropriate range of cells, enter the formula, and press Ctrl+Shift+Enter.

Returning an Array Result
in Either a Row or a Column

VBA considers a one-dimensional array to have the elements of the array in a row. Thus, if your custom function returns an array of three values, the user must select a horizontal range of three cells (1R × 3C) to receive the results. If the user selects a vertical range of three cells (3R × 1C), enters the function, and presses Ctrl+Shift+Enter, the first element of the array will be entered into all three cells.

You can use the **Caller** property to determine whether an array formula is entered in a row or a column. The **Caller** property of the **Application** object returns information about how a Visual Basic procedure was called; depending on the caller, different information is returned. If the procedure is called by a custom function, a **Range** object is returned, specifying the cell or cells in which the formula was entered. The following code (which assumes that the array *result* is one-dimensional) allows the user to select either a horizontal or a vertical range for the result.

```
If Application.Caller.Rows.Count > 1 Then _
    result = Application.Transpose(result)
DemoFunction = result
End Function
```

Figure 20-1. Part of a **Function** procedure that returns an array in either a row or a column.

Returning an Error Value

The case may arise in which a custom function is unable to calculate a result: for example, an argument supplied to the function is out of range, or a calculation within the procedure results in an error. In a case such as this, your custom function should return the appropriate Excel error value, such as #DIV/0!. You do this by using the CVErr function. The syntax is CVErr(*errornumber*), where *errornumber* can be one of the following *XlCVError* constants: xlErrDiv0, xlErrNA, xlErrName, xlErrNull, xlErrNum, xlErrRef, xlErrValue.

The following example illustrates the use of the CVErr function.

```
'Exit here if a root is found
If Abs(X3 - X1) < tolerance Then DemoFn = X3: Exit Function
Next J
'Exit here with error value if no root found after 100 iterations
DemoFn = CVErr(xlErrNA)
End Function
```

Figure 20-2. Part of a **Function** procedure that returns an error value.

A Custom Function that Takes an Optional Argument

A custom function can have optional arguments. Use the **Optional** keyword in the list of arguments to declare an optional argument. The optional argument or arguments must be last in the list.

You will need to determine whether the user provided or omitted optional arguments in a procedure by using the **IsMissing** keyword. As well, you may need to provide a default value if an argument is omitted.

In the example shown in Figure 20-1, the custom function MolWt calculates the formula weight of a compound from *formula* (a text string in a cell). The user can specify, by the optional argument *decimals*, the number of decimal places to be displayed in the returned value.

The MolWt function is described later in this chapter. It can be found in the workbook Molecular Weight Calculator in the CD-ROM that accompanies this book.

```
Function MolWt(formula As String, Optional decimals)
' If decimals is omitted, returns value as a number.
' If decimals is specified, returns value as text,
' formatted with specified number of decimals.
' Maximum number of decimal places allowed is determined in subroutine
' by decimal places of atomic weights.
:
:
:
```

Figure 20-3. Part of a **Function** procedure with an optional argument.

A Custom Function that Takes an Indefinite Number of Arguments

Some of Excel's built-in worksheet functions can accept an indefinite number of arguments. The SUM function is an example of such a function; its syntax is =SUM(number1,number2,...). To create a custom function with an indefinite number of arguments, use the **ParamArray** keyword. An argument declared with **ParamArray** is an array of **Variants**, and each element of the array can itself be an array.

The following example illustrates a custom function that accepts a number of worksheet ranges as arguments and combines them into a single array. This custom function is designed to be used with Excel's LINEST function (see Chapter 14). LINEST requires that the *known_x's* argument be a contiguous range of cells; you can't use a non-adjacent selection as the argument, even if the ranges are enclosed in parentheses. But when you perform multiple linear regression using several columns of x-values, oftentimes those columns of x-values are not contiguous, as illustrated in Figure 20-2. The expression =LINEST(E4:E13,(A4:A13,C4:C13),1,1) returns the #REF! error message. You could rearrange the worksheet to have all the columns of x values adjacent, but with the custom function shown in Figure 20-5 you can combine the separated ranges into a single array and use it as the known_x's argument of LINEST.

Since LINEST requires that there be the same number of values for each independent variable, we first check to make sure that each range in the array of arguments has the same number of rows. An interesting point: Although we specified **Option Base** 1, the **ParamArray** keyword produces an array whose lower array index is zero. Thus we must loop from zero to **UBound**(rng). Note that **UBound**(rng) is the number of elements in the array, while rng(z).**Columns.Count** returns the number of columns in an individual element in the array.

	A	B	C	D	E	F
3	x1	Intermediate Calculation	x2	Intermediate Calculation	x3	Y
4	0.39	etc.	0.15	etc.	0.04	19.84
5	0.64	etc.	0.40	etc.	0.33	37.07
6	2.43	etc.	5.89	etc.	0.50	65.64
7	1.44	etc.	2.07	etc.	1.45	103.47
8	1.45	etc.	2.11	etc.	0.86	151.04
9	4.05	etc.	16.39	etc.	10.97	209.41
10	6.52	etc.	42.56	etc.	13.76	277.28
11	5.41	etc.	29.24	etc.	8.35	355.57
12	7.33	etc.	53.70	etc.	20.54	443.44
13	5.56	etc.	30.94	etc.	21.57	541.07

Figure 20-4. LINEST fails when used with a non-adjacent selection of known_x's.

If an array element is not the correct size, we need to return an error value. Use the **CVErr** keyword to return a worksheet error value that Excel can handle appropriately. The error values are listed in Appendix F.

After checking all the elements in the array, we assemble them into the final array of dimensions XSize rows by YMax columns, and return the array to the worksheet.

Now the LINEST expression

{=LINEST(F4:F13,Arr(A4:A13,C4:C13,E4:E13),1,1)}

returns the correct result. You can use names for cell ranges in the Arr function, e.g.,

{=LINEST(YValues,Arr(XValues1,XValues2,XValues3),1,1)}

```
Option Explicit
Option Base 1
Function Arr(ParamArray range1())
'Combines individual 1-D or 2-D arrays into a final 2-D array.
'In this version all individual arrays must be "vertical".
'All individual arrays must have same number of rows.
Dim result()
Dim I As Integer, J As Integer, K As Integer
Dim TempX As Integer, TempY As Integer
Dim XDim As Integer, YDim As Integer
Dim YStart As Integer, YSize As Integer
'First, get sizes of individual arrays, check to make sure all are same size.
For J = 0 To UBound(range1)
'Handles either range, name or array constant arguments
If IsObject(range1(J)) = True Then     'reference is to a range or a name
     TempX = range1(J).Rows.Count
     TempY = range1(J).Columns.Count
ElseIf IsArray(range1(J)) Then
     TempX = UBound(range1(J), 1)
     TempY = UBound(range1(J), 2)
End If
If J = 0 Then XDim = TempX
If XDim <> TempX Then Arr = CVErr(xlErrRef): Exit Function
YDim = YDim + TempY
Next J
'Now combine each individual array into final array.
'I index is used to select within array of arrays.
'K and J are column & row indices of individual arrays.
ReDim Result(XDim, YDim)
YStart = 0
For I = 0 To UBound(range1)
  YSize = range1(I).Columns.Count
  For K = 1 To YSize
  For J = 1 To XDim
    result(J, YStart + K) = Application.index(range1(I), J, K)
  Next J, K
  YStart = YStart + YSize
Next I
Arr = result()
End Function
```

Figure 20-5. A Function procedure to combine arrays.

Excel Tip. A function argument name should not be too long, or the name will not be able to fit completely in the Function Arguments dialog box. The limit is about 10–12 characters, depending on character width.

Providing a Description for a Custom Function in the Paste Function Dialog Box

When you use the Function Wizard to enter a function, the function's description appears at the bottom of the dialog box. For a custom function, the default description is "Choose the Help button for Help on this function and its arguments", but you can provide a custom description for your custom function. There are two ways to do this: You can use the Macro dialog box (normally used only for **Sub** procedures), or you can write and run a simple one-line VBA **Sub** procedure that uses the **MacroOptions** property. Either way, the description becomes part of Excel and does not need to be entered each time you start Excel.

To use the Macro dialog box to enter a description, carry out the steps in the following box.

To Provide a Description for a Custom Function

1. Click on Macros in the Code section of the Developer tab (Excel 2007/2010) or choose **Macro** from the **Tools** menu and **Macros...** from the submenu (Excel 2003).

2. Type the name of the custom function in the Macro Name box.

3. Press the Options... button to display the Macro Options dialog box.

4. Type the description in the Description box, then press OK.

To enter a description by means of VBA code, type and run a **Sub** procedure similar to the one shown in Figure 20-4.

```
Sub CustomFunctionDescription()
Application.MacroOptions Macro:="FtoC", Description:=_
    "Converts Fahrenheit temperature to Celsius"
End Sub
```

Figure 20-4. A **Sub** procedure to add a function description.

Providing Descriptions for Function Arguments in the Paste Function Dialog Box (Excel 2010 Only)

For Excel's built-in functions, the Function Arguments dialog box provides help information about each argument as you begin to enter it. Excel 2010 augmented the **MacroOptions** Method to provide information about each argument in a custom function. The new argument of **MacroOptions**, ArgumentDescriptions, is an array of text values, one for each argument of the custom function. Figure 20-5 shows a **Sub** procedure that provides help information for the ALPHA custom function that is described in a following section.

```
Sub ALPHAFunctionDescription()
Dim ArgDesc(1 To 4) As String
ArgDesc(1) = "is the number of protons in the species to be calculated."
ArgDesc(2) = "is the pH value for the calculation."
ArgDesc(3) = "is a table of pKa (acid dissociation constant) or logK
(protonation constant) values."
ArgDesc(4) = "is a logical value: TRUE if eqm_constants are pKa values,
FALSE or omitted if they are logK values."
Application.MacroOptions macro:="ALPHA", _
    Description:="Returns the fraction of the total concentration of a
    polyprotic acid in a specified protonated form", _
    ArgumentDescriptions:=ArgDesc
End Sub
```

Figure 20-5. A **Sub** procedure to add a function description
and argument descriptions (Excel 2010 only).

For Excel 2003 or Excel 2007, there's no way to provide information about arguments for custom functions. But since the same description text appears in the Insert Function and the Function Arguments dialog boxes, you can provide information about the arguments in the description. Unfortunately, the limit for Description is 255 characters. You can provide line breaks in the text by using **Chr**(10) or **Chr**(13), but the Function Arguments dialog box can only display two lines of text.

Assigning a Custom Function to a Function Category

By default, your custom function will be located in the User Defined category in the Insert Function dialog box. You can specify the category in which the function will be located, by typing and running a **Sub** procedure similar to the one shown in Figure 20-6. The category values are listed in Appendix F.

```
Sub CustomFunctionDescription()
Application.MacroOptions macro:="FtoC", Category:=3
End Sub
```

Figure 20-6. A **Sub** procedure to change function category.

Excel Tip. Don't give a module containing a custom function the same name as the function. If you do, the function will return the #NAME? error value.

A Custom Function to Calculate Acid–Base Species Distribution Diagrams

To illustrate the variation in composition of an aqueous solution of a polyprotic weak acid species, it is useful to plot a species distribution curve such as the one for citric acid shown in Figure 20-7. The parameter plotted versus pH for each species is α, the fraction of the total citric acid concentration in the form of a particular species.

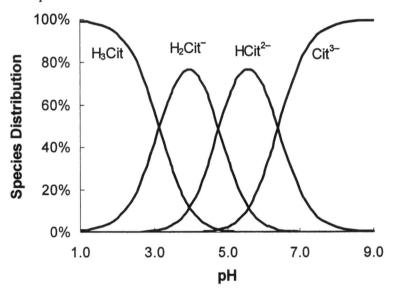

Figure 20-7. Species distribution diagram for citric acid.

For a weak acid H_JA, the fraction of the acid in the form containing j protons is

$$\alpha_j = \frac{[H_jA]}{[H_JA]_T} \tag{20-1}$$

For example, the expression for α_0 for the Cit^{3-} species is

$$\alpha_0 = \frac{[Cit^3]}{[H_3Cit]_T} \tag{20-2}$$

The equations that follow use protonation constants rather than dissociation constants. The stepwise protonation constant K^H of a weak base species B is the equilibrium constant for the following reaction (charges are omitted for clarity):

$$B + H^+ \rightleftharpoons HB \tag{20-3}$$

$$K_1^H = \frac{[HB]}{[B][H^+]} \tag{20-4}$$

and in general:

$$K_j^H = \frac{[H_jB]}{[H_{j-1}B][H^+]} \tag{20-5}$$

K^H values are typically much greater than 1 and are often reported as log K^H values. K_a values are typically very small, and are often reported as pK_a values ($-\log K_a$). Since K^H is the reciprocal of K_a, it follows that numerically, log K^H is identical to pK_a for a given conjugate acid/conjugate base equilibrium.

It is also convenient to use the cumulative protonation constant β_j for the overall reaction

$$B + jH^+ \rightleftharpoons H_jB \tag{20-6}$$

$$\beta j = \frac{[H_jB]}{[B][H^+]^j} \tag{20-7}$$

The relationship between K's and β_j is

$$\beta_j = K_1^H K_2^H ... K_j^H \tag{20-8}$$

The expression for the α_j value for a particular protonated species is easily calculated from the expression

$$\alpha_j = \frac{\beta_j[H^+]^j}{\sum\limits_{j=0}^{J}\beta_j[H^+]^j} \tag{20-9}$$

where $\beta_0 = 1$. Thus the expression for α_2 for citric acid is

$$\alpha_2 = \frac{K_1^H K_2^H [H^+]^2}{1 + K_1^H[H^+] + K_1^H K_2^H [H^+]^{2+} + K_1^H K_2^H K_3^H[H^{3+}]} \tag{20-10}$$

The custom function ALPHA provides a convenient way to calculate alpha values for polyprotic species, and thus to construct distribution diagrams. The macro is shown in Figure 20-7. The syntax of the function is ALPHA(*j, pH, eqm_constants, pKa_logical*). Either pK_a values or log K^H values can be used. If pK_a values are used, *pKa_logical* is TRUE; if log K^H values are used, *pKa_logical* is FALSE or omitted.

```
Function ALPHA(J, pH, eqm_constants, Optional pKa_logical)
'pKa_logical:      TRUE if constants are pKa values, FALSE or omitted if
they are logK values
'
Dim N As Integer, K As Integer
Dim logbeta As Double, denom As Double, numerator As Double

N = eqm_constants.Count
If (K < 0) Or (K > N) Or (pH = "") Or (N = 0) Then
    ALPHA = CVErr(xlErrNA)
    Exit Function
End If
If IsMissing(pKa_logical) Then pKa_logical = False
logbeta = 0: denom = 1: numerator = 1

If (pKa_logical = False) Then   'Calculation using protonation constants
    For K = 1 To N
    logbeta = logbeta + eqm_constants(K)
    denom = denom + 10 ^ (logbeta - pH * K)
    If (K = J) Then numerator = 10 ^ (logbeta - pH * K)
    Next
Else   'Calculation using dissociation constants
    For K = N To 1 Step -1
    logbeta = logbeta + eqm_constants(K)
    denom = denom + 10 ^ (logbeta - pH * (N - K + 1))
    If (N - K + 1 = J) Then numerator = 10 ^ (logbeta - pH * (N - K + 1))
    Next
End If
ALPHA = numerator / denom
End Function
```

Figure 20-8. The ALPHA custom function

A Custom Statistical Function

One of the underlying assumptions of the least squares method is that there is no error in the measurement of the independent variable (the y-values). This assumption is often not valid, and one of the most obvious cases of this is found in method-comparison analysis. A typical example of method-comparison analysis involves the comparison of two different instruments, a current production instrument and an improved model. Measurements made on a series of samples with the two instruments, and plotted by using current instrument readings as the x-values and new instrument readings as the y-values, should ideally result in a straight line of unit slope and zero intercept. The actual slope and intercept of the line can provide estimates of the proportional and constant error between the two methods.

It is obvious that there is error in both the x- and the y-values. Calculation of the least-squares slope and intercept by "standard" methods is clearly not valid. In the custom function that follows, the method of Deming is used to calculate the regression parameters for the straight line $y = mx + b$. The Deming regression calculation assumes Gaussian distribution of errors in both x- and y-values and uses duplicate measurements of x-values (and of y-values) to estimate the standard errors. A portion of a data table is shown in Figure 20-9.

	A	B	C
	Sample #	Xvalues	Yvalues
1			
2	1	24.70	25.30
3		25.50	26.75
4	2	25.10	25.10
5		25.75	27.70
6	3	24.75	24.90
7		28.30	27.55
8	4	25.10	24.95
9		26.50	26.55
10	5	20.60	21.70
11		23.55	22.20
12	6	45.30	44.60
13		44.85	44.80
14	7	28.45	29.60
15		29.45	30.00
16	8	46.70	47.45

Figure 20-9. Portion of data table to calculate Deming regression parameters.

The equations*for the Deming slope and intercept are

$$m = U + \sqrt{U^2 + (1/\lambda)} \tag{20-11}$$

and
$$b = \bar{y} - m\,\bar{x} \tag{20-12}$$

where
$$U = \frac{S_y^2 - (1/\lambda)S_x^2}{2rS_xS_y} \tag{20-13}$$

$$\lambda = \frac{S_{ex}^2}{S_{ey}^2} \tag{20-14}$$

---

* See, for example, P. J. Cornbleet and N. Gochman, *Clin. Chem.* **1979**, *25*, **432**.

$\bar{x}$ and $\bar{y}$ are the means and S_x^2 and S_y^2 are the variances of the MeanX and MeanY values, respectively, r is the Pearson correlation coefficient, and S_{ex}^2 and S_{ey}^2 are the error variances of the x-values and y-values, respectively, calculated from the equation

$$S_e = \sqrt{\frac{S(\text{difference between duplicates})^2}{2N}} \qquad (20\text{-}15)$$

The macro listing is shown in Figure 20-9. The two arguments used by the macro are a range of cells containing pairs of x measurements on a given sample and a corresponding range of pairs of y measurements.

Within the procedure, the major part of the code is virtually identical to the code that would be used to obtain the usual least-squares regression parameters for $y = mx + b$, namely, obtaining the sum of x-values, the sum of squares of x-values, etc. The difference is that pairs of values are used; the mean value of each pair of x-values or of y-values is used as the independent variable or dependent variable, respectively, and the estimate of the standard errors for the two sets of data is obtained from the differences between pairs.

Near the end of the code, some calculations that are specific to the Deming equations are performed. The function returns an array of two values, the slope and intercept of the Deming regression. Since an array is returned, the user must press Ctrl+Shift+Enter.

```
' Deming Regression
' Calculates Deming regression parameters for Y = mX + b
' Equations from Cornbleet & Gochman, Clin. Chem. 1979, 25, 432.
' Copyright 1997 by E. J. Billo
' Begun 5/10/97.  Last modified 10/3/00

Function Deming(XValues, Yvalues)
Dim MeanX(), MeanY()
'Get number of cells to use in calculation loop
Ncells = XValues.Count
ReDim MeanX(Ncells / 2), MeanY(Ncells / 2)

' Step thru pairs of cells, calculating sums for statistics calcs
For J = 2 To Ncells Step 2
    MeanX(J / 2) = (XValues(J - 1) + XValues(J)) / 2
    MeanY(J / 2) = (Yvalues(J - 1) + Yvalues(J)) / 2
    SumX = SumX + MeanX(J / 2): SumY = SumY + MeanY(J / 2)
    SumX2 = SumX2 + (MeanX(J / 2)) ^ 2
    SumY2 = SumY2 + (MeanY(J / 2)) ^ 2
    SumXY = SumXY + MeanX(J / 2) * MeanY(J / 2)
    SumDeltaX2 = SumDeltaX2 + (XValues(J - 1) - XValues(J)) ^ 2
```

```
     SumDeltaY2 = SumDeltaY2 + (Yvalues(J - 1) - Yvalues(J)) ^ 2
Next J

' Calculate some intermediate statistical quantities
XBar = SumX / N: YBar = SumY / N
Sx2 = (N * SumX2 - SumX ^ 2) / (N * (N - 1))
Sy2 = (N * SumY2 - SumY ^ 2) / (N * (N - 1))
Sdx2 = SumDeltaX2 / (2 * N)
Sdy2 = SumDeltaY2 / (2 * N)
rPearson = (N * SumXY - SumX * SumY) / _
Sqr((N * SumX2 - SumX ^ 2) * (N * SumY2 - SumY ^ 2))

' Calculate quantities that are specific to the Deming calculation
lambda = Sdx2 / Sdy2
U = (Sy2 - Sx2 / lambda) / (2 * rPearson * Sqr(Sx2) * Sqr(Sy2))
Slope = U + Sqr(U ^ 2 + 1 / lambda)
Intercept = YBar - Slope * XBar

Deming = Array(Slope, Intercept)
End Function
```

Figure 20-10. A **Function** procedure to calculate Deming regression parameters.

This function could be located in the Statistical category, rather than in the User Defined category.

A Custom Function
to Calculate Molecular Weights

The CD-ROM that accompanies this book contains the workbook Molecular Weight Calculator, with a useful custom function, MolWt, that calculates the molecular weight of a compound from its formula. The syntax of the function is MolWt(*formula,decimals*). *Formula* is a text string that can be interpreted as a chemical formula, e.g., H2SO4, a reference to a cell containing a formula, or a name. The number characters in the formula can be normal or subscripted. The optional *decimals* argument is used to specify the number of decimal places in the returned result. If *decimals* is larger than the number of decimal places allowed by the rules of significant figures, i.e., the smallest number of decimal places in the atomic weight values used, the returned value will be rounded to the allowed number of decimal places. If *decimals* is omitted, the returned value is obtained simply by adding the atomic weight values; not all of the figures are significant.

If *decimals* is specified, the returned value is text. Most numeric operations on the text value will operate correctly; if not, use the VALUE function to convert the text to a number. If *decimals* is omitted, the value returned is a number.

Examples. If cell A1 contains the text NaF, the formula =MolWt(A1,3) returns the text value 41.988; the formula =MolWt(A1) returns the number value 41.98817248 (see the first and second examples in Figure 20-10).

If cell A2 contains the text H2S, the formula =MolWt(A2,5) returns the text value 34.081, since the atomic weight of S, 32.065, determines the number of significant figures after the decimal.

Additional details. The workbook contains a database of atomic weights (IUPAC 2007 atomic weights). If necessary, you can update the atomic weights in the .xls backup copy and save it as a new Add-In

As well as the atomic weights of the elements, the database contains the molecular weights of a number of organic and amino acid residues. Thus, for example, you can use C_2H_6O, CH_3CH_2OH or EtOH for the formula of ethanol (see the third and fourth examples in Figure 20-10) or HGluCysGlyOH for the formula of a peptide. You can add your own abbreviations to the database; instructions can be found in the workbook.

You can use hyphens, dashes or the equal sign within a formula, as in $H_2C=CH_2$.

The function accepts parentheses, such as in $Ni(ClO_4)_2$ or $N(CH_2CH_2N(CH_3)_2)_3$. Up to seven levels of nested parentheses are possible.

The function also handles hydrates or similar species, such as $CuSO_4$ $5H_2O$, $CaSO_4 \cdot 1/2H_2O$ or $N(CH_2CH_2NMe_2)_3 \cdot 4HCl$. You can use the space character, the asterisk, the bullet or the centered dot to indicate a hydrate.

Error values. The function returns the following error values:

#N/A	*formula* argument missing (reference to an empty cell)
#VALUE!	unequal number of left and right parentheses
	more than seven levels of nested parentheses
	incorrect symbol, e.g., NaCL
	incorrect data for *formula* argument, e.g., a number
#NUM!	incorrect data for *decimals* argument, e.g., -1

It should be obvious that even if there is more than one error in the function, only one error message can be displayed.

Some examples of the use of the MolWt function are shown in Figure 20-11.

	A	B	C
1	Formula	MW	Formula entered in cell
2	NaF	41.988	=MolWt(A2,3)
3	NaF	41.98817248	=MolWt(A3)
4	CH$_3$CH$_2$OH	46.0684	=MolWt(A4,5)
5	EtOH	46.0684	=MolWt(A5,9)
6	N(CH$_2$CH$_2$NMe$_2$)$_3$·4HCl	376.237	=MolWt(A6,9)
7	HGluCysGlyOH	275.258	=MolWt(A7,3)
8	CaSO$_4$ 0.5H$_2$O	145.14824	=MolWt(A8)
9	CaSO$_4$·1/2H$_2$O	145.14824	=MolWt(A9)
10	H$_2$C=CH$_2$	28.05316	=MolWt(A10)
11	NaCL	#VALUE!	=MolWt(A11)
12	NaCl	#NUM!	=MolWt(A12,-1)

Figure 20-11. Some examples of the use of the MolWt custom function.

The function is most conveniently used when the Molecular Weight Calculator workbook is saved as an Add-In. See "Creating Add-In Function Macros" later in this chapter.

Keep the original Molecular Weight Calculator.xls document as back-up.

Creating Add-In Function Macros

Saving a custom function as an Add-In is by far the most convenient way to use it. Here are some of the advantages:

- An Add-In custom function is listed in the Paste Function list box without the workbook name preceding the name of the function, making it virtually indistinguishable from Excel's built-in functions.

- If the Add-In workbook is saved in the Add-Ins folder (the default location for Add-Ins) or in the Excel Startup folder, the Add-In will be available every time you start Excel.

- An Add-In workbook can be hidden so it can't be viewed by the user.

How to Create an Add-In Macro

To save a workbook as an Add-In using Excel 2007/2010 (the workbook containing the macro must be the active workbook), click on the Office Button (Excel 2007) or the File tab (Excel 2010), and click on Save As. In the Save As window, choose "Excel Add-In (*xlam)" in the Save As Type list box, and then press OK.

To save a workbook as an Add-In using Excel 2003 (the workbook containing the macro must be the active workbook): Choose **Save As...** from the **File** menu. Choose Microsoft Excel Add-In from the Save File As Type drop-down list box, then press OK. In Excel for Windows, Add-In macros are automatically given the filename extension .xla.

Add-Ins are saved in the Add-Ins folder.

The custom function will now be indistinguishable from Excel's built-in functions. (If you give them names using lowercase letters, you can distinguish them from Excel's built-in functions.)

Command macros can also be saved as Add-Ins.

How to Load/Unload an Add-In Macro

To load or unload an Add-In using Excel 2007/2010, use the following procedure: Press the Office button. In the Office window, press the Excel Options button at the bottom of the window. In the Excel Options window, click on Add-Ins to display the list of active and inactive Add-Ins. At the bottom of the window, press the "Manage Excel Add-Ins" button. Check or uncheck the box next to the desired Add-In, and then press OK.

If you're using Excel 2003, choose **Add-Ins** from the **Tools** menu. Check or uncheck the box next to the desired Add-In, then press OK.

How to Delete an Add-In Macro

To delete an Add-In using Excel 2007/2010, use the following procedure: Click on the Office button (Excel 2007) or the File tab (Excel 2010), choose Excel Options, choose Add-Ins, and press the "Manage Excel Add-Ins" button. In the list of Add-Ins, unload the add-in if it is loaded. Press the Browse... button to display the list of Add-Ins in the Add-Ins folder. Right-click on the Add-In to be deleted and choose **Delete** from the shortcut menu.

To delete an Add-In using Excel 2003, choose **Add-Ins** from the **Tools** menu. In the list of Add-Ins, unload the add-in if it is loaded. Press the Browse... button to display the list of Add-Ins in the Add-Ins folder. Right-click on the Add-In to be deleted and choose **Delete** from the shortcut menu.

The name of the add-in will remain in the Add-Ins list box until you exit from Excel.

How to Protect an Add-In Workbook

To protect an Add-In workbook, activate the Visual Basic Editor. Choose **Properties...** from the **Tools** menu and choose the Protection tab. Check the Lock Project For Viewing box, enter a password, and **Save** the workbook.

How to Edit an Add-In Workbook or Convert a .xla File to a .xls File

When you save a workbook as an Add-in workbook (.xla file in Excel 2003 or .xlam in Excel 2007/2010), the document can no longer be viewed or modified by the casual user. But it is possible to convert an add-in workbook to a normal Excel workbook, by using the Visual Basic Editor to change the IsAddin property of the workbook from True to False. Here's how to do it.

Make sure that the add-in file is open. Add-ins can be stored anywhere – in the Add-Ins folder, the XLSTART directory, on the desktop, or elsewhere.

Enter the Visual Basic Editor.

In the Project Explorer window, find the Add-in file.

Expand the Microsoft Excel Objects hierarchy tree by clicking on the plus sign to the left of the name of the Add-in file, and click on ThisWorkbook.

In the Properties window, scroll down to IsAddin and change the property from True to False. (When you click on the property box to the right of IsAddin, a drop-down list box will appear with the properties True or False; click on False to enter it.) If the workbook is protected with a password, you will be asked to enter the password. If you don't know the password, you will not be able to convert the add-in workbook.

Return to the Excel window. The file, now a normal Excel workbook, will be visible. Use **Save As...** to save it as a normal Excel workbook (the default file type will now be Microsoft Excel Workbook). The new workbook does not replace the Add-in workbook.

Alternatively, you can edit the Add-in workbook and then use **Save As...** to save it once again as an Add-In. The modified document will replace the existing file.

Advantages and Disadvantages of Using Function Macros

Some function macros perform calculations that could be performed by using a worksheet formula or formulas, while in other cases implementing the function on the worksheet is impossible (when looping or branching is required, for example).

The main advantage of using a custom function instead of worksheet formulas is minimization of errors occurring when formulas are entered. The main disadvantage is that calculations done by using a custom function are slower than the same calculations by means of worksheet formulas. The difference will probably only be apparent if a worksheet contains a large number of cells with custom functions, or if the custom functions involve a large number of repetitive calculations.

21

Automatic Procedures

Automatic procedures are VBA procedures that run when a specified event occurs. Usually the event is an action performed by the user, such as opening a document or clicking on a toolbutton, but a macro can also be set to run at a specified time, for example.

There are two kinds of automatic procedures. The original VBA introduced in Excel 5/95 included a number of VBA properties that allowed the user to create procedures that ran automatically. For example, an Auto_Open procedure runs automatically when the workbook is opened. We'll call these procedures OnEvent procedures.

In Excel 97, VBA introduced *event-handler procedures*. Event-handler procedures are more versatile than the earlier automatic procedures. But automatic procedures still function and in some cases are easier to create and use. Both kinds are described in the sections that follow.

OnEvent Procedures

OnEvent procedures are procedures that run automatically whenever a specified event occurs. For example, an Auto_Open procedure runs when a workbook is opened; an **OnCalculate** procedure runs when a worksheet is recalculated. Table 21-1 lists the some of the most useful procedures.

Auto_Open or Auto_Close Procedures

If you give a procedure the name Auto_Open or Auto_Close, that procedure will run whenever the workbook is opened or closed, respectively.

A simple Auto_Open procedure is shown in Figure 21-1.

Table 21-1. OnEvent Procedures

Procedure	Runs when ...
Sub Auto_Open	the workbook that contains the procedure is opened.
Sub Auto_Close	the workbook that contains the procedure is closed.
OnAction	an object (control, chart, menu command or toolbutton) is clicked.
OnCalculate	a worksheet is recalculated.
OnEntry	data is entered or edited.
OnKey	a specified key or key combination is pressed.
OnSheetActivate	the user switches to a sheet.
OnUndo	the user chooses **Undo** in the **Edit** menu.
OnWindow	the user activates a window.

After entering the code, close the workbook, saving the changes. When you re-open the workbook, the dialog box will be displayed and the value 12345 will be entered in cell A1.

Auto_Open procedures are useful for performing "housekeeping" tasks that need to be done each time a document is opened, or for installing custom menu commands (see Chapter 22, "Custom Menus, Menu Bars and Menu Commands"). Auto_Close procedures are also useful for housekeeping, for example, to remove a custom menu command when the workbook that uses it is closed.

```
Sub Auto_Open()
MsgBox "Hello"
Range("A1").Value = 12345
End Sub
```

Figure 21-1. An Auto_Open procedure.

OnEvent Procedures

The OnEvent keywords (see Table 21-1) are Properties, and thus always require an Object keyword. With the exception of the **OnUndo** procedure, which will be discussed later, **OnEvent** procedures have identical syntax:

> *object*.**OnEvent** = *ProcedureName*

Object allows you to specify the scope of the procedure. For example, for the **OnSheetActivate** procedure, if *object* is **Application**, the procedure runs when any sheet in any workbook is activated. If *object* is, for example, **Workbooks**("Deconvolution").**Worksheets**("Spectrum 2"), the procedure runs only when the specified sheet in the specified workbook is activated.

This one line of code must be executed once in order to set up the pointer to run the **Sub** procedure *ProcedureName* each time the event occurs.

For example, the code to run a specified procedure each time the user switches to a different worksheet is shown in Figure 21-2.

The line of code to set up the pointer (the Enable procedure in the example) could be in, for example, an Auto_Open procedure in the Personal Macro Workbook, so that it runs whenever Excel is started.

```
Sub Enable()
Application.OnSheetActivate = "CheckMargins"
End Sub

Sub CheckMargins()
(some code)
End Sub

Sub Disable()
Application.OnSheetActivate = ""
End Sub
```

Figure 21-2. OnSheetActivate procedures.

To remove the pointer to the procedure, set the **OnEvent** procedure to a null string, as illustrated in Figure 21-2.

Some Examples of OnEvent Code

OnSheetActivate.   The following example sets the **OnSheetActivate** property to the macro "Check_Margins". The procedure will run whenever the user activates a sheet in any workbook.

Application.OnSheetActivate = "Check_Margins"

OnWindow.   This example runs the procedure Install_Button when the user switches to the SCHEDULE.XLS window.

Application.Windows("SCHEDULE.XLS").OnWindow = "Install_Button"

OnEntry.   This example runs Data_Entry_Procedure when data is entered on the active worksheet.

ActiveSheet.OnEntry = "Data_Entry_Procedure"

OnAction. The following is an example of code that runs a procedure when the custom menu command **Add Data Labels...** in the **Chart** menu is clicked:

CommandBars("Chart Menu Bar").Controls("Chart"). _
    Controls("Add Data Labels...").OnAction ="DataLabeler"

Using OnUndo

You can install a custom **Undo** command in the Menu Commands group of the Add-Ins tab (Excel 2007/2010) or in the **Edit** menu (Excel 2003). The text of your custom **Undo** command replaces the built-in **Undo** command. For example, you could have a custom Undo procedure that will be installed each time a user chooses one of your custom toolbuttons.

Undo does not happen automatically; you must write code that saves the original situation, and later restores it.

The syntax of **OnUndo** is:

application.**OnUndo(***MenuText, ProcedureName***)**

MenuText specifies the text that appears in place of the **Undo** command on the **Edit** menu. *ProcedureName* specifies the name of the procedure that runs when you choose the custom Undo command.

Figure 21-3 illustrates part of an Undo procedure that restores text that was changed by a previous procedure that saved the original text in the array OldValue before changing it.

```
Sub RestoreText()
Range(OldAddr).Select 'selects range where text was formatted
y = 0
For Each CEL In Selection
CEL.Value = OldValue(y)
OldValue(y) = ""
y = y + 1
Next
End Sub
'----------------------------------------------------------------
Sub EnableUndo()
Application.OnUndo "Restore Original Text", "RestoreText"
End Sub
```

Figure 21-3. A custom Undo procedure.

Event-Handler Procedures

Event-handler procedures (Excel 97 *et seq*.) are not located in the usual VBA code modules, but in the module sheets associated with a worksheet, chart sheet or workbook object. In Figure 21-4, for Book1, that contains three worksheets and a chart sheet, the "Microsoft Excel Objects" folder contains code sheets for Chart1, Sheet1, Sheet2, Sheet3 and ThisWorkbook. These code sheets are intended for event-handler procedures only; if you create a normal **Sub** procedure in one of them it will not function.

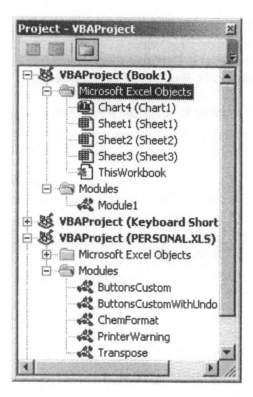

Figure 21-4. The VBA Project window.

The **ThisWorkbook** module sheet is used to contain event-handler procedures for workbook events. A few of the more useful ones are listed in Table 21-2. The **Workbook_Open** event-handler routine corresponds to the **Auto_Open** procedure.

Table 21-2. Some Workbook Event-Handler Procedures

Procedure	Runs when the user...
Open	opens the workbook that contains the procedure.
BeforeClose	closes the workbook that contains the procedure.
BeforeSave	saves the workbook that contains the procedure.
NewSheet	inserts a sheet in the workbook.
SheetActivate	activates any sheet in the workbook.
BeforePrint	prints anything in the workbook.

The Chart and Sheet code sheets are used to create event-handler procedures for a specific worksheet or chart sheet. Worksheet event-handler procedures include **Activate**, **Deactivate**, **Calculate** and **SelectionChange**. Chart events include **Activate**, **Deactivate**, **Resize** and **SeriesChange**

Creating a Workbook_Open Procedure

To create a **Workbook_Open** procedure, a procedure equivalent to an **Auto_Open** macro, follow the procedure in the following box

---

To Create a Workbook_Open Event-Handler Procedure

1. Access the Visual Basic Editor
2. In the Project window, locate the workbook for which you want to create an event-handler procedure.
3. Double-click on the node for **ThisWorkbook** to display the Code window.
2. Click on the drop-down button of the left-hand list box at the top of the code window and choose **Workbook.**
3. Click on the right-hand drop-down button to display the list of available event-handler procedures for Workbook (there are 28 of them in Excel 2003, 29 in Excel 2007 and 36 in Excel 2010).
4. In the drop-down list, click on **Open** (**Open** is the default event-handler procedure for **ThisWorkbook**). The shell of a **Workbook_Open** procedure will be automatically entered on the code sheet.
5. Enter the code.

A simple **Workbook_Open** procedure is shown in Figure 21-5.

```
Private Sub Workbook_Open()
MsgBox "Hello again"
Range("B1").Value = 6789
End Sub
```

Figure 21-5. A **Workbook_Open** procedure.

After entering the code, **Close** the workbook, saving the changes. When you re-open the workbook the dialog box will be displayed and the value 6789 will be entered in cell B1.

22

Custom Menus

In this chapter you'll learn how to add a new command to a menu, how to add a new menu to a menu bar, and how to create a whole new menu bar with separate menus and commands. In this way you can add new capabilities to Excel or even create a complete custom application. And even though Excel 2007/2010 no longer displays menus, you can still add a custom menu or a menu bar containing several menus in order to access built-in or custom commands.

In Excel 2003, menubars and toolbars are both members of the collection of CommandBar objects, and toolbuttons, drop-down menus and menu commands are all members of the class of Control objects. Command bars can contain buttons or menus or both. In general, however, mixing buttons and menu commands on the same toolbar doesn't seem to be a good idea, so this chapter deals exclusively with menus and Chapter 23 with toolbars.

In Excel 2007/2010, there are no built-in menus. But, using VBA, it's possible to create custom menu bars that are displayed in the Add-Ins ribbon tab. This chapter shows how to accomplish this.

Modifying Menu Bars, Menus or Menu Commands in Excel 2003

You can modify existing menu bars, menus or menu commands either manually or by means of VBA code. You modify them manually by choosing the **Customize...** command in the **Tools** menu and then dragging the menu command to the desired place on the menu bar.

Adding or Removing a Menu Command

You can add any of Excel's built-in menu commands to a menu, or remove menu commands that you don't use. The procedures in the following boxes cover these possibilities.

To Add a Built-in Menu Command to a Menu

1. The toolbar that contains the menu to be customized must be visible.

2. Choose **Customize...** from the **Tools** menu, or choose **Toolbars...** from the **View** menu and choose **Customize...**, or right-click on any toolbar and choose **Customize...** from the shortcut menu.

3. In the Customize dialog box, choose the Commands tab.

4. In the Categories box, choose the desired category, then choose the desired command from the list in the Commands box and drag it to the desired location in the copy of the menu. The mouse pointer will display a large "×" beside the mouse pointer image if the mouse pointer is over a location at which the command cannot be installed. If the command can be installed, the "×" will change to a "+".

5. You can also add a new menu to the menu bar: Choose "New Menu" in the Categories box.

Some commands in the list of commands in the Commands tab have an icon beside them, others don't. Commands with an icon are toolbuttons, and those without an icon are menu commands.

To Remove a Menu Command from a Menu

1. The toolbar that contains the menu to be customized must be visible.

2. Choose **Customize...** from the **Tools** menu, or choose **Toolbars...** from the **View** menu and choose **Customize...**, or right-click on any toolbar and choose **Customize...** from the shortcut menu.

3. Drag the menu command off the menu (when you click on it, the menu command is highlighted) and release the mouse button. The menu command will disappear. You can remove a complete menu from a menu bar in the same way.

Creating a New Menu Bar

In previous chapters you learned how to create useful **Sub** procedures. You may find that you use a particular macro so often that it would be more convenient to have it on one of Excel's drop-down menus, rather than having to "run" it by means of the **Run** command from the **Macro** menu. In addition, running a command macro by means of an Excel menu command makes the custom command accessible to Excel users who are not familiar with the use of macros.

If you have written a number of related macros, you may even decide to create a custom menu bar with several custom menus, each containing custom commands. While both "Custom Menu Command" and "New Menu" can be found in Customize, there's no "New Menu Bar". That's because menu bars are really toolbars. To create a new menu bar, you simply create a new toolbar and place menus on it instead of toolbuttons.

You can install either custom menus or any of Excel's built-in menus on your custom menu bar. To install custom menus, use the procedure in the following box; to install any of Excel's built-in menus, follow the same procedure but choose "Built-in Menus" from the Categories box. You can apply the same methods used for custom menus to add commands, remove commands, or rename the built-in menus. Menus installed on a custom menu bar are displayed with a drop-down arrow next to the menu name. Figure 22-1 shows such a custom menu bar.

Figure 22-1 A custom menu bar with three menus:
Excel's built-in **File** and **Edit** menus and a custom menu.

Adding a Custom Menu to a Menu Bar

To add a custom menu to a menu bar, use the procedure in the following box.

To Add a Custom Menu to a Menu Bar

1. The toolbar that contains the menu to be customized must be visible.

2. Choose **Customize...** from the **Tools** menu, or choose **Toolbars...** from the **View** menu, or right-click on any toolbar and choose **Customize...** from the shortcut menu.

3. In the Customize dialog box, choose the Commands tab.

4. Scroll through the Categories box and choose "New Menu".

5. Select New Menu and drag it to the desired location on the toolbar.

6. Right-click on the custom menu item to display the shortcut menu. Type the name of the new menu in the Name input box (the default text is "New Menu").

---

*Excel Tip: If you add a built-in menu to a toolbar and make changes to the menu, the changes will also appear in the built-in menu on the menu bar. Thus, if you want a modified **File** menu on a custom menu bar, you should add a custom menu, rename it **File**, and then add the desired menu commands.*

Adding a Custom Menu Command to a Menu

To add a custom menu command to a menu, use the procedure in the following box.

When you exit from Excel, any changes you made to the menu bars or toolbars are saved, and this menu bar/toolbar configuration will be displayed the next time you start Excel.

**To Add a Custom Menu Command to a Menu Bar
and Assign a Macro to It**

1. The toolbar that contains the menu to be customized must be visible.
2. Choose **Customize...** from the **Tools** menu, or choose **Toolbars...** from the **View** menu and choose **Customize...**, or right-click on any toolbar and choose **Customize...** from the shortcut menu.
3. In the Customize dialog box, choose the Commands tab.
4. Scroll through the Categories box and choose Macros.
5. Select Custom Menu Item and drag it to the desired location on the menu (a black bar will show where the command will be placed).
6. Right-click on the custom menu item to display the shortcut menu and choose Assign Macro...
7. Excel displays the Assign Macro dialog box. Choose the name of the macro to be assigned to the custom menu item (best located in the Personal Macro Workbook). Click the OK button.
8. In the shortcut menu, type the text for the new menu command in the Name input box (the default text is "&Custom Menu Command").
9. To insert a separator bar above the new menu command, choose Begin a Group from the shortcut menu; to remove a separator bar, uncheck Begin a Group from the shortcut menu.

Modifying Menus or Menu Bars by Using VBA

You can also modify existing menus or menu bars by using Visual Basic for Applications.

If you create a macro to be used by other people, you can make it easy for them to use the macro by installing a custom menu command that runs the procedure. As well, you can install the new menu command by means of an automatic procedure (see Chapter 21) so that all the users have to do is open the workbook containing the macro.

In contrast to changes to menus made by using the **Customize...** command, menus and menu commands installed by using VBA are not saved when you quit Microsoft Excel.

The Basic Structure of a Procedure to Install a New Menu Command

Use the **Add** method to add a drop-down menu, menu command or submenu. The syntax of the **Add** method is

*expression.***Add***(Type, Id, Parameter, Before, Temporary)*

The required argument *expression* specifies the menu bar and the menu where the new control will be added. Use the *Type* argument to specify the type of control to be added: *msoControlButton* to install a normal menu command, or *msoControlPopup* to install a drop-down menu on a menu bar or a submenu for a menu command.   The numbers equivalent to the built-in constants *msoControlButton* and *msoControlPopup* are 1 and 10, respectively.   Use the optional *Before* argument to specify the position of the new control on the menu bar or menu. The new control will be inserted before the control at this position; if omitted, the control is added at the end of the specified menu or menu bar. If the optional *Temporary* argument is set to **True**, the menu command is removed when you exit from Excel and must be re-installed when you start Excel; if set to **False**, the command will be present the next time you start Excel; the default is *Temporary* = **True**. The *Parameter* and *Id* arguments can be omitted; see VBA On-Line Help for information.

Creating code to install a custom menu command can be confusing: how to declare the necessary arguments, whether or not to use parentheses with arguments, whether to use **Set** to create an object, whether to use a **With...End With** structure, etc. The following procedures illustrate three approaches to install a new menu command in the Tools menu. The code in Figure 22-2 installs a command with the menu text "Test1" at the top of the **Tools** menu. The code in Figure 22-3 installs a command with the text "Test2" at the bottom of the **Tools** menu, with a separator bar above it.

```
Sub CommandDemo1()
Set NewCmd = CommandBars("Worksheet Menu Bar").Controls("Tools") _
     .Controls.Add (Type:=1, before:=1)
NewCmd.Caption = "Test1"
NewCmd.OnAction = "Procedure1"
End Sub
```

Figure 22-2. VBA code to install a custom menu command, using the **Set** keyword.

```
Sub CommandDemo2()
With CommandBars("Worksheet Menu Bar").Controls("Tools") _
     .Controls.Add (Type:= msoControlButton)
  .Caption = "Test2"
  .BeginGroup = True
  .OnAction = "Procedure2"
End With
End Sub
```

Figure 22-3. VBA code to install a custom menu command,
using the **With...End With** construction.

The procedure shown in Figure 22-4 shows how to add a new drop-down menu to the worksheet menu bar, with menu commands, submenu commands and separator bars. The custom menu is shown in Figure 22-5.

```
Sub CreateMenu()
' Adds a new menu to the right of the Window menu,
' with two menu commands and a submenu with two commands.
posn = CommandBars(1).Controls("Window").Index
Set NM = CommandBars("Worksheet Menu Bar").Controls.Add _
     (Type:=msoControlPopup, temporary:=True, before:=posn + 1)
NM.Caption = "Demo Menu"
'  Adds two menu items
Set NMC1 = NM.Controls.Add(Type:=msoControlButton)
With NMC1
   .Caption = "Menu Command 1"
   .OnAction = "MyProcedure1"
End With
Set NMC2 = NM.Controls.Add(Type:=msoControlPopup)
With NMC2
   .Caption = "Menu Command 2"
   .BeginGroup = True
End With
'  And two submenu commands
Set SMC1 = NMC2.Controls.Add(Type:=msoControlButton)
With SMC1
   .Caption = "Submenu Command 3"
   .OnAction = "MyProcedure3"
End With
Set SMC2 = NMC2.Controls.Add(Type:=msoControlButton)
With SMC2
   .Caption = "Submenu Command 4"
   .OnAction = "MyProcedure4"
End With
End Sub
```

Figure 22-4. Sample **Sub** procedure to install a custom menu
with two menu commands and a submenu.

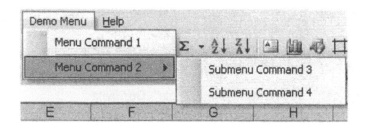

Figure 22-5 A custom menu and its menu commands.

Adding a Menu Command Automatically
By Means of an Event-Handler Procedure

If you've read the preceding chapter, it should be clear that you can create an automatic procedure to install a menu command when a particular workbook is opened (using either an Auto_Open procedure or a **Workbook_Open** event procedure) and remove the menu command when that workbook is closed (using an Auto_Close procedure or a **Workbook_BeforeClose** event procedure). Or you can install a command when a particular worksheet is activated and remove it when that worksheet is deactivated. The following example illustrates how to install a new menu command.

```
Private Sub Workbook_Open()
'This procedure installs a new menu command
'at the bottom of the Tools menu.

Set NMC = Application.CommandBars("Tools").Controls.Add _
(Type:=msoControlButton, temporary:=True)
  NMC.Caption = "New Menu Command..."
  NMC.OnAction = "MyProcedure"
End Sub
```

Figure 22-6. An event-handler procedure to install a custom menu command.

Figure 22-6 illustrates a **Workbook_Open** procedure that installs a new menu command in the **Tools** menu and assigns the **Sub** procedure named MyProcedure to it. To create your own procedure, replace "Tools", "New Menu Command..." and "MyProcedure" with suitable text. MyProcedure is a **Sub** procedure located in a module sheet.

You may want to install a new menu command immediately above or below a menu command that is already present in the menu. Since menus can be customized, you shouldn't use a numerical value for the position; instead, determine the position of the command programmatically. Use the **Index** property as shown in Figure 22-7. The BeginGroup property of a control inserts a separator bar above the menu command.

```
Private Sub Workbook_Open()
'This procedure installs a new menu command
'immediately below the Goal Seek command in the Tools menu
'with a separator bar above it.

posn = Application.CommandBars("Tools").Controls _
("Goal Seek...").Index

Set NMC = Application.CommandBars("Tools").Controls.Add _
(Type:=msoControlButton, before:=posn + 1, temporary:=True)
    NMC.Caption = "New Menu Command..."
    NMC.OnAction = "MyProcedure"
    NMC.BeginGroup = True
End Sub
```

Figure 22-7. An event-handler procedure to install a custom menu command
at a specified position.

Menu commands installed by this method are Temporary; that is, they remain in the menu until you exit from Excel.

Exiting Gracefully: Removing a Menu Command

The Solver Statistics workbook (on the CD-ROM that accompanies this book) contains three procedures: the **Sub** procedure that performs the calculations, a **Workbook_Open** procedure that installs a menu command in the **Tools** menu with a pointer to the calculation procedure, and a **Workbook_BeforeClose** procedure that removes the menu command when the user closes the Solver Statistics workbook. The single line of code required is shown in Figure 22-89.

```
Private Sub Workbook_BeforeClose(Cancel As Boolean)
Application.CommandBars("Tools").Controls("Solver Statistics...").Delete
End Sub
```

Figure 22-8. An event-handler procedure to remove a custom menu command.

Modifying a Built-In Menu Command

You can also modify the action of a built-in menu command. The following VBA statement runs the PrintWarning procedure when the user chooses the **Print...** command in the **File** menu.

```
CommandBars("Worksheet Menu Bar").Controls("File").Controls("Print...") _
    .OnAction = "PrintWarning"
```

Installing Menu Bars, Menus or Menu Commands in Excel 2007/2010

Although Excel 2007/2010 doesn't use menus, the VBA procedures described in the preceding sections will run successfully in Excel 2007/2010. The custom menus appear in the Add-Ins tab, in a new group labeled Menu Commands. Figure 22-9 shows the menu created by the procedure in Figure 22-5, as it appears in Excel 2007/2010. (Note that the procedure runs successfully, even though there is no **Tools** menu!)

You can use the procedure in Figure 22-4 as a framework to create a procedure to install your own custom menus in Excel 2007/2010.

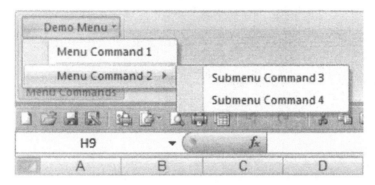

Figure 22-9. A custom menu and its menu commands, installed in Excel 2007/2010.

Displaying Excel 2003 Menus in Excel 2007/2010

Believe it or not, the Excel 2003 menus are hidden somewhere inside Excel 2007/2010. They can be displayed with a simple VBA procedure, which I found on the website of Excel guru John Walkenbach. You can find it by going to his website, www.j-walk.com, and searching for "menus".

The procedure shown in Figure 22-10 illustrates the principle. It displays only the **File** menu, which is shown in Figure 22-11.

```
Sub InstallMenuDemo()
On Error Resume Next
CommandBars("Classic Menus").Delete
CommandBars("Built-in Menus").Controls("File").Copy
CommandBars.Add("Classic Menus")
CommandBars("Classic Menus").Visible = True
End Sub
```

Figure 22-10. A demo procedure that displays the Excel 2003 File menu
in Excel 2007/2010.

As Walkenbach points out, "the menu isn't perfect. A few of the commands don't work, and the list of recent files in the **File** menu just shows placeholders."

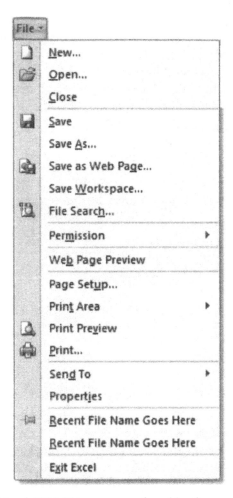

Figure 22-11. The Excel 2003 **File** menu produced by the procedure in Figure 22-10.

An example of a complete system of custom menu bar, custom menus and custom menu commands – the Classic Menus utility for Excel 2007/2010 – is provided on the CD that accompanies this book. The utility is discussed in "Displaying Classic Menus" in Chapter 1. The Classic Menus File menu, with its commands and a submenu, is shown in Figure 22-12.

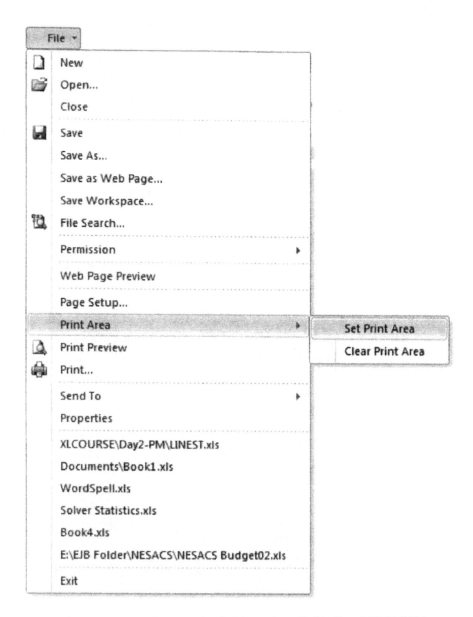

Figure 22-12. Excel 2003 Classic Menus installed in Excel 2007/2010.

23

Custom Toolbars and Toolbuttons

Excel 2003 allows the user to customize built-in toolbars and create new toolbuttons to simplify some of the operations that you perform often. Unfortunately, in Excel 2007/2010 these possibilities are severely curtailed.

Customizing Toolbars in Excel 2003

Some of the toolbuttons on Excel's toolbars are not very useful for scientists. You can remove the toolbuttons you don't use, giving a less cluttered workspace and providing room for other, more useful toolbuttons.

Moving and Changing the Shape of Toolbars

Excel's toolbars are usually located along the edges of the screen, most commonly along the top edge. Excel initially displays the Standard and Formatting toolbars, which provide tools for common Excel actions. There are a number of other toolbars provided with Excel, including the Chart, Drawing and Forms toolbars. You can also create custom toolbars, containing the toolbuttons you use most.

A toolbar can be moved from its position at the top of the screen and placed anywhere on the screen. Simply drag the *move handle* (located at the left side of a docked toolbar) to another location. A toolbar that is moved from a position along the edge of the screen is called a *floating toolbar*. It appears in its own window, with a title bar and a Close box. If you change the height or width of a floating toolbar, the tools are automatically rearranged to fit the new shape. If a floating toolbar (Figure 23-1) is dragged near the edge of the screen, Excel places it in a *toolbar dock*. There are four toolbar docks, one along each of the edges of the Excel application window.

If a toolbar does not fill the whole window (from left to right or vertically, depending on its orientation), the toolbar dock is visible as the extra blank space at either side of the toolbar, separated from the toolbar by the toolbar border.

Figure 23-1. The Excel 2003 Standard toolbar.

Excel Tip. *A toolbar that contains a drop-down list box, such as the Style box, cannot be placed in a vertical position in the left or right toolbar docks.*

Activating Other Toolbars

You can display a toolbar by choosing **Toolbars** from the **View** menu and then checking the box for the desired toolbar in the submenu. You can also display a toolbar by choosing **Customize...** from the **Tools** menu, choosing the Toolbars tab (Figure 23-2) and checking the box for the desired toolbar.

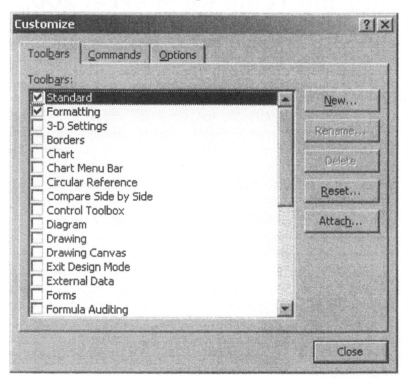

Figure 23-2. Displaying a built-in toolbar with the Toolbars dialog box.

To cancel the display of a toolbar, choose **Toolbars** from the **View** menu and uncheck the box for the toolbar, or drag the toolbar from the toolbar dock and then press the Close box.   You can have several toolbars in the Excel window at once.

Excel Tip. *Right-click on any toolbar to display the **Toolbar** shortcut menu.*

Adding or Removing Toolbuttons from Toolbars

Removing seldom-used toolbuttons from a toolbar provides a less cluttered working environment and makes room for other, more useful tools.   To delete a toolbutton from a toolbar or to add a built-in toolbutton to a toolbar, follow the procedures in the following boxes.

To Delete a Toolbutton from a Toolbar

1.   Choose **Toolbars** from the **View** menu and choose **Customize...**, or choose **Customize...** from the **Tools** menu, or right-click on a toolbar and choose **Customize...** from the shortcut menu.

2.   Drag the toolbutton off the toolbar (when you click on it, the toolbutton outline is highlighted) and release the mouse button.   The toolbutton will disappear.

To Add a Built-in Toolbutton to a Toolbar

1.   Choose **Toolbars** from the **View** menu and choose **Customize...**, or choose **Customize...** from the **Tools** menu, or right-click on a toolbar and choose **Customize...** from the shortcut menu.

2.   Choose the Commands tab.

3.   Select the desired toolbutton from the various categories (Figure 23-3).   Drag the toolbutton to the desired place on the toolbar. Separator bars between toolbuttons may be added or removed by using the procedure in the following box.

The built-in Microsoft Excel toolbars can be restored to their "factory-installed" condition by using the Reset button in the Toolbars tab of the Toolbars dialog box (Figure 23-2).   Individual toolbuttons can be restored by following the procedure described earlier.   Custom toolbars cannot be reset.

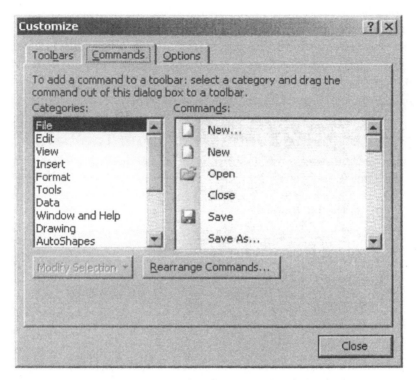

Figure 23-3. Using the Customize dialog box.

Toolbuttons can be organized into logical groups by grouping them together using *separator bars*.

To Insert or Remove a Separator Bar Between Toolbuttons

1. Choose **Toolbars** from the **View** menu and choose **Customize...**, or choose **Customize...** from the **Tools** menu, or right-click on a toolbar and choose **Customize...** from the shortcut menu.

2. Right-click on the button to the left of which you want to insert or remove a separator bar, to display the shortcut menu.

3. Choose Begin a Group.

Creating a New Toolbar

There are two ways to create a custom toolbar. One way is to modify an existing toolbar (such as the Standard toolbar). The other way is to create a new toolbar and then proceed to add built-in tools, as described earlier, or custom toolbuttons, as described later in this chapter. This way you can leave the built-in toolbars unmodified and display your own custom toolbar. To create a new toolbar, use the procedure in the following box.

To Create a Custom Toolbar

1. Choose **Toolbars** from the **View** menu and choose **Customize...**, or choose **Customize...** from the **Tools** menu, or right-click on a toolbar and choose **Customize...** from the shortcut menu.
2. Choose the Toolbars tab.
3. Press the New... button. Enter a name for the custom toolbar (see Figure 23-4) and then press OK. The (empty) custom toolbar will appear.

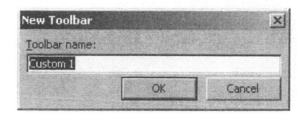

Figure 23-4. Naming a custom toolbar with the New Toolbars dialog box.

Creating Custom Toolbuttons

You may have created a number of macros. Some of these macros will be written for a very specific purpose, such as to prepare a specialized report. The workbook containing the macro will be opened only when you want to assemble the report, and you'll probably run the macro by using the Macro dialog box. Other macros automate tasks that you perform often, and you'll want to have them readily available whenever you're using Excel. These macros should be saved in the Personal Macro Workbook. To make these command macros even easier to use, you can add a custom toolbutton to a toolbar and assign the macro to it. The three macros described in this section — the NumberFormatConvert macro, the FullPage macro and the ChemicalFormat macro — are particularly convenient to use when they are assigned to a button.

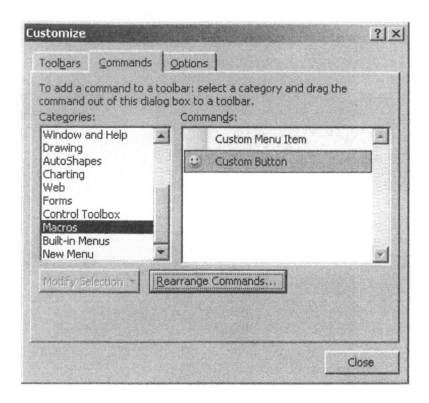

Figure 23-5. A custom toolbutton that can be placed on a toolbar.

The ToggleNumberFormat macro toggles a number between floating-point and scientific formats. The FullPage macro sets margins to zero and deletes the header and footer, to provide maximum space on a page for printing the spreadsheet. The ChemicalFormat macro formats text in a cell as a chemical formula (e.g., H2SO4 becomes H_2SO_4).

To Add a Custom Toolbutton to a Toolbar and Assign a Macro to It

1. Choose **Toolbars** from the **View** menu and choose **Customize...**, or choose **Customize...** from the **Tools** menu, or right-click on a toolbar and choose **Customize...** from the shortcut menu.

3. Scroll through the Categories box and choose Macros.

4. Select the "happyface" toolbutton (you'll create a new image for it later) and drag it to the desired location on the toolbar.

5. Now right-click on the custom button to display the shortcut menu, and choose **Assign Macro...** In the Assign Macro dialog box (Figure 23-6), select the name of the macro to be assigned to the toolbutton. Click the OK button.

The Macros category in the Commands tab (Figure 23-5) contains a custom toolbutton that can be assigned to a macro by following the procedure in the following box.

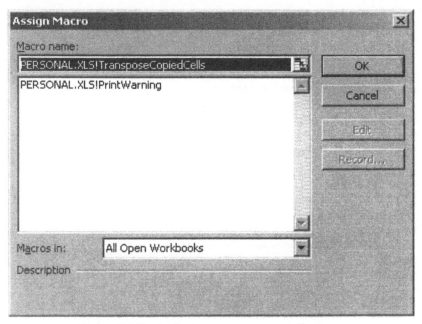

Figure 23-6. Assigning a macro to a toolbutton.

Excel Tip. When you place a custom toolbutton on a toolbar, you don't have to assign a macro to it right away. Later, if you click a custom toolbutton that doesn't have a macro assigned to it, the Assign to Tool dialog box will appear.

The ToggleNumberFormat Macro

In scientific spreadsheet computing, it's common to convert numbers from floating-point format to scientific, and vice versa. Although Excel provides toolbuttons in the Formatting toolbar to format numbers with Currency, Percent or Comma format, there isn't a toolbutton for Scientific format. The macro listing in Figure 23-7 toggles a number between Floating Point or General format and Scientific format.

The Macro. After getting the number format of the active cell (or top left cell of a range selection), the macro examines the number format. If characters corresponding to percent, currency, date or time formats are found, nothing is done. If an "E" is found, the range is formatted "General". If none of these characters is found, the number format is changed to the default Scientific format, 0.00E+00.

```
Sub ToggleNumberFormat()
'  Toggles between floating point and scientific number formats
'  Begun 3/21/97.  Last modified 03/09/11
'
'Check for a worksheet and make sure a range is selected
If TypeName(ActiveSheet) <> "Worksheet" Then Exit Sub
If TypeName(Selection) <> "Range" Then Exit Sub

'Get number format of cell, or top left cell of range
fmt = ActiveCell.NumberFormat
'Do nothing if format is percent, currency, date or time
If InStr(fmt, "%") Or InStr(fmt, "$") Then Exit Sub
If InStr(fmt, "m") Or InStr(fmt, "d") Or InStr(fmt, "y") Then Exit Sub
If InStr(fmt, "h") Or InStr(fmt, "s") Then Exit Sub

'Toggle the format
If InStr(fmt, "E") Then
   Selection.NumberFormat = "General"
Else
   Selection.NumberFormat = "0.00E+00"
End If
End Sub
```

Figure 23-7. Simple number-formatting macro to assign to a toolbutton.

Writing a macro that performs Floating Point/Scientific number formatting and returns a number with the same number of significant figures as in the original number is much more difficult. You may wish to try to write one.

The FullPage Macro

The FullPage macro maximizes the space on a page that is available for printing a worksheet, by eliminating margins, header and footer. To do this by using menu commands requires choosing **Page Setup** from the **File** menu, setting Left Margin, Right Margin, Top Margin and Bottom Margin to zero, choosing Header and deleting the header text, and then choosing Footer and deleting the footer text. The FullPage macro was written to do this at the click of a toolbutton; the listing is shown in Figure 23-8.

The Macro. Two custom toolbuttons, a Full Page Portrait toolbutton and a Full Page Landscape toolbutton, were created and positioned on the left side of the Standard toolbar. The FullPage macro was assigned to both buttons, as described earlier.

When either button is pressed, the macro uses **Application.Caller** to determine which button was pressed. **Application.Caller** returns a two-element array, the button position and the toolbar name. These are used to obtain the button caption text.

```
Sub FullPage ()

'Get info about which button pressed
WhichButton = Application.Caller
ButtonPosition = WhichButton(1)
BarName = WhichButton(2)
ButtonCaption = CommandBars(BarName). _
    Controls(ButtonPosition).Caption

'Use Excel4Macro command since it executes MUCH faster than the
VBA equivalent
If ButtonCaption = "Full Page Portrait" Then
ExecuteExcel4Macro ("PAGE.SETUP("""","""",0,0,0,0,,,1,1,1)")
'Double quotes needed
Else
ExecuteExcel4Macro ("PAGE.SETUP("""","""",0,0,0,0,,,1,1,2)")
End If

End Sub
```

Figure 23-8. FullPage Landscape and Portrait macros to assign to toolbar buttons.

This macro uses the **ExecuteExcel4Macro** method.   The Excel 4 Macro PAGE.SETUP executes much faster than its VBA equivalent.

The button caption text is used to set the orientation argument (the eleventh argument) in the Excel 4 Macro to 1 (for Portrait) or 2 (for Landscape). The first six arguments set the header and footer to null, and the margins to zero.

Creating a Custom Toolbutton Image

Use the Button Editor to edit an existing toolbutton image or create a new one. First, choose **Toolbars** from the **View** menu and choose **Customize...** from the shortcut menu. Right-click on the toolbutton image you wish to edit to display the **Toolbars** shortcut menu. Choose **Edit Button Image...** from the shortcut menu to display the Button Editor (Figure 23-9).

The button image will be displayed in the Picture area, 16 pixels wide by 15 pixels high. Click on any pixel in the Picture area to add a pixel of a selected color. Click a second time if you want to remove that color. The background color "Erase" provides a gray background identical to the rest of the toolbar.

Toolbuttons in Excel 2003 are much more artistic than the ones in Excel 2000. The hue, saturation and luminosity of the colors can be modified. To modify colors, press the Color Picker button in the Button Editor to display the Color dialog box (Figure 23-10).

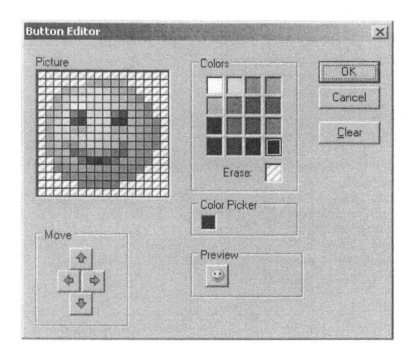

Figure 23-9. The Button Editor dialog box.

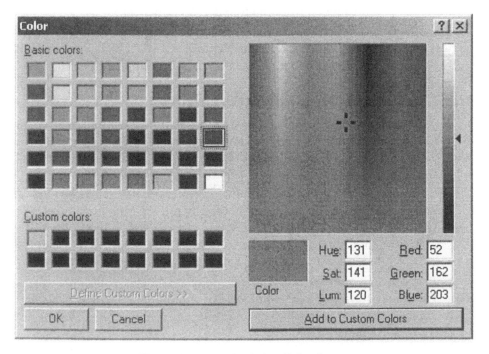

Figure 23-10. The Color dialog box.

In the Button Editor, the Move buttons allow you to shift the image up, down, left or right, but only if the background is "Erase". If you want to create a new picture, press the Clear button before you begin. You will start with a complete "Erase" background. The Preview window shows the appearance of the toolbutton image.

To edit large toolbuttons, choose **Toolbars** from the **View** menu and check the Large Buttons check box. Then display the button image to be edited. The pixel area displayed is 24 pixels wide by 23 high.

Custom toolbuttons were created for the Full Page Portrait ▦ , Full Page Landscape ▦ , Chemical Format H₂0 and Toggle Number Format ▦ macros. The ChemicalFormat macro is described in Chapter 19. Module sheets containing these macros are best saved in the Personal Macro Workbook.

How to Add a ToolTip to a Custom Button

If you create a custom toolbutton by dragging the Custom button onto a toolbar, the ToolTip message is simply "Custom". You can add your own ToolTip text. Figure 23-11 shows a custom ToolTip message displayed with the Toggle Number Format toolbutton. To add or change ToolTip text, use the procedure in the following box.

To Add a ToolTip to a Custom Toolbutton

1. Choose **Toolbars** from the **View** menu and choose **Customize...**, or choose **Customize...** from the **Tools** menu, or right-click on a toolbar and choose **Customize...** from the shortcut menu.

2. Right-click on the custom button to display the shortcut menu.

3. In the Name box, type the ToolTip text.

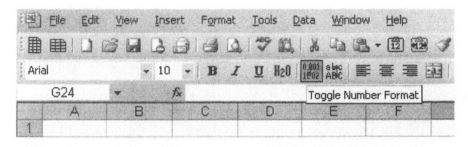

Figure 23-11. A ToolTip for the Toggle Number Format custom toolbutton.

Customizing in Excel 2007/2010

Although the Excel 2007 Ribbon can't be customized by the casual user, the Quick Access Toolbar can be customized. You can customize it with the buttons that represent the commands that you use most frequently. For example, you can create a toolbar similar to the one that's displayed in Excel 2003 when you uncheck the box "Show Standard and Formatting toolbuttons on two rows" in the Options tab of the Excel 2003 **Customize...** menu command.

You can also assign a macro to a custom button and place it on the Quick Access Toolbar. What you can't do, however, is create a custom button image. You can select an image from a gallery of approximately 181 button images, but there's no built-in way to change the image.

Excel 2010 introduced the ability to customize the Ribbon, not just the Quick Access Toolbar. New tabs can be added to the Ribbon, and built-in commands can be added to custom tabs, as described in a following section.

Moving the Quick Access Toolbar

There are two possible locations for the Quick Access Toolbar. The default location is above the ribbon. You can move the Quick Access Toolbar to the other location, below the ribbon, which you may find more convenient, especially if you customize the Quick Access Toolbar to make it resemble the "Standard and Formatting toolbars share one row" of Excel 2003.

To move the Quick Access Toolbar, click on the drop-down button at the right-hand end of the Quick Access Toolbar to display the Customize Quick Access Toolbar shortcut menu, or simply right-click on the Ribbon menu bar. Click on "Show Quick Access Toolbar below the ribbon".

Adding a Built-In Command Button to the Quick Access Toolbar

There are two ways to customize the Quick Access Toolbar. The first way is the following: click on the appropriate ribbon tab to display the button that you want to add to the Quick Access Toolbar. Right-click on the button and choose "Add to Quick Access Toolbar" in the shortcut menu. If you use this method, the command is always added to the end of the Quick Access Toolbar.

A better way to customize the Quick Access Toolbar is to right-click on it and choose "Customize the Quick Access Toolbar" from the shortcut menu. (Alternatively, in Excel 2007, click on the Office button and choose Excel Options, then in the Excel Options window, choose Customize. In Excel 2010, click on the File tab, choose Options, and then choose Customize Ribbon.) This will display the Customize window, shown in Figure 23-12. The left window shows available commands, while the right window shows the current configuration of the Quick Access Toolbar.

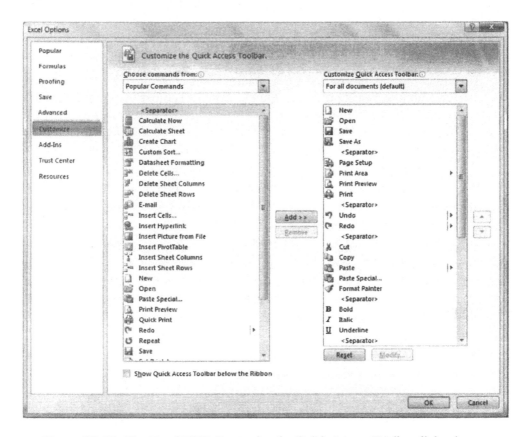

Figure 23-12. The Excel 2007 Customize the Quick Access Toolbar dialog box.

If you customize the Quick Access Toolbar using the Customize window, you can add toolbuttons that are not found on any tab of the ribbon, you can group related commands together rather than simply adding them at the end of the Quick Access Toolbar, and you can add separator bars between groups of related commands.

To add a command to the Quick Access Toolbar, click on the desired command in the list of available commands (I recommend that you choose "All Commands" in the "Choose commands from:" drop-down list above the list of available commands.) Press the "Add" button. The command will be added at the bottom of the current configuration of the Quick Access Toolbar (the list on the right), or immediately below the highlighted command in the list on the right. You can use the Up or Down buttons to move the command. You can separate groups of commands with the <separator> command, found at the top of the list of available commands. When you press OK, the new configuration of the Quick Access Toolbar will be set.

Adding a Custom Toolbutton
to the Quick Access Toolbar

To add a custom button to run a macro, right-click on the Quick Access Toolbar and choose "Customize the Quick Access Toolbar" from the shortcut menu. In the "Choose commands from:" drop-down list box, select Macros. In the list of open macros, choose the macro that you want to assign to a custom button, and then click Add to place the command in the list of current Quick Access Toolbar commands.

The default button image for a macro is ⬚. The button will be added at the right-hand end of the Quick Access Toolbar; you can move it elsewhere if you wish, using the Customize the Quick Access Toolbar dialog box.

You can't create a custom button image as you can in Excel 2003, using the Button Editor. You can only choose a different image from a gallery of button images. To change the button image, while the command is still highlighted in the list of Quick Access Toolbar commands, press the Modify button to display the Modify Button window. Select the button image that you want to use and press OK.

Even though there are 180 button images, most of them don't seem to be too useful.

Figure 23-13. The Excel 2007/2010 gallery of button images.

To change the ToolTip text, type the text in the "Display name:" box in the Modify Button window. Click OK to add the macro button to the Quick Access Toolbar.

Adding Custom Toolbuttons to the Ribbon

You can also add custom toolbuttons to the ribbon, in the same way that you can add custom menus (in the previous chapter). When added, custom buttons appear in the Add-Ins tab in a new group labeled Toolbar Commands. The procedure shown in Figure 23-14 installs a custom button labeled "Chemical Format" and assigns it to the procedure ChemicalFormat, located in the Personal Macro Workbook. Interestingly, you must specify the Excel 2003 toolbar name (in this case "Standard") even though there is no such toolbar in Excel 2007/2010. The name of any of the built-in toolbars of Excel 2003 seems to work.

```
Sub InstallButton()
Dim NewButton As Object

Set NewButton = Application.CommandBars("Standard").Controls.Add _
    (Type:=msoControlButton, ID:=2950)
With NewButton
  .Caption = "Chemical Format"
  .OnAction = "Personal.xlsb!ChemicalFormat"
End With
End Sub
```

Figure 23-14. Procedure to install a custom toolbutton in the Excel 2007/2010 ribbon.

In the above procedure, the ID number, 2950, is the number for the Custom button; it results in the button image illustrated in Figure 23-15. Obviously, if you install a number of custom buttons in the Toolbar Commands group, you will need a way to change the button images.

Figure 23-15. A custom toolbutton in the Add-Ins tab of the Excel 2007/2010 ribbon.

Excel 2007/2010 does not have a built-in way to modify toolbutton images, but there is a way to use the Excel 2003 Button Editor to create a custom button image for a custom button in Excel 2007/2010. This is described in the following section.

How to Use the Excel 2003 Button Editor in Excel 2007/2010

In order to use the Excel 2003 Button Editor to create a custom button image for a custom button in Excel 2007/2010, you must have both Excel 2003 and Excel 2007 or Excel 2010 installed on your computer.

Use the Excel 2003 Button Editor to create the desired button image. Choose **Customize...** from the **Tools** menu, right-click on the button image and choose Copy Button Image from the shortcut menu. Now activate Excel 2007/2010 and run the PasteFace procedure shown in Figure 23-16. You must specify the name of the custom button, in this case "Chemical Format", in the procedure.

```
Sub PasteFace()
Application.CommandBars("Standard"). _
      Controls("Chemical Format").PasteFace
End Sub
```

Figure 23-16. VBA **Sub** procedure to paste a button image.

Figure 23-17. Two custom toolbuttons (Chemical Formatting and Quick Chart) in the Add-Ins tab of the Excel 2007/2010 ribbon.

Removing Custom Toolbuttons from the Ribbon

To remove a custom toolbutton from the Toolbar Commands group in the Add-Ins tab of the ribbon, simply right-click on the button. This will display the shortcut menu shown in Figure 23-18.

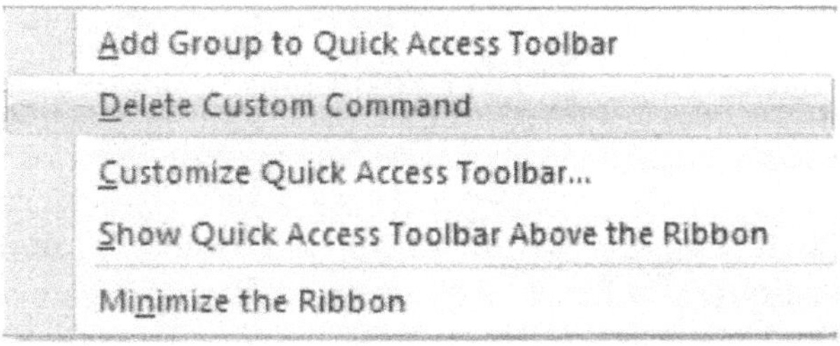

Add Group to Quick Access Toolbar
Delete Custom Command
Customize Quick Access Toolbar...
Show Quick Access Toolbar Above the Ribbon
Minimize the Ribbon

Figure 23-18. Using a shortcut menu to delete a custom toolbutton.

When you click on "Delete Custom Command", you will be asked to confirm that you want to delete the toolbutton.

You can also put a copy of the Toolbar Commands group on the Quick Access Toolbar. To do this, right-click on the bar containing the text "Toolbar Commands" in the Toolbar Commands group to display the shortcut menu shown in Figure 23-19, and choose Add Group to Quick Access Toolbar. This will install a toolbutton at the extreme right-hand end of the Quick Access Toolbar. Clicking on the toolbutton will display a drop-down like the one shown in Figure 23-20.

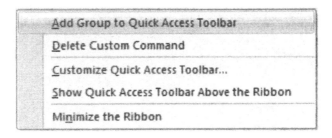

Figure 23-19. Adding Toolbar Commands group to the Quick Access Toolbar

Figure 23-20. The Toolbar Commands group added to the Quick Access Toolbar

Customizing the Ribbon (Excel 2010 Only)

Excel 2010 added the possibility of customizing the Ribbon. You can add new tabs or groups, or change the built-in tabs. In the Excel 2010 Excel Options dialog box, the Customize group contains both Customize Ribbon and Quick Access Toolbar possibilities.   Click on Customize Ribbon to display the Customize the Ribbon dialog box, shown in Figure 23-21.

Custom buttons must be added to a new Custom Group. To add a new group, in the hierarchy tree on the left, choose the tab in which you want to add the group. Press the New Group button to add a custom Group. You can use the Up and Down buttons on the right side to position the new group.

With the new group still highlighted, choose Macros from the drop-down list in the top left. In the list of macros, click on the one you want to assign to a new button, and press Add. This will add a button with the default button image for a macro, 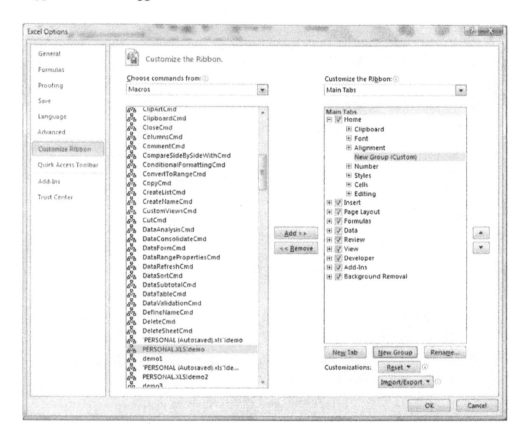. Finally, use the Rename button to change the name of the Group and the name of the macro, as well as to choose a suitable button image from the gallery of button images shown in Figure 23-13. Figure 23-22 shows part of the Home tab, with a new group, Toggle Formats or Case, with two custom buttons, Toggle Case and Toggle Number Format.

Figure 23-21. The Excel 2010 Customize the Ribbon dialog box.

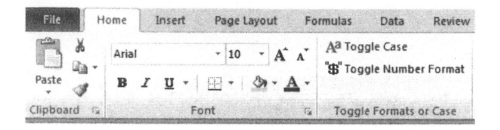

Figure 23-22. A custom group added to the Home tab of the Excel 2010 ribbon.

APPENDICES

Appendix A

What's Where
in Excel 2007/2010

Excel 2003 **File** menu	Excel 2007/2010 location
New...	Office Button (Excel 2007), File tab (Excel 2010) → New
Open...	Office Button (Excel 2007), File tab (Excel 2010) → Open
Close	Office Button (Excel 2007), File tab (Excel 2010) → Close
Save	Office Button (Excel 2007), File tab (Excel 2010) → Save
Save As...	Office Button (Excel 2007), File tab (Excel 2010) → Save As
Page Setup...	Office Button → Print → Print Preview → Page Setup (Excel 2007)
	File tab → Print → Print Preview hyperlink (Excel 2010)
	or Page Layout tab → Page Setup
Print Area ▶	Page Layout tab → Page Setup → Print Area
Print Preview	Office Button → Print → Print Preview (Excel 2007)
	File tab → Print → Print Preview (Excel 2010)
Print...	Office Button (Excel 2007), File tab (Excel 2010) → Print
Recent Documents	Office Button → Recent Documents (Excel 2007)
	File tab → Recent (Excel 2010)
Exit	Office Button (Excel 2007), File tab (Excel 2010) → Exit

Excel 2003 **Edit** menu	Excel 2007/2010 location
Cut	Home tab → Clipboard → Cut
Copy	Home tab → Clipboard → Copy
Office Clipboard	Home tab → Clipboard → Office Clipboard
Paste	Home tab → Clipboard → Paste
Paste Special...	Home tab → Clipboard → Paste → Paste Special
Paste as Hyperlink	Home tab → Clipboard → Paste → Paste as Hyperlink
Fill ▶	Home tab → Editing → Fill▼
Clear ▶	Home tab → Editing → Clear▼
Delete	Home tab → Cells → Delete
Delete Sheet	Home tab → Cells → Delete → Delete Sheet
Move or Copy Sheet...	Home tab → Cells → Format → Move or Copy Sheet
Find...	Home tab → Editing → Find & Select → Find
Replace...	Home tab → Editing → Find & Select → Replace
Go To...	Home tab → Editing → Find & Select → Go To
Links...	Data tab → Connections → Edit Links
Object	(double-click an object to edit it)

Excel 2003 **View** menu	Excel 2007/2010 location
Normal	View tab → Workbook Views → Normal
Page Break Preview	View tab → Workbook Views → Page Break Preview
Formula Bar	View tab → Show/Hide → Formula Bar
Header and Footer...	Insert tab → Text → Header & Footer
Comments	Review tab → Comments
Custom View tabs	View tab → Workbook Views → Custom Views
Full Screen	View tab → Workbook Views → Full Screen
Zoom...	View tab → Zoom

Excel 2003 **Insert** menu	Excel 2007/2010 location
Cells...	Home tab → Cells → Insert
Rows	Home tab → Cells → Insert → Insert Sheet Rows
Columns	Home tab → Cells → Insert → Insert Sheet Columns
Worksheet	Home tab → Cells → Insert → Insert Sheet
Chart	Insert tab → Charts
Symbol	Insert tab → Text → Symbol
Page Break	Page Layout tab → Page Setup → Breaks → Insert Page Break
Reset All Page Breaks	Page Layout tab → Page Setup → Breaks → Reset All Page Breaks
Function...	Formulas tab → Function Library → Insert Function
Name → Define	Formulas tab → Defined Names → Name Manager
Name → Paste	Formulas tab → Defined Names → Use in Formula → Paste Names
Name → Create	Formulas tab → Defined Names → Create from Selection
Name → Apply	Formulas tab → Defined Names → Define Name → Apply Names
Comment	Review tab → Comments → New Comment
Picture → Clip Art	Insert tab → Illustrations → Clip Art
Picture → From File	Insert tab → Illustrations → Picture
Picture → From File	Chart Tools → Layout → Insert → Picture
Picture → From File	PivotChart Tools → Layout → Insert → Picture
Picture → AutoShapes	Insert tab → Illustrations → Shapes
Picture → WordArt	Insert tab → Illustrations → WordArt
Diagram...	Insert tab → Illustrations → SmartArt
Object...	Insert tab → Text → Object
Hyperlink...	Insert tab → Links → Hyperlink

Excel 2003 **Format** menu	Excel 2007/2010 location
Cells	Home tab → Cells → Format
Row	Home tab → Cells → Format
Column	Home tab → Cells → Format
Sheet	Home tab → Cells → Format
Sheet → Background	Page Layout tab → Page Setup → Background
AutoFormat	Home tab → Styles → Format as Table
Conditional Formatting	Home tab → Conditional Formatting
Style	Home tab → Style → Cell Styles

Excel 2003 **Tools** menu	Excel 2007/2010 location
Spelling...	Review tab → Proofing → Spelling
Research...	Review tab → Proofing → Research
Error Checking...	Formulas tab → Formula Auditing → Error Checking
Shared Workspace...	Office Button → Publish → Create Document Workspace
Share Workbook...	Review tab → Changes → Share Workbook
Track Changes	Review tab → Changes → Track Changes
Protection	Review tab → Changes
Protection → Protect Sheet	Home tab → Cells → Format → Protect Sheet
Goal Seek...	Data tab → Data Tools → What-If Analysis → Goal Seek
Scenarios...	Data tab → Data Tools → What-If Analysis → Scenario Manager
Formula Auditing	Formulas tab → Formula Auditing
Macro ▶	Developer tab → Code → Macros
Add-Ins...	Office Button → Excel Options → Add-Ins (Excel 2007) File tab → Options → Add-Ins (Excel 2010)
AutoCorrect Options...	Office Button → Excel Options → Proofing → AutoCorrect Options (Excel 2007) File tab → Options → Proofing → AutoCorrect Options (Excel 2010)
Customize...	Office Button → Excel Options → Customize (Excel 2007) File tab → Options → Customize Ribbon or Quick Access Toolbar (Excel 2010)
Options...	Office Button → Excel Options (Excel 2007) File tab → Options (Excel 2010)

Excel 2003 **Data** menu	Excel 2007/2010 location
Sort...	Data tab → Sort & Filter → Sort
Filter ▶	Data tab → Sort & Filter → Filter
Form...	(add the Form button to the Quick Access Toolbar)
Subtotals...	Data tab → Outline → Subtotal
Validation...	Data tab → Data Tools → Data Validation
Table...	Data tab → Data Tools → What-If Analysis → Data Table
Text to Columns...	Data tab → Data Tools → Convert Text to Table
Consolidate...	Data tab → Data Tools → Consolidate
Group and Outline ▶	Data tab → Outline
PivotTable...	Insert → Tables → PivotTable → PivotTable/PivotChart
Import External Data	Data tab → Get External Data
List → Create List	Insert tab → Tables → Table
List ▶	Table Tools → Design
XML ▶	Developer → XML
Refresh Data	PivotChart Tools → Analyze → Data → Refresh
Refresh Data	Table Tools → Design → External Table Data → Refresh
Refresh Data	Data tab → Connections → Refresh → Refresh

Excel 2003 **Chart** menu	Excel 2007/2010 location
Chart Type...	Chart Tools → Design tab → Type → Change Chart Type
Source Data...	Chart Tools → Design tab → Data → Select Data
Chart Options...	Chart Tools → Layout tab
Location...	Chart Tools → Design tab → Location → Move Chart
Add Data...	Chart Tools → Design tab → Data → Edit Data Source
Add Trendline...	Chart Tools → Layout tab → Analysis → Trendline
3-D View...	Chart Tools → Layout tab → Background → 3-D View

Excel 2003 **Window** menu	Excel 2007/2010 location
New Window	View tab → Window → New Window
Arrange...	View tab → Window → Arrange All
Compare Side by Side with	View tab → Window → View Side by Side
Hide	View tab → Window → Hide
Unhide...	View tab → Window → Unhide
Split	View tab → Window → Split
Freeze Panes	View tab → Window → Freeze Panes
Open Workbooks	View tab → Window → Switch Windows

Appendix B

Selected Worksheet Functions by Category

Excel 2003 and 2007/2010 provide more than 300 worksheet functions, organized in ten categories: Database, Date and Time, Engineering, Financial, Information, Logical, Lookup and Reference, Math and Trigonometry, Statistical and Text. This appendix lists selected worksheet functions by category. In Appendix C, the functions are listed alphabetically, with comments, examples and related functions.

DATABASE FUNCTIONS

DAVERAGE	Returns the average of selected values.
DCOUNT	Returns the number of cells that contain numbers in a specified field, according to specified criteria.
DCOUNTA	Returns the number of cells that contain values in a specified field, according to specified criteria.
DGET	Returns a single record that matches the specified criteria.
DMAX	Returns the maximum value in the range *Database*.
DMIN	Returns the minimum value in the range *Database*.
DPRODUCT	Returns the product of values in a specified field that match specified criteria in a database.
DSTDEV	Returns the standard deviation of values in a specified field that match specified criteria in a database.
DSTDEVP	Returns the standard deviation of values in a specified field that match specified criteria in a database.
DSUM	Returns the sum of numbers in the field column of records in the database that match the criteria.
DVAR	Returns the variance of values in a specified field that match specified criteria in a database.
DVARP	Returns the variance of values in a specified field that match specified criteria in a database.
SUBTOTAL	Returns a subtotal in a database.

DATE & TIME FUNCTIONS

DATE	Returns the serial number of a date.
DATEVALUE	Converts a date in the form of text to a serial number.
DAY	Converts a serial number to a day of the month.
MONTH	Converts a serial number to a month.
NOW	Returns the serial number of the current date and time.
TODAY	Returns the serial number of today's date.
WEEKDAY	Converts a serial number to a day of the week.
YEAR	Converts a serial number to a year.

INFORMATION FUNCTIONS

CELL	Returns information about the formatting, location or contents of a cell.
COUNTBLANK	Counts the number of blank cells within a range.
ERROR.TYPE	Returns a number corresponding to an error value.
INFO	Returns information about the current operating environment.
ISBLANK	Returns TRUE if the cell is blank.
ISERR	Returns TRUE if the argument is any error value except #N/A.
ISERROR	Returns TRUE if the argument is an error value.
ISEVEN	Returns TRUE if the argument is even.
ISLOGICAL	Returns TRUE if the argument is a logical value.
ISNA	Returns TRUE if the argument is #N/A.
ISNONTEXT	Returns TRUE if the argument is not text.
ISNUMBER	Returns TRUE if the argument is a number.
ISODD	Returns TRUE if the argument is odd.
ISREF	Returns TRUE if the argument is a reference.
ISTEXT	Returns TRUE if the argument is text.
N	Returns a value converted to a number. Excel usually does this automatically when necessary.
NA	Returns the error value #N/A
TYPE	Returns a number indicating the data type of a value.

LOGICAL FUNCTIONS

AND	Returns TRUE if all arguments are true, otherwise FALSE.
FALSE	Returns the logical value FALSE.
IF	Returns one value if *logical_test* is TRUE, another value if *logical_test* is FALSE.
IFERROR	Returns *value_if_error* if *expression* evaluates to an error, otherwise returns *expression* (Excel 2007/2010 only).

NOT   Reverses the logic of its argument.
OR   Returns TRUE if any argument is TRUE.
TRUE   Returns the logical value TRUE.

LOOKUP & REFERENCE FUNCTIONS

ADDRESS   Returns a reference in the form of text.
AREAS   Returns the number of areas in a multiple selection.
CHOOSE   Chooses a value from a list of values, based on index number.
COLUMN   Returns the column number of a reference.
COLUMNS   Returns the number of columns in a reference.
HLOOKUP   Finds the value in the first row of an array that is equal to or less than *lookup_value*. Returns the associated value in the same column, as determined by *offset_num*.
INDEX   Returns a value from a reference or array, using a specified index.
INDIRECT   Returns a reference specified by a text value.
LOOKUP   Looks up values in an array.
MATCH   Looks up a value in an array and returns its relative position.
OFFSET   Returns a reference offset from a base reference by specified number of rows and columns.
ROW   Returns the row number of a reference.
ROWS   Returns the number of rows in a reference.
TRANSPOSE   Returns the transpose of an array.
VLOOKUP   Finds the value in the first column of an array that is equal to or less than *lookup_value*. Returns the associated value in the same row, as determined by *offset_num*.

MATH & TRIG FUNCTIONS

ABS   Returns the absolute value of a number.
ACOS   Returns the angle corresponding to a cosine value.
ACOSH   Returns the inverse hyperbolic cosine of a number.
ASIN   Returns the angle corresponding to a sine value.
ASINH   Returns the inverse hyperbolic sine of a number.
ATAN   Returns the angle corresponding to a tangent value.
ATAN2   Returns the angle defined by a pair of x- and y coordinates.
ATANH   Returns the inverse hyperbolic tangent of a number.
COS   Returns the cosine of a given angle.
COSH   Returns the hyperbolic cosine of a number.
COUNTIF   Returns the number of non-blank cells within a range

	that meet the given criteria.
DEGREES	Converts a value in radians to degrees.
EXP	Returns the value of *e* raised to the power *number*.
FACT	Returns the factorial of a number, i. e., 1*2*3*... *number*.
INT	Rounds a number down to the nearest integer.
LN	Returns the base-*e* logarithm of a number.
LOG	Returns the logarithm of a number to the specified base.
LOG10	Returns the base-10 logarithm of a number.
MDETERM	Returns the determinant of an array.
MINVERSE	Returns the inverse of a matrix.
MMULT	Returns the product of two matrices.
MOD	Returns the remainder of the division of number by divisor.
MROUND	Returns a number down to the nearest specified level.
PI	Returns π.
PRODUCT	Returns the product of the specified arguments.
RADIANS	Converts an angle in degrees to radians.
RAND	Returns a random number between 0 and 1.
RANDBETWEEN	Returns a random number within a specified range.
ROMAN	Converts an Arabic number to Roman numerals.
ROUND	Rounds a number to a specified number of digits.
ROUNDDOWN	Rounds a number down.
ROUNDUP	Rounds a number up.
SIGN	Returns the sign of a number.
SIN	Returns the sine of a given angle.
SINH	Returns the hyperbolic sine of a number.
SQRT	Returns the square root of a number.
SUM	Returns the sum of all the numbers in a range.
SUMIF	Returns the sum of all the numbers in a range that satisfy the specified criteria.
SUMIFS	Returns the sum of the cells specified by multiple criteria (Excel 2007/2010 only).
SUMPRODUCT	Returns the sum of the products of corresponding array components.
SUMSQ	Returns the sum of the squares of arguments.
TAN	Returns the tangent of a given angle.
TANH	Returns the hyperbolic tangent of a number.
TRUNC	Truncates a number.

STATISTICAL FUNCTIONS

AVEDEV	Returns the average absolute deviation of data points from the mean.
AVERAGE	Returns the average of all the numbers in a range.
AVERAGEIF	Returns the average of all values in a range that satisfy a specified criterion (Excel 2007/2010 only).
AVERAGEIFS	Returns the average of all values in a range that satisfy multiple criteria (Excel 2007/2010 only).
CORREL	Returns the correlation coefficient between two data sets.
COUNT	Returns the number of numbers in a range.
COUNTA	Returns the number of non-blank values in a range.
COUNTIF	Returns the number of non-blank values in a range within a range that meet a specified criterion.
COUNTIFS	Returns the number of non-blank values in a range within a range that satisfy multiple criteria (Excel 2007/2010 only).
INTERCEPT	Returns the intercept of the linear regression line $y = mx + b$.
LARGE	Returns the kth largest value in a list of values.
LINEST	Returns the parameters of multiple linear regression.
MAX	Returns the maximum value in a list of arguments.
MEDIAN	Returns the median value in a list of arguments.
MIN	Returns the minimum value in a list of arguments.
RSQ	Returns the square of the correlation coefficient.
SLOPE	Returns the slope of the linear regression line $y = mx + b$.
SMALL	Returns the kth smallest value in a list of values.
STDEV	Returns the standard deviation of a sample.
VAR	Returns the variance of a sample.

TEXT FUNCTIONS

CHAR	Returns the character corresponding to the character code number.
CLEAN	Removes all non-printable characters from a text string.
CODE	Returns the numeric code corresponding to a character.
CONCATENATE	Concatenates text values into a single text string.
DOLLAR	Converts a number to a text value in currency format.
EXACT	Compares two text strings; returns TRUE if they are the same.
FIND	Returns the position at which one text string occurs within another text string.

FIXED	Rounds a number to the specified number of decimal places and returns the result as text.
LEFT	Returns the specified number of characters from a text string, beginning at the left.
LEN	Returns the number of characters in a text string.
LOWER	Converts a text string to lowercase.
MID	Returns the specified number of characters from a text string, beginning at the specified position.
PROPER	Capitalizes the first letter in each word of a text string.
REPLACE	Replaces characters at a specified position within a text string.
REPT	Repeats a text value a specified number of times.
RIGHT	Returns the specified number of characters from a text string, beginning at the right.
SEARCH	Finds the position of a text value within a text string.
SUBSTITUTE	Finds and substitutes characters within a text string.
T	Converts a value to text. Excel usually does this automatically when necessary.
TEXT	Formats a number and returns it as text.
TRIM	Removes spaces from a text string, except for single spaces between words.
UPPER	Converts a text string to uppercase.
VALUE	Converts a text argument to a number. Excel usually does this automatically when necessary.

Appendix C

Alphabetical List of Selected Worksheet Functions

This appendix lists selected worksheet functions that will be useful in creating advanced worksheet formulas for chemical applications. For each function, a description of the function and its syntax are given; most cases include an example of the use of the function, and any related functions are listed. Function arguments are in italics; arguments are in boldface type if the argument is required, and in plain text if the argument is optional. The required data type for an argument is indicated by the suffixes *_num*, *_ref*, *_logical*, *_text*. If a function can accept an argument of more than one type, then no suffix is attached.

ABS
Returns the absolute value of a number
Syntax: ABS(***number***)
Example: =ABS(-7.3) returns 7.3
Related Function: SIGN

ACOS
Returns the angle corresponding to a cosine value.
Syntax: ACOS(***number***)
Number must be from -1 to $+1$. The returned angle is in radians, in the range 0 to π. To convert the result to degrees, multiply by $180/\pi$.
Example: =ACOS(0) returns 1.570796327, or 90 degrees.
Related Functions: COS, other trigonometric functions

ADDRESS
Returns a reference in the form of text.
Syntax: ADDRESS(***row_num,column_num***,abs_num,a1,sheet_text)
Abs_num specifies the type of reference to return. If *abs_num* is 1 or omitted, returns an absolute reference; if 2, absolute row and relative column; if 3, relative row and absolute column; if 4, relative. A1 is a logical value that

specifies the A1 or R1C1 reference style. If A1 is TRUE or omitted, ADDRESS returns an A1-style reference; if FALSE, ADDRESS returns an R1C1-style reference. Sheet_text is text specifying the name of the worksheet to be used as the external reference. If sheet_text is omitted, no sheet name is used.

Examples:
>   =ADDRESS(2,3) returns "$C$2"
>   =ADDRESS(2,3,2) returns "C$2"
>   =ADDRESS(2,3,2,FALSE) returns "R2C[3]"
>   =ADDRESS(2,3,1,FALSE,"[Book1]Sheet1") returns "[Book1]Sheet1!R2C3"

Related Functions: COLUMN, INDIRECT, ROW

AND

Returns TRUE if all arguments are TRUE, otherwise returns FALSE.
Syntax: AND(*logical1*,*logical2*,...)
Up to 30 logical conditions can be tested.
Example: If A1 contains 0 and A2 contains 110, then =AND(A1=0,A2>100) returns TRUE.
Related Functions: NOT, OR

AREAS

Returns the number of areas in a multiple selection.
Syntax: AREAS(*reference*)
Related Functions: COLUMN, COLUMNS, ROW, ROWS

ASIN

Returns the angle corresponding to a sine value.
Syntax: ASIN(*number*)
Number must be from -1 to $+1$. The returned angle is in radians, in the range $-\pi/2$ to $+\pi/2$. To convert the result to degrees, multiply by $180/\pi$.
Example: If A1 contains 0.7071, then =ASIN(A1) returns 0.785388573, or 45 degrees.
Related Functions: SIN, other trigonometric functions

ATAN

Returns the angle corresponding to a tangent value.
Syntax: ATAN(*number*)
The returned angle is in radians, in the range 0 to π. To convert the result to degrees, multiply by $180/\pi$.
Example: =ATAN(0) returns 0.785388573 or 45 degrees.
Related Functions: ATAN2, TAN, other trigonometric functions

ATAN2

Returns the angle defined by a pair of x- and y coordinates. The angle is between the x-axis and the line connecting the origin $(0,0)$ and the point (x,y).

Syntax: ATAN2(***x_num,y_num***)
The returned angle is in radians, in the range $-\pi$ to π. To convert the result to degrees, multiply by $180/\pi$. A negative result represents a clockwise angle from the *x*-axis.
Example: =ATAN2(3,4) returns 0.927295218, or 53.13 degrees.
Related Functions: ATAN, TAN, other trigonometric functions

AVERAGE
Returns the average of all the numbers in a range.
Syntax: AVERAGE(***number1***,*number2*,...)
The arguments may be numbers, or names, arrays or references that contain numbers. Up to 30 separate arguments can be listed. Only numbers in the array or range are counted.
Related Function: MEDIAN

AVERAGEIF
Returns the average of all values in a range that satisfy a specified criterion (available in Excel 2007/2010 only).
Syntax: AVERAGEIF(***range, criterion,*** *average_range*))
Range is the range of cells to be averaged. *Criterion* is the number, reference, text or expression that specifies which cells are to be averaged. *Average_range* is the range of cells to be averaged; if omitted, values in range are averaged.
Example: =AVERAGEIF(Major, "Biology", Grade)
Related Functions: COUNT, COUNTA, COUNTIF, COUNTBLANK, SUMIF

AVERAGEIFS
Returns the average of all values in a range that satisfy multiple criteria (available in Excel 2007/2010 only).
Syntax: AVERAGEIFS(***average_range, criterion_range1, criterion1,*** *criterion_range2, criterion2,...*)
Average_range is the range of cells to be averaged. *Criterion_range* is the range of values to be compared with *criterion*. Up to 127 *criterion_range, criterion* pairs are allowed. Note the difference in order of arguments between AVERAGEIF and AVERAGEIFS; to return the same result as in the example shown above for AVERAGEIF, the expression would be AVERAGEIFS(Grade, Major, "Biology")
Example: =AVERAGEIFS(Grade, Major, "Biology", Year_of_Graduation, 97)
Related Functions: COUNT, COUNTA, COUNTIF, COUNTBLANK, SUMIFS

CELL
Returns information about location, formatting and contents of a cell.
Syntax: CELL(***info_type_text***,*reference*)
Info_type_text specifies the cell information to be returned. If *reference* is omitted, information about the active cell is returned. For further information, see On-Line Help.

Example: If the cell format is 0.00E+00, =CELL("format") returns S2.
Related Function: INFO

CHAR
Returns the character corresponding to the character code number.
Syntax: CHAR(***number***)
Number must be between 1 and 255.
Example: =CHAR(65) returns A.
Related Function: CODE

CHOOSE
Chooses a value from a list of values, based on index number.
Syntax: CHOOSE(***index_number, value1***,*value2*,...)
Index_number must be a number between 1 and 29, or a formula or reference that evaluates to same. The arguments may be numbers, names, formulas or references.
Example: If cell H8 contains 1, then
=CHOOSE(H8+1,$F$2,$G$2,$H$2,$I$2)/SUM($F$2:$I$2) calculates the fraction $G$2/SUM($F$2:$I$2).
Related Functions: INDEX, MATCH

CLEAN
Removes all non-printable characters from a text string.
Syntax: CLEAN(***text***)
Use on text imported from other applications that may contain non-printing characters.
Related Function: TRIM

CODE
Returns the numeric code for the first character of text,
Syntax: CODE(***text***)
Example: =CODE ("A") returns 65.
Related Function: CHAR

COLUMN
Returns the column number of reference.
Syntax: COLUMN(*reference*)
If *reference* is a range of cells, returns the column number of the upper left cell of the range. *Reference* cannot be a multiple selection. If *reference* is omitted, it is assumed to be the reference in which the COLUMN function appears.
Example: =COLUMN($C$3:$N$10) returns 3.
Related Functions: COLUMNS, ROW, ROWS

COLUMNS
Returns the number of columns in reference.
Syntax: COLUMNS(*reference*)
If *reference* is a multiple selection, use INDEX to select a specified area within the selection.
Example: =COLUMNS($C$3:$N$10) returns 12.
Related Functions: COLUMN, ROW, ROWS

CORREL
Returns the correlation coefficient between two data sets.
Syntax: CORREL(*array1,array2*)
Array1 and *array2* must have the same number of data points.
Related Functions: SLOPE, INTERCEPT, COVAR, PEARSON, RSQ

COS
Returns the cosine of a given angle.
Syntax: COS(*number*)
Number is the angle in radians. To convert an angle in degrees to one in radians, multiply by $\pi/180$.
Related Functions: ACOS, SIN, TAN, other trigonometric functions

COUNT
Returns the number of numbers in the list or range.
Syntax: COUNT(*value1,value2,...*)
Up to 30 arguments are allowed. Any value, other than an empty cell, is counted.
Related Functions: COUNTA, COUNTBLANK, COUNTIF

COUNTA
Returns the number of non-blank values in the list or range.
Syntax: COUNTA(*value1,value2,...*)
Up to 30 arguments are allowed. Nulls (""), logical values, numbers formatted as dates, or text values that can be converted to numbers, are counted.
Related Functions: COUNT, COUNTBLANK, COUNTIF

COUNTBLANK
Returns the number of blank cells within a range.
Syntax: COUNTBLANK(*reference*)
Cells containing null values are counted, but cells containing zero values are not counted.
Related Functions: COUNT, COUNTA, COUNTIF

COUNTIF
Returns the number of cells within a range that meet the specified criteria.
Syntax: COUNTIF(*reference,criteria*)

Example: =COUNTIF(year_of_graduation>2000)
Related Functions: COUNT, COUNTA, COUNTBLANK, SUMIF

COUNTIFS
Counts the number of times that multiple criteria are met.
*Syntax:*COUNTIFS(**criterion_range1, criterion1,** *criterion_range2, criterion2,…*)
Criterion_range is the range of cells to be examined. *Criterion* is the range of values to be compared with *criterion_range*. Up to 127 *criterion_range, criterion* pairs are allowed. Note the difference in order of arguments between COUNTIF and COUNTIFS.
Example: =COUNTIFS(year_of_graduation>2000, Major="Biology")
Related Functions: COUNT, COUNTA, COUNTIF, COUNTBLANK, SUMIF

DATE
Returns the serial number of a particular date.
Syntax: DATE(**year, month, day**)
Year is a number from 1900 to 9999. *Year* can be specified using two digits but four digits are recommended. *Month* or *day* can be greater than 12 or 31 (see example).
Example: DATE(94,7,4) corresponds to July 4, 1994. DATE(94,7,90) corresponds to Sept. 28, 1994.
Related Functions: DATEVALUE, DAY, MONTH, YEAR, NOW, TODAY

DATEVALUE
Converts a date in the form of text to a serial number.
Syntax: DATEVALUE(**text**)
Example: =DATEVALUE("3 Aug 1938") returns 14095, or 12633 if the 1904 Date System is in effect.
Related Functions: NOW, TODAY

DAY
Converts a serial number to a day of the month.
Syntax: DAY(**serial_number**)
Serial_number can also be given as text, such as "Jul-4-1994". Returns a value in the range 1 to 31.
Related Functions: NOW, TODAY, WEEKDAY, YEAR, MONTH

DEGREES
Converts an angle in radians to degrees
Syntax: DEGREES(**number**)
Related Functions: RADIANS, trigonometric functions

EXACT

Compares two text strings; returns TRUE if they are the same.
Syntax: EXACT(*text1,text2*)
EXACT is case-sensitive.
Related Functions: FIND, SEARCH

EXP

Returns the value of *e* raised to a power.
Syntax: EXP(*number*)
Returns the value of *e* raised to the power *number*.
Related Functions: LN, LOG

FACT

Returns the factorial of a number.
Syntax: FACT(*number*)
Returns the factorial of a number, i.e., 1*2*3*... *number*. *Number* must be positive. If *number* is not an integer, it is truncated.

FIND

Returns the position at which one text string occurs within another text string.
Syntax: FIND(*find_text, within_text, start_number*)
Find_text is the text string you want to find. *Within_text* is the string you are searching.   *Start_number* is the character at which to start searching.   If *start_number* is omitted, it is assumed to be 1. FIND is case-sensitive.
Example: If cell A3 contains BILLO, E JOSEPH, the formula =FIND(",",A3) returns 6.
Related Functions: EXACT, SEARCH

FIXED

Rounds a number to the specified number of decimal places and returns the result as text.
Syntax: FIXED(*number, decimals, no_commas_flag*)
Number is the number to be converted. *Decimals* is the number of decimal places desired; if *decimals* is omitted, it is assumed to be 2. If *decimals* is negative, *number* is rounded to the left of the decimal point. If *no_commas_flag* is TRUE, commas are not included in the formatted number.
Example: =FIXED(12345,3,1) returns 12345.000. =FIXED(12345) returns 12,345.00. =FIXED(PI(),4) returns 3.1416.
Related Functions: ROUND, TEXT

HLOOKUP

Finds the value in the first row of an array that is equal to or less than *lookup_value*. Returns the associated value in the same column, as determined by *offset_num*.

Syntax: HLOOKUP(***lookup_value, array, offset_num***, *match_logical*)

The values in the first row must be in ascending order. If *match_logical* is TRUE or omitted, returns the largest value that is less than or equal to *lookup_value*. If *match_logical* is FALSE, returns #N/A if an exact match is not found.

Example: See example under VLOOKUP.

Related Functions: INDEX, MATCH, VLOOKUP

IF

Returns one value if *logical_test* is TRUE, another value if *logical_test* is FALSE.

Syntax: IF(***logical_test, value_if_true***, *value_if_false*)

Value_if_true and/or *value_if_false* can be IF functions; up to seven IF functions can be nested.

Example: =IF(result=0, 0.08*estimate, 0.08*result)

Related Functions: AND, NOT, OR, IS functions

IFERROR

Returns *value_if_error* if *expression* evaluates to an error, otherwise returns *expression*.

Syntax: IFERROR(***value, value_if_error***)

INDEX

Chooses a value from an array, based on row and column number pointers.

Syntax: INDEX(***array, row_number, column_number***)

If the array is one-dimensional, only one pointer argument need be specified.

Example: If the reference $C$2:$C$102 contains the atomic weights of the elements, arranged in order of atomic number, then =INDEX($C$2:$C$102,51) returns the atomic weight of element number 51.

Related Functions: CHOOSE, HLOOKUP, MATCH, VLOOKUP

INDIRECT

Returns a reference specified by a text string.

Syntax: INDIRECT(***reference_text,***A1_logical***)

Reference_text can be an A1- or R1C1-style reference, or a name. *A1_logical* specifies the form of *reference_text*: if *reference_text* is TRUE or omitted, *reference_text* is interpreted as an A1-style reference; if *reference_text* is FALSE, *reference_text* is interpreted as an R1C1-style reference.

Example: If cell B3 contains "Sheet2!A5" then the formula =INDIRECT(B3) displays the contents of Sheet2!A5.

Related Functions: OFFSET

INFO
Returns information about the operating environment.
Syntax: INFO(***type_text***)
Type_text specifies what information is to be returned. Some useful values of *type_text*: "directory" returns path of the current directory or folder, "memavail" returns the amount of memory available, "release" returns Microsoft Excel version, "system" returns mac or pcdos. See On-Line Help for details.
Example: The formula =INFO("release") returns 12.0 if Excel 2007 is being used.
Related Function: CELL

INT
Rounds a number down to the nearest integer.
Syntax: INT(***number***)
INT rounds downward. Use TRUNC to return the integer part of a number, irrespective of sign.
Example: =INT(3.897) returns 3. =INT(-0.05) returns −1.
Related Functions: ROUND, TRUNC

INTERCEPT
Returns the intercept of the linear regression line $y = mx + b$.
Syntax: INTERCEPT(***known_ys, known_xs***)
See LINEST for details.
Related Functions: LINEST, SLOPE

ISBLANK
Returns TRUE if the value is blank.
Syntax: ISBLANK(***value***)
Related Functions: other IS functions

ISERR
Returns TRUE if the value is an error value other than #N/A.
Syntax: ISERR(***value***)
Related Functions: other IS functions

ISERROR
Returns TRUE if the value is any error value.
Syntax: ISERROR(***value***)
Related Functions: other IS functions

ISNA
Returns TRUE if the value is #N/A.
Syntax: ISNA(***value***)
Related Functions: other IS functions

ISNUMBER
Returns TRUE if the value is a number.
Syntax: ISNUMBER(***value***)
Does not convert text representation of a number.
Example: =ISNUMBER(5) returns TRUE. =ISNUMBER("9") returns FALSE.
Related Functions: other IS functions

ISTEXT
Returns TRUE if the value is text.
Syntax: ISTEXT(***value***)
Related Functions: other IS functions

LARGE
Returns the *k*th largest value in a list of values.
Syntax: LARGE(***array,k***)
Related Functions: MAX, MIN, SMALL

LEFT
Returns the specified number of characters from a text string, beginning at the left.
Syntax: LEFT(***text***, *num_chars*)
If *num_chars* is omitted, it is assumed to be 1.
Example: =LEFT("CHEMISTRY",4) returns CHEM.
Related Functions: LEN, MID, RIGHT

LEN
Returns the number of characters in a text string.
Syntax: LEN(***text***)
Example: =LEN("CHEMISTRY") returns 9.
Related Functions: LEFT, MID, RIGHT

LINEST
Returns an array of linear regression parameters.
Syntax: LINEST(***known_ys, known_xs***, *const_logical, stats_logical*)
See Chapter 13 for details.
Related Functions: CORREL, INTERCEPT, SLOPE

LN
Returns the natural (base-*e*) logarithm of a number.
Syntax: LN(***number***)
Related Functions: EXP, LOG

LOG

Returns the logarithm of a number to the specified base.
Syntax: LOG(***number, base***)
If *base* is omitted, returns the base-10 logarithm.
Related Functions: LN, LOG10

LOG10

Returns the base-10 logarithm of a number.
Syntax: LOG10(***number***)
Related Functions: LN, LOG

LOOKUP

Looks up a value in an array.
The LOOKUP function has two syntax forms: vector and array. The vector form of LOOKUP looks in a one-row or one-column range for a value and returns a value from the same position in a second one-row or one-column range. See the On-Line Help for information about the array form of LOOKUP.
Syntax: LOOKUP(***lookup_value,lookup_vector,result_vector***)
The values in *lookup_vector* must be placed in ascending order; if LOOKUP can't find the *lookup_value*, it matches the largest value in *lookup_vector* that is less than or equal to *lookup_value*.
Related Functions: HLOOKUP, MATCH, VLOOKUP

LOWER

Converts a text string to lowercase.
Syntax: LOWER(***text***)
Related Functions: PROPER, UPPER

MATCH

Looks up a value in an array and returns its relative position.
Syntax: MATCH(***lookup_value, array***, match_type)
If *match_type* = 0, the function finds the first value that is exactly equal to *lookup_value*; *array* can be in any order. If *match_type* = 1, finds the largest value that is less than or equal to *lookup_value*; *array* must be in ascending order. If *match_type* = −1, finds the smallest value that is greater than or equal to *lookup_value*; *array* must be in descending order.
Example: =MATCH(MAX(Spectrum),Spectrum,0) returns the relative position in the array *Spectrum* of the maximum value in the array.
Related Functions: HLOOKUP, INDEX, VLOOKUP

MAX

Returns the maximum value in a list of arguments.

Syntax: MAX(***number1***, *number2,...*)

There may be up to 30 arguments. If an argument is a reference, only numbers in the reference are examined.

Related Function: MIN

MDETERM

Returns the determinant of an array.

Syntax: MDETERM(***array***)

Array must have an equal number of rows and columns. If any cells in *array* do not contain numbers, returns #VALUE!.

Example: See Chapter 12 for details.

Related Functions: MINVERSE, MMULT, SUMPRODUCT, TRANSPOSE

MEDIAN

Returns the median value in a list of arguments.

Syntax: MEDIAN(***number1***, *number2,...*)

If there are an even number of numbers in the set, returns the average of the two median values.

Related Function: AVERAGE

MID

Returns the specified number of characters from a text string, beginning at the specified position.

Syntax: MID(***text, start_num, num_chars***)

If *num_chars* extends beyond the end of text, returns characters to the end of text.

Example: If cell A4 contains H2SO4, =MID(A4,3,1) returns "S".

Related Functions: LEFT, LEN, RIGHT

MIN

Returns the minimum value in a list of arguments.

Syntax: MIN(***number1***, *number2,...*)

There may be up to 30 arguments. If an argument is a reference, only numbers in the reference are examined.

Related Function: MAX

MINVERSE

Returns the inverse of a matrix.

Syntax: MINVERSE(***array***)

Array must have an equal number of rows and columns. If any cells in *array* do not contain numbers, MINVERSE returns #VALUE!. If MDETERM for the array returns 0, the array cannot be inverted; MINVERSE will return #NUM! error.

Example: See Chapter 12 for details.
Related Functions: MDETERM, MMULT, SUMPRODUCT, TRANSPOSE

MMULT
Returns the product of two matrices.
Syntax: MMULT(**array1, array2**)
COLUMNS for *array1* must equal ROWS for *array2*.
Example: See Chapter 12 for details.
Related Functions: MDETERM, MINVERSE, SUMPRODUCT, TRANSPOSE

MOD
Returns the remainder of the division of *number* by *divisor*.
Syntax: MOD(**number, divisor**)
If *divisor* is greater than *number*, returns *number*.
Use MOD(number,1) to return the decimal part of a floating-point number.
Example: =MOD(2.3333,2) returns 0.3333. =MOD(2.3333,3) returns 2.3333.
=MOD(10.37,1) returns 0.37.
Related Functions: INT, ROUND, TRUNC

MONTH
Converts a serial number to a month.
Syntax: MONTH(**serial_number**)
Serial_number can also be given as text, such as "Jul-4-1994". Returns a number
between 1 and 12.
Related Functions: YEAR, DAY, WEEKDAY

N
Returns a value converted to a number
Syntax: N (**value**)
Provided mostly for compatibility with other spreadsheet programs. Use N to
convert TRUE and FALSE to 1 and 0.

NA
Returns the error value #N/A
Syntax: NA()

NOT
Reverses the logical value of its argument.
Syntax: NOT(**logical**)
Example: =IF(NOT(ISERROR(F2/G2)),F2/G2,"")
Related Functions: AND, OR

ROMAN
Converts an Arabic number to Roman numerals.
Syntax: ROMAN(**number, form**)
If *form* = 0, TRUE or omitted, the "classic" form is returned. *Form* can also be 1, 2, or 3, in which case successively more "concise" forms are returned. If *form* = 4 or FALSE, a "simplified" form is returned.
Example: =ROMAN(1997,0) returns MCMXCVII.

ROUND
Rounds a number to a specified number of digits.
Syntax: ROUND(**number, digits**)
If the result ends in a zero, the zero is not displayed.
Example: If cell A1 contains 0.001736, the formula =ROUND(A1,5) returns 0.00174. If cell A1 contains 0.007702, the same formula returns 0.0077, i.e., the number is not padded out with trailing zeros.
Related Functions: INT, TRUNC, TEXT

ROW
Returns the row number of reference.
Syntax: ROW(*reference*)
If *reference* is a range of cells, returns the row number of the upper left cell of the range. *Reference* cannot be a multiple selection.
Example: =ROW(B5:C7) returns 5.
Related Functions: COLUMN, ROWS

ROWS
Returns the number of rows in reference.
Syntax: ROWS(**reference**)
If *reference* is a multiple selection, can use INDEX to select a specified area within the selection.
Example: =ROWS($B$5:$C$7) returns 3.
Related Functions: ROW, COLUMNS

RSQ
Returns the correlation coefficient R^2.
Syntax: RSQ(**known_ys, known_xs**)
This is the correlation coefficient returned by LINEST. For more details, see LINEST.
Related Functions: LINEST, CORREL

SEARCH
Finds the position of a text value within a text string.
Syntax: SEARCH(**find_text, within_text**, *start_num*)
Find_text is the text to be found. Include the ? wildcard character in *find_text* to

match any single character or * to match any sequence of characters within the source string *within_text*.   *Start_num* is the position in *within_text* to begin searching. SEARCH is not case-sensitive, FIND is case-sensitive.
Example: See example under FIND.
Related Functions: FIND, REPLACE, SUBSTITUTE

SIGN
Returns the sign of a number.
Syntax: SIGN(**number**)
Returns 1, 0, −1 if the number is positive, zero or negative, respectively.
Related Function: ABS

SIN
Returns the sine of a given angle.
Syntax: SIN(**number**)
Number is the angle in radians. To convert an angle in degrees to one in radians, multiply by $\pi/180$.
Related Functions: ASIN, COS, TAN, other trigonometric functions

SLOPE
Returns the slope of the linear regression line $y = mx + b$.
Syntax: SLOPE(**known_ys, known_xs**)
See LINEST for details.
Related Functions: INTERCEPT, LINEST

SMALL
Returns the kth smallest value in a list of values.
Syntax: SMALL(**array,k**)
Related Functions: LARGE, MAX, MIN

SQRT
Returns the square root of a number.
Syntax: SQRT(**number**)

STDEV
Returns the standard deviation of a sample.
Syntax: STDEV(**value1**, value2,...)
The sample can consist of up to 30 separate arguments or arrays.
Related Functions: AVERAGE, AVDEV

SUBSTITUTE
Finds and substitutes characters within a text string.
Syntax: SUBSTITUTE(**text, old_text, new_text**, instance_num)
Text is the string in which characters are to be substituted. *Old_text* is the text to be replaced. *New_text* is the text to replace *old_text*. *Instance_num* specifies

which occurrence of *old_text* is to be replaced; if omitted, all are replaced.
Example: =SUBSTITUTE(A3," ","",2) replaces the second occurrence of a space in the string in cell A3 with a null .
Related Function: REPLACE

SUM
Returns the sum of all the numbers in a range.
Syntax: SUM(*number1, number2,...*)
The arguments may be numbers, names, arrays or references that contain numbers. Up to 30 separate arguments. Only numbers in the array or range are counted.
Related Functions: AVERAGE, COUNT, COUNTA

SUMIF
Returns the sum of all the numbers in a range that satisfy the specified criteria.
Syntax: SUMIF(*range, criteria, sum_range*)
Range is the range of cells in which criteria will be evaluated. *Sum_range* contains the cells that will be summed. If *sum_range* is omitted, cells in *range* are summed.
Example: =SUMIF(B3:B553,>100,D3:D553)
Related Functions: AVERAGE, COUNT, COUNTA

SUMIFS
Returns the sum of the cells specified by multiple criteria (available in Excel 2007/2010 only).
Syntax: SUMIFS(*sum_range, criterion_range1, criterion1,* criterion_range2, criterion2,...)
Sum_range is the range of cells to be summed. *Criterion_range* is the range of values to be compared with *criterion*. Up to 127 *criterion_range*, *criterion* pairs are allowed. Note the difference in order of arguments between SUMIF and SUMIFS.

SUMPRODUCT
Returns the sum of the products of corresponding array components.
Syntax: SUMPRODUCT(*array1, array2,...*)
Up to 30 arrays may be included. All arrays must have the same dimensions.
Related Functions: MMULT, TRANSPOSE

TAN
Returns the tangent of a given angle.
Syntax: TAN(*number*)
Number is the angle in radians. To convert an angle in degrees to one in radians, multiply by $\pi/180$.
Related Functions: ATAN, ATAN2, other trigonometric functions

TEXT
Formats a number and returns it as text.
Syntax: TEXT(*value, format_text*)
Format_text is a number format, similar to those in the Number format dialog box. TEXT converts a number to text; the result will sometimes fail to be calculated as a number.
Example: The formula =TEXT(PI(),"0.000") in cell B1 returns 3.142. The formula =2*B1 returns 6.284, but the formula =SUM(B1) returns 0.
Related Function: FIXED

TIME
Returns the serial number of a time.
Syntax: TIME(*hour, minute, second*)
Hour is a number from 0 to 23; *minute* is a number from 0 to 59; *second* is a number from 0 to 59.
Related Function: NOW

TODAY
Returns the serial number of today's date.
Syntax: TODAY()
TODAY returns the integer part of the serial number returned by NOW.
Related Functions: DATE, DAY, NOW

TRANSPOSE
Returns the transpose of an array
Syntax: TRANSPOSE(*array*)
Must be entered as an array formula in a range with number of columns equal to number of rows in *array*, and number of rows equal to number of columns in *array*.
Example: Does the same thing as **Paste Special** (Transpose) from the **Edit** menu.
Related Functions: MDETERM, MINVERSE, MMULT, SUMPRODUCT

TRIM
Removes spaces from a text string, except for single spaces between words.
Syntax: TRIM(*text*)
Related Function: CLEAN

TRUNC
Truncates a number.
Syntax: TRUNC(*number, num_digits*)
Num_digits is the number of digits after the decimal point. If omitted, *num_digits* is taken to be zero. TRUNC does not round up.
Example: The formula =TRUNC(-8.913) returns -8. The formula =TRUNC(PI(),3)

returns 3.141.
Related Functions: INT, ROUND

TYPE
Returns a number indicating the data type of a value.
Syntax: TYPE(**value**)
Returns 1 (if *value* is a number), 2 (text), 4 (logical value), 16 (error value) or 64 (array).
Related Functions: CELL, IS functions

UPPER
Converts a text string to uppercase.
Syntax: UPPER(**text**)
Related Functions: LOWER, PROPER

VALUE
Converts a text argument to a number.
Syntax: VALUE(**text**)
Not usually necessary, since Excel automatically converts text to number.
Example: =ISNUMBER(MID("H2SO4",2,1)) returns FALSE but
=ISNUMBER(VALUE(MID("H2SO4",2,1))) returns TRUE.
Related Functions: FIXED, TEXT

VAR
Returns the variance of a sample.
Syntax: VAR(**number1**,number2,...)
The sample can consist of up to 30 separate arguments or arrays.
Example: See On-Line Help for details.
Related Function: STDEV

VLOOKUP
Finds the value in the first column of *array* that is equal to or less than *lookup_value*. Returns the associated value in the same row, as determined by *offset_num*.
Syntax: VLOOKUP(**lookup_value, array, offset_num**, match_logical)
If *match_logical* is FALSE, returns an exact match, or #N/A if an exact match is not found. If *match_logical* is TRUE or omitted, returns the largest value that is less than or equal to *lookup_value*. In the latter case, the values in the first column must be in ascending order.
Example: See example under HLOOKUP.
Related Functions: HLOOKUP, INDEX, MATCH

WEEKDAY
Converts a serial number to a day of the week.

Syntax: WEEKDAY(*serial_number, return_type_num*)
Returns a number indicating the day of the week. If *return_type_num* = 1 or omitted, returns 1 (Sunday) to 7 (Saturday). If *return_type_num* = 2, returns 1 (Monday) to 7 (Sunday). If *return_type_num* = 3, returns 0 (Monday) to 6 (Sunday).
Related Functions: DAY, NOW, TODAY

YEAR
Converts a serial number to a year.
Syntax: YEAR(*serial_number*)
Related Function: DAY, MONTH

Appendix D

Renamed Functions
in Excel 2010

The following Statistical functions were renamed "*so that they are more consistent with the function definitions of the scientific community and with other function names in Excel.*"

Original function	Renamed function
BETADIST	BETA.DIST
BETAINV	BETA.INV
BINOMDIST	BINOM.DIST
CRITBINOM	BINOM.INV
CHIDIST	CHISQ.DIST.RT
CHIINV	CHISQ.INV.RT
CHITEST	CHISQ.TEST
CONFIDENCE	CONFIDENCE.NORM
COVAR	COVARIANCE.P
EXPONDIST	EXPON.DIST
FDIST	F.DIST.RT
FINV	F.INV.RT
FTEST	F.TEST
GAMMADIST	GAMMA.DIST
GAMMAINV	GAMMA.INV
HYPGEOMDIST	HYPGEOM.DIST
LOGNORMDIST	LOGNORM.DIST
LOGINV	LOGNORM.INV
MODE	MODE.SNGL
NEGBINOMDIST	NEGBINOM.DIST
NORMDIST	NORM.DIST
NORMINV	NORM.INV
NORMSDIST	NORM.S.DIST
NORMSINV	NORM.S.INV

Original function	Renamed function
PERCENTILE	PERCENTILE.INC
PERCENTRANK	PERCENTRANK.INC
POISSON	POISSON.DIST
QUARTILE	QUARTILE.INC
RANK	RANK.EQ
STDEVP	STDEV.P
STDEV	STDEV.S
TDIST	T.DIST.2T
TDIST	T.DIST.RT
TINV	T.INV.2T
TTEST	T.TEST
VARP	VAR.P
VAR	VAR.S
WEIBULL	WEIBULL.DIST
ZTEST	Z.TEST

Appendix E

Selected
Visual Basic Keywords
by Category

This appendix lists selected VBA keywords for functions, statements, methods and properties. See Excel's On-Line Help for a complete list of keywords.

FUNCTIONS

Abs	Returns the absolute value of a number.
Array	Returns a **Variant** containing an array.
Asc	Returns the numeric code for the first character of text.
Atn	Returns the angle corresponding to a tangent value.
Chr	Returns the character corresponding to a code.
Cos	Returns the cosine of an angle.
Exp	Returns *e* raised to a power.
Fix	Truncates a number to an integer.
Format	Formats a value according to a formatting code expression.
InputBox	Displays an input dialog box and waits for user input.
Int	Rounds a number to an integer.
IsArray	Returns **True** if the variable is an array.
IsDate	Returns **True** if the expression can be converted to a date.
IsEmpty	Returns **True** if the variable has been initialized.
IsMissing	Returns **True** if an optional argument has not been passed to a procedure.
IsNull	Returns **True** if the expression is null (i.e., contains no valid data).
IsNumeric	Returns **True** if the expression can be evaluated to a number.
LBound	Returns the lower limit of an array dimension.
LTrim	Returns a string without leading spaces.
LCase	Converts a string into lowercase letters.
Left	Returns the leftmost characters of a string.
Len	Returns the length (number of characters) in a string.

Log	Returns the natural (base-e) logarithm of a number.
Mid	Returns a specified number of characters from a text string, beginning at a specified position.
MsgBox	Displays a message box.
Now	Returns the current date and time.
Right	Returns the rightmost characters of a string.
Rnd	Returns a random number between 0 and 1.
RTrim	Returns a string without trailing spaces.
Sgn	Returns the sign of a number.
Sin	Returns the sine of an angle.
Sqr	Returns the square root of a number.
Str	Converts a number to a string.
Tan	Returns the tangent of an angle.
Trim	Returns a string without leading or trailing spaces.
UBound	Returns the upper limit of an array dimension.
UCase	Converts a string into uppercase letters.
Val	Converts a string to a number.

STATEMENTS (COMMANDSS)

Beep	Makes a "beep" sound.
Call	Transfers control to a **Function** or **Sub** procedure.
Dim	Declares an array and allocates storage for it.
Do...Loop	Delineates a block of statements to be repeated.
Else	Optional part of **If...Then** structure.
ElseIf	Optional part of **If...Then** structure.
End	Terminates a procedure or block.
Exit	Exits a **Do...**, **For...**, **Function**... or **Sub**... structure.
For Each...Next	Delineates a block of statements to be repeated.
For...Next	Delineates a block of statements to be repeated.
Function	Marks the beginning of a **Function** procedure.
GoSub	Branches to a subroutine within a procedure.
GoTo	Unconditional branch within a procedure.
If...Then...End If	Delineates a block of conditional statements.
On...GoSub	Branches to one of several specified subroutines, depending on the value of an expression.
On...GoTo	Branches to one of several specified lines, depending on the value of an expression.
Option Base	Used at module level to declare lower bound for an array.
Preserve	Preserves data in an existing array when using **ReDim**.
Private	Indicates that the procedure is available to all other procedures.
Public	Indicates that the procedure is available only to procedures in the same module.

ReDim	Allocates or re-allocates dynamic array storage.
Return	Delineates the end of a subroutine within a procedure.
Select Case	Executes one of several blocks of statements, depending on the value of an expression.
Set	Assigns an object reference to a variable.
Static	Preserves a procedure's local variables between calls.
Stop	Stops execution, but does not close files or clear variables.
Sub	Marks the beginning of a **Sub** procedure.
Until	Optional part of **Do...Loop** structure
While	Optional part of **Do...Loop** structure
With...End With	Delineates a block of statements to be executed on a single object.

METHODS

Activate	Activates an object.
Address	Returns a reference, as text
Cells	Returns a single cell by specifying the row and column.
Clear	Clears formulas and formatting from a range of cells.
Close	Closes a window, workbook or workbooks.
Columns	Returns a **Range** object that represents a single column or multiple columns.
Copy	Copies the selected object to the Clipboard or to another location.
Cut	Cuts the selected object to the Clipboard or to another location.
Delete	Deletes the selected object.
FillDown	Copies the contents and format(s) of the top cell(s) of a specified range into the remaining rows.
FillRight	Copies the contents and format(s) of the leftmost cell(s) of a specified range into the remaining columns.
InputBox	Displays an input dialog box and waits for user input.
Insert	Inserts a range of cells in a worksheet.
MacroOptions	Sets options in the Macro Options dialog box.
Paste	Pastes the contents of the Clipboard onto a worksheet.
Quit	Quits Microsoft Excel.
Range	Returns a **Range** object that represents a cell or range of cells.
Rows	Returns a **Range** object that represents a single row or multiple rows.
Save	Saves changes to active workbook.
SaveAs	Saves changes to active workbook or other document with a different filename.
Select	Selects an object.
Sort	Sorts a range of cells.

PROPERTIES

ActiveCell	Returns the active cell of the active window.
ActiveSheet	Returns the active sheet of the active workbook.
Bold	Returns **True** if the font is Bold. Sets the Bold font.
Column	Returns a number corresponding to the first column in the range.
Count	Returns the number of items in the collection.
Font	Returns the font of the object.
Formula	Returns or sets the formula associated with an object.
FontStyle	Returns or sets the font of the object.
Italic	Returns **True** if the font is Italic. Sets the Italic font.
Name	Returns or sets the name of an object.
NumberFormat	Returns or sets the number format code of a cell.
Row	Returns a number corresponding to the first row in the range.
Selection	Returns the selected object.
Text	Returns or sets the text associated with an object.
Value	Returns the value of an object.

OTHER KEYWORDS AND OPERATORS

False	Boolean keyword.
True	Boolean keyword.
And	Logical operator.
Or	Logical operator.

Appendix F

Alphabetical List of Selected VBA Keywords

This listing of VBA objects, properties, methods, functions and other keywords will be useful when creating your own VBA procedures. The list is not exhaustive, but contains mainly those keywords that are used in the procedures shown in this book.

For each VBA keyword, the required syntax is given, along with some comments on the required and optional arguments, one or more examples and a list of related keywords. See Excel's On-Line Help for further information.

Abs Function
Returns the absolute value of a number.
Syntax: **Abs(number)**
Example: **Abs(-7.3)** returns 7.3
See also: **Sgn**

Activate Method
Activates an object.
*Syntax: object.***Activate**
Object can be **Chart, Worksheet** or **Window**.
Example: **Workbooks**("BOOK1.XLS").**Worksheets**("Sheet1").**Activate**
See also: **Select**

ActiveCell Property
Returns the active cell of the active window. Read-only.
Syntax: **ActiveCell** and **Application.ActiveCell** are equivalent.
See also: **Activate, Select**

ActiveSheet Property
Returns the active sheet of the active workbook. Read-only.
*Syntax: object.***ActiveSheet**
Object can be **Application, Window** or **Workbook**.
Example: **Application.ActiveSheet.Name** returns the name of the active sheet of the active workbook. Returns **None** if no sheet is active.
See also: **Activate, Select**

Address Property
Returns a reference, as text
*Syntax: object.***Address** *(rowAbsolute,columnAbsolute, referenceStyle, external, relativeTo)*
All arguments are optional. If *rowAbsolute* or *columnAbsolute* are **True** or omitted, returns that part of the address as an absolute reference. *ReferenceStyle* can be xlA1 or xlR1C1. If *external* is **True,** returns an external reference. See On-Line Help for information about the *relativeTo* argument.
See also: **Offset**

And Operator
Logical operator. (expression1 **And** expression2) evaluates to **True** if both expression1 and expression2 are **True.** Also can be used to perform bitwise comparison of two numerical values: (13 **And** 6) evaluates to 4. (13 = 00001101, 6 = 00000110, 4 = 00000100).
See also: **Or, Not, Xor**

Application Object
Represents the Microsoft Excel application.

Array Function
Returns a **Variant** containing an array.
Syntax: **Array** *(arglist)*
Example: **Array** *(31,28,31,30,31,30,31,31,30,31,30,31)*
See also: **Dim**

As Keyword
Used with **Dim** to specify the data type of a variable.

Asc Function
Returns the numeric code for the first character of text.
Syntax: **Asc***(character)*
Example: **Asc** ("A") returns 65.
See also: **Chr**

Atn Function
Returns the angle corresponding to a tangent value.
Syntax: **Atn***(number)*
Number can be in the range $-\infty$ to $+\infty$. The returned angle is in radians, in the range $-\pi/2$ to $+\pi/2$ ($-90°$ to $90°$). To convert the result to degrees, multiply by $180/\pi$.
Example: **Atn**(1) returns 0.785388573 or 45 degrees.
See also: **Cos, Sin, Tan**

Bold Property
Returns **True** if the font is Bold. Sets the Bold font. Read-write.
*Syntax: object.***Bold**
Object must be **Font**.
Example: **Range**("A1:E1").**Font.Bold = True** makes the cells bold.
See also: **Italic**

Boolean Data Type
Use to declare a variable's type as **Boolean (True** or **False)**, either in a **Dim** statement, or in a **Sub** or **Function** statement. Two bytes required per variable. When number values are converted to Boolean values, 0 becomes **False** and all other values become **True**. When Boolean values are converted to numbers, **False** becomes 0 and **True** becomes -1.
See also: **Dim, As, Double, Integer, String, Variant**

Call Command
Transfers control to a **Sub** procedure.
Syntax: **Call** *name (argument1, ...)*
Name is the name of the procedure. *Argument1*, etc., are the names assigned to the arguments passed to the procedure. **Call** is optional; if omitted, the parentheses around the argument list must also be omitted.
Example: **Call** Task1(argument1,argument2)
See also: **Sub, Function**

Case Keyword
See: **Select Case**

Cells Method
Returns a single cell by specifying the row and column.
*Syntax: object.***Cells***(row, column)*
Object is optional; if not specified, **Cells** refers to the active sheet.
Example: **Cells**(2,1).**Value** = 5 enters the value 5 in cell A2.
See also: **Range**

Characters Object
Represents characters in any object containing text. Use the **Characters** object to format characters within a text string.
*Syntax: expression.***Characters***(start, length)*
Example: **Selection.Characters(Start:=x, Length:=1).Font.Subscript = True**

Clear Method
Clears formulas and formatting from a range of cells.
*Syntax: object.***Clear**
Object can be **Range** (or **ChartArea**).
Example: **Range**("A1:C10").**Clear**
See also: **ClearContents, ClearFormats** in Excel's On-Line Help.

Close Method
Closes a window, workbook or workbooks.
Syntax: For workbooks, use *object.***Close**. For a workbook or window, use
*object.***Close**(*SaveChangesLogical, FileName*).
Object can be **Window, Workbook** or **Workbooks**. If *SaveChangesLogical* is
False, does not save changes; if omitted, displays a "Save Changes?" dialog box.
Example: **Workbooks**("BOOK1.XLS").**Close**
See also: **Open, Save, SaveAs**

Column Property
Returns a number corresponding to the first column in the range. Read-only.
*Syntax: object.***Column**
Object must be **Range**.
See also: **Columns, Row, Rows**
Columns Method
Returns a **Range** object that represents a single column or multiple columns
*Syntax: object.***Columns**(*index*)
Object can be **Worksheet** or **Range**. *Index* is the name or number (column A =
1, etc.) of the column.
Example: **Selection.Columns.Count** returns the number of columns in the
selection.
See also: **Range, Rows**

ColumnWidth Property
Returns or sets the width of all columns in the range. If columns in the range
have different widths, returns **Null**.
Example: **Worksheets**("Sheet1").**Columns**("C").**ColumnWidth** = 30
See also: **RowHeight**

ConvertFormula Method
Converts cell references between A1-style and R1C1-style, and between absolute
and relative. On-Line Help states that *Formula* must begin with an equal sign,
but references in a string that does not begin with an equal sign are also
converted.
Syntax: **expression.ConvertFormula**(*Formula, FromReferenceStyle,*
ToReferenceStyle, ToAbsolute, RelativeTo)
Example:

FormulaString = **Application.ConvertFormula**(FormulaString, **xlA1,   xlA1, xlAbsolute**)
See also: **Address**

Copy Method
Copies the selected object to the Clipboard or to another location.
*Syntax: object.***Copy**(*destination*)
Object can be **Range, Worksheet, Chart** and many other objects. *Destination* specifies the range where the copy will be pasted. If omitted, copy goes to the Clipboard.
Example: **Worksheets**("Sheet1").**Range**("A1:C50").**Copy**
See also: **Cut, Paste**

Cos Function
Returns the cosine of an angle.
Syntax: **Cos**(*number*)
Number is the angle in radians; it can be in the range $-\infty$ to $+\infty$. To convert an angle in degrees to one in radians, multiply by $\pi/180$. Returns a value between -1 and 1.
See also: **Atn, Sin, Tan**

Count Property
Returns the number of items in the collection. Read-only.
*Syntax: object.***Count**
Object can be any collection.
Example: The statement N = array.**Count** counts the number of values in the range array.

Cut Method
Cuts the selected object and pastes to the Clipboard or to another location.
*Syntax: object.***Cut**(*destination*)
Object can be **Range, Worksheet, Chart** or one of many other objects. *Destination* specifies the range where the copy will be pasted. If omitted, copy goes to the Clipboard.
Example: **Worksheets**("Sheet1").**Range**("A1:C50").**Cut**
See also: **Copy, Paste**

CVErr Function
Returns a Variant containing an error value specified by the user.
Syntax: **CVErr**(*number*)
CVErr can return either Excel's built-in worksheet error values, or a user-defined error value. The values of *number* for built-in worksheet error values are xlErrDiv0, xlErrNA, xlErrName, xlErrNull, xlErrNum, xlErrRef, xlErrValue.
See also: **IsError**

Delete Method
Deletes the selected object.
*Syntax: object.***Delete***(shift)*
Object can be **Range**, **Worksheet, Chart** and many other objects. *Shift* specifies
how to shift cells when a range is deleted from a worksheet (xlToLeft or xlUp).
Can also use *shift* = 1 or 2, respectively. If *shift* is omitted, Excel moves the cells
without displaying the "Shift Cells?" dialog box.
Example: **Worksheets**("Sheet12").**Range**("A1:A10").**Delete** (xlToLeft) deletes
the indicated range and shifts cells to left.

Dim Keyword
Declares an array and allocates storage for it.
Syntax: **Dim** *variable (subscripts)*
Variable is the name assigned to the array. *Subscripts* are the size dimensions of
the array; an array can have up to 60 size dimensions. Each size dimension has a
default lower value of zero; a single number for a size dimension is taken as the
upper limit. Use *lower* **To** *upper* to specify a range that does not begin at zero.
Use **Dim** with empty parentheses to specify an array whose size dimensions are
defined within a procedure by means of the **ReDim** statement.
Example: **Dim** Matrix (5,5) **As Double** creates a 6×6 array of double-precision
variables.
See also: **ReDim**

Do...Loop Command
Delineates a block of statements to be repeated.
Syntax: The beginning of the loop is delineated by **Do** or **Do Until** *condition* or
Do While *condition*. The end of the loop is delineated by **Loop** or **Loop Until**
condition or **Loop While** *condition*. *Condition* must evaluate to **True** or **False.**
Example: See examples of **Do...Loop** structures in Chapter 16.
See also: **Exit, For, Next, Wend, While**

Double Data Type
Use to declare a variable's type as double-precision floating-point (15 significant
digits), either in a **Dim** statement, or in a **Sub** or **Function** statement. Eight
bytes required per variable.
Example: **Dim** tolerance **As Double**
See also: **Dim, As, Boolean, Integer, String, Variant**

Else Keyword
Optional part of **If...Then** structure.

ElseIf Keyword
Optional part of **If...Then** structure.

End Command

Terminates a procedure or block.

Syntax: **End** terminates a procedure. **End Function** is required to terminate a **Function** procedure. **End If** is required to terminate a block **If** structure. **End Select** is required to terminate a **Select Case** structure. **End Sub** is required to terminate a **Sub** procedure. **End With** is required to terminate a **With** structure.

Example: See examples under **Select Case**.

See also: **Exit, Function, If, Then, Else, Select Case, Sub, With**

EndIf Keyword

Optional part of **If...Then** structure.

Err Function

Returns a run-time error number. Use in error-handling routine to determine the error and take appropriate corrective action.

Example: **If Err.Number** = 13 Then
                (code for corrective action here)
                Resume pt1
                End If

See also: **Error, On Error, Resume**

Evaluate Method

Converts a name or formula to a value.

Syntax: **Evaluate**(expression)

Expression must be a string, maximum length 255 characters. An initial equal sign is not necessary.

Example: F$ = "2*3"
                MsgBox Evaluate(F$)

See also: **Formula**

Exit Command

Exits a **Do...**, **For...**, **Function...** or **Sub...** structure.

Syntax: **Exit Do, Exit For, Exit Function, Exit Sub**

From a **Do** or **For** loop, control is transferred to the statement following the **Loop** or **Next** statement, or, in the case of nested loops, to the loop that is one level above the loop containing the **Exit** statement. From a **Function** or **Sub** procedure, control is transferred to the statement following the one that called the procedure.

Example: See examples of **Exit** procedures in Chapter 16.

See also: **Do, For...Next, Function, Stop, Sub**

Exp Function
Returns *e* raised to a power.
Syntax: **Exp**(*number*)
Returns the value of *e* raised to the power *number*.
See also: **Log**

False Keyword
Use the keywords **True** or **False** to assign the value **True** or **False** to Boolean (logical) variables.
When other numeric data types are converted to Boolean values, 0 becomes **False** while all other values become **True**. When Boolean values are converted to other data types, **False** becomes 0 while **True** becomes -1.
Example: **If** SubFlag = **False Then**...
See also: **True**

FillDown Method
Copies the contents and format(s) of the top cell(s) of a specified range into the remaining rows.
*Syntax: object.***FillDown**
Object must be **Range**.
Example: **Worksheets**("Sheet12").**Range**("A1:A10").**FillDown**
See also: **FillLeft, FillRight, FillUp** in Excel's On-Line Help.

FillRight Method
Copies the contents and format(s) of the leftmost cell(s) of a specified range into the remaining columns.
*Syntax: object.***FillDown**
Object must be **Range**.
Example: **Worksheets**("Sheet12").**Range**("A1:A10").**FillRight**
See also: **FillDown, FillLeft, FillUp** in Excel's On-Line Help.

Fix Function
Truncates a number to an integer.
Syntax: **Fix**(*number*)
If *number* is negative, **Fix** returns the first negative integer greater than or equal to *number*.
Example: **Fix**(-2.5) returns -2.
See also: **Int**

Font Property
Returns the font of the object. Read-only.
*Syntax: object.***Font**
Example: ActiveCell.Font.Bold = **True** makes characters in the active cell bold.
See also: **FontStyle**

FontStyle Property
Returns or sets the font of the object. Read-write.
*Syntax: object.***FontStyle**
Example: **Range**("A1:E1").**Font.FontStyle** = "Bold"
See also: **Font**

For...Next Command
Delineates a block of statements to be repeated.
Syntax: **For** *counter* = *start* **To** *end* **Step** *increment*
                        (statements)
                        Next *counter*
Step *increment* is optional; if not included, the default value 1 is used.
Increment can be negative, in which case *start* should be greater than *end*.
Example: See examples of **For...Next** procedures in Chapter 16.
See also: **Do...Loop, Exit, For Each...Next, While...Wend**

For Each...Next Command
Delineates a block of statements to be repeated.
Syntax: **For Each** *element* **In** *group*
                        (statements)
                        Next *element*
Group must be a collection or array. *Element* is the name assigned to the
variable used to step through the collection or array. *Group* must be a collection
or array.
Example: See examples of **For Each...Next** procedures in Chapter 16.
See also: **Do...Loop, Exit, For...Next, While...Wend**

Format Function
Formats a value according to a formatting code expression.
Syntax: **Format**(*expression,formattext*)
Expression is usually a number, although strings can also be formatted.
Formattext is a built-in or custom format. Additional information can be found
in *Microsoft Excel/Visual Basic Reference*, or VBA On-Line Help.
Example: **Format**(TelNumber,"(###) ###-####") formats the value TelNumber in
the form of a telephone number.

Formula Property
Returns or sets the formula in a cell.
If a cell contains a value, returns the value; if the cell contains the formula,
returns the formula as a string.
See also: **Text, Value**

IsDate Function
Returns **True** if the expression can be converted to a date.
Syntax: **IsDate**(*expression*)
See also: other **Is** functions

IsEmpty Function
Returns **True** if the variable has been initialized.
Syntax: **IsEmpty**(*expression*)
See also: other **Is** functions

IsMissing Function
Returns **True** if an optional argument has not been passed to a procedure.
Syntax: **IsMissing**(*name*)
See also: other **Is** functions

IsNull Function
Returns **True** if the expression is null (i.e., contains no valid data).
Syntax: **IsNull**(*expression*)
See also: other **Is** functions

IsNumeric Function
Returns **True** if the expression can be evaluated to a number.
Syntax: **IsNumeric**(*expression*)
See also: other **Is** functions

Italic Property
Returns **True** if the font is Italic. Sets the Italic font. Read-write.
*Syntax: object.***Italic**
Object must be **Font**.
Example: **Range**("A1:E1")**.Font.Italic = True** makes the cells italic.
See also: **Bold**

LBound Function
Returns the lower limit of an array dimension.
Syntax: **LBound**(*array,dimension*)
Array is the name of the array. *Dimension* is an integer (1, 2, 3, etc.) specifying
the dimension to be returned; if omitted, the value 1 is used.
Example: If the array table was dimensioned using the statement **Dim** table (1 **To**
3, 1000), **LBound**(table,1) returns 1, **LBound**(table,2) returns 0.
See also: **Dim, UBound**

LCase Function
Converts a string into lowercase letters.
Syntax: **LCase**(*string*)
See also: **UCase**

LTrim Function
Returns a string without leading spaces.
Syntax: **LTrim**(*string*)
See also: **RTrim**

Left Function
Returns the leftmost characters of a string.
Syntax: **Left**(*string,number*)
If *number* is zero, a null string is returned. If *number* is greater than the number
of characters in *string*, the entire string is returned.
Example: **Left**("CHEMISTRY",4) returns CHEM
See also: **Len, Mid, Right**

Len Function
Returns the length (number of characters) in a string.
Syntax: **Len**(*string*)
Example: **Len**("CHEMISTRY") returns 9.
See also: **Left, Mid, Right**

Log Function
Returns the natural (base-e) logarithm of a number.
Syntax: **Log**(*number*)
Number must be a value or expression greater than zero. VBA does not provide
base-10 logarithms; use **Log**(value)/**Log**(10).
See also: **Exp**

MacroOptions Method
Sets options in the Macro Options dialog box.
Syntax: **Application.MacroOptions**(*macro, description, hasMenu, menuText,
hasShortcutKey, shortcutKey, category, statusbar, helpContext, helpFile*)

macro is the name of the macro. *description* is the description that appears in
the dialog box. *category* is the function category that the macro appears in:
Financial, 1; Date & Time, 2; Math & Trig, 3; Statistical, 4; Lookup &
Reference, 5; Database, 6; Text, 7; Logical, 8; Information, 9; User Defined, 14;
Engineering, 15.
Example: **Application.MacroOptions** macro:="FtoC", Description:= "Converts
Fahrenheit temperature to Celsius", Category:=3
provides a description for the macro FtoC and assigns it to the Math & Trig
category.

Mid Function
Returns the specified number of characters from a text string, beginning at the
specified position.
Syntax: **Mid**(*string,start,number*)

If *start* is greater than the number of characters in *string*, returns a null string. If *number* is omitted, all characters from *start* to the end of the string are returned.
Example: **Mid**("H2SO4",2,1) returns 2.
See also: **Left, Len, Right**

Mod Operator
Returns the remainder resulting from the division of two numbers.
Syntax: result = number1 **Mod** *number2*

MsgBox Function
Displays a message box.
Syntax: **MsgBox(*prompt,***buttons,title,helpfile,context*)*
See *Microsoft Excel/Visual Basic Reference* or On-Line Help for details.
See also: **InputBox**

Name Property
Returns or sets the name of an object.
Example: SeriesName = **Selection.Name** assigns the name of the selected chart series to the variable SeriesName.
See also: **NameLocal, Names**

Next Keyword
Delineates the end of a **For...Next** or **For Each...Next** block of statements.
Not Operator
Logical operator. Performs logical negation: **True** becomes **False, False** becomes **True**.
See also: **And, Or**

Now Function
Returns the current date and time.
Syntax: **Now**
See also: other date and time functions.

NumberFormat Property
Returns or sets the number format code of a cell.
Example: **Range**("A1:A10").**NumberFormat=** "0.00" sets the number format of the specified range of cells.
See also: **GoSub, GoTo, Return, Select Case**

On...GoTo Command
Branches to one of several specified lines, depending on the value of an expression.
Syntax: **On** *expression* **GoTo** *label1, ...*
See explanation under **On...GoSub** command.

Example: See examples of **On...GoTo** procedures in Chapter 16.
See also: **GoSub, GoTo, Return, Select Case**

On Error GoTo Command
Enables an error-handling routine and specifies the action to be taken in event of an error.
Examples: **On Error GoTo** *line* (enables the error-handling routine at the specified location in the procedure)
On Error Resume Next (execution resumes with the statement immediately following the statement that caused the error)
On Error GoTo 0 (disables any enabled error handler in the current procedure)

Open Method
Opens a workbook.
*Syntax: object.***Open**(*filename, ...*)
Object must be **Workbooks**. *Filename* is required. See On-Line Help for the remaining arguments.
Example: **Workbooks.Open**("SOLVSTAT.XLS")
See also: **Close, Save, SaveAs**

Option Base Keyword
Use at module level to declare lower bound for an array.
Can be **Option Base** 0 or 1. The statement can appear only once in a module and must precede all **Dim** or equivalent declaration.
See also: **Dim, LBound, ReDim**

Option Explicit Statement
Use at module level to force explicit declaration of all variables in that module.
See also: **Option Base, Option Compare**

Optional Keyword
Indicates that an argument in a function is not required. All arguments following the **Optional** keyword must be optional. All optional arguments are **Variant**.
Syntax: **Function** name(*argument1,...* **Optional** *argument*)
See also: **Function, ParamArray**

Or Operator
Logical operator. (expression1 **Or** expression2) evaluates to **True** if either expression1 or expression2 is **True**. Also can be used to perform bitwise comparison of two numerical values: (13 **Or** 6) evaluates to 15. (13 = 00001101, 6 = 00000110, 15 = 00001111).
See also: **Or, Not, Xor**

ParamArray Keyword
Allows the use of an indefinite number of arguments for a function. The argument becomes an array of **Variant** elements. The array has lower array index of zero, even if **Option Base 1** is declared.
Syntax: **Function** name(*argument1*,... **ParamArray** *argument*() **As Variant)**
Example: **Function** test **(ParamArray** rng() **As Variant)**
See also: **Dim, Function, Variant**

Paste Method
Pastes the contents of the Clipboard onto a worksheet.
*Syntax: object.***Paste**(*destination*)
Object must be **Worksheet.** There are other **Paste** methods, with different syntax, for **Chart** and many other objects. *Destination* specifies the range where the copy will be pasted. If omitted, copy is pasted to the current selection.
Example: **Worksheets**("Sheet1")**.Range**("A1:C50")**.Copy**
            ActiveSheet.Paste
See also: **Copy, Cut**

Preserve Command
Preserves data in an existing array when using **ReDim.**

Private Command
Indicates that the procedure is available only to procedures in the same module.

Public Command
Indicates that the procedure is available to all other procedures.

Quit Method
Quits Microsoft Excel.
*Syntax: object.***Quit**
Object must be **Application**.
Example: **Application.Quit**
See also: **Close, Save**
Range Method
Returns a **Range** object that represents a cell or range of cells.
*Syntax: object.***Range**(*reference*)
Object is required if it is **Worksheet**. *Reference* must be an A1-style reference, in quotes, or the name of the reference.
Example: **Worksheets**("Sheet12")**.Range**("A1")**.Value** = 5
See also: **Cells**

ReDim Keyword
Allocates or re-allocates dynamic array storage.
Syntax: **ReDim** *variable* (*subscripts*)
For discussion of *variable* and *subscripts*, see comments under the entry for **Dim**.

You can use **ReDim** repeatedly to change the number of elements in an array, or the number or dimensions.

Example: **Dim** Matrix()
              (statements)
              ReDim Matrix (5,5)
              (statements)
              ReDim Matrix (15,25)

See also: **Dim**

Resume Command
Resumes execution after an error-handling routine is finished.
Examples: **Resume 0**
              Resume Next (execution resumes with the statement immediately following the statement that caused the error)
              Resume *label* (Execution resumes at the specified location in the procedure)
See *also:* **On Error GoTo**

Return Command
Delineates the end of a subroutine within a procedure.

Right Function
Returns the rightmost characters of a string.
Syntax: **Right**(*string,number*)
If *number* is zero, a null string is returned. If *number* is greater than the number of characters in *string*, the entire string is returned.
Example: **Right**(303585842,4) returns 5842.
See also: **Left, Len, Mid**

Rnd Function
Returns a random number between 0 and 1.
Syntax: **Rnd**

Row Property
Returns a number corresponding to the first row in the range. Read-only.
Syntax: *object*.**Row**
Object must be **Range**.
Example: **If ActiveCell.Row** = 10 **Then ActiveCell.Interior.ColorIndex** = 27
changes the interior color of the active cell to yellow if it is in row 10.
See also: **Column, Columns, Rows**

RowHeight Property
Returns or sets the height of all rows in the range.
Example: **Worksheets**("Sheet1").**Rows**(1).**RowHeight** = 15
See also: **ColumnWidth**

Rows Method
Returns a **Range** object that represents a single row or multiple rows.
*Syntax: object.***Rows**(*index*)
Object can be **Worksheet** or **Range**. *Index* is the name or number of the row.
Example: **Selection.Rows.Count** returns the number of rows in the selection.
See also: **Columns, Range**

RTrim Function
Returns a string without trailing spaces.
Syntax: **RTrim**(*string*)
See also: **LTrim, Trim**

Save Method
Saves changes to active workbook.
*Syntax: object.***Save**(*filename*)
Object must be **Workbook**. If *filename* is omitted, uses a default name.
Example: **ActiveWorkbook.Save**
See also: **Close, Open, SaveAs**

SaveAs Method
Saves changes to active workbook or other document with a different filename.
*Syntax: object.***SaveAs**(*filename, ...*)
Object can be **Worksheet, Workbook, Chart** or other document types. See
Microsoft Excel/Visual Basic Reference or On-Line Help for details.
Example: NewChart.**SaveAs**("New Chart")
See also: **Close, Open, Save**

Select Method
Selects an object.
*Syntax: object.***Select**
Object can be **Chart, Worksheet** or one of many other objects.
Example: Range("A1:C50").**Select**
See also: **Activate**

Select Case Command
Executes one of several blocks of statements, depending on the value of an
expression.
Syntax: **Select Case** *expression*
                Case *expression1*
                (statements)
                Case *expression2*
                (statements)
                End Select
You can also use the **To** keyword in *expression*, e.g., **Case** "A" **To** "M".
Expression can also be a logical expression. Use **Case Else** (not required) to

handle all cases not covered by the preceding **Case** statements.
Example: See examples of **Select Case** procedures in Chapter 16.
See also: **If...Then...Else, On...GoSub, On...GoTo**

Selection Property
Returns the selected object.   The object returned depends on the type of
selection.
See also: **Activate, ActiveCell, Select**

Set Command
Assigns an object reference to a variable.
See also: **Dim, ReDim**

Sgn Function
Returns the sign of a number.
Syntax: **Sgn**(*number*)
Returns 1, 0 or −1 if *number* is positive, zero or negative, respectively.
Example: **Sgn**(-7.3) returns −1.
See also: **Abs**

Sin Function
Returns the sine of an angle.
Syntax: **Sin**(*number*)
Number is the angle in radians; it can be in the range −∞ to +∞. To convert an
angle in degrees to one in radians, multiply by $\pi/180$. Returns a value between
−1 and 1.
See also: **Atn, Cos, Tan**

Sort Method
Sorts a range of cells.
Syntax: *object*.**Sort**(*sortkey1,order1,sortkey2,order2, ...*)
Object must be **Range**. See *Microsoft Excel/Visual Basic Reference* or On-Line
Help for details.

Sqr Function
Returns the square root of a number.
Syntax: **Sqr**(*number*)
Number must be greater than or equal to zero.

Step Keyword
Stops execution, but does not close files or clear variables.
See also: **End**
Stop Command
Stops execution, but does not close files or clear variables.
See also: **End**

Str Function
Converts a number to a string.
Syntax: **Str**(*number*)
A leading space is reserved for the sign of the number; if the number is positive, the string will contain a leading space.
See also: **Format**

String Data Type
Use to declare a variable's type as **String**, either in a **Dim** statement, or in a **Sub** or **Function** statement. One byte/character required per variable.
Example: **Dim J As Integer**
See also: **Dim, As, Boolean, Double, String, Variant**

Sub Keyword
Marks the beginning of a **Sub** procedure.
Syntax: **Sub** *name* (*argument1*, ...)
Name is the name of the procedure. *Argument1*, etc., are the names assigned to the arguments passed from the caller to the procedure. The end of the procedure is delineated by **End Sub**
Example: See examples of **Sub** procedures in Chapter 18.
See also: **Call, Function**

Tan Function
Returns the tangent of an angle.
Syntax: **Tan**(*number*)
Number is the angle in radians; it can be in the range $-\infty$ to $+\infty$. To convert an angle in degrees to one in radians, multiply by $\pi/180$. Returns a value between $-\infty$ and $+\infty$.
See also: **Atn, Cos, Sin**

Text Property
Returns or sets the text associated with an object.
The text can be associated with a chart, button, textbox, control or range. For all except range, this property is read-write, but for a range, it is read-only.
Example: **Worksheets**("Sheet1").**Buttons**(1).**Text** = "Undo"
See also: **Formula, Value**

Trim Function
Returns a string without leading or trailing spaces.
Syntax: **Trim**(*string*)
See also: **LTrim, RTrim**

True Keyword
Use the keywords **True** or **False** to assign the value **True** or **False** to Boolean

(logical) variables.

When other numeric data types are converted to Boolean values, 0 becomes **False** while all other values become **True**. When Boolean values are converted to other data types, **False** becomes 0 while **True** becomes –1.

Example: **If** FirstFlag = **True Then GoTo** 2000

UBound Function

Returns the upper limit of an array dimension.

Syntax: **UBound(*array, dimension*)**

Array is the name of the array. *Dimension* is an integer (1, 2, 3, etc.) specifying the dimension to be returned; if omitted, the value 1 is used.

Example: If the array table was dimensioned using the statement **Dim** table (1 **To** 3, 1000), **UBound**(table,3) returns 1, **UBound**(table,2) returns 1000.

See also: **Dim, LBound**

UCase Function

Converts a string into uppercase letters.

Syntax: **UCase(*string*)**

See also: **LCase**

Union Method

Returns a **Range** object that represents the union of two or more ranges, i.e., performs the same function as the comma character in the worksheet expression SUM(A1, B2, C3).

Syntax: **Union (*range1, range2*)**

See also: **Intersect, Areas, Caller**

Until Command

Optional part of **Do...Loop** structure.

Syntax: See explanation under **Do...Loop.**

Val Function

Converts a string to a number.

Syntax: **Val(*string*)**

Val stops at the first non-numeric character other than the period.

Example: **Val**("21 Lawrence Avenue") returns 21.

See also: **Str**

Value Property

Returns the value of an object.

*Syntax: object.***Value**

If *object* is **Range**, returns or sets the value(s) of the cell(s). Read-write.

If **Range** contains more than one cell, returns an array of values.

Example: **Worksheets**("Sheet12").**Range**("A1").**Value** = "Volume, mL"

Variant Data Type
Use to declare a variable's type as **Variant**, either in a **Dim** statement, or in a **Sub** or **Function** statement. **Variant** is the default data type, so usually not required. It is required when using the **ParamArray** keyword. Sixteen bytes + one byte/character required per variable.
Example: **Function** test **(ParamArray** rng() **As Variant)**
See also: **Dim, As, Boolean, Double, Integer, String**

Wend Command
Delineates the end of a **While...Wend** procedure.
Syntax: See explanation under **Do...Loop.**
See also: **Do...Loop, While...Wend**

While...Wend Command
Executes a series of statements as long as a specified condition is true.
Syntax: See explanation under **Do...Loop.**
See also: **Do...Loop, Wend**

With...End With command
Delineates a block of statements to be executed on a single object.
Syntax: **With** *object*
                 (statements)
                 End With
See also: **Do...Loop, While...Wend**

XOr Operator
Exclusive Or operator.
Use to perform bitwise comparison of two numerical values: (13 **XOr** 6) evaluates to 11. (13 = 00001101, 6 = 00000110, 11 = 00001011).
See also: **Or, Not, Or**

Appendix G

Selected Excel 4 Macro Functions

DOCUMENTS

Returns a horizontal array of the names of the specified open workbooks. Includes hidden workbooks.

Syntax: DOCUMENTS*(type_num, match_text)*

Type_num is a number specifying whether to include add-in workbooks in the array of workbooks, according to the following table. If *type_num* = 1 or omitted, returns names of all open workbooks except add-in workbooks; if *type_num* = 2, returns names of add-in workbooks only; if *type_num* = 3, returns names of all open workbooks.

Match_text specifies the workbooks whose names you want returned and can include wildcard characters. If *match_text* is omitted, DOCUMENTS returns the names of all open workbooks.

Comments: When used as a named formula, does not update if a new workbook is opened or an open workbook is closed.

Example: If there are three open workbooks – an Add-In workbook Classic Menus.xla, the hidden workbook Personal.xls and the open, and not yet saved, workbook Book1 – and the named formula

    =TRANSPOSE(DOCUMENTS(1))

has been given the name OpenWorkbooks, then the expression

    {=OpenWorkbooks}

entered as an array formula in a vertical range of cells, returns the values Personal.xls and Book1.

GET.DOCUMENT

Returns information about a sheet in a workbook.

Syntax: GET.DOCUMENT*(type_num, name_text)*

Type_num is a number (1-88) that specifies what information is returned. The following are some useful values of *type_num* and the corresponding result.

Name_text is the name of the workbook, about which information will be returned; the workbook must be open. If *name_text* is omitted, information about the active workbook is returned

Type_num	*Returns*
1*	Name of the worksheet, as text, in the form "[Book1]Sheet1". If there is only one sheet in the workbook, returns only the name of the workbook.
9	Number of the first used row. If the document is empty, returns 0.
10	Number of the last used row. If the document is empty, returns 0.
11	Number of the first used column. If the document is empty, returns 0.
12	Number of the last used column. If the document is empty, returns 0.
50	Total number of pages that would be printed based on current settings, excluding notes, or 1 if the document is a chart. Does not update is a page break is added.
76*	Name of the active worksheet
87	Position number of the sheet. The first sheet is position 1. Hidden sheets are included in the count.
88*	Name of the active workbook in the form "Book1".

Example: If the names FR, FC, LR, LC were assigned to the formulas

   =GET.DOCUMENT(9,SheetName)

and so on., the names BookName and SheetName to cells containing the values Book1 and Sheet2, and the name UsedRange to the formula

   ="'["&BookName&"]"&SheetName&"'!R"&FR&"C"&FC&":R"&LR&"C"&LC

then the worksheet formula

   =INDIRECT(UsedRange,0)

entered on Sheet3 returns a reference to all of the values entered on Sheet2. (Note the use of apostrophes enclosing the external reference, to handle cases where book or sheet name may include a space character.)

GET.WORKBOOK

Returns information about a workbook.

Syntax: GET.WORKBOOK(***type_num***, *name_text*)

Type_num is a number (1-38) that specifies what information is returned.

Name_text is the name of an open workbook. If omitted, it is assumed to be the active workbook.

Type_num	*Returns*
1	Names of all sheets in the workbook, as a horizontal array. Does not update if a sheet is added or deleted.

3   Names of the currently selected sheets in the workbook, as a
           horizontal array.
4   Number of sheets in the workbook.
38*   Name of the active worksheet.

* This information can be accessed using the worksheet function CELL.

Appendix H

Selected Shortcut Keys by Keystroke

Key	Applies to	Action
arrow key	Worksheet	Move one cell in direction of arrow
arrow key	Dialog box	Move between options in a group of options
arrow key	Object	Move selected object in one-pixel increments
arrow key	Chart	Move between elements in a group of chart elements
arrow key	Dialog box	Move one character to the right or left in the entry in an edit box (right/left arrows only)
arrow key	Dialog box	Move between options in a list box (up/down arrows only)
Backspace	Formula bar	Delete character to the left of the insertion point, or selected character(s)
Ctrl	Worksheet	Make a non-adjacent selection
Ctrl	Worksheet	When using **AutoFill,** copy a cell or range rather than create a series
Delete	Formula bar	Delete character to the right of the insertion point, or selected character(s)
Delete	Worksheet	Delete contents of selected cells
End	Dialog box	Move to the end of the entry in an edit box
End	Formula bar	Extend selection to last-used cell (lower-right corner)
End	Worksheet	Toggle END mode on/off
Enter	Worksheet, Formula bar	Enter data and move down one cell
Enter	Dialog box	Perform the action assigned to the default button (usually the OK button)
Esc	Dialog box	Cancel and close the dialog box or shortcut menu

Key	Applies to	Action
Esc	Worksheet, VBE	Halt execution of a macro
Esc	Formula bar	Undo editing
Home	Dialog box	Move to the beginning of the entry in an edit box
Home	Formula bar	Move insertion point to beginning
Home	Worksheet	Select cell in column A in same row as active cell
Insert	Worksheet	Toggle between Insert and Typeover modes
Shift	Worksheet	Select multiple objects
Tab	Dialog box	Move to the next option
Tab	Dialog box	Move to the next input box
Tab	Worksheet	Move right one cell, or move to next unlocked cell in a protected worksheet
Tab	Worksheet	Move to the next toolbutton (when toolbar is active)
(space)	Dialog box	Perform the action assigned to the active button, or select or clear the active check box
F1	Worksheet	**Help**
F2	Worksheet	Begin Edit. Position the insertion point at the end of the line
F3	Worksheet	Display the Paste Name dialog box
F4	Worksheet	**Repeat**
F4	Formula bar	Toggle Absolute/Relative references
F4	VBE	Display Properties window
F5	Worksheet	Display the **Go To** dialog box
F5	VBE	Run a macro
F6	Worksheet	Move to the next pane in a split workbook
F7	Worksheet	Display the Spell Check dialog box
F7	VBE	Display Code window
F8	Worksheet	Activate/deactivate Extend mode (then use arrow keys)
F8	VBE	Step through macro
F9	Worksheet	Recalculate all sheets in all open workbooks
F9	Formula bar	Calculate and display result of selection

Key	Applies to	Action
F9	VBE	Toggle breakpoint
F10	Worksheet	Activate menu bar
F11	Worksheet	Create a new chart that uses the current selection
F12	Worksheet	Display the Save As dialog box

Ctrl + (key)	Applies to	Action
arrow key	Worksheet	Jump to end of block of data in direction of arrow
right/left arrow	Formula bar	Move to beginning of next element in formula, in direction of arrow
right/left arrow	Dialog box	Move one word to the right or left in the entry in an edit box
Backspace	Worksheet	Display active cell
Break	VBE	Halt macro execution
Delete	Formula bar	Clear from insertion point to end
Home	VBE	Jump to beginning of module
End	Worksheet	Select bottom right cell
End	VBE	Jump to end of module
Enter	Worksheet	Enter the contents of the formula bar in the selected range (references are adjusted)
Page Down	Worksheet	Select next sheet in workbook
Page Down	Dialog box	Switch to next tab
Page Up	Worksheet	Select previous sheet in workbook
Page Up	Dialog box	Switch to previous tab
Tab	Worksheet	Move to the previous tool button (when toolbar is active)
' (apostrophe)	Worksheet	Copy value or formula from cell above (references are not adjusted)
(space)	VBE	Display global list of Methods and Properties (with cursor on blank line)
(space)	VBE	Jump to end of next element in line of code
(space)	Worksheet	Select entire column
&	Worksheet	Applies outline border to the selected cells.

Ctrl + (key)	Applies to	Action
*	Worksheet	Select block of data containing the active cell (use numeric keypad)
- (minus)	Worksheet	Display the Delete… dialog box
+ (plus)	Worksheet	Display the Insert… dialog box
(period)	Worksheet	Move clockwise to the next corner of the selection
/ slash)	Worksheet	Select all cells of the array containing the active cell
: colon)	Worksheet	Insert current time
; semicolon)	Worksheet	Insert current date
[	Worksheet	Select cells that are direct precedents of cells in selection
]	Worksheet	Select cells that are direct dependents of cells in selection
~ (tilde)	Worksheet	Toggle between values and formulas
0	Worksheet	**Hide** column(s) containing selection
1	Worksheet, Chart	Display the Format Cells dialog box
2	Worksheet, Chart	Apply or remove Bold formatting
3	Worksheet, Chart	Apply or remove Italic formatting
4	Worksheet, Chart	Apply or remove Underline formatting
5	Worksheet, Chart	Apply or remove Strikethrough formatting
6	Worksheet	Toggle between displaying objects, placeholders for objects or hiding objects
7	Worksheet	Toggle display of Standard toolbar (Excel 2003 only)
8	Worksheet	Toggle display of outline symbols
9	Worksheet	**Hide** row(s) containing selection
A	Formula bar	Display the Formula Palette (the Insert Function Step 2 dialog box) after typing a function name in a formula
A	Worksheet	Select entire worksheet
B	Worksheet, Chart Formula bar	Apply or remove Bold formatting
C	Worksheet, Chart Formula bar	**Copy**

Ctrl + (key)	Applies to	Action
D	Worksheet	**Fill Down**
F	Worksheet, VBE	Display the Find dialog box
G	VBE	Display Immediate window
G	Worksheet	Display the Go To dialog box
H	Worksheet, VBE	Display the Replace dialog box
I	Worksheet, Formula bar	Apply or remove Italic formatting
J	VBE	Display list of Methods and Properties for selected object. With cursor on blank line, display global list of Methods and Properties
K	Worksheet	Display the Insert Hyperlink dialog box
N	Worksheet	**New** (Open a new workbook)
O	Worksheet, VBE	Display the Open dialog box
P	Worksheet	Display the Print dialog box
R	VBE	Display the Project Explorer window
R	Worksheet	**Fill Right**
S	Worksheet	**Save** the active workbook
T		
U	Worksheet	Apply or remove Underline formatting
V	Worksheet, VBE, Formula bar	**Paste**
W	Worksheet, Formula bar	**Close**
X	Worksheet, VBE, Formula bar	**Cut**
Y	VBE	Delete current line in a macro
Y	Worksheet, Chart	**Repeat**
Z	Worksheet, Chart	**Undo**
F1	Worksheet	Display/Hide the Ribbon (Excel 2007/2010)
F1	Worksheet	Display Startup Task Pane
F2	Worksheet	Displays the Print Preview window
F3	Worksheet	Display the Define Name dialog box
F4	Worksheet	Close the active window
F5	Worksheet	Restore window size

Ctrl + (key)	Applies to	Action
F6	Worksheet	Move to the next workbook active window
F7	Worksheet	Move the window (use arrow keys)
F8	VBE	Run to cursor
F8	Worksheet	Re-size the window (use arrow keys)
F9	VBE	Select statement to execute next
F9	Worksheet	Minimize the window
F10	Worksheet	Maximize the window
F11	Worksheet	Insert an Excel 4 Macro sheet
F12	Worksheet	Display **Open** dialog box

Shift+(key)	Applies to	Action
mouse click	Chart	Selects chart as an object. Can then use arrow keys to move it, one point at a time.
arrow key	Worksheet	Extend selection by one row/column
arrow key	Worksheet	With END mode on, extend selection to end of block in direction of arrow
right/left arrow	Dialog box	Select or unselect one character to the right or left in the entry in an edit box
(space)	Worksheet	Select entire row
End	Dialog box	Select from the insertion point to the end of the entry in an edit box
End	Formula bar	Extend selection from insertion point to end of current line in formula
End	Worksheet	Extend selection to end of row
Enter	Worksheet	Enter data and move up one cell
Home	Dialog box	Select from the insertion point to the beginning of the entry in an edit box
Home	Formula bar	Extend selection from insertion point to beginning of current line in formula
Home	Worksheet	Extend selection to beginning of row
Home	Worksheet	With END mode on, extend selection to last-used cell (lower-right corner)
Return	Worksheet	With END mode on, extend selection to last cell in row
Tab	Dialog box	Move to the previous option
Tab	Worksheet	Enter data and move left one cell

Shift+(key)	Applies to	Action
F1	Worksheet	**Help**
F1	VBE	Get Help on a selected VBA keyword
F2	Worksheet	Insert or edit a **Cell Comment**
F2	VBE	Display the Object Browser window
F3	Worksheet	Display the **Insert Function** dialog box
F4	Worksheet	Display the **Find Next** dialog box
F5	Worksheet	Display the **Find** dialog box
F6	Worksheet	Move to the previous pane in a workbook that has been split
F8	Worksheet	Toggle Modify Selection mode
F9	Worksheet	Recalculate the active worksheet
F9	VBE	Display Quick Watch window
F10	Worksheet	Display a shortcut menu
F11	Worksheet	Insert a new worksheet
F12	Worksheet	Display the **Save** As dialog box

Ctrl + Shift+ (key)		Action
arrow key	Formula bar	Select to end of block of data in direction of arrow (inserts reference in formula)
arrow key	Worksheet	Select to end of block of data in direction of arrow
right/left arrow	Formula bar	Extend selection, one term at a time
right/left arrow	Dialog box	Select or unselect one word to the right or left in the entry in an edit box
End	Formula bar	Extend selection from insertion point to end of formula
End	Worksheet	Extend selection to end of worksheet (bottom right corner)
Enter	Worksheet	Enter an array formula in the selected range
Home	Formula bar	Extend selection from insertion point to beginning of formula
Home	Worksheet	Extend selection to beginning of worksheet (upper left corner)

Ctrl + Shift+ (key)		Action
Page Down	Worksheet	Select current and next sheet in workbook
Page Up	Worksheet	Select current and previous sheet in workbook
Tab	Dialog box	Switch to previous tab
!	Worksheet	Apply Decimal number format (two decimal places, thousands separator, minus sign for negative numbers)
#	Worksheet	Apply Date number format (d-mmm-yy)
$	Worksheet	Apply Currency number format (two decimal places, negative numbers in parentheses)
%	Worksheet	Apply Percent number format (no decimal places)
&	Worksheet	Applies border to selection
_ (underscore)	Worksheet	Removes border from selection
(	Worksheet	**Unhide** rows
)	Worksheet	**Unhide** columns
* (asterisk)	Worksheet	Select block of data containing the active cell
+	Worksheet	Display the Insert dialog box
@	Worksheet	Apply Time number format (hours and minutes, AM and PM)
^	Worksheet	Apply Scientific number format (two decimal places)
{	Worksheet	Select cells that are direct or indirect precedents of cells in selection
}	Worksheet	Select cells that are direct or indirect dependents of cells in selection
~ (tilde)	Worksheet	Apply General number format
"(quote)	Worksheet	Copy value from cell above
(space)	Worksheet	Select entire worksheet

Alt+(key)	Applies to	Action
A	Formula bar	Insert placeholder arguments and closing parenthesis after typing a function name in a formula
F	Worksheet	Activate the **Font** list box
J	VBE	Display list of constants
O	Worksheet	Select all cells with comments
P	Worksheet	Activate the **Font Size** list box
U	Worksheet	Expand/Collapse Formula Bar (Excel 2007/2010 only)
down arrow	Dialog box	Open the selected drop-down list box
Enter	Formula bar	Insert a line break
Backspace	Worksheet	**Undo**
' (apostrophe)	Worksheet	Display the **Style** dialog box
- (minus)	Worksheet	Display the Workbook Icon menu to Maximize/ Minimize/Restore a window
;(semicolon)	Worksheet	Select only visible cells in the current selection
=	Worksheet	Insert the SUM function in a cell
(letter)	Dialog box	Select the menu or dialog box option indicated by the underlined letter
F1	Worksheet	Create a new chart that uses the current selection
F2	Worksheet	Display the **Save As** dialog box
F4	Worksheet	**Close** the active workbook
F8	Worksheet	Display **Macro** dialog box
F11	Worksheet	Toggle between **Visual Basic Editor** and **Excel**

Ctrl + Alt + (key)		Action
right/left arrow	Worksheet	Move between non-adjacent selections

Ctrl + Alt + (function key)		
F9	Worksheet	Recalculate all sheets in the active workbook

Ctrl + Shift + (function key)

F3	Worksheet	Display the **Create Names** dialog box
F4	Worksheet	Display the **Find Next** dialog box
F6	Worksheet	Move to the previous workbook active window
F12	Worksheet	Display **Print** dialog box

Alt + Shift + (function key)

F1	Worksheet	Insert a new worksheet
F2	Worksheet	Display the **Save As** dialog box

Appendix I

Selected Shortcut Keys by Category

Cut, Copy, Paste, Undo

Undo	Ctrl+Z or Alt+Backspace
Cut	Ctrl+X
Copy	Ctrl+C
Paste	Ctrl+V
Clear contents of selection (formula bar or worksheet)	Delete
Repeat	Ctrl+Y or F4

Moving and Selecting

Jump to end of block of data in direction of arrow	Ctrl+arrow key
Select to end of block of data in direction of arrow	Ctrl+Shift+arrow
Select entire row	Shift+(space)
Select entire column	Ctrl+(space)
Select all cells of the array containing the active cell	Ctrl+/(slash)
Extend selection by one row/column	Shift+arrow key
Extend selection in a formula, one term at a time	Ctrl+Shift+ right or left arrow
Hide column(s) containing selection	Ctrl+0
Hide row(s) containing selection	Ctrl+9
Unhide rows	Ctrl+Shift+(
Unhide columns	Ctrl+Shift+)
Select block of data containing the active cell	Ctrl+Shift+* or Ctrl+* (numeric keypad)
Select only visible cells in the current selection	Alt+;(semicolon)
Select all cells with comments	Ctrl+Shift+O
Select entire worksheet	Ctrl+A
Move between non-adjacent selections	Ctrl+Alt+arrow

Entering or Editing a Formula

Begin Edit. Position the insertion point at the end of the line	F2
Enter the contents of the formula bar in the selected range (references are adjusted)	Ctrl+Enter
Insert current time	Ctrl+: (colon)
Insert current date	Ctrl+; (semicolon)
Insert a line break	Alt+Enter
Toggle between values and formulas	Ctrl+~ (tilde)
Display the Function Arguments dialog box after typing a function name in a formula	Ctrl+A
Insert placeholder arguments and closing parenthesis after typing a function name in a formula	Ctrl+Shift+A
Enter data and move down one cell	Enter
Enter data and move up one cell	Shift+Enter
Enter data and move right one cell	TAB
Enter data and move left one cell	Shift+Tab
Enter an array formula in the selected range	Ctrl+Shift+ Enter
Toggle between relative and absolute reference	F4
Calculate and display result of selection	F9
Fill Down	Ctrl+D
Fill Right	Ctrl+R

Formatting

Apply or remove Bold formatting	Ctrl+B
Apply or remove Italic formatting	Ctrl+I
Apply or remove Underline formatting	Ctrl+U
Apply General number format	Ctrl+Shift+~ (tilde)
Apply Decimal number format (two decimal places, thousands separator, minus sign for negative numbers)	Ctrl+Shift+!
Apply Currency number format (two decimal places, negative numbers in parentheses)	Ctrl+Shift+$

Apply Percent number format (no decimal places)   Ctrl+Shift+%
Apply Scientific number format (two decimal places)   Ctrl+Shift+^
Apply Date number format (d-mmm-yy)   Ctrl+Shift+# or
                                                         Ctrl+Shift+&
Apply Time number format (hh:mm AM/PM)   Ctrl+Shift+@
Apply border to selected cells   Ctrl + &

Menu Commands

New (Open a new workbook)	Ctrl+N
Display the **Open...** dialog box	Ctrl+O or Ctrl+F12
Close	Alt+F4
Save the active workbook	Ctrl+S or F12 or Alt+F2
Display the **Save** As dialog box	Shift+F12 or Alt+Shift+F2
Display the **Print...** dialog box	Ctrl+P or Ctrl+Shift+F12
Display the **Delete...** Rows/Columns dialog box	Ctrl+-(minus)
Display **Insert Cells...** dialog box	Ctrl+Shift++(plus)
Display the **Find...** dialog box	Ctrl+F or Shift+F5
Display the **Find Next** dialog box	Shift+F4
Display the **Replace...** dialog box	Ctrl+H
Display the **Go To...** dialog box	Ctrl+G or F5
Insert a **Cell Comment**	Shift+F2
Display the **Paste Function** dialog box	Shift+F3
Display the **Define Name** dialog box	Ctrl+F3
Display the **Create Names** dialog box	Ctrl+Shift+F3
Display the **Paste Name** dialog box.	F3
Display **Format Cells** dialog box	Ctrl+1
Display the **Macro** dialog box.	Alt+F8
Display the **Spell Check** dialog box	F7
Help	F1 or Shift+F1

Workbooks and Worksheets

Move to the previous workbook active window	Ctrl+Shift+F6
Select next sheet in workbook	Ctrl+Page Down
Select previous sheet in workbook	Ctrl+Page Up
Switch to next tab	Ctrl+Page Down
Switch to previous tab	Ctrl+Page Up
Create a new chart that uses the current selection	F11 or Alt+F1
Recalculate the active worksheet	Shift+F9
Recalculate all sheets in the active workbook	Ctrl+Alt+F9
Insert a new worksheet	Shift+F11 or Alt+Shift+F1
Toggle Window Maximize	Ctrl+F10

Visual Basic

Halt execution of a macro	Esc
Run a macro	F5
Step through a macro	F8
Toggle breakpoint	F9
Toggle between Visual Basic Editor and Excel	Alt+F11

Appendix J

ASCII Codes

The following table lists the ASCII codes for some useful non-printing keyboard characters, the keyboard characters and the alternate character set. The alternate characters can be printed by holding down the Alt key while typing 0###, e.g., for ±, type Alt+0177, *using the numeric keypad.*

Non-Printing Keyboard Characters:

8	backspace	10	line feed	27	escape
9	horizontal tab	13	carriage return		

Keyboard Characters:

32	(space)	64	@	96	`
33	!	65	A	97	a
34	"	66	B	98	b
35	#	67	C	99	c
36	$	68	D	100	d
37	%	69	E	101	e
38	&	70	F	102	f
39	'	71	G	103	g
40	(	72	H	104	h
41	)	73	I	105	i
42	*	74	J	106	j
43	+	75	K	107	k
44	,	76	L	108	l
45	-	77	M	109	m
46	.	78	N	110	n
47	/	79	O	111	o
48	0	80	P	112	p
49	1	81	Q	113	q
50	2	82	R	114	r
51	3	83	S	115	s
52	4	84	T	116	t
53	5	85	U	117	u
54	6	86	V	118	v
55	7	87	W	119	w
56	8	88	X	120	x
57	9	89	Y	121	y
58	:	90	Z	122	z
59	;	91	[	123	{
60	<	92	\	124		
61	=	93	]	125	}
62	>	94	^	126	~
63	?	95	_	127	*(bksp)*

Alternate Characters:

128	€	178	²	228	ä
129	*(NP)**	179	³	229	å
130	,	180	´	230	æ
131	ƒ	181	µ	231	ç
132	„	182	¶	232	è
133	…	183	·	233	é
134	†	184	¸	234	ê
135	‡	185	¹	235	ë
136	ˆ	186	º	236	ì
137	‰	187	»	237	í
138	Š	188	¼	238	î
139	‹	189	½	239	ï
140	Œ	190	¾	240	ð
141	*(NP)**	191	¿	241	ñ
142	Ž	192	À	242	ò
143	*(NP)**	193	Á	243	ó
144	*(NP)**	194	Â	244	ô
145	`	195	Ã	245	õ
146	'	196	Ä	246	ö
147	"	197	Å	247	÷
148	"	198	Æ	248	ø
149	•	199	Ç	249	ù
150	–	200	È	250	ú
151	—	201	É	251	û
152	˜	202	Ê	252	ü
153	™	203	Ë	253	ý
154	š	204	Ì	254	þ
155	›	205	Í	255	ÿ
156	œ	206	Î		
157	*(NP)**	207	Ï	**non-printing
158	ž	208	Ð	
159	Ÿ	209	Ñ	
160		210	Ò	
161	¡	211	Ó	
162	¢	212	Ô	
163	£	213	Õ	
164	¤	214	Ö	
165	¥	215	×	
166	¦	216	Ø	
167	§	217	Ù	
168	¨	218	Ú	
169	©	219	Û	
170	ª	220	Ü	
171	«	221	Ý	
172	¬	222	Þ	
173		223	ß	
174	®	224	à	
175	¯	225	á	
176	°	226	â	
177	±	227	ã	

*non-printing

Appendix K

Contents of the CD-ROM

This appendix describes the worksheets, macros and other files that are on the CD-ROM that accompanies this book. Most of the workbooks are in 2003 format (.xls); they can be read by users with Excel 97/2000/2003 or Excel 2007/1010. There are a few, intended only for Excel 2007/2010, that are in .xlsx, .xlsm or .xlam format.

The documents described below are listed in the order in which they appear in each chapter.

Chapter 3   Excel Formulas and Functions

Intersection Operator demo.xls illustrates the use of the intersection operator in a formula.

IF demo.xls illustrates a simple use of the IF function to prevent the display of error values.

COUNTIFS demo.xlsm illustrates the use of an Excel 2007 function to count with multiple criteria. The workbook can be viewed using Excel 2003 but the function will not be available.

SUMIF demo.xls illustrates the use of the SUMIF function.

Conditional Formatting demo.xls shows one use of conditional formatting: how to color-band a column of dates so as to distinguish one month from another.

Megaformula.xls illustrates the use of "megaformulas", in this case to parse text into separate columns.

Chapter 4   Excel 2007 Charts

Charts 2007 demo.xlsx shows examples of the different kinds of chart described in the chapter.

Chapter 5   Excel 2003 Charts

Charts 2003 demo.xls shows examples of the different kinds of chart described in the chapter.

Chapter 6   Advanced Worksheet Formulas

Using Names demo.xls is a simple illustration of the use of names in formulas: in this case, the equation of a straight line, using m, b and x as names for cells or ranges.

Lookup demo1.xls illustrates the use of VLOOKUP in a one-way table.

Lookup demo2.xls illustrates the use of VLOOKUP in a two-way table. The data table used in this example is part of the Steam Tables and is reprinted from *ASME International Steam Tables for Industrial Use*. The information in the table was provided by the National Institute for Standards and Technology (NIST) and is in the public domain.

Circular reference demo.xls is a simple example using intentional circular references. When the workbook is opened, the "circular reference" error message will be displayed; press Cancel to close the box.

Chapter 7   Array Formulas

Array Demo.xls illustrates a non-array-formula approach and three different array formulas for the calculation of the sum of squares of deviations.

PowerSeries.xls shows how to create a series of integers for use in array formulas.

Multiple Criteria demo.xls shows how to use an array formula to extract information from a table using multiple criteria.

Auto Alphabetize demo.xls shows how to sort text values in alphabetical order by using a formula.

CAS Numbers demo.xls shows how to validate a CAS# by means of a formula.

CAS# Conditional Formatting.xls shows how to validate a CAS# by means of a formula using Conditional Formatting.

Chapter 8   Advanced Charting Techniques

Methane Hydrate chart.xls illustrates techniques for formatting a chart and producing a smooth curve through data points.

Secondary Y-Axis demo.xls illustrates an XY Scatter chart with a secondary Y-axis.

Secondary X- & Y-Axis chart.xls illustrates an XY Scatter chart with a secondary X Axis and a secondary Y Axis.

Column Chart with Secondary Axis.xls illustrates a data layout to produce a column chart with a secondary Y Axis.

Error Bars demo.xls illustrates a chart with error bars.

Overlapping Zero on Scale demo.xls illustrates how to use custom number formatting to improve the look of a chart axis.

Additional Axis chart.xls is an example of using simple plotting and formatting techniques to produce a non-standard chart type.

AutoUpdateChart.xls shows how to use named formulas to create a chart that updates automatically when new data are added.

Chapter 9   Using Excel's Database Features

Database demo.xls is a sample database to illustrate Excel's database capabilities.

Chapter 10 Importing Data into Excel

NISPEC.DAT is a comma-delimited text file to be used with Excel's **Text to Columns** or **Import External Data** menu commands.

Spectral data (imported).xls is the scanned spectral data from a monograph (raw data) that can be converted into a table in usable form by means of a recorded macro.

Nth.xls illustrates four different worksheet formulas to select every Nth data point from a data table.

Chapter 11 Adding Controls to a Spreadsheet

List Box demo.xls illustrates two drop-down list boxes that display selected names, addresses and telephone numbers from a database.

Dropdown List Box demo.xls illustrates using a drop-down list box to display a sublist from an alphabetized list of names. Uses named formulas.

Option Buttons & List Box demo.xls illustrates using option buttons and a list box. Uses named formulas.

Hyperlink demo.xls contains the hyperlink examples shown in the chapter.

Chapter 12 Other Language Versions of Excel

EUR Date fmt to US.xls illustrates one method to convert dates in a non-standard European format to US date format.

FunctionNameTranslator.xls is a worksheet that displays a table of worksheet function names in a specified language and their equivalents in another specified language.juikgkg,

Chapter 13 Mathematical Methods for Spreadsheet Calculations

Interpolation demos.xls illustrates how to perform linear and cubic interpolation.

Matrix demo.xls illustrates the tools available for matrix mathematics, with examples.

Derivs of Titration Data demo.xls illustrates how to obtain the first and second derivatives of a data set.

Derivs of a Function demo.xls illustrates how to calculate the first derivative of a function.

Area Under a Curve demo.xls illustrates three worksheet formulas that can be used to obtain the area under a curve.

Root Finding demos.xls illustrates three ways to find a real root of a function: by successive approximations, by the secant method, and by using Goal Seek.

Simultaneous Equations demo.xls illustrates ways to solve sets of simultaneous linear equations by using matrices.

Three-Component Spec demo.xls illustrates methods for the analysis of the spectrum of a mixture.

Polar to Cartesian demo.xls illustrates how to convert from polar to Cartesian coordinates.

Chapter 14 Linear Regression

Calibration Curve demo.xls is an example of finding the least-squares slope and intercept of a linear calibration curve using the SLOPE, INTERCEPT and RSQ worksheet functions.

LINEST.xls illustrates how to obtain the least-squares slope, intercept, plus R^2 and some other regression statistics of a linear calibration curve using the LINEST worksheet functions.

Amino Acids.xls contains the data and results for the multiple linear regression example of Figure 14-10.

Power Series example.xls illustrates the use of LINEST to fit data with a cubic fitting function.

LINEST with Non-contiguous Ranges demo.xls shows how to handle data in non-contiguous ranges.

LINEST Collinearity demo.xls illustrates how LINEST handles collinearity.

T Table.xls shows how to calculate values of Student's t by using a worksheet function.

Chapter 15 Nonlinear Regression Using the Solver

A-B-C example.xls is the Solver example shown in Figures 15-1, 15-4 and 15-5.

Deconvolution.xls illustrates deconvolution of a UV-visible spectrum consisting of four overlapping bands, using the Solver.

NMR example.xls is an example of the calculation of a binding constant from NMR data, using the Solver.

Solver Statistics.xls is a command macro that returns the standard deviations for non-linear regression analysis performed by the Solver. See "Instructions for Using Solver Statistics" at the end of this appendix. The macro can be used with Excel 2007/2010 as well as with Excel 2003 and earlier versions.

Chapter 17 Programming with VBA

Abs and Rel reference demo.xls contains three macros; a macro recorded using absolute references, a macro recorded using relative references, and a modified version of the second procedure that eliminates the **Select** keyword.

MsgBox demo.xls provides some examples of the built-in dialog box to display a message. **MsgBox** returns a value indicating which button the user pressed.

InputBox demo.xls provides some examples of the built-in dialog box for user input.

VBA to Open a File.xls illustrates several ways to open a file within a procedure.

Chapter 18 Working with Arrays in VBA

Array Demos.xls contains 11 **Sub** or **Function** procedures that illustrate various features of using arrays in VBA.

Chapter 19 Command Macros

Chemical Format demos.xls contains three **Sub** procedures that apply chemical formatting. ChemicalFormatDemo1 and ChemicalFormatDemo2 illustrate the development of the procedure. The third **Sub** procedure, ChemicalFormat, applies chemical formatting to text in a cell, in a range of cells, in a chart or in a textbox.

Chart Labeler demo.xls is the simple **Sub** procedure shown in Figure 19-9 that adds user-specified data labels to a chart.

Chart Labeler 2.xls is a more advanced version of the Chart Labeler macro. It uses a custom dialog box (not discussed in the text). See "Instructions for Using Labeler2" at the end of this appendix.

Classic Menus Chart Labeler.xlsm is the Excel 2007/2010 version of **ChartLabeler2.xls**.

Install the QuickChart toolbutton.xlsm installs the QuickChart toolbutton and connects it to a dummy procedure.

Chapter 20 Custom Functions

Array Maker.xls is a **Function** procedure that combines non-contiguous worksheet ranges into an array. This **Function** procedure illustrates the use of the **ParamArray** keyword.

Alpha.xls is a custom function that returns α values for a polyprotic acid species. **Alpha.xlsm** is the Excel 2007/2010 version.

Deming.xls is a **Function** procedure that returns the slope and intercept calculated by the method of Deming.

Molecular Weight Calculator.xls is a **Function** procedure that returns the formula weight from text that can be interpreted as a chemical formula; See "Instructions for Using MolWt" at the end of this appendix.

Molecular Weight Calculator.xla is the Add-in version.

"Instructions for Using MolWt" is a Word document that gives more details about using the custom function.

Chapter 22 Custom Menus and Menu Bars

Menu Workbook.xls is a **Sub** procedure that illustrates how to install a new menu in the Worksheet menu bar, with menu commands and sub-menus.

Classic Menus.xlsm is an Excel 2007/2010 utility that displays a version of the Excel 2003 Worksheet Menu Bar and Chart Menu Bar in Excel 2007/2010. The menu is located in the Add-Ins tab of the Ribbon. Most of the commands found in the Excel 2003 Worksheet Menu Bar are available. Figure 1-68 shows the **Tools** menu with the **Macro** submenu.

A few menu commands are absent from the menus: the **Toolbars** command in the **View** menu and the **Customize...** command in the **Tools** menu, are omitted because they are not applicable to Excel 2007/2010. The box in Chapter 1 following Figure 1-68 gives a more complete list of the differences between the Excel 2003 menus and Classic Menus.

.   Two add-ins, the Solver and the Analysis ToolPak, have commands that are installed in the **Tools** menu when they are loaded. Because of certain limitations in programming, these menu commands do not appear in the **Tools** menu after they are loaded until you click on the **File** menu.

Chapter 23 Custom Toolbars and Toolbuttons

NumFmt.xls is a simple command macro that toggles between floating-point and scientific number formats. The macro can be easily assigned to a toolbutton.

FullPage.xls is a simple command macro that can be used to obtain the maximum amount of space on a page for printing a worksheet. It sets either portrait or landscape orientation, sets margins to zero, and removes header and footer text. The macro assumes that you have created two custom toolbuttons on the Standard toolbar and have labeled them "Full Page (Portrait)" and "Full Page (Landscape)". The macro can be easily assigned to a toolbutton.

Instructions for Using Classic Menus

The CD contains a macro-enabled workbook (Classic Menus.xlsm). If you open the workbook, it will display the Classic Menus and automatically Hide itself. You can use **SaveAs...** to save the workbook as an Add-in workbook in the Add-ins folder, so that you can use the **Add-ins...** command in the **Tools** menu to load or unload the Add-in as desired. Classic Menus is an Auto_Open workbook that installs the menus automatically.

The VBA code is not available for viewing

To save Classic menus as an Add-in workbook, follow these steps:
1. Start Excel 2007 or 2010.
2. Open Classic Menus.xlsm. The Classic Menus will be displayed in the Add-ins tab of the Ribbon.
3. Use Unhide in the Window group in the View tab of the Excel 2007/2010 Ribbon to display the Classic Menus.xlsm workbook.
4. Use Save As (Excel 2007/2010) to save the workbook as an Add-in (.xlam). The default location is in the Add-ins folder.
5. Exit Excel.
6. Restart Excel.
7. Click on Excel 2007 or 2010 Options, click on the Add-ins button and check Classic Menus in the Add-ins dialog box.

Excel 2010 users: Classic Menus loads the Excel 2003 version of the Solver. If you want to use the Excel 2010 Solver in Classic Menus, you must unload Classic Menus and then use the Options button in the File tab of the Excel 2010 ribbon to load Solver. Once you have done this, you can load/unload the Solver using the **Add-ins...** command in the **Tools** menu of Classic Menus.

Instructions for Using Solver Statistics

This command macro returns the standard deviations of regression coefficients obtained by using the Solver, plus the correlation coefficient and the RMSD; these statistical parameters are not available from the Solver. The array of values returned is in a format similar to that returned by LINEST.

The sheet must contain y_{calc} values. The y_{obsd} and y_{calc} values must each be in either a single column or row. The regression coefficients returned by the Solver do not have to be in adjacent cells.

To use the macro, simply **Open** SolvStat.xls; it will appear on screen and then **Hide** itself. It installs a new menu command, **Solver Statistics...**, immediately under the **Solver...** command in the **Tools** menu. If the Solver Add-in has not been loaded, the **Solver Statistics...** command will be at the top of the menu. The command will remain in the menu until you exit from Excel.

Activate the document in which the **Solver** has already been used to obtain regression coefficients by minimizing the sum of squares of deviations between observed and calculated y-values. Choose **Solver Statistics...** from the menu. Follow the directions in the dialog boxes.

Instructions for Using MolWt

To install the function, simply open the Excel document Molecular Weight Calculator.xls. The document automatically hides itself. The function is available even though the document is hidden. You can examine the workbook by choosing **Unhide** from the **Window** menu. The VBA code is not available for viewing.

Syntax: MolWt(*formula, decimals*). *Formula* will usually be a reference to a cell containing a text string that can be interpreted as a chemical formula, e.g., H2SO4. *Decimals*, an optional argument, is the number of decimal places to be displayed in the returned value.

Instructions for Using ChartLabeler2

To use the macro, simply **Open** ChartLabeler2.xls; it will appear on screen and then **Hide** itself. It installs a new menu command, "**Add Data Labels...**" in the **Chart** menu. To view the workbook, containing the macro and two sample charts, **Unhide** the workbook.

To add data labels to a data series, click on the chart (you can pre-select the data series or choose it later in the dialog box). Choose "**Add Data Labels...**" in the **Chart** menu. This will display the Apply Data Labels dialog box. Choose the desired options and the range of cells on the worksheet that contain the data labels and then press OK.

The custom dialog box is designed for XY Scatter plots only.

Index